Edited by
Nicola Pinna and Mato Knez

Atomic Layer Deposition of Nanostructured Materials

Related Titles

Nicolais, L., Carotenuto, G.

Nanocomposites

In Situ Synthesis of Polymer-Embedded Nanostructures

Hardcover

ISBN: 978-0-470-10952-6

Kumar, C. S. S. R. (ed.)

Semiconductor Nanomaterials

2010

Hardcover

ISBN: 978-3-527-32166-7

Kumar, C. S. S. R. (ed.)

Nanostructured Thin Films and Surfaces

2010

Hardcover

ISBN: 978-3-527-32155-1

Mathur, S., Singh, M. (eds.)

Nanostructured Materials and Nanotechnology III, Volume 30, Issue 7

Hardcover

ISBN: : 978-0-470-45757-3

Kumar, C. S. S. R. (ed.)

Nanostructured Oxides

2009

Hardcover

ISBN: 978-3-527-32152-0

Zehetbauer, M. J., Zhu, Y. T. (eds.)

Bulk Nanostructured Materials

2006

Hardcover

ISBN: 978-3-527-31524-6

Morkoc, H.

Handbook of Nitride Semiconductors and Devices

Hardcover

ISBN: 978-3-527-40797-2

Irene, E. A.

Electronic Material Science and Surfaces, Interfaces, and Thin Films for Microelectronics

Hardcover

ISBN: 978-0-470-22478-6

Eftekhari, A. (ed.)

Nanostructured Materials in Electrochemistry

2009

Hardcover

ISBN: 978-3-527-31876-6

Atwood, J. L., Steed, J. W. (eds.)

Organic Nanostructures

2008

Hardcover

ISBN: 978-3-527-31836-0

Edited by
Nicola Pinna and Mato Knez

Atomic Layer Deposition of Nanostructured Materials

WILEY-VCH Verlag GmbH & Co. KGaA

The Editors

Prof. Dr. Nicola Pinna
Department of Chemistry and CICECO
University of Aveiro
Campus Universitario de Santiago
3810-193 Aveiro
Portugal

and

School of Chemical and Biological Engineering
College of Engineering
Seoul National University (SNU)
151-744 Seoul
Korea

Dr. Mato Knez
MPI für Mikrostrukturphysik
Experimental Dept. II
Weinberg 2
06120 Halle
Germany

Library of Congress Card No.: applied for

British Library Cataloguing-in-Publication Data
A catalogue record for this book is available from the British Library.

Bibliographic information published by the Deutsche Nationalbibliothek
The Deutsche Nationalbibliothek lists this publication in the Deutsche Nationalbibliografie; detailed bibliographic data are available on the Internet at http://dnb.d-nb.de.

Typesetting Thomson Digital, Noida, India
Printing and Binding Fabulous Printers Pte Ltd, Singapore

Cover Design Grafik-Design Schulz, Fußgönheim

Printed in Singapore
Printed on acid-free paper

Print ISBN: 978-3-527-32797-3
ePDF ISBN: 978-3-527-63993-9
oBook ISBN: 978-3-527-63991-5
ePub ISBN: 978-3-527-63992-2
mobi ISBN: 978-3-527-63994-6

Foreword

Atomic layer deposition (ALD) is a relatively new and low-temperature growth method capable of depositing a variety of thin films on virtually any substrate. Although it is a vapor-based technique like chemical vapor deposition (CVD), there are two main differences that make it a particularly powerful method for very thin and conformal film growth. First, precursor molecules react only with the surface (not with themselves) under deposition temperatures. Second, growth is achieved by sequential introduction of typically two precursors, separated by a thorough inert gas purge, so that different precursors are not present together in the gas phase. These characteristics are in fact responsible for generating much interest from researchers belonging to scientific areas that had thus far been considered unrelated, thus creating a stimulating interdisciplinary environment. These are gas-phase chemistry, surface science, solid-state chemistry, kinetics of thin film growth, and engineering of gas flow reactors.

The community that is responsible for the development of ALD came from *organometallic chemists*. They were responsible for identifying or synthesizing precursors capable of reacting at moderate temperatures with specific surface groups while remaining stable in the gas phase (no self-reaction), which is key for the ALD process. The challenge for this community has been to produce complementary precursors that react with complete ligand exchange at moderate temperatures, essential to grow pure films. Incomplete reactions lead to the incorporation of carbon or other impurities into the films.

Surface scientists have quickly been drawn to ALD because the process is based on surface reactions. As is now recognized, surface processes can be substantially different from gas-phase chemistry because the surface structure and steric interactions dramatically affect chemical reactions at surfaces. *In situ* characterization is therefore necessary to verify choices based on gas-phase chemistry and to quantify surface reactivity with precursors. Furthermore, it has now been shown in several instances that the substrate can also play an active role. Specifically, at typical process temperatures reactions with the substrate can take place, such as formation of SiO_2 during deposition of metal oxides on oxide-free Si substrates. Therefore, much thought is being given to developing diffusion barriers to prevent oxidation of and/or ion penetration into the substrate, including surface functionalization and even self-assembled monolayers. Given the diversity of approaches and complexity of the mechanisms, the whole arsenal of surface techniques has been brought to bear: vibrational spectroscopy (infrared, Raman), electron spectroscopy (X-ray photoelectron, photoemission, Auger), ion scattering (Rutherford backscattering, medium- and low-energy ion spectroscopies), imaging (transmission electron microscopy, atomic and in some cases scanning tunneling spectroscopies), mass spectroscopy, and quartz crystal mass analysis. Much needed fundamental work continues to be performed as new precursors and systems (e.g., metal films) are considered.

Solid-state chemists and modelers are also engaged in ALD processes because the simple picture of alternative deposition of element A and element B to form an amorphous A_xB_y film is not strictly correct. Atoms A and B require substantial rearrangement and rebonding as their ligands are removed, which is not always consistent with layer-by-layer growth and is certainly more complicated than the simple ALD models typically presented. Kinetics play a central role. A striking example of complex, kinetically controlled surface mechanism is the agglomeration of metal atoms upon thin metal film deposition, making it virtually impossible to initiate a uniform thin metal film growth. A combination of measurements and advanced modeling is needed to develop a mechanistic understanding of the processes with some predictive power.

Chemical engineers have played an important role in addressing the gas flow requirements that are much more stringent in ALD than in any other vapor-phase growth method. Indeed, the complete removal of one precursor gas before the other is introduced into the reactor requires that flow patterns are optimized such that the purge gas can efficiently remove all traces of the precursor gas. For industrial applications, uniform supply of precursor gas over large wafers also demands careful optimization of precursor delivery rates and flow.

The chapters in the first part of this book involve authors from this wide community and address many of the issues outlined above, including the growth of organic thin films. The focus of the book, however, is related to the extraordinarily diverse ALD-based applications that have emerged over the past few years. While the initial work was motivated by metal oxide film deposition for microelectronic applications (ALD was initially adopted by industry for the growth of high-*k* dielectrics, for instance), the ALD technique has begun to contribute to a wide range of systems. It is now used for sensor fabrication, fiber coating, and biomedical device fabrication, with a focus on highly structured material and nanomaterials. The second part of the book brings together a diversity of topics that underscore the value and importance of the ALD technique and foreshadows its growing popularity.

With such an explosion of applications and a growing community using ALD, it is clear that fundamental studies of ALD processes are needed to provide the necessary understanding for progress and success. This book is therefore a motivation for researchers in other fields to bring their discipline and expertise to critically evaluate and address the complex issues arising in all these applications. It constitutes an excellent reference book for students and young scientists interested in ALD, and its applications and challenges. Much remains to be done, well beyond precursor development, to fully implement this powerful technique for the current and future applications.

Dallas, Texas
January 2011

Prof. Yves J. Chabal

Contents

Preface

Atomic layer deposition (ALD) is a coating technology that in the past two decades rapidly developed from a niche technology to an established method. The method itself is not too difficult to understand and apply. The basic requirements are a vacuum chamber, at least two reactive precursors, and valves for alternate dosing of the precursors, since the different precursors must never be present in the chamber at the same time. For most of the processes, one metal-containing chemical is used as metal source and subsequently reacted with another chemical, which is the source of, for example, oxygen, nitrogen, sulfur, and so on. The chemistry is normally very simple and often shows hydrolysis, oxidation, reduction, or an organic coupling reaction in the particular case of molecular layer deposition (MLD). There are some restrictions for the chemicals: One has to ensure that the precursors have a reasonably high vapor pressure in order to allow saturation of the chamber volume upon dosing and a good thermal stability to avoid decomposition prior to the next step. A further requirement is that the precursor must chemisorb onto the substrate to be coated. Once these preconditions are fulfilled, the coating process, consisting of alternating pulse and purge steps for each precursor, can be started. The timely separation of the precursors is necessary in order to achieve a saturation (self-termination) of the substrate surface with the precursor and thus to enable a very precise thickness control, which is normally on the angstrom scale. Since ALD is not a line-of-sight coating technology, even complicated 3D structures with not easily accessible surfaces (e.g., trenches, grooves, pores, aerogels, etc.) can be uniformly and conformally coated.

The chemistry used for ALD is often derived from the chemical approaches used for chemical vapor deposition (CVD) and one may argue that it is a modified form of CVD. This comparison is indeed correct as demonstrated from the fact that all ALD precursors can be used for CVD; however, not necessarily vice versa. On the other hand, (i) the timely separation of the precursors and the resulting film thickness control with the number of cycles instead of processing time, (ii) the often much higher compactness of the deposited films compared to CVD, and (iii) the exceptionally good step coverage make ALD an outstanding subset of CVD that surely deserves particular attention as an individual coating technology.

The rapid increase of popularity of ALD in the past two decades is clearly demonstrated by the number of articles published every year (Figure P.1). An almost

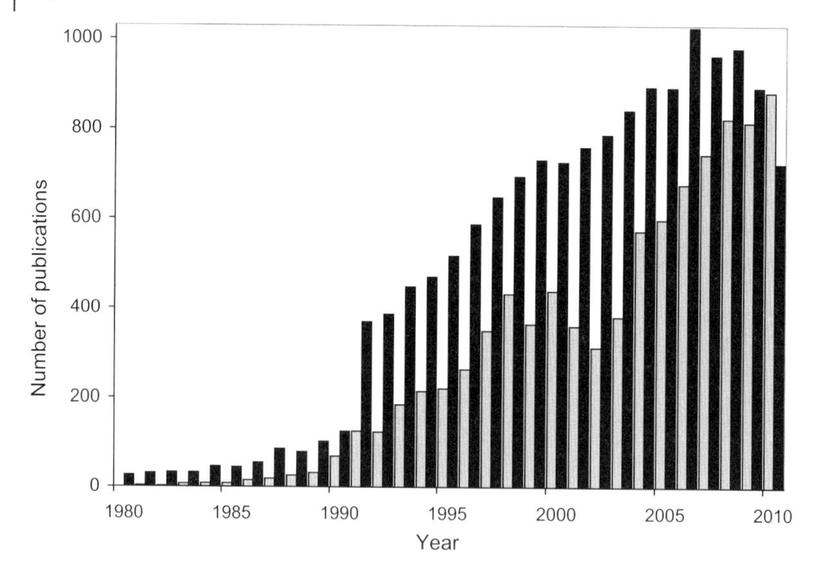

Figure P.1 Number of articles per year published on ALD (gray bars) and CVD divided by five (black bars). *Source*: Web of Science accessed on December 20, 2010.

linear increase is observed, reaching a maximum of around 900 articles published in 2010. Although the number of articles per year is around five times larger, a similar trend can be observed for CVD. The only difference is that since 2004 the number of papers has remained somehow constant. The development of publications related to ALD promises a bright future for research and development.

The book is divided in two parts: the first part (Chapters 1–7) deals with all the basic aspects of the technique, while the second part focuses on ALD-based nanostructured materials and their fields of application (Chapters 8–17).

The introductory chapter, written by J. Niinistö and L. Niinistö is an introduction to ALD and its development since its discovery in 1976. The following two chapters describe theoretical modeling of ALD processes (Chapter 1) and step coverage (Chapter 2). Chapter 3 describes the precursors used in ALD and their requirements. Chapter 4 describes the soft chemistry routes to oxides and the comparison of the chemistry taking place in ALD and in solution. Chapter 5 introduces molecular layer deposition, which is equivalent to ALD for the deposition of hybrid organic–inorganic thin films. The features and the relevance of low-temperature processes are discussed in Chapter 6. Chapter 7 describes the plasma-enhanced ALD.

The second part of the book (Chapter 8) starts with the applications of ALD in microelectronics; these are without any doubt the reasons why ALD became so popular nowadays. Chapter 9 introduces area-selective ALD and shows how it can be used for the formation of nanopatterns. Chapters 10, 11, and 14 describe the coatings of high aspect ratio nanostructures, nanoparticles and nanowires, and carbon nanotubes. Chapters 12 and 13 describe the coatings of soft materials such as polymers and biological materials. Chapter 15 describes the coating of nanostructures for optical applications such as opals. Chapter 16 describes the fabrication and properties of multilayers or nanolaminates. Finally, Chapter 17 discusses the challenges ALD is currently facing and possible novel directions and applications.

The book is structured in such a way to fit both the need of the expert reader (due to the systematic presentation of the results at the forefront of the technique and their applications) and the ones of students and newcomers to the field (through the first part detailing the basic aspects of the technique).

Halle, Germany
December 21, 2010

Dr. Mato Knez
Prof. Dr. Nicola Pinna

Introduction: Basic Features and Historical Development of Atomic Layer Deposition

Jaakko Niinistö and Lauri Niinistö

1
Introduction

Atomic layer deposition (ALD) has gained wide interest as an advanced thin film growth technique for various applications in modern technology. The strength of the method relies on its unique growth process where alternate, self-limiting surface reactions of the precursors, separated by inert gas purging, form a growth cycle whereupon thin, up to one atomic layer of material is grown. Precise thickness control can be achieved by repeating the growth cycle a desired number of times. The unique growth mode leads to perfectly conformal films [1–4].

The ALD technology was developed and patented almost 40 years ago in Finland by Suntola and coworkers [5]. The goal in the early development of the ALD technology was strictly application-oriented, namely, the need to develop flat panel displays for mass production. However, the breakthrough of the technology was driven by the semiconductor industry some 10 years ago and industrial applications for ALD are now emerging.

This chapter describes briefly the basic principles of the ALD method, including the benefits and limitations. As it is quite interesting to understand how the ALD technology has evolved over the years, a short historical perspective to ALD technology is also given. In addition, we try to emphasize the development of the ALD community, and novel processes and applications studied by a large number of research groups in industry and academia. Especially, answer is sought for the question: Which research areas of ALD were in main focus in the past and which seem to be emerging in the future?

2
Basic Features of ALD

ALD processes and their possible as well as industrial applications have been reviewed numerous times in the past and even very extensive reviews have been published. [1, 6, 7] ALD is a special variant of the well-known chemical vapor

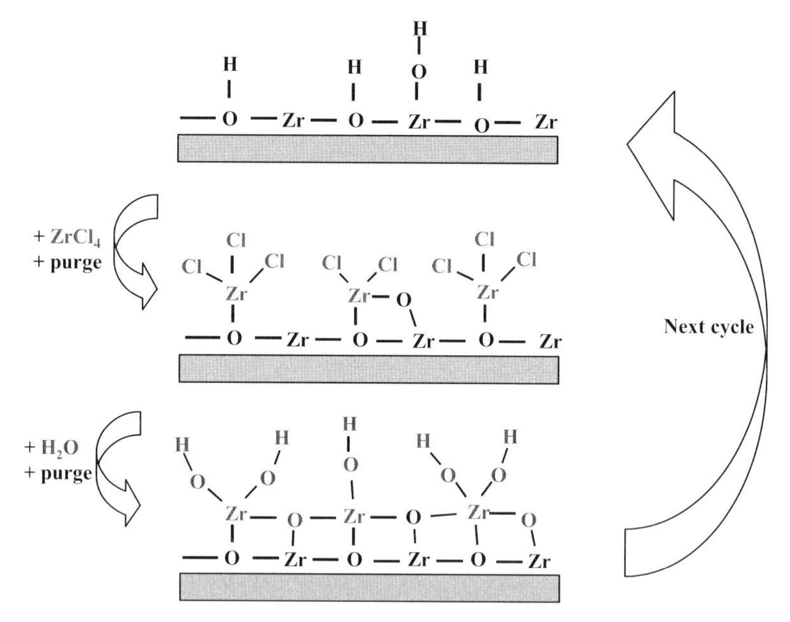

Figure 1 Sketch of a typical ALD growth cycle leading to ZrO_2 from $ZrCl_4$ and H_2O precursors. Reproduced with permission from ref. [3].

deposition (CVD) method. However, the differences between ALD and CVD are obvious. Whereas in CVD the precursors react at the same time on the surface or in the gas phase and precursors can decompose, in ALD the highly reactive precursors react separately by saturating surface reactions without self-decomposition. The method is surface controlled rather than process parameter controlled as in the case of CVD. In ALD the growth process proceeds in a cyclic manner and can be described as follows (exemplified with Figure 1).

An ideal ALD growth cycle is characterized by

- exposure of substrate surface to pulse of the first gaseous precursor;
- chemisorption of the first precursor onto the substrate;
- inert gas purge;
- introduction of the second precursor;
- surface reaction to produce the thin film;
- inert gas purge to remove gaseous reaction by-products.

In order to achieve a surface saturative ALD-type process, the growth rate has to be independent of the precursor dose provided that the dose is sufficiently large so that all the available surface sites have been occupied (Figure 2a). In other words, the precursor decomposition leading to a CVD-type growth mode should be avoided.

Often, but not always, a region with a constant deposition rate, also known as ALD window, is observed [2, 8, 9]. The ALD window certainly is not a requirement for an ALD-type growth mode, but it is a desirable feature that leads to the reproducibility of the film growth. Especially, if a ternary material is to be deposited, overlapping ALD windows of the constituent binary processes offer a good starting point for the development of a ternary process. The observed growth rates *vs.* temperature in ALD processes are shown in Figure 2b.

2.1 Limitations and Benefits of ALD

Compared to CVD, the deposition rate in ALD is rather low: An ALD cycle requires some time, typically few seconds, and the resulting thickness after one cycle is ideally one monolayer but in practice, due to steric effects, only a distinct factor of a monolayer. On the other hand, in modern technology the required film thickness is often from a couple of nanometers up to few tens of nanometers, thus the low growth rate is not seen as severe problem any more. In addition, large batch processing is a rather straightforward way to increase throughput [10].

Another limitation often considered is the limited materials selection, that is, the available materials that can be grown effectively by ALD. Several technologically interesting materials such as Si, Ge, Cu, multicomponent oxide superconductors, and many transition metals are lacking effective and production-worthy processes. However, the list of available processes [1, 6] is expanding continuously and process development through precursor chemistry is increasingly studied by many research groups, in academia and the chemical industry. In addition, with plasma processing the materials selection is likely to be increased [11].

During the recent years, many research groups have been studying growth characteristics during the early stages of growth in order to understand the growth mechanism and to better control the growth process (Figure 2). In many cases, poor nucleation has been observed that can lead to rough surfaces and in the case of many metals a discontinuous film when the thickness is only couple of nanometers. Also, in the case of high-*k* dielectric growth on bare silicon can suffer from inhibited growth and it can be difficult to obtain a sharp interface without a mixed interfacial layer between the material and the Si substrate [12, 13]. Such process-specific deviations are not characteristic to the method itself, but rather the precursor

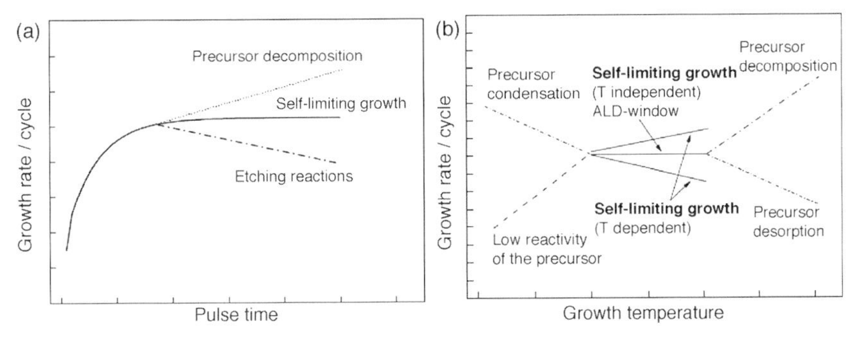

Figure 2 Growth per cycle as a function of the pulse time (a) and temperature (b) for different growth modes.

chemistry should be considered as one of the reasons. Other deviations related to precursor chemistry include reactive by-products that could etch the growing film or readsorb on the surface. Even small partial decomposition of the precursor can change the growth mechanism substantially. However, it should be noted that some applications can tolerate some of these deviations and "an ideal ALD growth" is not always needed.

The list of advantages of ALD is long. The characteristic feature of ALD, the self-limiting growth mode, leads to a precise control of the film thickness, a large area capability, and uniformity also in batch processes as well as to a good reproducibility. The excellent conformality is the main feature that distinguishes ALD from other thin film deposition methods and possible applications related to that are numerous. Separate dosing of highly reactive precursors allows production of high-quality materials (e.g., pinhole-free materials) at relatively low temperatures. The growth temperatures can often be reduced using, for example, plasma-generated radicals. Further benefits include the possibility to produce multilayer structures and the straightforward doping.

2.2 The Impact of Precursor Chemistry

As noted previously, a successful ALD process relies on the underlying chemistry. The alternately introduced precursors undergo saturative surface reactions. Thermal decomposition of the applied precursor destroys this self-limiting growth mode and thus the thermal stability of the precursor is the key issue in the ALD process. Naturally, being a vapor-phase method, ALD requires sufficient volatility of the precursors. Fast and complete reactions are mandatory; first of all, precursors must adsorb or react with the surface sites and these intermediates must be reactive toward the other precursor, for example, H_2O in the case of many oxide processes. To maintain film uniformity, precursor molecules or reaction by-products should not etch the growing film.

In general, volatile metal-containing ALD precursors can be divided into five categories: elements (only Zn and Cd), halides, oxygen-coordinated precursors (alkoxides and beta-diketonates), true organometallics (carbon-coordinated precursors, such as alkyls and cyclopentadienyls), and nitrogen-coordinated precursors (alkylamides, silylamides, and acetamidinates). For nonmetal precursors, hydrides are most widely used. Many successful oxide processes apply water as an oxygen source. Usually O_2 is found too inert in thermal ALD but it is used as a radical source in plasma-enhanced ALD and, for example, in growth of noble metals. Ozone has gained increasing interest as a powerful oxygen source and some industrial oxide processes rely on the use of ozone. The other most commonly used nonmetal precursors include H_2 and NH_3, also as a radical source. For sulfide film growth, H_2S should be mentioned.

As the precursor chemistry is extremely important factor in ALD processing, more detailed description on different aspects is given elsewhere in this book.

3
Short History of the ALD Technology

The purpose of the development of ALD technology in the 1970s by T. Suntola and coworkers was to meet the needs of producing improved thin films and structures based thereupon for thin film electroluminescent (TFEL) flat panel displays. The first Finnish patent application was filed in November 1974 and the first US patent was granted in 1977 [5]. The name of the technology then was named by T. Suntola as atomic layer epitaxy (ALE). The word epitaxy can be translated as "on arrangement" and thus Dr. Suntola used this broad definition even though usually ALD is used to grow nonepitaxial films. Until the end of the twentieth century, the method was mostly called ALE, which was not universally accepted because mostly polycrystalline and amorphous films are deposited rather than single crystalline growth on single crystalline substrate. Other names were used as well, for example, atomic layer CVD. For the past 15 years, the term ALD became generally accepted.

The above-mentioned patents showed the basis of reactor and process technology of ALD. It should be noted that in Russia, V. Aleskowski and others had published papers in Russian language about "molecular layering," describing alternate surface reactions characteristic to ALD. This work has been reviewed extensively by R. Puurunen earlier [6]. Here we focus on the development of ALD technology, especially reactor and process development for various applications.

3.1
ALD Reactor Development

The strength of T. Suntola's work inventing the ALD technology is clearly based on the development of feasible reactors for the processing. The aim of the first patent by Suntola and Antson was to describe a method for producing compound thin films using pure elements as reactants. The reactor configuration is shown in Fig. 3. In a rotating substrate reactor, the substrates are moved from one precursor flux to another and thus high-speed valving of pulses is not needed. This kind of concept is getting more interest again when roll-to-roll ALD or spatial ALD reactors are being developed.

Usually ALD reactors are based on valving the precursors and purges into the reaction chamber. Pulsing of the precursors separately, especially if the precursor has high vapor pressure, is simple with pneumatic or solenoid valves. Thus, it is possible to use CVD reactors as ALD reactors and many ALD reactor configurations are modified pulsed CVD reactors. The development of an inert gas valving system in early 1980s can be considered as a breakthrough for introducing low-pressure precursors. This invention led to the introduction of first commercial ALD research-scale reactors few years later. The reaction chamber and thus ALD reactors can be roughly divided into cross-flow (traveling-wave) or perpendicular-flow (e.g., showerhead) type reactors. The benefit of the cross-flow reactor is the higher throughput, and both precursor pulsing and inert gas purging are fast. However, the film thickness uniformity can be difficult to control, as CVD-type growth components may be present. Thus, careful optimization of the gas flow dynamics

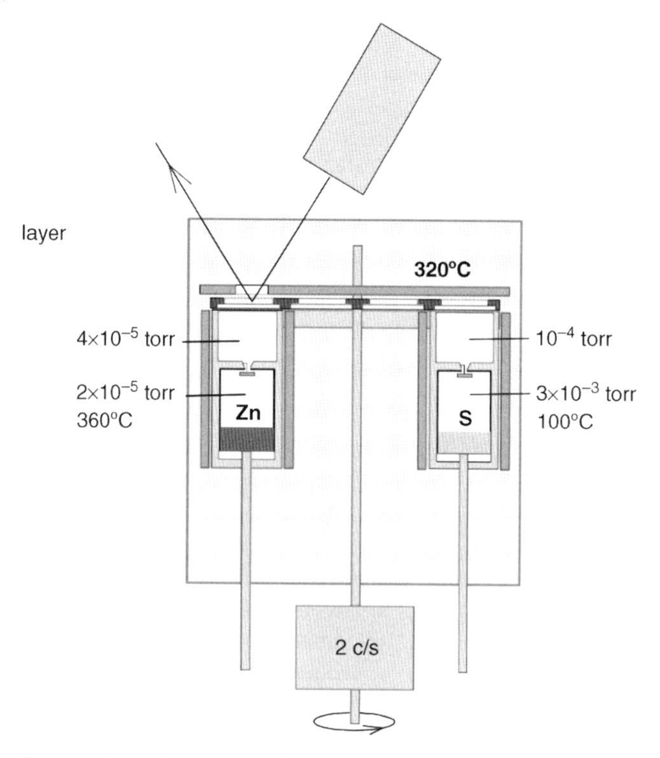

Figure 3 An early reactor configuration. Courtesy of Tuomo Suntola.

is needed. Batch reactors were introduced in the 1990s to overcome the major limitation of ALD technology, the low throughput. Currently, large batch reactors are used for various applications in volume production [14]. Carefully optimizing the configuration with multiple flow channels, it is possible even to reach nearly as fast cycle times as in single-wafer systems.

As the plasma became popular giving additional energy in CVD processing, the first reports of plasma-enhanced ALD became available in the late 1990s [15]. Many reactor manufacturers are introducing their PEALD reactors into the market. Using remote hydrogen plasma, for example, can overcome some problems related to ALD of transition metals. It should be noted that the PEALD techniques can be roughly divided to two categories, direct and remote PEALD. Among those, remote PEALD has gained wider interest, as mainly radicals are reaching the growing film during plasma exposure and ion bombardment can be avoided.

3.2
The Early Years – ALE

As mentioned earlier, TFEL displays were the key issue when inventing ALD technology. The research in ALE (the definition used originally) was focused on developing these devices and patents were mainly applied rather than research articles published. ALD showed its significance as a method producing dense and pinhole-free films for TFEL displays, where insulator–luminescent layer–insulator films should withstand high electric fields. The ALD grown layers for this rather thick (even >1 μm) structure were Al_2O_3 or Al_xTi_yO, ZnS:Mn, and Al_2O_3 [1]. The commercial ALD-based production for TFEL displays started already in the early 1980s [16]. In 1984, the first scientific meeting, First Symposium on Atomic Layer Epitaxy, was organized in Finland (Figure 4a). The contributed talks concentrated mainly on TFEL displays and only half a dozen groups participated in the meeting [17].

At the same time, besides the ALE research related to EL displays, interest toward polycrystalline II–VI semiconductors, such as CdTe, also rose.

3.3 Novel Method Still Unknown – ALE/ALD in the 1980s and 1990s

The epitaxial growth of III–V and II–VI compounds began to gain interest internationally from mid-1980s until mid-1990s. The research was extremely active in this area in Japan and the United States. The First International Symposium on ALE was arranged in summer 1990 in Finland (Figure 4b). The majority of contributed papers focused on III–V compounds, especially the growth of epitaxial GaAs [18].

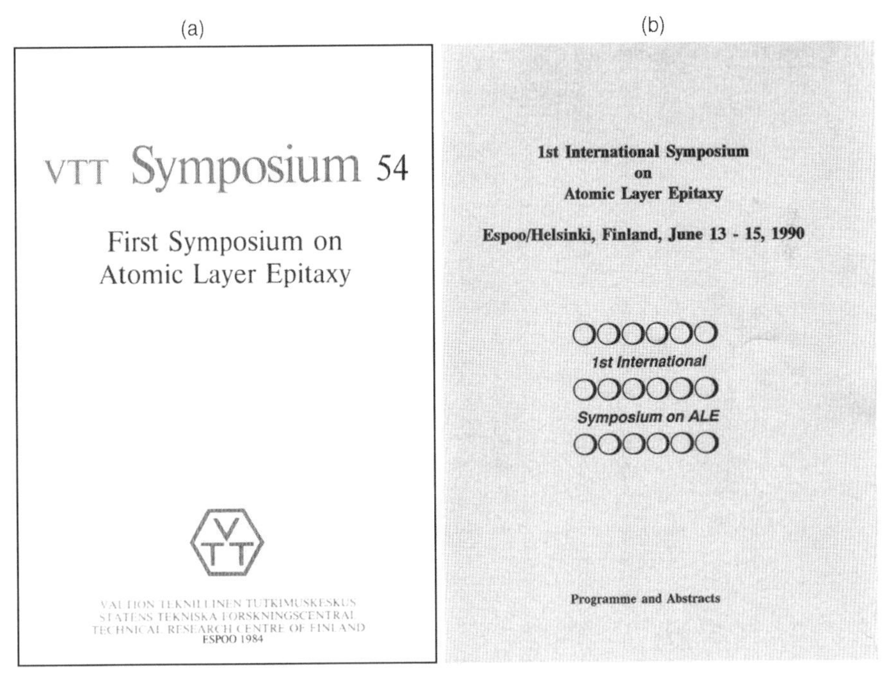

Figure 4 First pages of the book of the abstracts of the first ALE symposium held in Espoo in 1984 (a) and the first international symposium on ALE held in Espoo in 1990 (b).

In addition, during that time one major research focus was the deposition of Si and Ge. Superconductor research was also of interest.

At the same time, research efforts focusing on polycrystalline and amorphous oxides increased. This research was extremely important in the later stage, when microelectronics industry realized the potential of ALD. In a review published in 1996, growth processes for a dozen of binary oxide materials such as HfO_2 and TiO_2 were listed [19]. In addition, surface chemistry of ALD processes began to gain interest [20]. However, in the 1990s the full recognition for ALD was still to be seen and the interest was growing rather slowly.

3.4 Breakthrough for ALD – Microelectronics

The final breakthrough for the ALD technology can be timed to approximately the beginning of the twenty-first century. The reason for the increase in interest was the continuous scaling of microelectronic devices. The major semiconductor companies started their research already in the end of the 1990s and soon more and more research groups in industry and academia established ALD research in their laboratories. The expanding ALD community is discussed in more detail in Section 4.

The first microelectronic application was related to memories, to be more specific, an ALD Al_2O_3 insulating layer in DRAM capacitors [14]. The need for rather thin conformal coating on high aspect ratio structures is perfectly suited for ALD. The technology is used in high-volume production, current state-of-the-art ALD material as DRAM high-k dielectric materials being ZrO_2-based [21, 22]. Future materials include $SrTiO_3$ in a metal–insulator–metal (MIM) capacitor structure and ALD is the method of choice for conformal coatings with such materials.

However, largest publicity around ALD was reached somewhat later, when the semiconductor company Intel announced adaptation of ALD in manufacturing HfO_2-based high-k gate oxide in CMOS transistors [23]. The research efforts for this application started more than 10 years earlier and the most likely ALD process for this gate oxide process was published in 1994 [24]. Implementing novel manufacturing technology, such as ALD, in microelectronics industry took time, but it seems that now novel ALD processes can be adapted into production somewhat faster.

Due to the obvious need for ALD processed materials, a large number of ALD journal articles were related to those applications. High-k oxides, especially HfO_2, seemed to dominate the field in the beginning of the 2000s. Wide research efforts have been put not only on oxides, but also on metals for microelectronic applications. Many novel processes were developed, such as Ru processes. Copper for interconnects was also studied intensively.

In spite of the numerous research groups studying ALD for microelectronic applications, there are still many goals to be achieved. Better processes for future materials need to be developed and integration problems need to be solved. For example, $SrTiO_3$ for DRAM applications is extensively studied, and whether the industry will use ALD $SrTiO_3$ in high-volume production is to be seen. It should be noted that not only volatile memories but also nonvolatile memories offer a lot of possibilities for ALD to solve many problems, phase-change memories with ALD GST being a good example.

3.5 Emerging Applications for the Future

As microelectronic applications dominated the ALD technology field in the past decade, the current research efforts are going into the direction of much wider application areas. Nanotechnology applications seem to be well suited for ALD and more and more research papers are covering these topics. Photovoltaics is certainly an emerging field for ALD. Also, energy-related applications are research fields where ALD is applied. Other possible breakthrough fields for ALD include MEMS, biotechnological applications, and applications related to growth on nanotubes and graphene. Various interesting topics are covered later in this book. It seems that ALD has become a mainstream technology, and the possible application areas are very broad.

4 The ALD Community in the Academia and Industry

As mentioned several times earlier, the ALD community, groups concentrating on various ALD research topics, precursor and reactor manufacturers, and industrial "end users" of ALD technology, has increased dramatically during the past 15 years. In the early days of ALD, only few groups were more or less active. For example, the first ALD patent was applied in 1974 and the first research paper was published in 1980 [25]. In spite of the small number of groups in the field, the first ALD symposium was held in 1984 (Figure 4a). This national symposium in Finland mainly consisted of talks related to EL displays. As mentioned, the interest rose slowly, but already in 1990 the first international conference was held in Finland [26] and it acted as a start for first series of ALD conferences. This international symposium series was held four times, latest in 1996 [27]. At the same time, longest lasting conference series was established, first as a small symposium between Finnish and Estonian ALD groups and then expanding as a Baltic ALD conference series, latest organized in Germany in 2010. The international ALD conference series was established in 2001, and the first international conference on ALD was held in Monterey, CA [28], and it gathers the expanding ALD community every summer. In addition, many large international materials science conferences do have special ALD sessions annually.

A review article from the year 1998 shows the number of ALD groups at that time more or less actively using ALD and publishing ALD-related papers or patents [29]. Out of those groups (about 40), which were mainly concentrated in Finland, the United States, and Japan, some pioneering groups are still strongly studying ALD. However, from the end of the 1990s, the number of groups and ALD reactors in use has increased at such a pace that it would be impossible to gather such data. Hundreds of ALD reactors have been distributed and groups working in industry and academia are constantly expanding. This can be easily seen in Figure 5, where the

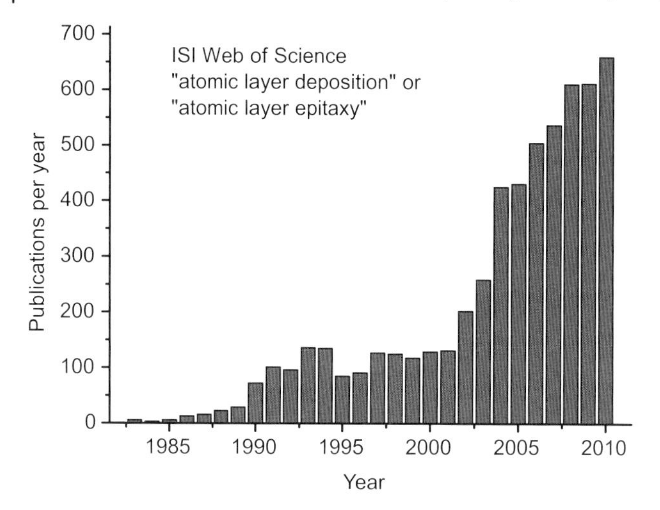

Figure 5 Number of articles per year published on ALD (keywords: "atomic layer deposition" or "atomic layer epitaxy"). Source Web of Science accessed on 2011.

number of scientific papers per year is plotted. From the early years, where only couple of ALD publications appeared, number approaching 1000 publications is the current status. For patents and patent applications, the increase has been even more intense during the recent years.

5
Conclusions

As a method for producing high-quality thin films, ALD has in the recent 35 years increased to a mainstream technology. Its future seems bright, and more and more materials scientists are using this method for numerous applications. Its importance for modern technology has been proved, and the number of industrial ALD applications is constantly increasing. However, there are many topics where ALD need to be further improved. Especially, the development of novel ALD processes through precursor chemistry is important. In addition, the development of ALD reactor technology should provide more and more advantages in terms of productivity for ALD compared to many other film deposition techniques.

References

1 Ritala M. and Leskelä, M., Atomic layer deposition in Handbook of Thin Film Materials, ed. Nalwa, H.S., Academic Press, New York 2002, vol. 1, pp. 103–159.

2 Niinistö, L., Päiväsaari, J., Niinistö, J., Putkonen M., and Nieminen, M. (2004) Advanced electronic and optoelectronic materials by atomic layer deposition: an overview with special emphasis on recent progress in processing of high-k dielectrics and other oxide materials, *Phys. Stat. Sol. A* **201**, 1443–1452.

3 Leskelä, M. and Ritala, M. (2003) Atomic layer deposition chemistry: recent developments and future challenges, *Angew. Chem. Int. Ed.*, **42**, 5548–5554.

4 M. Ritala, and J. Niinistö, Atomic Layer Deposition, in Chemical Vapour Deposition: Precursors and Processes, Eds. Jones, A.C. and Hitchman, M.L., *Royal Society of Chemistry*, 2009, pp. 158–206.

5 Suntola, T. and Antson, J., Method for producing compound thin films, *US Patent* **4** 058 430 (1977).

6 Puurunen, R.L. (2005) Surface chemistry of atomic layer deposition: A case study for the trimethylaluminum/water process, *J. Appl. Phys.*, **97** 121301–52.

7 George, S.M. (2009) Atomic layer deposition: an overview, *Chem. Rev.*, **110**, 111–131.

8 Suntola, T. (1996) Surface Chemistry of Materials Deposition of Atomic Layer Level *Appl. Surf. Sci.*, **100/101** 391–398.

9 Suntola, T., Atomic layer epitaxy. In Handbook of Crystal Growth, vol. 3, Edited by Hurle D.T.J., Amsterdam, Elsevier. (1994) 601–663.

10 Granneman, E., Fischer, P., Pierreux, D., Terhorst, H., and Zagwijn, P. (2007) Batch ALD: Characteristics, comparison with single wafer ALD, and examples, *Surf. Coat. Technol.*, **201**, 8899.

11 Kessels, W.M.M., Heil, S.B.S., Langereis, E., van Hemmen, J.L., Knoops, H.C.M., Keuning, W., and van de Sanden, M.C.M. (2007) Opportunities for plasma-assisted atomic layer deposition, *ECS Trans.*, **3**, 183–190.

12 Copel, M., Gibelyuk M., and Gusev, E. (2000), Structure and stability of ultra-thin zirconium oxide layers on Si(001), *Appl. Phys. Lett.*, **76**, 436.

13 Ritala, M., Atomic layer deposition, in High-k Gate Dielectrics, ed. Houssa, M. Institute of Physics Publishing, Bristol, United Kingdom, 2004, pp. 17–64.

14 Ritala, M. and Niinistö J. (2009) Industrial applications of atomic layer deposition, *ECS Transactions*, **25** (8), 641–652.

15 Sherman, A., Atomic layer deposition for nanotechnology (2008) Ivoryton Press, Ivoryton, USA, 239 p.

16 Pakkala, A., AVS 4th International Conference on Atomic Layer Deposition, Helsinki, Finland, 2004, presentation available on CD-ROM.

17 Paananen, R. (Ed.) Proceedings of the First Symposium on Atomic Layer Epitaxy, VTT, Espoo, 1984, 46 p.

18 Niinistö L., (Ed.): Proceedings of the First International Symposium on Atomic Layer Epitaxy, *Acta Chem. Scand.*, Helsinki, 1990, 209 p.

19 Niinistö, L., Ritala, M., and Leskelä, M. (1996) Synthesis of oxide thin films and overlayers by atomic layer epitaxy for advanced applications, *Mater. Sci. Eng. B*, **41**, 23–29.

20 George, S.M., Ott, A.W., and Klaus, J.W. (1996) Surface chemistry for atomic layer growth, *J. Phys. Chem.*, **100**, 13121–13131.

21 Kuesters, K. H., *et al.* (2009) New materials in memory development sub 50 nm: trends in flsh and DRAM, *Adv. Eng. Mater.*, **11**, 241–248.

22 Niinistö, J., Kukli, K., Heikkilä, M., Ritala, M., and Leskelä, M. (2009) Atomic layer deposition of high-k oxides of the group 4 metals for memory applications, *Adv. Eng. Mater.*, **11**, 225–234.

23 Bohr, M. T., Chan, R. S., Ghani, T., and Mistry, K. (2007) The high-k solution, *IEEE Spectrum*, **44** (10), 29–35.

24 Ritala, M., Leskelä, M., Niinistö, L., Prohaska, T., Friedbacher, G., and Grasserbauer, M. (1994) Develoment of crystallinity and morphology in hafnium dioxide thin films grown by atomic layer epitaxy, *Thin Solid Films*, **250**, 72.

25 Ahonen, M, Pessa, M., and Suntola, T. (1980) *Thin Solid Films*, **65**, 301.

26 Niinistö, L. and Karlemo, T. (Eds.) Abstract book the 1st International Symposium on ALE, Espoo, Finland, June 13–15, 1990.

27 Sitter H. and Heinrich, H. (Eds.): Proceedings of the Fourth International Symposium on Atomic Layer Epitaxy and Related Surface Processes, Amsterdam, Elsevier, 1997

28 AVS 1st International Conference on Atomic Layer Deposition, Monterey, USA, May 14–15, 2001, Extended Abstracts available on CD-ROM.

29 Niinistö L. (1998) Atomic layer epitaxy, Curr. Opin. *Solid State & Mater. Sci.*, **3**, 147–152.

List of Contributors

Aziz I. Abdulagatov
University of Colorado
Departments of Chemistry and Biochemistry and Chemical and Biological Engineering
Boulder, CO 80309-0215
USA

Stacey F. Bent
Stanford University
Department of Chemical Engineering
381 North-South Mall
Stanford, CA 94305
USA

Guylhaine Clavel
University of Aveiro
Department of Chemistry and CICECO
Campus Universitario de Santiago
3810-193 Aveiro
Portugal

Jeffrey W. Elam
Argonne National Laboratory
Communications & Public Affairs
9700 S. Cass Avenue
Argonne, IL 60439
USA

Hong Jin Fan
Nanyang Technological University
School of Physical & Mathematical Sciences
Division of Physics and Applied Physics
21 Nanyang Link
Singapore 637371
Singapore

Davy P. Gaillot
University of Lille 1 (USTL Lille 1)
TELICE Group
IEMN/UMR 8520, Bâtiment P3
59655 Villeneuve d'Ascq Cedex
France

Steven M. George
University of Colorado
Departments of Chemistry and Biochemistry and Chemical and Biological Engineering
Boulder, CO 80309-0215
USA

Zachary M. Gibbs
University of Colorado
Departments of Chemistry and Biochemistry and Chemical and Biological Engineering
Boulder, CO 80309-0215
USA

Robert A. Hall
University of Colorado
Departments of Chemistry and
Biochemistry and Chemical and
Biological Engineering
Boulder, CO 80309-0215
USA

Cheol Seong Hwang
Seoul National University
Department of Materials Science and
Engineering and Inter-university
Semiconductor Research Center
151-744 Seoul
Korea

Erwin Kessels
Eindhoven University of Technology
Department of Applied Physics
5600 MB Eindhoven
The Netherlands

Mato Knez
Max-Planck-Institute of Microstructure
Physics
Weinberg 2
06120 Halle
Germany

Byoung H. Lee
University of Colorado
Departments of Chemistry and
Biochemistry and Chemical and
Biological Engineering
Boulder, CO 80309-0215
USA

Han-Bo-Ram Lee
Stanford University
Department of Chemical Engineering
381 North-South Mall
Stanford, CA 94305
USA

Younghee Lee
University of Colorado
Departments of Chemistry and
Biochemistry and Chemical and
Biological Engineering
Boulder, CO 80309-0215
USA

Markku Leskelä
University of Helsinki
Department of Chemistry
00014 Helsinki
Finland

Catherine Marichy
University of Aveiro
Department of Chemistry and CICECO
Campus Universitario de Santiago
3810-193 Aveiro
Portugal

Jens Meyer
Princeton University
Department of Electrical Engineering
Olden Street
Princeton, NJ 08544
USA

Charles B. Musgrave
University of Colorado at Boulder
Department of Chemical and Biological
Engineering
Boulder, CO
USA

Kornelius Nielsch
University of Hamburg
Institute of Applied Physics
Jungiusstrasse 11
20355 Hamburg
Germany

Jaakko Niinistö
University of Helsinki
Department of Chemistry
00014 Helsinki
Finland

Lauri Niinistö
Aalto University School of Chemical Technology
Laboratory of Inorganic Chemistry
02015 Espoo
Finland

and

Picosun Oy
Tietotie 3
02150 Espoo
Finland

Sovan Kumar Panda
School of Advanced Materials Engineering and Center for Materials and Processes of Self-Assembly
Kookmin University
Jeongneung-Gil 77, Seoul, 136702
South Korea

Gregory N. Parsons
North Carolina State University
Department of Chemical and Biomolecular Engineering
Raleigh, NC 27695
USA

Nicola Pinna
University of Aveiro
Department of Chemistry and CICECO
Campus Universitario de Santiago
3810-193 Aveiro
Portugal

and

Seoul National University (SNU)
School of Chemical and Biological Engineering
College of Engineering
World Class University (WCU) Program of Chemical Convergence for Energy & Environment (C2E2)
151-744 Seoul
Korea

Stephen Potts
Eindhoven University of Technology
Department of Applied Physics
5600 MB Eindhoven
The Netherlands

Harald Profijt
Eindhoven University of Technology
Department of Applied Physics
5600 MB Eindhoven
The Netherlands

Andrea Pucci
University of Aveiro
Department of Chemistry and CICECO
Campus Universitario de Santiago
3810-193 Aveiro
Portugal

Matti Putkonen
Beneq Oy
01511 Vantaa
Finland

and

Aalto University
School of Science and Technology
Laboratory of Inorganic Chemistry
00076 Aalto, Espoo
Finland

Thomas Riedl
University of Wuppertal
Institute of Electronic Devices
Rainer-Gruenter-Str. 21
42119 Wuppertal
Germany

and

Chemnitz University of Technology
Institute of Physics
Chemnitz
Germany

Dragos Seghete
University of Colorado
Departments of Chemistry and Biochemistry and Chemical and Biological Engineering
Boulder, CO 80309-0215
USA

Hyunjung Shin
School of Advanced Materials Engineering and Center for Materials and Processes of Self-Assembly
Kookmin University
Jeongneung-Gil 77, Seoul, 136702
South Korea

Christopher J. Summers
Georgia Institute of Technology
School of Materials Science and Engineering
Atlanta, GA 30338-0245
USA

Adriana V. Szeghalmi
Max-Planck-Institute of Microstructure Physics
Weinberg 2
06120 Halle
Germany

Richard van de Sanden
Eindhoven University of Technology
Department of Applied Physics
5600 MB Eindhoven
The Netherlands

Marc-Georg Willinger
University of Aveiro
Department of Chemistry and CICECO
Campus Universitario de Santiago
3810-193 Aveiro
Portugal

and

Fritz Haber Institute of the Max Planck Society
Department of Inorganic Chemistry
Faradayweg 4-6
14195 Berlin
Germany

Byunghoon Yoon
University of Colorado
Departments of Chemistry and Biochemistry and Chemical and Biological Engineering
Boulder, CO 80309-0215
USA

Part One
Introduction to ALD

1
Theoretical Modeling of ALD Processes

Charles B. Musgrave

1.1
Introduction

This chapter describes simulations of atomic layer deposition (ALD) using quantum chemical electronic structure methods. Section 1.2 provides a brief overview of the quantum chemistry methods useful in the study of chemical processes and materials behavior. Although this section includes a summary of quantum chemical methods, it is not meant to be a comprehensive review of quantum chemistry and the reader is encouraged to examine the many excellent textbooks and reviews of quantum chemistry [1–3]. Section 1.3 overviews the use of quantum chemistry for predicting the properties of molecules and materials, while Section 1.4 specifically overviews the use of these methods to study ALD mechanisms, and Section 1.5 provides several examples of the determination of ALD mechanisms using quantum chemical methods. Again, this section is not meant to provide a comprehensive review of the use of electronic structure theory to study ALD and the motivated reader can examine the literature to explore what has specifically been done in this area. In addition to the various manuscripts our group has published on this topic [4–24], the groups of Esteve and Rouhani [25–29] and Raghavachari [30–35] have published many excellent articles on using quantum chemical methods to study ALD. While the overview of simulations presented here is meant to be helpful in understanding the approaches used to theoretically study the chemistry of ALD, it can be skipped for those who are already familiar with these methods or who do not seek a deeper understanding of electronic structure theory.

1.2
Overview of Atomistic Simulations

While simulations can be aimed at developing a description and understanding of the ALD process or at predicting the properties of the resulting ALD film, we here specifically focus on simulations intended to explain the chemistry of the ALD process. The unique features of ALD specifically rely on the nature of the half-

Atomic Layer Deposition of Nanostructured Materials, First Edition. Edited by Nicola Pinna and Mato Knez.

reactions that describe the chemistry of each half-cycle. Specifically, ALD relies on the self-limiting nature of the surface reactions that result from the use of reagents that each do not self-react and that are introduced into the ALD reactor in separate pulses, temporally separated from each other by a reactor purge [36]. During each precursor pulse, the precursor reacts selectively with the functional groups remaining from the previous precursor pulse, but not with itself. For a successful ALD process, each half-reaction must produce a surface functional group reactive toward the subsequent precursor and deposit at most one atomic layer. Ideally, because reactants do not self-react and are introduced into the reactor separately, chemistry and transport are decoupled in ALD reactors. Thus, the nature of the surface reactions is the central feature of the ALD process that provides it with its ideal attributes – uniformity, conformality, and nanometer-scale control of film thickness and composition. Consequently, a fundamental understanding of an ALD process involves a detailed description of the ALD chemical mechanism, and accordingly, we focus on summarizing methods that describe these surface reactions that provide ALD with its unique advantages.

1.2.1
Quantum Simulations

The most fundamental approach to describing chemical reactions is based on using quantum mechanics to describe the electronic structure of the reacting system. During a chemical reaction, the reactants transform into products by rearranging their atomic coordinates from that of the reactants to that of the products. The actual trajectory followed by the reacting species is variable as atomic motion is a highly dynamical process. Each intermediate structure along the trajectory followed during the reaction involves a unique ground-state electronic wave function. That is, the electron density of the system is dynamically redistributed as the atomic structure evolves from that of the reactants to that of the products. Because the electrons have much lower mass than the nuclei, the reacting system typically stays in its electronic ground state during the course of reaction. In other words, the redistribution of electron density along the trajectory of atomic configurations is adiabatic. Nonetheless, the key point is that this complex process is intrinsically quantum mechanical, despite being electronically adiabatic because the properties along the trajectory of the reaction, and specifically the energy, depend exclusively on the wave function at each atomic configuration.

Because chemical reactions are inherently quantum mechanical, any method used to simulate chemical reactions must either be quantum mechanical or somehow empirically incorporate the quantum mechanical nature of the process of reaction into its description. Classical molecular dynamics (MD) potentials are generally not capable of accurately describing chemical reactions, except in the very few cases where potentials have been designed and trained specifically to describe a particular chemical reactivity. Fortunately, over the past few decades the meteoric rise in computational power together with significant advances in electronic structure methods and algorithmic progress has made the direct application of quantum mechanics to describing atomistic systems of reasonable size practical, although nowhere as efficient as classical MD. In fact, high-quality quantum calculations that required supercomputers to execute just 20 years ago can now be performed on desktop computers. Another major development that enabled the ongoing revolution in quantum chemistry is the development of high-quality quantum mechanical density functional theory (DFT) methods that have been implemented in reasonably priced simulation packages with relatively simple graphical interfaces. This, combined with the availability of affordable modern computers, has enabled researchers on even the most modest budgets access to the power of quantum chemical calculations. The result is an explosion in the quantity of research conducted using quantum chemical simulations. Thus, the computational approaches we focus on are quantum chemical methods, both because of their now widespread use and because they are the most appropriate and reliable approaches for describing the chemistry of ALD.

A large number of different quantum chemical methods, including DFT, are now available in a variety of software packages. These can be categorized as being either semi-empirical or *ab initio* (from first principles). Semi-empirical methods approximately solve the Schrödinger equation, but use empirically derived parameters to compute the effects of terms that are either neglected or approximated in the specific approximations used within the method. These methods have the advantage of being extremely fast relative to *ab initio* methods. They can also be relatively accurate if the system and property of interest are within or not too different from the training set used to parameterize the method. Unfortunately, semi-empirical methods are generally not reliable for the prediction of activation barriers, which of course are central to predicting the chemistry of a reacting ALD system as the energies of the various possible transition states (TS) determine whether a reaction is active. Consequently, we do not discuss the application of semi-empirical methods to simulate ALD.

In contrast to semi-empirical methods, first principles methods are by definition not empirically fit to any experimental data set. *Ab initio* methods start with a fundamental quantum mechanical description of the system and then employ various approximations to make the solution of the quantum mechanical problem tractable. While approximations are employed, they do not involve fitting to any data. As we will see shortly, many DFT methods involve an empirical component in the development of the exchange correlation (XC) functionals that define each DFT method. Nevertheless, DFT methods are still typically called *ab initio* methods. Despite the fact that most DFT methods do not strictly conform to the definition of *ab initio*, we will also adhere to the general convention of calling these methods first principles methods.

1.2.2
Wave Function-Based Quantum Simulations

Quantum simulation methods are classified as either wave function or density functional methods. *Ab initio* wave function methods involve directly solving the Schrödinger equation using various approximations with the quality of the method depending on the degree of approximation. Thus, *ab initio* wave function methods can be ranked within a hierarchy that depends on the extent of the approximations used, ranging from the mean field approximation of the Hartree–Fock (HF) method [37–39] to exact methods that include *n*th-order perturbation theory [40] (in the limit of large *n* and if the perturbation is small), full configuration interaction (CI), and quantum Monte Carlo methods [41]. The most approximate *ab initio* wave function method is the HF method. The HF wave function is the lowest energy single Slater determinant wave function for an *n*-electron system, where the wave function is the *determinant* of the *n*-by-*n matrix* where each column is a different electron orbital and each row a different electron. A determinant form for the wave function guarantees that it is antisymmetric, a requirement for fermions, and that the electrons are treated as identical, indistinguishable particles. Unfortunately, despite being *ab initio*, the mean field approximation that defines the HF method causes it to severely overestimate activation barriers in the vast majority of reactions, among other inaccuracies, which thus makes it unsuitable as a method for accurately describing chemical reactivity. On the other hand, HF serves as the basis for almost all wave function-based methods. Generally, these approaches combine the *n*-electron HF ground-state wave function with excited HF wave functions constructed by replacing occupied orbitals of the ground state with up to *n* unoccupied (virtual) orbitals to improve the quality of the wave function. The mixing in of excited HF wave functions with the HF ground state to improve the description of the wave function is usually accomplished using perturbation theory or variationally using configuration interaction.

Because a detailed description of wave function methods is beyond the scope of this chapter and because the application of these methods to simulate ALD has been much more limited than the application of density functional methods, we will forgo a more extensive account of these methods. However, we note that *ab initio* wave function methods have the advantage that they can be systematically improved. That is, we can methodically climb the hierarchical ladder of methods from the base Hartree–Fock method up toward the exact methods by, for example, increasing the order of the many-body perturbation theory (MBPT) or increasing the order of excitations in a configuration interaction method, which include coupled cluster and complete active space methods. Thus, we can robustly determine whether a method accurately describes the system by systematically improving it and analyzing its approach to convergence. Unfortunately, the accurate wave function methods are generally prohibitively expensive for many systems of interest. This is especially true for the case of ALD because the systems of interest are surface reactions, and thus involve a model of the reacting surface containing too many atoms for practical simulation using the higher quality wave function methods. However, they can still play the role of calibrating more approximate methods practical enough for simulating ALD processes to validate the choice of method and provide error bars on the predicted reaction energetics.

1.2.3
Density Functional-Based Quantum Simulations

An alternative to using the electronic wave function as the basis of a quantum mechanical method is to use the electron density. Methods based on this approach are called DFT methods. Hohenberg and Kohn first showed that the electron density uniquely determines the number of electrons and the location and identity of the atoms, that is, the potential, and thus the system Hamiltonian and its associated Schrödinger equation [42]. Consequently, the electron density uniquely determines the wave function and thus all information contained in the wave function must also be contained in the density! The key is to construct an energy functional equivalent to the Hamiltonian that includes all important contributions to the electronic energy. A functional is essentially a function of a function; in this case, the energy is a function of the electron density, which is itself a function of position. Kohn and Sham proved that the exact energy functional exists [43], but unfortunately proof of its existence did not prescribe an approach to its determination. However, knowing that an exact energy functional exists motivated the development of DFT where the key challenge has been to derive energy functionals that accurately describe the system.

The general approach to developing DFT energy functionals has been to exploit the fact that energy is an extensive property in order to separate the energy functional into its component energy contributions, including the kinetic energy, electron–electron repulsions, electron–nuclear attractions, the exchange energy, and the correlation energy. The explicit inclusion of the exchange and correlation energy through the exchange and correlation density functionals, called the exchange correlation functional when combined, may appear strange here because they are not explicitly included in the Hamiltonian of wave function methods. This is because no antisymmetry requirement is imposed on the density, so exchange does not naturally arise in DFT as it does when using a Slater determinant form of the wave function and so it must be explicitly added. On the other hand, correlation is an opportunistic addition to the energy functional used to improve the quality of a density functional to help it reproduce the properties of the system. This explicit addition of correlation energy is remarkably computationally efficient compared to the approaches for incorporating additional correlation energy into wave function methods that rely on mixing large numbers of excited states into the ground-state wave function. Furthermore, density functional methods are computationally efficient because they not only use a single Slater determinant description of the electronic structure, but also do not involve calculating any "two-electron integrals," which is a requirement of all orbital-based wave function methods, making the wave function methods scale with at least the fourth power of the system size.

One drawback of DFT methods is that they are generally not improved systematically; whereas wave function methods calculate exchange and correlation based on expansions, the exchange and correlation density functionals that define the DFT method are based on specific functions. Another drawback is that these functionals often involve parameters that are empirically determined by a fitting process. Although virtually all DFT methods involve XC functionals that are empirically fit and are thus not strictly first principles methods, they are commonly referred to as "*ab initio*." Because DFT functionals are often empirically determined, great care should be taken in conducting simulations using DFT to ensure that the functional is appropriate for the property and system of interest. Unfortunately, it is common for DFT results to be published in the literature that do not confirm that the choice of DFT method is appropriate for the property of interest and the reader is left with the question of whether the results are reliable. This issue is often not fully appreciated. This might be partially caused by the confidence in the method instilled in the nonexpert by the use of the term "*ab initio*" or "first principles" to describe DFT methods. A similar caution could be given for wave function methods where the level of approximation within the method may make it incapable of accurately describing the system or phenomena of interest. Again, the user should confirm that the method is appropriate for the property being predicted.

1.2.4 Finite and Extended Quantum Simulations

Computational quantum mechanical methods can also be categorized by whether the method is applied in a finite or extended fashion. A finite calculation involves simulation of a localized system, such as an atom, molecule, or cluster of atoms, whereas an extended calculation models systems such as a surface or bulk material by periodically repeated unit cells. Wave function methods are almost exclusively only applied to finite systems, whereas DFT methods are routinely applied to both finite and extended systems. Extended systems are modeled using periodic boundary conditions (PBC), a good review of which can be found in Ref. [3], even if the underlying system is itself not periodic. The application of PBC requires that a supercell be defined that will be periodically repeated along the lattice vectors. This model naturally applies to periodic systems, such as crystals and their surfaces. However, it can also be applied to systems that do not exhibit periodicity by defining a relatively large supercell containing a locally nonperiodic structure that is repeated on the longer length scale of the chosen supercell. Quantum simulations of surface reactions, such as those of ALD on the growing surface, can be performed using either a finite cluster or periodically extended supercell model of the reacting surface site. Thus, for simulating ALD surface reactions the researcher must decide whether to use a finite cluster or extended model of the reacting surface.

If a cluster is chosen to model the reacting surface, it should be large enough that it includes the changes in the system occurring during the chemical reaction, including all significant interactions with the surrounding material. In other words, any changes induced on the substrate by the ALD surface reactions should be insignificant at the cluster edge. If a cluster model is chosen that is too small to model the ALD surface reaction, the edge of the cluster can introduce nonphysical interactions with the reacting atoms. For example, for a material such as GaAs the broken bonds at the cluster edges should be terminated in such a way as to preserve the electronic structure of the surface. However, a material such as GaAs involves bonds that are partially covalent and bonds that are partially ionic. When a surface is created, the broken bonds preferentially become lone pairs (in the case of As) and empty orbitals (in the case of Ga). Thus, in the case of a cluster, the surfaces that define the edge of the cluster are more difficult to "terminate" as the terminating groups, such as a H atom, will not form a bond to the Ga or the As atom that accurately mimics the Ga–As bonds of the bulk material. Of course, these artifacts can be reduced by using large clusters so that the edges are far from the reactive site of the cluster. Unfortunately, most quantum chemical methods scale with at least the third power of the system size. It is then important to define cluster models that are just large enough to accurately describe the effects at play during the reaction. Fortunately, the effects involved in chemical reactivity are relatively local phenomena and so the assumption that reactions on a cluster mimic reactions on the surface is usually valid. This is true except in cases of cluster models that are obviously too small because the gas-phase reagent reacts with more than a single surface reactive site, or because of edge effects due to the cluster being too small or designed with poor edge termination.

If a supercell model of the reacting surface is chosen to model the reactive surface site, then a supercell must be defined that is large enough to include the relevant interactions within the cell and where self-interactions resulting from interactions with periodic images in neighboring cells are small. Although the underlying substrate may be periodic, surface reactions break this periodicity as does the growth of the film, which is rarely epitaxial with the underlying substrate. Spurious periodic interactions with images in neighboring unit cells caused by the surface reactions and imposed by PBC can lead to artifacts, and thus like cluster models, care must be taken to avoid edge effects when supercell size is insufficient to isolate the reacting surface site from neighboring reacting sites. Although both approaches have their drawbacks, they have both been successfully applied to simulations of ALD and generally agree with each other when the choices made in model design and method are appropriate so as to not introduce significant artifacts [14].

1.2.5 Basis Set Expansions

In both finite and extended methods, the electronic structure is expanded as linear combinations of basis functions. For finite systems, including clusters of atoms designed to model regions of an extended structure such as a surface, basis functions are localized atomic orbital-like functions with their radial component described using either Gaussians, $\exp(-\zeta r^2)$, or Slater functions, $\exp(-\zeta r)$, where the exponent ζ determines the spatial extent of the basis function. On the other hand, extended systems modeled using supercells within PBC generally expand the electron density over a series of plane waves up to a chosen cutoff energy. This expansion is analogous to a Fourier expansion of a periodic function, in this case the density, and involves the usual difficulties of Fourier expansions, such as describing functions with discontinuities in their derivatives. This challenge can be met by including very high frequency plane waves in the expansion or describing the nucleus and core electrons of the system using pseudopotentials. This also has the effect of reducing the number of electrons that are treated explicitly, greatly reducing the computational cost for systems with large numbers of electrons.

Large basis sets include more basis functions or plane waves to generally allow for a better description of the wave function or the electron density, which is the square of the wave function. When using localized or plane wave basis functions, the number of unknown coefficients to be determined, and thus the computational demand, grows with the size of the basis set. Fortunately, the quality of the calculations typically converges at reasonably sized basis sets, although it is good practice to confirm this convergence. Methods have also been developed to estimate the energy predicted by infinitely large basis sets using what is called a basis set extrapolation.

1.3 Calculation of Properties Using Quantum Simulations

Quantum mechanical calculations are generally relatively computationally intensive. However, their ability to describe esoteric phenomena and accurately predict system properties, often beyond the capability of experiment, motivates their use. A wide variety of properties useful in understanding an ALD process can be computed using quantum simulations. The most basic and widely calculated property is the system energy. Although the system energy itself has little intrinsic value, differences in energies form the basis for calculating a wide range of properties. These include bond energies, adsorption energies, reaction energies, intermediate energies, activation barriers, electron affinities, ionization potentials, excitation energies, band gaps, and free energies. Furthermore, first and second derivatives of the energy with respect to the atomic coordinates allow one to calculate the forces and vibrational frequencies of the system, and thus optimize geometries. Calculation of forces allows one to predict stationary states (where all forces are zero) including stable structures and transition states, which are first-order saddle points on the system potential energy surface (PES) $E(R_A, \ldots, R_{3N})$, where the set $\{R_i\}$ contains the $3N$ coordinates of the N atoms. Furthermore, the calculation of frequencies enables the prediction of experimental vibrational spectra, such as infrared and Raman frequencies. Calculated frequencies also allow for the determination of zero-point energies and vibrational partition functions, from which entropies, enthalpies, heat capacities, and thermal corrections can be calculated, which then enable the calculation of free energies and other temperature-dependent properties.

Additional properties that can be calculated and that are useful for understanding ALD include the electron density, from which valuable insight into the electronic structure of the reacting system can be gleaned, HOMO–LUMO gaps, which are valuable in understanding the interactions involved during reaction, and band gaps and band offsets, which are often useful in predicting the electronic properties of the film and its interface with the substrate or an overlying metal electrode.

1.3.1 Calculation of Transition States and Activation Barriers

Our focus on simulations of the ALD process, specifically the reactions involved in ALD, naturally motivates the desire to predict properties related to the chemical reactivity. These properties include the active chemical mechanism of the process and the associated energetics and kinetics of the mechanism. We have already mentioned that the calculation of forces allows us to optimize structures to determine the energies of reactants, intermediates, products, and TS connecting these stable structures. If we optimize a series of stable structures from reactants and products along with a series of TS connecting the various intermediate structures, we have a candidate mechanism for a chemical reaction. The differences in energy between the transition states and the structures they connect are the forward and reverse barriers for that step in the mechanism. Finding a low-energy path from reactants to products allows us to state that the pathway is viable. However, we cannot ignore the possibility that a lower energy pathway may exist. This generally motivates the researcher to explore all reasonable pathways, meaning pathways that are chemically rational. Unfortunately, although the optimization of stable structures along the minimum energy path (MEP) of the reaction is straightforward in modern simulation codes, TS searches are still challenging despite the development of various algorithms for locating saddle points on the potential energy surface. One challenge is the sheer number of saddle points in a system with even only a moderate number of atoms and the possibility that incorrect saddle points are located. Another is that saddle points themselves are more subtle topological features on the potential energy surface than the minima of stable structures and involve the more complex constraint that all second derivatives, except the one that corresponds to the reaction coordinate, are zero. Thus, the forces on the atoms near transition states tend to be smaller than those on the atoms near stable structures for equal displacements from the stationary point.

Best practices for determining reaction mechanisms include:

1) using chemical intuition to initiate TS searches at starting structures near the TS structure,
2) ensuring that the right number of imaginary modes are found,
3) visualizing the vibrational mode corresponding to the reaction coordinate to confirm that it moves the geometry toward the two states the TS is intended to connect, and
4) performing an intrinsic reaction coordinate calculation where the MEP is followed in both directions from the TS to identify what structures the TS connects.

These efforts locate and confirm the nature of the TS, but unfortunately do not ensure that this TS is actually that of the MEP. And even if the correct TS is found, it is possible that a competitive pathway exists at the reaction temperature. Consequently, it is important to locate all low-energy pathways that might be active at the reactor conditions. A second issue involves the possibility that even if the reactants follow the MEP through the lowest energy TS, they may not follow the MEP on the product side of the potential energy surface because as the potential energy of the system at the TS is converted to kinetic energy as the system proceeds toward the products, that kinetic energy can enable the system to follow a trajectory different from the MEP along the exit channel. This possibility is relatively rare, but is more probable in cases where the kinetic energy is not quickly thermalized (redistributed to modes other than the reaction coordinate) or where low-energy TS within the exit channel exist. The above issues make locating of TS structures one of the hardest challenges for the user of simulation methods. Fortunately, experience, understanding of the chemical nature of the system, and creative use of the various techniques for finding TS structures will reward the persistent researcher.

Once sufficient confidence in a predicted mechanism is obtained, one can then use the schematic potential energy surface of this mechanism as a basis for understanding the relevant ALD process. While this is valuable, a number of additional useful properties related to the mechanism can also be calculated.

1.3.2
Calculation of Rates of Reaction

In addition to predicting the activation barriers for each step in a mechanism, one can also use the vibrational frequencies of the intermediate (or reactant) and transition states to calculate their partition functions and thus the preexponential factor of the rate constant using transition state theory (TST). Consequently, a quantitative prediction of the rates of reaction for the individual steps and an overall reaction can be calculated. Although the prediction of rate constants using TST with input calculated from quantum calculations is enormously valuable, several words of caution should be provided. The most important is that because the rate law depends exponentially on the ratio of the activation barrier to the thermal energy $k_B T$, where k_B is the Boltzmann constant, through the Boltzmann factor $\exp(-E_A/k_B T)$, a small error in the barrier can lead to considerable errors in the predicted rate. This is exacerbated at low temperatures where $\exp(-\Delta E_A/k_B T)$ is large for relatively small errors in the barrier, ΔE_A. For example, at 298 K an approximately 1.36 kcal/mol error in an activation barrier leads to a factor of 10 error in the rate constant. Unfortunately, most methods do not achieve an error of less than 1.5 kcal/mol in the activation barrier. For example, for most DFT methods errors of 3–4 kcal/mol are relatively typical making errors in the rate constant at 298 K near a factor of 10 000 common. The good news is that most methods tend to be relatively systematic in their under- or overprediction of barriers so that while the predicted rates may be off by a substantial amount, the relative rates between different steps will generally be in error by a significantly lower amount. For example, DFT methods tend to systematically underestimate barriers. Thus, the prediction of branching ratios can be expected to be relatively good compared to the absolute rates. When the competing pathways involve barriers that are different by more than 3–4 kcal/mol, DFT methods can reliably predict the correct active reaction. Of course, more accurate methods can be used to calibrate and verify the predictions of DFT or to provide a correction factor. For example, the CCSD(T) coupled cluster method can predict barriers to within $\sim$1 kcal/mol if the reaction possesses what is called "single-reference" character. Because CCSD(T) is computationally demanding, it is usually just used on small systems, for example, small models of the reaction including less than 10 heavy atoms, that are analogous to that of the actual ALD reaction. Another kinetic property that can be calculated is the kinetic isotope effect (KIE) because it is just the ratio of the rate constants of a reaction where in one case the reacting species has been isotopically substituted. Prediction of KIEs for the rate-limiting steps of a reaction thus allows simulation to predict experimental KIEs, providing a powerful approach to confirming the rate-limiting step, and thus the mechanism of a reaction.

A less common quantum simulation approach for investigating ALD involves *ab initio* molecular dynamics. While classical MD is relatively efficient, as mentioned above, it does not usually provide an accurate description of the bond dissociation or formation process. *Ab initio* MD, on the other hand, allows one to follow dynamics trajectories with energies and forces described using various flavors of quantum mechanics, most commonly DFT. Unfortunately, the computational expense of DFT, although low relative to methods of comparable or greater accuracy, still prohibits the simulation of large systems or for processes that take place over relatively long timescales compared to the vibrational timescale of the system. Despite these limitations, DFT-based MD can be quite useful in elucidating the details of an ALD process. The two main varieties of *ab initio* MD are the Born–Oppenheimer (BOMD) and Car–Parrinello (CPMD) methods, reviewed in Ref. [3]. BOMD solves the electronic structure problem explicitly at each time step and uses the resulting energy to calculate forces on the ions. In contrast, CPMD treats the electron coordinates explicitly as dynamical variables in an extended Lagrangian. Both approaches have proven quite useful, although the BOMD approach does not suffer from the electronic drag and nonadiabaticity often encountered with CPMD. These problems are due to how CPMD propagates the electron density forward where it is no longer the correct electronic structure for the "external potential" presented by the ion arrangement.

This brief treatment of quantum simulation methods is only intended to remind the reader of the basic ideas related to these simulation methods. For readers who are inclined to delve into the field of quantum simulations, a thorough review of quantum mechanics and quantum simulations is highly recommended [1–3].

1.4 Prediction of ALD Chemical Mechanisms

As described generally above, electronic structure methods such as DFT are uniquely capable of providing a first principles description of the chemical reactivity of reacting systems. This involves determining the active reaction pathways by identifying and characterizing the intermediates and transition states leading to the products of the ALD reaction. The key challenges are (i) locating the low-lying transition states, (ii) confirming that they indeed connect the states along the reaction path, and (iii) establishing confidence that these transition states are indeed the active pathways. Transition states are first-order saddle points, meaning all forces are zero (they are stationary points) and all second derivatives except one of the energy with respect to the nuclear coordinates are positive. The negative second derivative is the normal mode corresponding to the reaction coordinate for this step. Locating these often subtle topological features in a hyperdimensional surface can be difficult. This is exacerbated by the fact that many low-lying first-order saddle points may reside on this energy landscape, including hindered rotors (e.g., methyl groups) and skeletal modes of larger molecules. Furthermore, at higher temperatures pathways that involve higher lying transition states may become active. Although these obstacles can make determining ALD mechanisms using quantum chemical methods problematic, it is usually the only practical way to obtain a detailed description of the chemistry, and especially of the nature of the transition states as this type of information is not readily available from experiment. Despite these challenges, researchers have successfully used quantum chemical calculations to explain the mechanisms of a variety of ALD chemistries. We will next discuss the general types of ALD reactions that arise in ALD chemistry and then go through the specific example of the ALD of Al_2O_3 using trimethylaluminum (TMA) and water to illustrate a simple case of determining an ALD mechanism with DFT [20].

1.4.1 Gas-Phase Reactions in ALD

The first possible reaction in the ALD system might involve gas-phase self-reactions that could occur when the precursor is introduced into the ALD chamber. Although ALD precludes self-reactions, one could first calculate the barriers to possible gas-phase self-reactions to estimate the extent of the CVD component in this ALD chemistry as a function of temperature. Ideally, the CVD component is very low, but prediction of these barriers could provide evidence to show this. Often, experiment has shown that the precursors of the ALD chemistry being investigated do not self-react at reasonable temperatures and this step is skipped.

Other gas-phase reactions that could be studied as a prelude to the ALD surface chemistry are the reactions between the complementary ALD precursors. Although these species are introduced into the reactor in pulses temporally separated by intervening reactor purges to prevent them from reacting anywhere but on the substrate, the predicted reaction energies and TS energies can provide a reasonable estimate for the thermodynamics and kinetics of the ALD surface reactions. This is because these reactions are often isodesmic to the surface reactions. That is, they involve breaking and forming the same number and types of bonds and often proceed through a similar TS. Consequently, they are small analogues of the surface reactions and thus their chemistry can be calculated relatively quickly to screen and develop understanding of the ALD surface chemistry. In addition to providing insight into the driving forces and rates of the ALD reactions, they can also be used to facilitate the search for the TS of the analogous surface reactions.

1.4.2 Surface Reactions in ALD

1.4.2.1 Adsorption Reactions in ALD

The ALD process can be generally treated as a series of surface chemical reactions. Although the surface reactions that define ALD are self-limiting, they can be treated in the same fashion as the surface reactions of chemical vapor deposition, organic functionalization of surfaces, and even heterogeneous catalysis. This entails exploring the individual steps that may be active in the reacting system. A typical first step in an ALD mechanism is adsorption, which can be either dissociative or molecular.

Usually adsorption of the ALD precursor on the reactive substrate is molecular because of the self-limiting requirement of the surface chemistry. To calculate the adsorption energy, one calculates the energy of adsorbed state and subtracts the energies of the initial state, which is the isolated precursor and the substrate, with the structures of each of these states optimized to their minimum energy geometries. Adsorption may or may not involve going over a barrier. In the majority of cases, molecular adsorption either is barrierless or has an insignificant barrier [5, 8, 9, 14, 18–21]. In either case, the PES that describes the energy as a function of the adsorption reaction coordinate can be calculated.

Although adsorption is usually one of the simplest steps in an ALD mechanism to calculate, a few points of caution should be exercised. One is that the molecularly adsorbed state might involve van der Waals interactions, which are poorly described by most DFT methods not specifically designed to describe dispersive interactions. A second issue relates to the possibility that a precursor may adsorb to more than one surface site and that these sites might be different. If this is the case, the model of the reacting substrate must include these sites. This point becomes even more important when the precursor is relatively large or possesses ligands that can bond to various possible surface sites. Similarly, this can also be important in cases where different substrate conditions provide various possible surface reactive sites. For example, in the ALD of metal oxides using water as the oxygen source, a purge subsequent to the water pulse may leave the substrate with different relative concentrations of OH groups, –O– bridge sites, and even molecularly adsorbed H_2O [14, 44]. Finally, caution should be exercised when the interaction between the adsorbed precursor and the substrate can lead to nonlocal effects. For example, in the ALD of metal oxides and nitrides, metal precursors can act as Lewis acids and dative bond to surface Lewis base groups to form Lewis acid–base complexes. Similarly, oxygen and nitrogen sources can act as Lewis bases and dative bond to metal atoms of the surface, which act as Lewis acids. In the latter case, the donation of an electron lone pair from the adsorbing species to the surface can lead to charge transfer to neighboring sites, which requires that the model include those neighboring sites to describe this effect.

Although adsorption may appear less interesting than the reactions directly involved in removing precursor ligands to deposit the atoms of interest, it can play an important role in both establishing the initial structures from which ligand exchange reactions follow and often determining the energetics of the ALD process. For example, if the initial ALD adsorbed precursor–substrate complex is too stable, it can act as a trapped intermediate [18, 19]. Furthermore, the stability of this complex, together with the stability of the subsequent intermediate, affects the barrier connecting these states. If the TS to proceed to the following intermediate lies above the energy of the entrance channel (i.e., if the forward barrier is higher than the barrier to desorption), the ALD reaction may be relatively slow and have inefficient precursor utilization [18, 19].

1.4.2.2 Ligand Exchange Reactions

A key step in any ALD process is replacing the functional groups that cover the reacting surface with the functional groups of the reacting species. The surface reactions that are often responsible for this transformation are ligand exchange reactions. For example, the adsorbed precursor complex often involves a dative bond between the initial surface functional group and the precursor. If this initial functional group is not a component of the deposited film, it must be transformed to remove any atoms that are not constituents of the final material. For example, if the reactive surface site is an OH group and the ALD process deposits a metal oxide, the H atom must be removed by the reactions of the ALD mechanism. In this example, this can be achieved by a hydrogen transfer from the OH to one of the ligands of the adsorbed precursor to form a volatile by-product to carry away the H together with one of the precursor ligands. This ligand exchange reaction thus converts the OH functionalized surface to one functionalized by the ligands of the precursor. Similarly, ligand exchange can replace these functional groups by OH groups via a H transfer from absorbed or absorbing H_2O and the surface ligand.

Although a ligand exchange reaction replaces the ligands deposited by the previous half-cycle with those of the pulsed precursor, the by-product of that reaction must be volatile. Often, this species is physisorbed to the surface, for example, through a dative bond arising from a dipolar interaction. This binding must not be so strong as to severely inhibit the desorption of the product, otherwise these species may require long purges at relatively high temperatures to remove them. A purge that does not completely remove these by-products can lead to film contamination and nonuniform growth through site blocking. Although the by-product should have a reasonably low desorption barrier, ALD does not require that each half-cycle be exothermic. In fact, each half-reaction can be relatively endothermic because ALD is run under nonequilibrium flow conditions. Thus, the chemical potential of the reactants remains relatively high because the pulse continues to supply the precursors while the chemical potential of the by-products is kept relatively low by the fact that as the by-product is formed and desorbed from the surface, it is removed from the reactor chamber by the flow, providing the thermodynamics to drive the reaction forward toward film growth despite the endothermicity of the reaction.

1.5
Example of a Calculated ALD Mechanism: ALD of Al_2O_3 Using TMA and Water

The ALD of Al_2O_3 using trimethylaluminum (TMA, $Al(CH_3)_3$) and water (H_2O) is one of the most commonly employed ALD processes, both because of the wide number of applications of thin films of Al_2O_3 and because this ALD process is one of the most likely to succeed for any given substrate. In fact, the ALD of Al_2O_3 using TMA and water is sometimes considered the prototypical ALD process [45–47].

As with any ALD process, the reactions between the precursors and the surface can be separated into two half-reactions:

$$\mathrm{Al-OH^* + Al(CH_3)_3 \rightarrow Al-O-Al-CH_3{}^* + CH_4} \tag{1.1}$$

$$\mathrm{Al-CH_3{}^* + H_2O \rightarrow Al-OH^* + CH_4} \tag{1.2}$$

where the asterisks denote the surface species. Although this reaction appears to be not balanced, it is written to illustrate the new bonds that form and the surface termination during each half-cycle. To determine the atomistic detail of the ALD mechanism represented by these two half-cycles, and to predict the associated thermochemistry and kinetics of the TMA and H_2O half-reactions, the B3LYP [48] gradient-corrected exchange DFT method combined with the 6-31 + G(d,p) double-zeta plus polarization and diffuse functions basis set was used by Widjaja and Musgrave [20]. The geometry of each species was optimized by finding stationary points followed by frequency calculations needed to identify the nature of the stationary points on the potential energy surface and to calculate the zero-point energy corrections.

The $Al(OH)_2{-}OH$, $Al(OH){-}(CH_3)_2$, and $Al(OH_2){-}CH_3$ clusters were used to represent the $Al{-}OH^*$ and $Al{-}CH_3^*$ reactive surface sites. In addition, the larger $Al[O{-}Al(OH)_2]_2{-}OH$, $Al[O{-}Al(OH)_2]_2{-}CH_3$, and $Al[O{-}Al(OH)_2]{-}(CH_3)_2$ clusters were used to determine the effect of cluster size by mimicking the effects of the surrounding material of a model Al_2O_3 surface site. Because the clusters were fully relaxed, the calculations may underestimate the strain energy of the system. The clusters used are shown in Figure 1.1. The PES of the $Al(OH)_2{-}OH + Al(CH_3)_3$ reaction, representing reactions during the TMA pulse at $Al{-}OH^*$ surface sites, is shown in Figure 1.2. TMA first adsorbs molecularly at an $Al{-}OH^*$ with an adsorption energy of 14.1 kcal/mol. The adsorbed complex involves a Lewis acid–base interaction, with TMA acting as the Lewis acid and OH^* acting as the Lewis base. Next, one H atom from the surface $-OH^*$ group transfers to the methyl group of the adsorbed $TMA{-}OH^*$ complex to form CH_4, which then desorbs. This transition state lies below the entrance channel: the TS energy is lower than that of the reactants, although it lies 12.0 kcal/mol above the chemisorbed TMA. Overall, the TMA half-reaction is exothermic by 25.1 kcal/mol and results in adding an Al layer to the surface

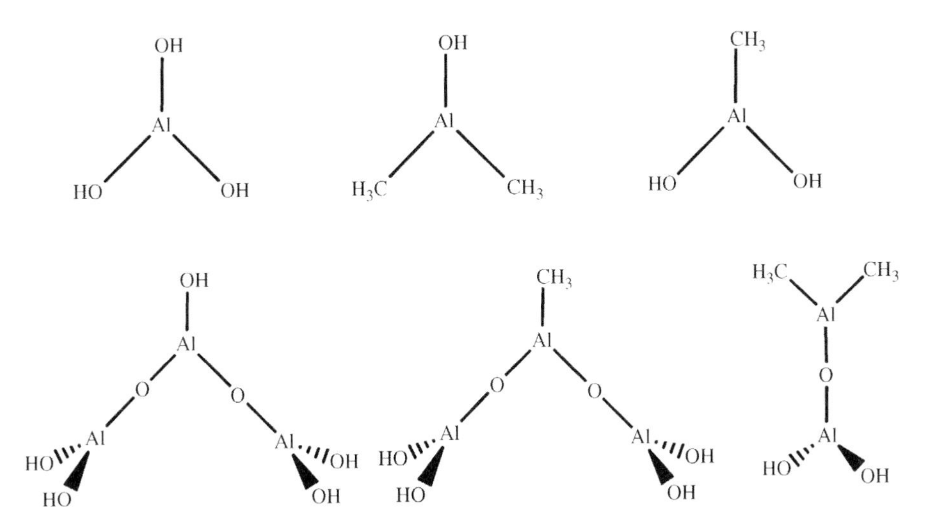

Figure 1.1 Schematic illustration of the clusters used to model reactive sites on the Al_2O_3 surface.

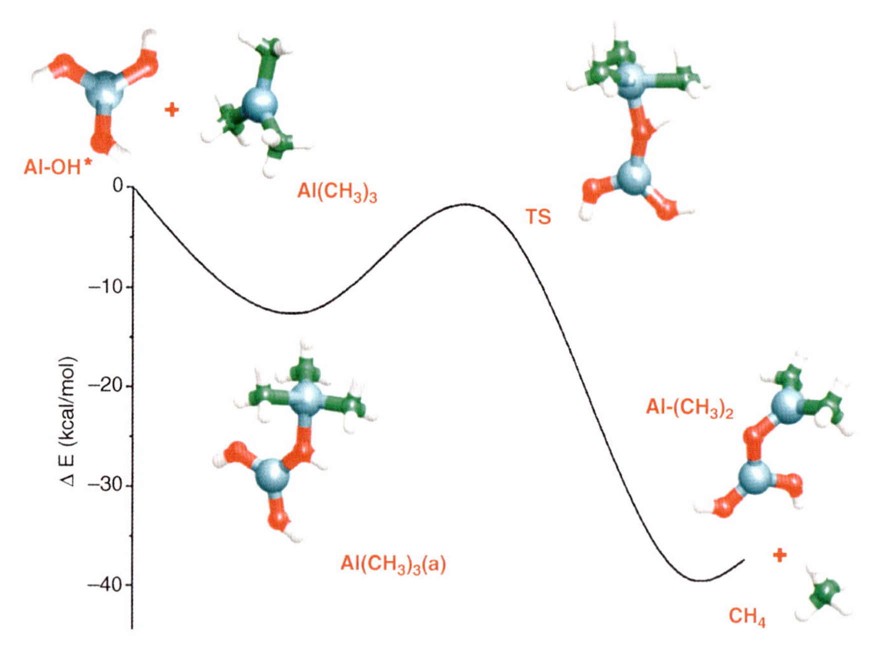

Figure 1.2 Schematic potential energy surface for the first half-reaction of TMA + an OH* site on the growing Al_2O_3 surface. Note that the barrier lies ~2 kcal/mol below the reactants. See the text and Ref. [20] for more details.

while replacing surface hydroxyl groups with surface methyl groups. Following a reactor purge, H_2O is then introduced into the ALD chamber.

The water pulse reacts with methyl-terminated Al_2O_3 to replace methyl groups with hydroxyl groups while adding a layer of oxygen to the surface. Determining the mechanism of this half-reaction is analogous to the calculations described above to study the TMA half-cycle. In fact, not only is the approach similar, but the reaction is also isodesmic to that of the first half-cycle. In the water half-cycle, the OH bond of water and the Al−C bond of the surface $Al-(CH_3)_2^*$ site are broken while an Al−O bond and a C−H bond are formed to make $Al-(CH_3)(OH)^*$ and CH_4. This involves forming and breaking the same bonds as the TMA half-reaction, and consequently, the energetics are different to the degree that effects beyond the active atoms in the reaction are important, which is a general expectation of isodesmic reactions. To predict the reaction mechanism of H_2O with $Al-(CH_3)_2^*$ sites, we calculated the states along a reaction path similar to that of the first half-reaction. The resulting PES is shown in Figure 1.3. As Figure 1.3 shows, H_2O first adsorbs molecularly by forming a Lewis acid–base complex with the $Al-(CH_3)_2^*$ surface site. The complex energy is 13.1 kcal/mol below the energy of the reactants. The adsorption of water is followed by CH_4 formation and desorption, which involves a barrier of 16.1 kcal/mol and an overall reaction energy of −21.0 kcal/mol, both relative to the adsorbed H_2O complex. As expected for the ALD process, reaction of $Al-(CH_3)_2^*$ with H_2O results in $Al-(CH_3)(OH)^*$ sites and a second reaction with water is required to remove the

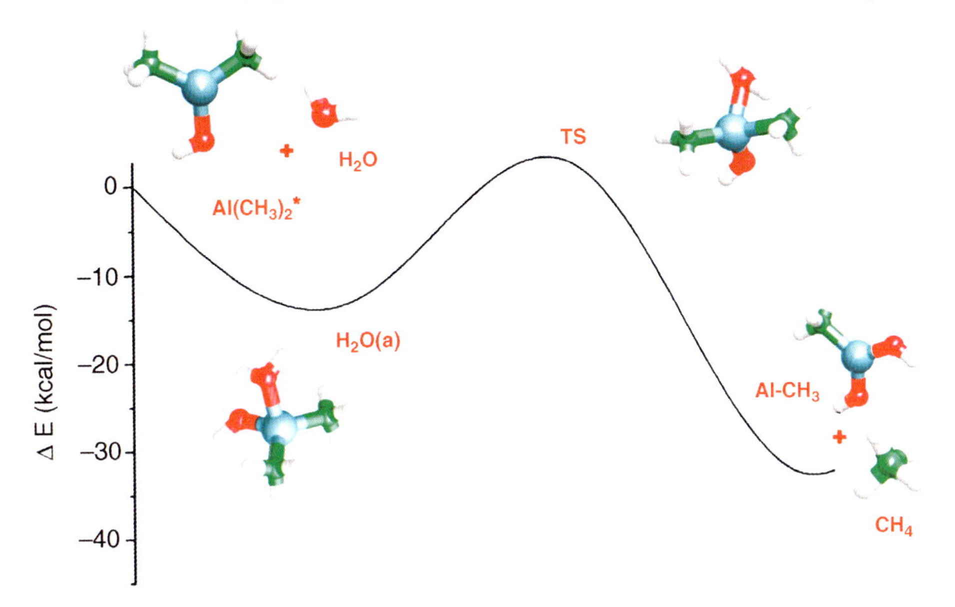

Figure 1.3 Schematic PES of the reaction of water with the $Al-CH_3^*$ sites on the methyl-terminated surface. Unlike the reaction of TMA with $Al-OH^*$ surface sites, the TS lies above the entrance channel. Note that the water reaction is slightly less exothermic than the TMA reaction and because the barrier lies above the entrance channel, it is kinetically slower than the TMA reaction. See the text and Ref. [20] for more details.

remaining methyl group. Reaction of the $Al-CH_3$ group with H_2O is subsequently investigated. Similar to the reaction of $Al-(CH_3)_2$ with H_2O, the reaction follows a trapping-mediated pathway. Although the reaction is analogous to reaction at the site with two methyl groups, replacing one methyl by a hydroxyl group affects the energetics of the reaction. We found that the adsorbed water complex energy is −17.1 kcal/mol, 4.0 kcal/mol more stable than for the $Al-(CH_3)_2^*$ surface site. We then calculated that the ligand exchange reaction barrier was 21.0 kcal/mol relative to the H_2O adsorbed state, ~5 kcal/mol larger than for reaction at the $Al-(CH_3)_2^*$ site. Finally, we saw that the reaction was exothermic by 12.9 kcal/mol, which was about 8 kcal/mol less exothermic than for water reacting at the $Al-(CH_3)_2^*$ surface site. Despite these differences, the qualitative nature of the chemistry is quite similar. Although the differences in the reactivity may suggest that the effects are nonlocal, it should be remembered that the two cases shown here differed by exchanging a methyl group directly bound to the active Al surface atom with an OH group. In the case of using different clusters to model surface reactions, the effects are less direct as the clusters are modified further from the reactive site of the cluster, as shown next.

We also calculated the reaction energies using the larger $Al(O-Al(OH)_2]_2-OH$, $Al[O-Al(OH)_2]-(CH_3)_2$, and $Al[O-Al(OH)_2]_2-CH_3$ clusters. The reaction mechanisms are identical to those found using the analogous smaller clusters. Furthermore, the calculated energies are very similar to those calculated using the smaller clusters, with energy differences of less than 2.5 kcal/mol. This showed that the ALD surface reactions for the case of TMA and H_2O on the Al_2O_3 surface do not exhibit significant nonlocal effects and that relatively small clusters are good models of the surface active sites because they are sufficient to describe the surface reactions.

Similar to the ALD of Al_2O_3 using TMA and water, formation of complexes has also been predicted for the ALD of ZrO_2 and HfO_2 with H_2O and $ZrCl_4$ and $HfCl_4$ precursors [5, 12, 14, 18, 19]. In those cases, an oxygen lone pair is donated to an empty d orbital of the metal center to form a Lewis acid–base complex with an energy below that of the final products of the half-cycle. In contrast to the complexes formed in the cases of $ZrO_2$and HfO_2 ALD, the energy of the TMA–water complex is above the energy of the final product. As a result, Al_2O_3 ALD is exothermic relative to both the initial state and the complex, so the H_2O complex should not be observed at room temperature, unlike the cases of ZrO_2 and HfO_2. This is confirmed experimentally for water reacting with methyl-terminated Al_2O_3 with infrared spectroscopy where the H_2O bending mode in the region between 1380 and 1650 cm^{-1} is not observed [49]. In the case of Al_2O_3 deposition, long purge sequences to drive the reaction toward the products are probably not necessary, because equilibrium favors the reaction products, in contrast to ALD of ZrO_2 and HfO_2. However, if excess water is introduced beyond that necessary to remove methyl groups, then a long purge may be required.

This example illustrates the basic use of quantum chemical methods to investigate an atomistic mechanism, namely, the simple chemistry of Al_2O_3 ALD using TMA and H_2O. It is quite common to find that the ALD half-reactions for various metal oxides and nitrides proceed through a stable intermediate complex before the reaction removing the H atom of the OH or NH_2 surface reactive site. These complexes result from the interaction between the oxygen or nitrogen lone pair electrons and the empty p orbital of the metal atom.

References

1 Szabo, A. and Ostlund, N.S. (1996) *Modern Quantum Chemistry: Introduction to Advanced Electronic Structure Theory*, Dover, Mineola, NY.
2 Cramer, C. (2004) *Essentials of Computational Chemistry*, John Wiley & Sons, Ltd, Chichester.
3 Payne, M.C., Teter, M.P., Allan, D.C., Arias, T.A., and Joannopoulos, J.D. (1992) *Rev. Mod. Phys.*, **64**, 1045.
4 Ardalan, P., Musgrave, C.B., and Bent, S.F. (2009) *Langmuir*, **25**, 2013.
5 Han, J.H., Gao, G.L., Widjaja, Y., Garfunkel, E., and Musgrave, C.B. (2004) *Surf. Sci.*, **550**, 199.
6 Heyman, A. and Musgrave, C.B. (2004) *J. Phys. Chem. B*, **108**, 5718.
7 Kang, J.K. and Musgrave, C.B. (2002) *J. Appl. Phys.*, **91**, 3408.
8 Kelly, M.J., Han, J.H., Musgrave, C.B., and Parsons, G.N. (2005) *Chem. Mater.*, **17**, 5305.
9 Mui, C. and Musgrave, C.B. (2004) *J. Phys. Chem. B*, **108**, 15150.
10 Mui, C., Senosiain, J.P., and Musgrave, C.B. (2004) *Langmuir*, **20**, 7604.
11 Mui, C., Widjaja, Y., Kang, J.K., and Musgrave, C.B. (2004) *Surf. Sci.*, **557**, 159.
12 Mukhopadhyay, A.B. and Musgrave, C.B. (2006) *Chem. Phys. Lett.*, **421**, 215.
13 Mukhopadhyay, A.B. and Musgrave, C.B. (2007) *Appl. Phys. Lett.*, **90**, 173120.
14 Mukhopadhyay, A.B., Musgrave, C.B., and Sanz, J.F. (2008) *J. Am. Chem. Soc.*, **130**, 11996.

15 Mukhopadhyay, A.B., Sanz, J.F., and Musgrave, C.B. (2007) *J. Phys. Chem. C*, **111**, 9203.
16 Musgrave, C.B., Han, J.H., and Gordon, R.G. (2003) *Abstr. Pap. Am. Chem. Soc.*, **226**, U386.
17 Musgrave, C.B., Mukhopadhyay, A., and Sanz, J.F. (2007) *Abstr. Pap. Am. Chem. Soc.*, **234**.
18 Widjaja, Y. and Musgrave, C.B. (2002) *J. Chem. Phys.*, **117**, 1931.
19 Widjaja, Y. and Musgrave, C.B. (2002) *Appl. Phys. Lett.*, **81**, 304.
20 Widjaja, Y. and Musgrave, C.B. (2002) *Appl. Phys. Lett.*, **80**, 3304.
21 Xu, Y. and Musgrave, C.B. (2004) *Chem. Mater.*, **16**, 646.
22 Xu, Y. and Musgrave, C.B. (2005) *Surf. Sci.*, **591**, L280.
23 Xu, Y. and Musgrave, C.B. (2005) *Appl. Phys. Lett.*, **86**, 192110.
24 Xu, Y. and Musgrave, C.B. (2005) *Chem. Phys. Lett.*, **407**, 272.
25 Dkhissi, A., Esteve, A., Mastail, C., Olivier, S., Mazaleyrat, G., Jeloaica, L., and Rouhani, M.D. (2008) *J. Chem. Theory Comput.*, **4**, 1915.
26 Dkhissi, A., Mazaleyrat, G., Esteve, A., and Rouhani, M.D. (2009) *Phys. Chem. Chem. Phys.*, **11**, 3701.
27 Esteve, A., Rouhani, M.D., Jeloaica, L., and Esteve, D. (2003) *Comput. Mater. Sci.*, **27**, 75.
28 Fetah, S., Chikouche, A., Dkhissi, A., Esteve, A., Rouhani, M.D., Landa, G., and Pochet, P. (2010) *Thin Solid Films*, **518**, 2418.
29 Jeloaica, L., Esteve, A., Rouhani, M.D., and Esteve, D. (2003) *Appl. Phys. Lett.*, **83**, 542.
30 Fenno, R.D., Halls, M.D., and Raghavachari, K. (2005) *J. Phys. Chem. B*, **109**, 4969.
31 Halls, M.D. and Raghavachari, K. (2003) *J. Chem. Phys.*, **118**, 10221.
32 Halls, M.D. and Raghavachari, K. (2004) *J. Phys. Chem. A*, **108**, 2982.
33 Halls, M.D. and Raghavachari, K. (2004) *J. Phys. Chem. B*, **108**, 4058.
34 Halls, M.D., Raghavachari, K., Frank, M.M., and Chabal, Y.J. (2003) *Phys. Rev. B*, **68**, 161302.
35 Raghavachari, K. and Halls, M.D. (2004) *Mol. Phys.*, **102**, 381.
36 George, S.M., Ott, A.W., and Klaus, J.W. (1996) *J. Phys. Chem.*, **100**, 13121.
37 Hartree, D.R. (1928) *Proc. Camb. Philos. Soc.*, **24**, 89.
38 Slater, J.C. (1951) *Phys. Rev.*, **81**, 385.
39 Slater, J.C. (1930) *Phys. Rev.*, **35**, 0210.
40 Bartlett, R.J. (1981) *Annu. Rev. Phys. Chem.*, **32**, 359.
41 Grimm, R.C. and Storer, R.G. (1971) *J. Comput. Phys.*, **7**, 134.
42 Hohenberg, P. and Kohn, W. (1964) *Phys. Rev. B*, **136**, B864.
43 Kohn, W. and Sham, L.J. (1965) *Phys. Rev.*, **140**, 1133.
44 Mukhopadhyay, A.B., Sanz, J.F., and Musgrave, C.B. (2006) *Chem. Mater.*, **18**, 3397.
45 George, S.M., Sneh, O., Dillon, A.C., Wise, M.L., Ott, A.W., Okada, L.A., and Way, J.D. (1994) *Appl. Surf. Sci.*, **82–83**, 460.
46 Dillon, A.C., Ott, A.W., Way, J.D., and George, S.M. (1995) *Surf. Sci.*, **322**, 230.
47 Ott, A.W., McCarley, K.C., Klaus, J.W., Way, J.D., and George, S.M. (1996) *Appl. Surf. Sci.*, **107**, 128.
48 Stephens, P.J., Devlin, F.J., Chabalowski, C.F., and Frisch, M.J. (1994) *J. Phys. Chem.*, **98**, 11623.
49 Hakim, L.F., Blackson, J., George, S.M., and Weimer, A.W. (2005) *Chem. Vapor Depos.*, **11**, 420.

2
Step Coverage in ALD

Sovan Kumar Panda and Hyunjung Shin

2.1
Introduction

Conformal coatings are an exclusive matter in microelectronics and in the fabrication of nanostructured materials. Conducting, insulating, and dielectric films, commonly used in microelectronics, need to be fabricated in an ultraprecise manner for electronic devices to provide satisfactory performance. The coating of flat (two-dimensional) surfaces is not a big issue anymore in modern technology. The advanced technology and miniaturization of the devices require the uniform coating of the complex features with high aspect ratios (ratio of length to diameter) in the regime of nanometer scale. That is a really challenging problem in our scientific and technological community. The term "conformal coverage" is considered for the uniform coating of functional materials on the horizontal and vertical surfaces. Microsteps, trenches, grooves, pores, and so on are common features in micro- and nanoelectronics that form during microfabrication processes such as lithography, etching, material deposition, and so on. In some process steps of integrated circuits (VLSI and ULSI technology), microelectromechanical systems (MEMS), and nanoelectromechanical systems (NEMS) fabrication, a conformal coating is essential for diffusion barriers [1–3], adhesive buffer layers [4–6], etch stops [7, 8], and sacrificial layers [9, 10]. For these purposes, the coating should be continuous and uniform throughout the structure even inside the complex features. Figure 2.1a and b shows the schematic diagrams of a step structure and uniform coating on the feature that shows 100% step coverage, respectively. Nonuniform coating causes poor step coverage that in turn generates kink, microcracks, and pores in the features. Figure 2.1c and d illustrates schematically how the poor step coverage creates kinks and pores in the complex features, respectively. Inadequate step coverage leads to an increase in electrical breakdown and mechanical failure at the crack region, which have been shown to be a major source of failure in device reliability testing. The kink on the conducting strip enhances electrical resistivity that exhibits higher Joule heating and sometimes accelerates thermal burnout of the device. Therefore, highly

Atomic Layer Deposition of Nanostructured Materials, First Edition. Edited by Nicola Pinna and Mato Knez.

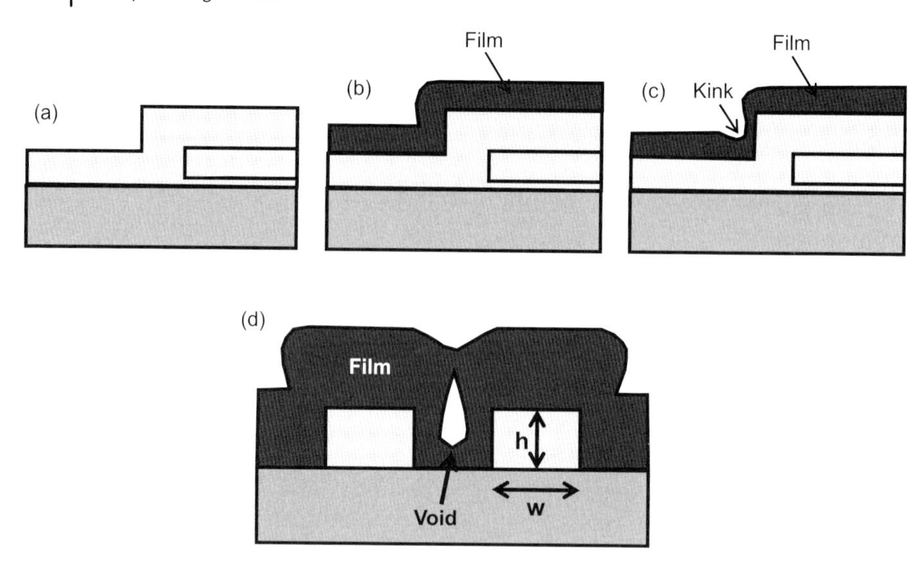

Figure 2.1 Schematic diagrams of (a) microstep, (b) conformal coating of a microstep (100% step coverage), (c) nonuniform coating of a microstep (step coverage <100%) causing the formation of kinks or microcracks, and (d) formation of voids due to poor step coverage (<100%).

conformal films with excellent step coverage are desirable for microelectronic device fabrications.

In particular for the dynamic random access memory (DRAM) of three-dimensional cells, conformal coatings of high aspect ratios (>100) will be preferred in the near future [11]. Therefore, a considerable attention has been paid to the conformal coating and step coverage of the micrometer- and nanometer-scale features for the sake of advanced microelectronic devices. The use of conformal coatings is not limited to semiconductor memories, but can be applied to reduce pore size in a highly controlled way in filtration and gas separation systems [12, 13]. The conformal coating process is also useful to spread catalysts efficiently over the surface of the supports with large surface area [14].

2.2 Growth Techniques

The selection of deposition technique is vital for conformal coating of complex structures with high aspect ratios. Generally, well-known thin film deposition techniques such as physical vapor deposition (PVD) [15–18], chemical vapor deposition (CVD) [19–22], solution techniques [23, 24], electrochemical deposition [25], and atomic layer deposition (ALD) [26–30] are used for this purpose. Among them, PVD shows very poor step coverage and limited ability to coat high aspect ratio features due to the fast surface reactions, limited rearrangement, and line-of-sight deposition (shadowing by surface topography). Film deposition by PVD occurs via

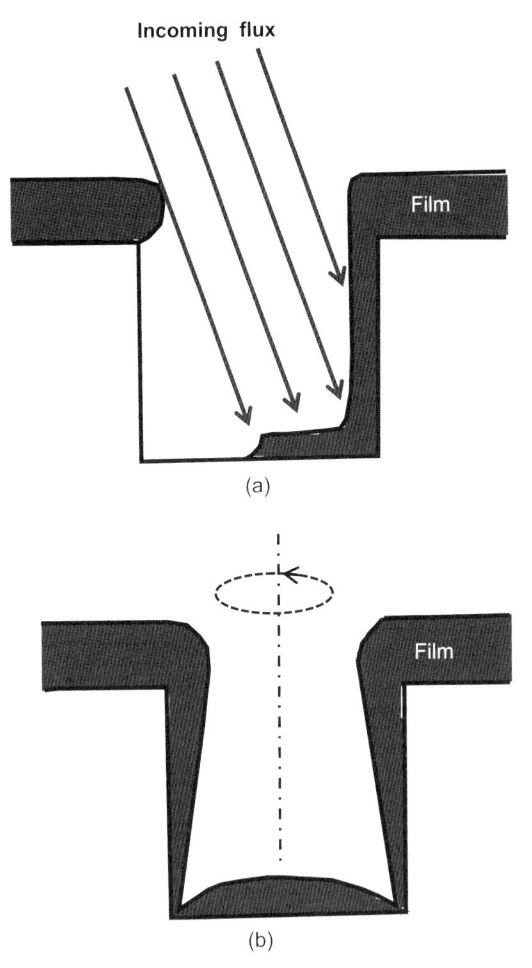

Figure 2.2 Schematic diagrams of (a) high aspect ratio well coated nonuniformly by PVD due to line-of-sight deposition and (b) improvement of film uniformity by using substrate rotation in PVD.

the line-of-sight method obeying the cosine law of emission that states that the emission is favored in the directions normal to the emitting surface. Therefore, large surface area or higher aspect ratio substrates could not be coated uniformly. Figure 2.2a shows the schematic diagram of a high aspect ratio well coated non-uniformly by PVD due to its line of sight. There are different ways to improve the film uniformity in the PVD technique. Best performance could be achieved by mounting the substrates on a curved surface and placing the point source at the center of the curve so that the distance between each point of the substrates and the source remains same and the flux approach the substrate perpendicularly to confirm maximum and uniform deposition. The coating of large-area substrates in this technique is not practical. The substrate rotation is another scheme in the PVD system to deposit large-area substrates with moderate uniformity. The improvement

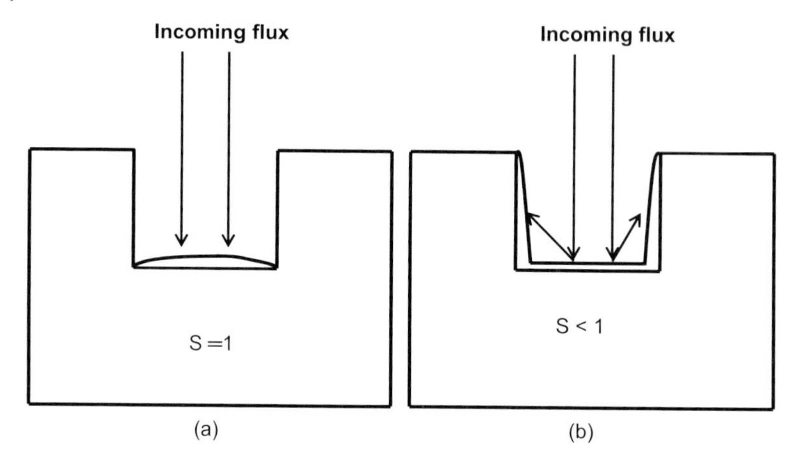

Figure 2.3 Dependence of the step coverage on sticking coefficient in PVD for (a) a very high sticking coefficient ($S = 1$) and (b) a low sticking coefficient ($S < 1$). A lower sticking coefficient provides better step coverage in PVD by resputtering of the deposited film.

of film uniformity by using the substrate rotation technique is shown schematically in Figure 2.2b. When PVD is applied to coat submicron features, particularly for trenches, pores, and vias above an aspect ratio of 0.5 or close, void formation takes place due to poor coverage. Use of collimated sputter deposition is the solution for this limitation. The application of collimated sputter deposition for conformal coating is limited due to its high cost. Therefore, an efficient technique called ionized PVD (I-PVD) is thought of as an alternative for collimated sputter deposition. In case of I-PVD, films are mostly deposited from ions that are directed perpendicularly on the substrate by plasma sheath. Improved film conformality compared to collimated PVD can be achieved because of the resputtering of the deposited film that causes coating of lower side walls of the high aspect ratio features more uniformly. In this case, the sticking coefficient ($S = F_{reacted}/F_{incident}$) plays an important role for uniform coverage. A lower sticking coefficient provides better step coverage by resputtering of the deposited film, which is shown schematically in Figure 2.3a and b. From the above-mentioned limitations, it can be concluded that PVD is not suitable for highly conformal coating and 100% step coverage of narrow and arbitrary shaped features.

Chemical vapor deposition has been widely used in microelectronics, which can coat high aspect ratio features conformally. Laminar gas flow is necessary in CVD in order to have good uniformity. The precursors have to diffuse through the boundary layer to reach on the substrate surfaces. Due to the presence of this boundary layer, a velocity gradient perpendicular to the flow direction is generated in the horizontal reactor. The precursor concentration in the CVD reactor is also dependent on the distance from the gas inlet. The concentration is maximum at the source end and decreases gradually toward the exit. Therefore, large-area uniform coating is tricky in the horizontal CVD reactor. This difficulty could be overcome by tilting the substrates with respect to the gas flow direction (Figure 2.4), which compensate the reduced precursor concentration with distance by the velocity gradient on the sloped surface and facilitate uniform film growth. Further improvement in terms of uniformity of

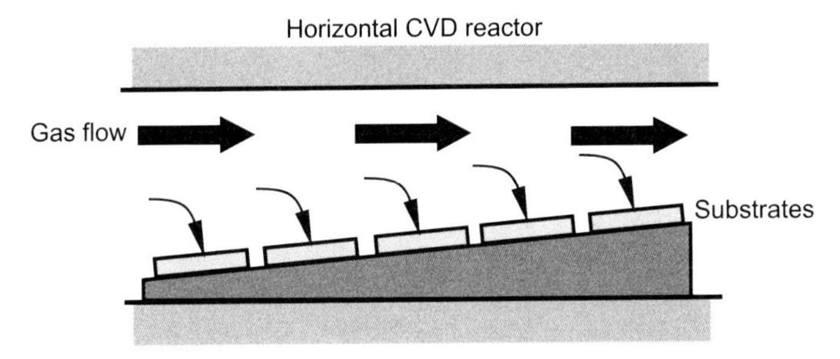

Figure 2.4 Schematic diagram of the horizontal CVD reactor with tilted substrate with respect to the gas flow direction to improve film uniformity.

the large-area substrates in the CVD reactor is possible by using shower-type gas flow together with substrate rotation arrangements. For these flexibilities, CVD is being increasingly used in microelectronics. However, all the CVD reactions are not conformal. To have a conformal coating in CVD, the reaction rate should be very slow and rate-limiting steps should be on the surface of the growing films, rather than the gas phase. In fact, high aspect ratios can be coated conformally by CVD by supplying two complementary reactant vapors in alternating pulses. In this way, the reactions are forced to be entirely on the surface, since they never meet in the gas phase. This method of deposition is well known as atomic layer deposition.

Although there are so many similarities between ALD and CVD, ALD shows some attractive rather unique characteristics over CVD that are related to its self-controlling features. ALD comprises of a binary sequence of self-limiting chemical reactions between gas-phase precursor molecules and a solid surface that is solely dependent on the properties of the surface. This self-limiting nature of ALD gives rise to a conformal growth behavior and an additional control over the total stack thickness. The film thickness can be determined precisely by the number of coating cycles. In comparison to CVD, where film growth generally takes place by vapor-phase reaction of the precursors, ALD occurs via saturating surface reactions, allowing full or partial monolayer formation per cycle. In addition, ALD helps to avoid powder formation since no reaction and no nucleation takes place in the vapor phase. Another advantage of ALD over CVD is its ability to grow high-quality films – that is, dense or pinhole free – at lower temperature [31, 32], which is mostly important for polymer-based device applications and for coating of heat-sensitive materials, for example, organic and/or biomaterials [33, 34]. It is a common misconception that ALD always proceeds through layer-by-layer deposition. In some cases, island-type growth also occurs at the initial growth stage depending on the surface energies [35, 36]. The growth rate of ALD is affected by the formation of islands, which causes a nonlinear growth and disturbs the smoothness of the surface. This island structure is usually observed for metal ALD on oxide substrates [37]. However, the island formation is not limited to metal ALD, but extended to a wide range of materials. Nevertheless, the ALD process helps in reduction of island formation compared to other methods. Figure 2.5 shows schematically the step coverage properties of different thin film deposition techniques that indicate that ALD is the best method for step coverage.

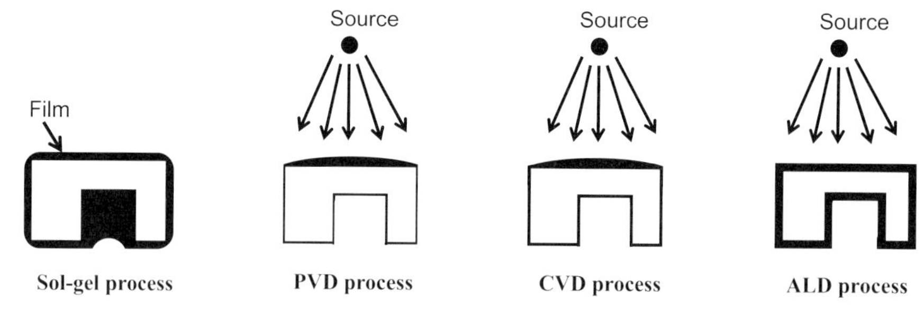

Figure 2.5 Schematic representation of step coverage properties of different thin film deposition techniques: (a) sol–gel; (b) PVD; (c) CVD; (d) ALD.

2.3 Step Coverage Models in ALD

Even in ALD, it is difficult to obtain uniform film thickness as the aspect ratio of the micro- and/or nanofeatures increases in dramatic manner. To support this task, theoretical understanding of the film deposition inside the micro/nanostructure is necessary. There have been several reports that deal theoretically and/or experimentally with the film deposition characteristics inside high aspect ratio features.

2.3.1 Gordon's Model for Step Coverage

Gordon *et al.* reported a kinetic model for step coverage using ALD inside cylindrical holes [38]. This model discusses about the exposure time and partial pressure of the precursor species that limit the conformal coating in ALD. The basic assumptions for Gordon's model for 100% step coverage in ALD are as follows:

1) Each of the two reactants must undergo a self-limiting reaction with the surface covered with the reaction products from the other reactant and the sticking probability for the reactants on the surface should be 1. The reaction must be fast and irreversible, and after completion of each reaction, no further reaction should take place with the excess reactant or with the by-products. The reaction by-products must be volatile enough at the deposition temperature to avoid the physisorption.
2) The dose of the reactants should be large enough so that stoichiometric amount of materials is available for deposition over the entire surface.
3) The reactants with proper doses must be present on the entrance of the holes for a long time so that the reactants get sufficient time to diffuse and react with the entire interior surface of the holes.

2.3.1.1 Parameters for Step Coverage

The parameters that mainly affect the step coverage are dose of the reactants, partial pressure (P) of the reactants ($Pa = N/m^2$), exposure time or pulse duration (t) of the reactants (s), molecular mass (m) (kg), temperature (T) during exposure (K), and aspect ratio (a) of the features. These parameters need to be optimized to have reasonable step coverage.

Stoichiometric Requirement Precursor dose is an important factor for step coverage. In ALD, a certain maximum amount of the supplied dose of each reactant is chemisorbed onto the surface. The minimum amount of each reactant needed to coat entire surface area can be calculated by multiplying the total surface area (flat surface + interior surface of holes + interior surface area of the reactor chamber) by saturated surface density per cycle (S) assuming the sticking probability to be 1. It is very simple to measure the area of the flat surface and reactor surface area, but for 100% coverage, the bottom surface of the pores will also be covered uniformly by the film. Therefore, the bottom of the holes should be included in the calculation. The interior area of the hole is perimeter (p) times the depth (L).

Therefore, number of atoms per cycle is

$$N = S(Lp) = S\left(4A\frac{Lp}{4A}\right) = 4SAa, \tag{2.1}$$

where A is the cross-sectional area of holes, and aspect ratio (a) can be defined as

$$a = \frac{Lp}{4A}.$$

Equation (2.1) indicates that N increases with increase in the aspect ratio a. Therefore, the stoichiometric amount of reactant required to coat a hole conformally increases linearly with increase in aspect ratio. In practice, a slightly excess dose (~10%) compared to its saturated value is provided to make sure that entire surface is saturated.

Exposure Time Estimation Minimum exposure for 100% step coverage by ALD can be estimated from the fundamental principles. In case of the coating of porous sample, the reaction occurs both on the flat surface and inside the holes. The reaction on the flat surface is faster compared to that inside the holes because of the diffusion of molecules down the holes.

Flat Surface First we will consider the kinetics on a flat surface. According to the kinetic theory, the impingement flux (J) ($m^{-2}\,s^{-1}$) onto a surface, that is, the number of molecules crossing a plane from one side per unit time and unit area, is [39]

$$J = \frac{P}{\sqrt{2\pi mkT}}, \tag{2.2}$$

where P is the partial pressure of the precursor near the surface, m is the molecular mass, k is the Boltzmann's constant ($1.38 \times 10^{-23}\,JK^{-1}$), and T is the temperature.

Therefore, the time (t) required for the vapor to reach to the saturation dose (S) is

$$t = \frac{S}{J} = \frac{S\sqrt{2\pi mkT}}{P}. \tag{2.3}$$

The product of partial pressure (P) and time (t) is a useful measure for the exposure (Pa s), which can be expressed as

$$Pt = S\sqrt{2\pi mkT}. \tag{2.4}$$

Equation (2.4) suggests that the reaction could be completed either by increasing the time and decreasing the partial pressure or vice versa, keeping the product term (Pt) constant. Equation (2.4) has been estimated by assuming the reaction probability to be 1, which is an ideal case. In practice, all the reactions will have reaction probability less that 1. Therefore, to have 100% step coverage, the exposure should be higher than the calculated value. This calculated value of the exposure can be considered as the lower limit of the required exposures.

Tubular Structure To estimate the minimum exposure required for 100% step coverage of tubular hole, the following assumptions should be taken into account:

1) There is a negligible depletion of each precursor at the entrance of the holes.
2) The precursors diffuse into the holes by molecular flow.

During coating of tubular holes, the reactants with stoichiometric doses have to reach to the entrance of the holes. From the entrance, the molecules diffuse down in order to reach the lower point of the hole by molecular flow. The upper portion of the hole that remains close to the entrance gets coated and saturates first and then the reaction proceeds further and further down the hole. Therefore, the sidewalls of the hole get coated first followed by the bottom surface. Schematic representation of a cylindrical hole is shown in Figure 2.6.

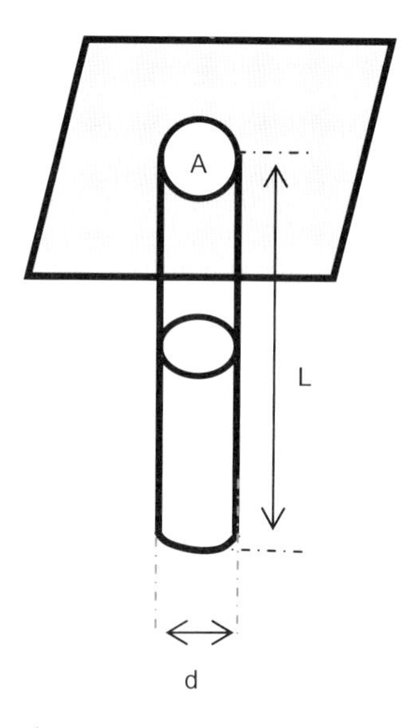

Figure 2.6 Geometry of a tubular hole. Reproduced with permission from Ref. [38]. Copyright 2003, Wiley-VCH Verlag GmbH & Co.KGaA.

The exposure required for complete coating of the sidewalls of the hole is

$$(Pt)_{sidewalls} = S\sqrt{2\pi mkT}[4a + (3/2)a^2]. \quad (2.5)$$

The required exposure for complete coating of the bottom area is

$$(Pt)_{bottom} = S\sqrt{2\pi mkT}[1 + (3/4)a]. \quad (2.6)$$

Therefore, exposure required for 100% step coverage of a tubular hole is the sum of the exposures for sidewalls and bottom surface, which is expressed as

$$(Pt)_{total} = S\sqrt{2\pi mkT}[1 + (19/4)a + (3/2)a^2]. \quad (2.7)$$

Comparing Eqs. (2.7) and (2.4), we can easily tell that the term $S\sqrt{2\pi mkT}$ is due to the exposure required for 100% coating of a flat surface and the additional term $[1 + (19/4)a + (3/2)a^2]$ is a factor by which the exposure has been increased to coat a hole conformally with aspect ratio a. The exposure increases much more rapidly with aspect ratio, and for larger aspect ratio, the exposure is proportional to the square of aspect ratio. Equation (2.7) shows that the exposure required for complete coating of a hole is proportional to the square root of molecular weight for a particular aspect ratio and temperature. This relation can be used to determine the precursor that limits conformality or penetration into the holes. Larger pressure and exposure time help the reactant to penetrate deeper into the pores, but the larger molecular mass hinders its ability to enter a pore.

2.3.2
Kim's Model for Step Coverage

The model proposed by Gordon *et al.* is useful for the determination of the exposure in the saturation region but not suitable to apply in the nonsaturation region. The model proposed by Kim *et al.* is applicable for conformal coating of microfeatures in both the saturation and nonsaturation regions [40]. This model is based on the chemisorption rate of precursors at a certain position along the depth of a microfeature that is determined by the total precursor flux and sticking probability.

The model is based on the following assumptions:

1) The entrance of the hole is exposed to an ideal gas with precursor concentration n_v (m^{-3}) and has a Maxwellian distribution of velocities. The depletion of the precursors just outside the hole is negligible due to the sufficient concentration, fast gas-phase diffusion, and rapid flow over the flat surface.
2) The precursors diffuse inside the hole by molecular flow because of the submicron diameter of holes and a gas pressure of 1–5 Torr.
3) The absorption rate is much higher than the desorption rate. This implies that the bonding between the reactants with the surface is stronger enough to give a lower desorption rate. The substrate temperature should not be very high to validate this assumption.
4) All the reactants chemisorbed on the surface are converted into solid films.

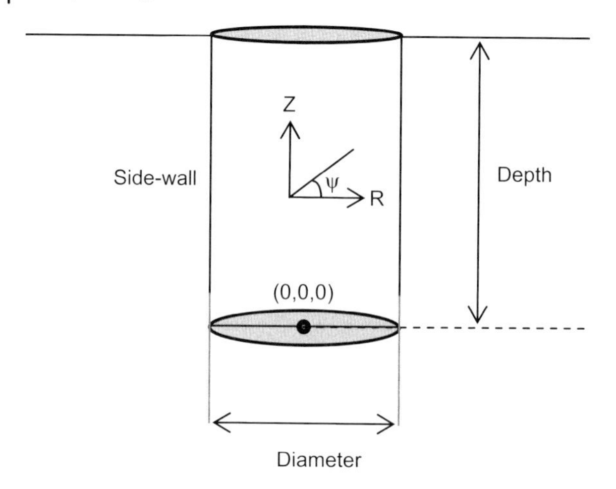

Figure 2.7 Schematic cross section of a cylindrical hole with origin of the coordinate system fixed at the center of the bottom surface.

According to the kinetic theory, the impingement flux on the flat surface is [41]

$$J_{\mathrm{f}}^{q} = \frac{n_{\nu}\bar{c}}{4}, \tag{2.8}$$

where J_{f}^{q} is the precursor flux ($\mathrm{m}^{-2}\,\mathrm{s}^{-1}$) that strikes the flat surface in the qth cycle and $\bar{c}$ is the average velocity of the precursors (m/s).

Kim *et al.* considered a cylindrical hole with origin of the coordinate system at the center of the bottom, which is shown in Figure 2.7. The total flux at a point (r, φ, z) inside the hole is the sum of the flux coming from the entrance of the hole and the flux reflected from other positions inside the hole. When a precursor reacts with the surface, a portion of it is chemisorbed on the surface and the rest is reflected. The sticking probability $[S(\theta)]$ is defined as the ratio of the chemisorbed precursors on the surface to the incident precursors. The probability for the reflecting precursors is $[1 - S(\theta)]$. In this model, a reemission mechanism has also been taken into account. Various reemission mechanisms such as diffuse elastic reemission, cosine reemission, and specular reemission have been illustrated in Figure 2.8. In diffuse elastic and cosine reemission mechanisms, the incoming molecules and the surface strongly interact with each other. Therefore, the molecules lose their all information about the previous trajectory. In diffuse elastic reemission, the probability of the molecules being reflected in any direction above the surface is equal. In cosine reemission, probability of the molecules being reflected with an angle β is proportional to its cosine function. Specular reemission shows negligible interaction of the molecules with the surface, so the angle of emission is equal to the angle of incidence.

Therefore, the precursor flux that strikes position (r, φ, z) inside the hole on xth collision in the qth cycle is given by

$$^{x}J_{\mathrm{h}(r,\varphi,z)}^{q} = \int_{A} \left\{1 - S(\theta_{\mathrm{h}(r',\varphi',z')}^{q})\right\}^{x-1} J_{\mathrm{h}(r',\varphi',z')}^{q} f(s,\alpha,\beta)\mathrm{d}A, \tag{2.9}$$

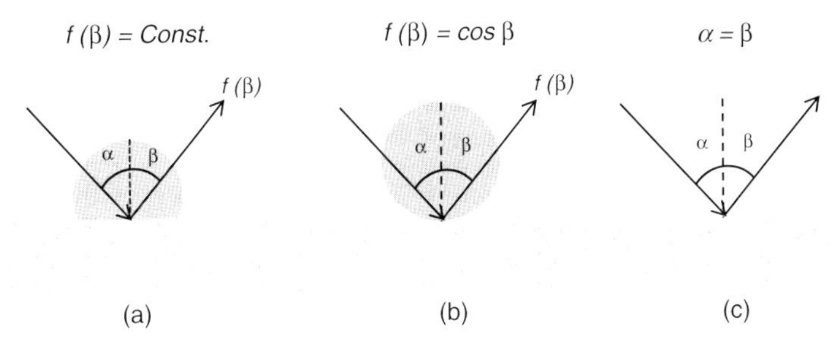

Figure 2.8 Reemission mechanisms of incident molecules impinging on the surface: (a) diffuse reemission; (b) cosine reemission; (c) specular reemission. Reproduced with permission from Ref. [40]. Copyright 2007, American Institute of Physics.

where ${}^{x-1}J^q_{\mathrm{h}(r',\varphi',z')}$ is the precursor flux that strikes position (r', φ', z') inside the hole on the ($x-1$)th collision in the qth cycle ($\mathrm{m}^{-2}\,\mathrm{s}^{-1}$), $S(\theta^q_{\mathrm{h}(r',\varphi',z')})$ is the sticking probability at position (r', φ', z') inside a hole in the qth cycle, $f(s, \alpha, \beta)$ is a function related to the reemission mechanism, each position inside a hole is connected by a line segment of length s (m), the normal vector at each position inside a hole forms the angles of α and β with respect to this connecting line, and dA is the differential surface area inside a hole.

Therefore, the total precursor flux that strikes position (r, φ, z) inside a hole in the qth cycle is given by

$$J^q_{\mathrm{h}(r,\varphi,z)} = {}^1J^q_{\mathrm{h}(r',\varphi',z)} + {}^2J^q_{\mathrm{h}(r',\varphi',z)} + {}^3J^q_{\mathrm{h}(r,\varphi,z)} + \cdots . \tag{2.10}$$

The surface coverage of chemisorbed precursor at the flat surface or at position (r, φ, z) inside a hole in the qth cycle is given by

$$\left[\theta^q_{\mathrm{f\ or\ h}(r,\varphi,z)}\right] = \left\{1-\exp\left[-\frac{S(0)J^q_{\mathrm{f\ or\ h}(r,\varphi,z)}}{K_{\max}}t_{\mathrm{pt}}\right]\right\} \quad \text{for } n=1, \tag{2.11}$$

$$\left[\theta^q_{\mathrm{f\ or\ h}(r,\varphi,z)}\right] = \left\{1-\left[(n-1)\frac{S(0)J^q_{\mathrm{f\ or\ h}(r,\varphi,z)}}{K_{\max}}t_{\mathrm{pt}}+1\right]^{1/(1-n)}\right\} \quad \text{for } n \neq 1, \tag{2.12}$$

where $K_{\max}$ is the maximum number of chemisorbed precursors per unit area (m^{-2}), t_{pt} is the precursor injection time (s), $S(0)$ is the initial sticking probability, and n is the adsorption order.

The film thickness at the flat surface or at position (r, φ, z) inside a hole in the qth cycle is given by

$$\left[T^q_{\mathrm{f\ or\ h}(r,\varphi,z)}\right] = r\left[\theta^q_{\mathrm{f\ or\ h}(r,\varphi,z)}\right] = r\left\{1-\exp\left[-\frac{S(0)J^q_{\mathrm{f\ or\ h}(r,\varphi,z)}}{K_{\max}}t_{\mathrm{pt}}\right]\right\} \quad \text{for } n=1, \tag{2.13}$$

$$\left[T^{q}_{\mathrm{f\ or\ h}(r,\varphi,z)}\right] = r\left[\theta^{q}_{\mathrm{f\ or\ h}(r,\varphi,z)}\right] = r\left\{1-\left[(n-1)\frac{S(0)J^{q}_{\mathrm{f\ or\ h}(r,\varphi,z)}}{K_{\max}}t_{\mathrm{pt}}+1\right]^{1/(1-n)}\right\} \quad \text{for } n \neq 1, \tag{2.14}$$

where r is the saturated film thickness per cycle (in monolayer).

The model parameters can be obtained by fitting the experimentally obtained film thickness data at the flat surface or at the bottom inside of a hole per ALD cycle depending on the precursor injection time. The maximum number of chemisorbed precursors per unit area ($K_{\max}$) and the saturated film thickness per cycle (r) can be obtained from the measured film thickness per cycle in the saturation region. r is the maximum film thickness during one deposition cycle and $K_{\max}$ is obtained from Rutherford backscattering analysis.

2.4 Experimental Verifications of Step Coverage Models

Elam *et al.* studied the ALD of Al_2O_3 and ZnO inside the anodic aluminum oxide (AAO) nanopores using alternating pulses of trimethylaluminum (TMA) and deionized water [28]. They used a nanoporous AAO membrane with a diameter of 65 nm and a length of 50 μm with open end on either side of the membrane. The AAO membrane was coated by Al_2O_3 at 177 °C using 100 cycles. The SEM study showed that the top, bottom, and middle of the AAO membrane have different degrees of coating and the film conformality depends on the exposure time. They also coated AAO membranes by ZnO at 177 °C using 64 cycles and the conformality study of the AAO pores was performed by EPMA mapping. The distribution of Zn along the cross section of the AAO in the direction of the long axis of the pores was quantified using EPMA line scan. EPMA study revealed that the Zn concentration is not uniform throughout the pore but decreases radially toward the center for a short exposure time (~1 s). For a long exposure time (~30 s), the Zn concentration became almost uniform throughout the pore, which indicates a uniform coverage. They also showed that the ZnO film coverage depends on the pore diameter. The normalized integrated ZnO coverage is linearly proportional to the pore diameter. Elam *et al.* further studied the ZnO coating of an extremely high aspect ratio pore by ALD [28]. The AAO template had a diameter of 19 nm and a depth of 92 μm, that is, an aspect ratio of ~5000. To coat such high aspect ratio pores, they used a viscous flow ALD reactor operated in the quasistatic mode. The coating was performed at 177 °C and 5 Torr reactant pressure using 19 ALD cycles for 120 s exposure time and the resulting EPMA study represented a continuous coating throughout the pores. These experimental results are in good agreement with the proposed step coverage models.

Gordon *et al.* verified the kinetic model for step coverage in ALD by hafnium oxide deposition using tetrakis(dimethylamino)hafnium [$Hf(NMe_2)$] and water vapor as two reactants [38]. They calculated the exposure value using Eq. (2.4) for complete coverage of a flat surface at a saturated dose (S) of $Hf(NMe_2)$ $\sim 2.5 \times 10^{18}$ molecules, molecular mass ~5.88×10^{-25} kg, and temperature ~473 K, which was 3.9×10^{-4} Pa s or 3×10^{-6} Torr s. This result indicates that 100% step coverage of the flat surface could be possible by a pressure of 1 Torr for 3 μs, 1 mTorr for 3 ms, and so on. The exposure was also calculated for pattern substrates having pores with a depth of 7.3 μm and an aspect ratio of 36. After coating the pores with HfO_2 for a thickness of 0.25 μm, the aspect ratio increased to 43 due to the reduction of internal diameter. From Eq. (2.7), the calculated exposure for 100% step coverage of the pore was 3000 times longer than that for the flat surface. If the flat surface requires 3×10^{-6} Torr s exposure to be covered conformally, the pores require 9×10^{-3} Torr s for the same. Gordon *et al.* experimentally proved that below 9×10^{-3} Torr s exposure 100% step coverage is not possible. They further showed that pores coated using 2×10^{-3} Torr s exposure were only coated the top quarter along their length. The pores coated using 4×10^{-3} Torr s exposure were coated over about three-quarters of their length. Therefore, exposure less than 9×10^{-3} Torr s causes step coverage less than 100%. These experimentally observed evidences are consistent with the results from Gordon's model.

Using Gordon's equation, Perez *et al.* calculated the aspect ratio of the pores in AAO that could be coated conformally by ALD using two separate precursors such as water and tetrakis(ethylmethylamino)hafnium (TEMAH) [42]. The estimated aspect ratio of a pore for which TEMAH can fully coat is about 30 : 1, while the required aspect ratio for water is 90: 1. In this case, TEMAH is limiting the depth up to which the ALD film can penetrate the pores. TEMAH has higher molecular mass compared to water; therefore, the larger molecular mass reduces deeper penetration inside the narrow hole. Perez *et al.* have also showed the wall thickness of the HfO_2 nanotubes by measurement from TEM images using a Matlab routine. For TEM study, the AAO membrane was dissolved into NaOH and free HfO_2 nanotubes were transferred on the carbon-coated copper grid. The nanotubes have an approximate length of ~1500 nm and a diameter of ~60 nm, that is, an aspect ratio of 25 : 1, which is consistent with the calculated aspect ratio (30 : 1) for conformal coating of a tubular structure from Gordon's equation. Variation of the inner diameter with length of the nanotubes was also observed for long tubes. The thinning was rapid near the deepest end inside the pores. The variation of the wall thickness of the nanotubes with distance was shown (Figure 2.9) mainly in two zones: a saturated precursor zone where wall thickness is constant with distance and a depleted precursor zone where wall thickness decreases with distance. The experimentally observed average slope value for several nanotubes (AAO pore diameter 60 nm and length 4 μm) in the depletion region is 0.020, which is close to the expected slope from Gordon's model in the depletion ALD region (0.0196).

Kim *et al.* experimentally verified their proposed model for the ALD coating of a 0.3 μm diameter hole with an aspect ratio 10 by Al_2O_3 using trimethylaluminum as an aluminum source and oxygen plasma as the oxidant at 225 °C and 3 Torr deposition pressure [40]. In this experiment, they observed that the Al_2O_3 film thickness per cycle on both the flat surface and bottom of the hole depends on the precursor injection time. The film thickness at the bottom of the hole is smaller compared to the flat surface for shorter precursor injection time and conformal

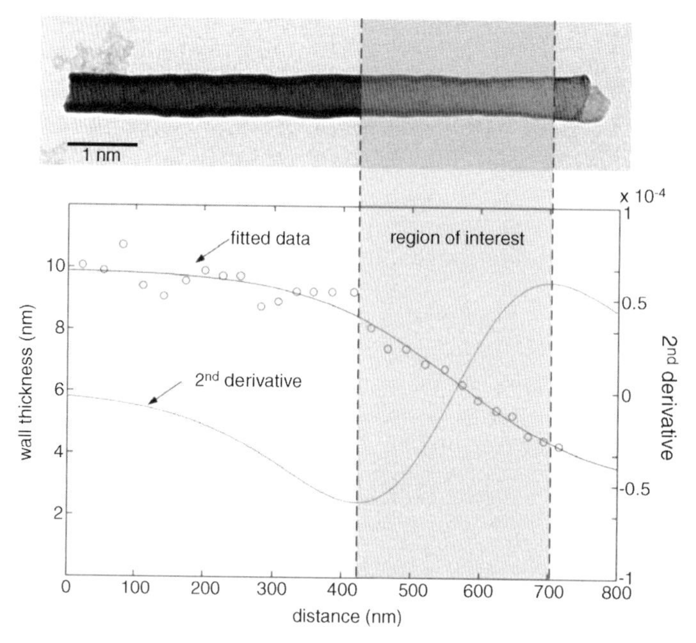

Figure 2.9 Plot of wall thickness of the nanotube obtained from TEM image with respect to the distance from top to the bottom of the tube is fitted with a sinusoidal curve. Reproduced with permission from Ref. [42]. Copyright 2008, Wiley-VCH Verlag GmbH & Co. KGaA.

coating can be achieved for relatively longer precursor injection time. The experimental data for Al_2O_3 film thickness per cycle on both the flat surface and bottom of the hole match well with the simulation results from the proposed model. The experimental data for step coverage on TMA injection time were in good agreement with the model predictions at different partial pressures of TMA.

Kim *et al.* further applied the step coverage model for the deposition of TiO_2 film inside a pore with a diameter of ~0.15 μm and an aspect ratio of 8.7 using titanium tetraisopropoxide (TTIP) and oxygen plasma [43]. TiO_2 film deposited with large enough precursor doses at a pulse time of oxygen plasma of 10 s showed 100% step coverage. To study the variation of film thickness at the bottom of the hole, TiO_2 films were deposited at 225 °C for TTIP injection time varied from 0.3 to 100 s keeping the number of cycles fixed at 500. The thickness of the TiO_2 film at the bottom of the hole was measured by cross-sectional TEM. It was found that the TiO_2 film thicknesses at the bottom of the hole for the TTIP injection times of 1, 5, and 100 s are 33, 91, and 160 Å, respectively. They also studied the thickness profile of the TiO_2 film along the depth of the hole for 5 s TTIP injection time (Figure 2.10), which is less than the required time for 100% step coverage. In this case, the number of cycles was also fixed at 500. The step coverage of the TiO_2 film was about 52% and the experimental data for the step coverage along the depth of a hole were consistent with the simulated results obtained using Kim's step coverage model. It was found that the thickness of the TiO_2 film at the bottom of the hole varied nonlinearly with number of cycles,

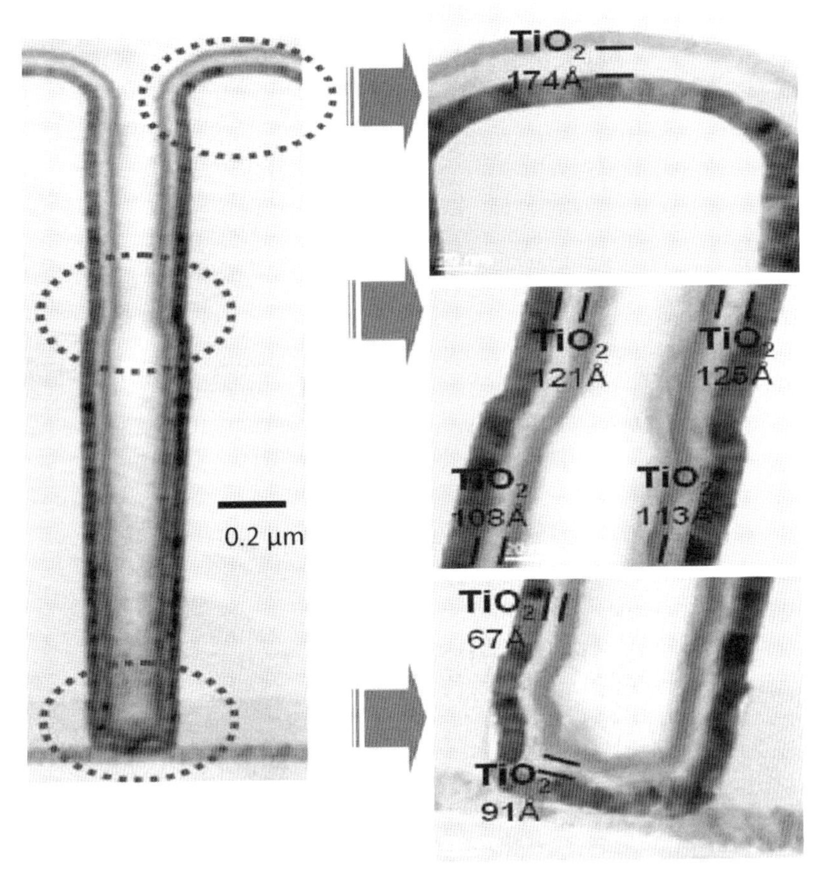

Figure 2.10 Study of the thickness profile of TiO_2 film along the depth of the hole for 5 s TTIP injection time and 500 ALD cycles. The step coverage of the TiO_2 film was about 52%. Reproduced with permission from Ref. [43]. Copyright 2007, the Electrochemical Society.

which is inconsistent with ALD (film thickness should vary linearly with ALD cycles in the flat surface). Therefore, the step coverage gradually decreases as the number of cycles increases. As the number of cycles increases, the area of the pore entrance reduces and as a result precursor diffusion through the pore reduces, which causes poor step coverage for higher number of ALD cycles. In this case, the precursor injection time was fixed at 5 s, which is not sufficient for the precursors to saturate the entire pore. Therefore, to achieve reasonable step coverage, the precursor injection time is required to increase. However, the increase in precursor injection time may cause a problem in throughput. Another solution for improving step coverage is to increase the precursor partial pressure. The precursor injection time required to achieve reasonable step coverage shortens because film thickness per cycle in the saturation region is not affected and the number of precursors entering inside a hole per unit time increases due to the increase in the precursor partial pressure in ALD.

2.5 Summary

ALD is the best technique for conformal coating and step coverage due to its self-limiting surface reaction. Conformal coating of high aspect ratio micro- or nanometer features is possible by selecting a suitable pair of reactants supplied by alternate doses. The precursor exposure time must be sufficiently long in order to coat high aspect ratio features conformally. The increase in the precursor partial pressure shortens the precursor exposure time because exposure is the product of precursor partial pressure and time, which is an important factor for step coverage. Larger molecular mass hinders its ability to enter into the pore, and therefore requires longer exposure for step coverage. The minimum exposure required for 100% step coverage for a flat surface and a tubular pore can be estimated by Gordon's model, which is useful for the determination of the exposure in the saturation region but tricky to apply in the nonsaturation region. The model proposed by Kim *et al.* is applicable for conformal coating of microfeatures in both the saturation and non-saturation regions. In comparison to Gordon's model, where the sticking probability was assumed to be 100%, the precursor injection time for the coating of a flat surface and the bottom of a hole based on Kim's model is longer because any real absorption will have less than 100% sticking probability.

References

1 Kaloyeros, A.E., Chen, X., Lane, S., Frisch, H.L., and Arkles, B. (2000) Tantalum diffusion barrier grown by inorganic plasma-promoted chemical vapor deposition: performance in copper metallization. *J. Mater. Res.*, **15**, 2800–2810.

2 Shin, Y.H. and Shimogaki, Y. (2004) Diffusion barrier property of TiN/Al/TiN films deposited with FMCVD for Cu interconnection in ULSI. *Sci. Technol. Adv. Mater.*, **5**, 399–405.

3 Sherman, A. (1991) Growth and properties of low pressure chemical-vapor-deposited TiN for ultra large scale integration. *Jpn. J. Appl. Phys.*, **30**, 3553–3557.

4 Yap, Y.K., Aoyama, T., Wada, Y., Yoshimura, M., Mori, Y., and Sasaki, T. (2000) Growth of adhesive c-BN films on a tensile BN buffer layer. *Diamond Relat. Mater.*, **9**, 592–595.

5 Ma, Z., Zhou, G.L., Morkoc, H., Allen, L.H., and Hsieh, K.C. (1994) Solid-state reaction-mediated low-temperature bonding of GaAs and InP wafers to Si substrates. *Appl. Phys. Lett.*, **64**, 772–774.

6 De Carlo, F., Song, J.J., Mancini, D.C. (1998) Enhanced adhesion buffer layer for deep x-ray lithography using hard x-rays. J. Vac. Sci. Technol. B, **16**, 3539–3542.

7 Shin, J.U., Kim, D.J., Park, S.H., Han, Y.T., Sung, H.K., Kim, J., and Park, S.J. (2002) An etch-stop technique using Cr_2O_3 thin film and its application to silica PLC platform fabrication. *ETRI J.*, **24**, 398–400.

8 Broekaer, T.P.E. and Fonstao, C.G. (1991) High power 980nm ridge waveguide lasers with etch-stop layer. *Electron. Lett.*, **27**, 2032–2033.

9 Navarro, M., Lopez-Villegas, J.M., Samitier, J., Morantea, J.R., and Bausells, J. (1997) Improvement of the porous silicon sacrificial-layer etching for micromachining application. *Sens. Actuators A*, **61**, 676–679.

10 Birol, H., Maeder, T., and Ryser, P. (2007) Application of graphite-based sacrificial layers for fabrication of LTCC (low

temperature co-fired ceramic) membranes and micro-channels. *J. Micromech. Microeng.*, **17**, 50–60.
11 International Technology Roadmap for Semiconductors, 2005.
12 Berland, B.S., Gartland, I.P., Ott, A.W., and George, S.M. (1998) *In situ* monitoring of atomic layer controlled pore reduction in alumina tubular membranes using sequential surface reactions. *Chem. Mater.*, **10**, 3941–3950.
13 Cameron, M.A., Gartland, I.P., Smith, J.A., Diaz, S.F., and George, S.M. (2000) Atomic layer deposition of SiO_2 and TiO_2 in alumina tubular membranes: pore reduction and effect of surface species on gas transport. *Langmuir*, **16**, 7435–7444.
14 Haukka, S., Kytokivi, A., Lakomaa, E.L., Lehtovirta, U., Lindblad, M., Lujala, V., and Suntola, T. (1995) The utilization of saturated gas–solid reactions in the preparation of heterogeneous catalysts. *Stud. Surf. Sci. Catal.*, **91**, 957–966.
15 Taylor, D.S., Jain, M.K., and Cale, T.S. (1998) Deposition rate dependence of step coverage of sputter deposited aluminium–(1.5%) copper films. *J. Vac. Sci. Technol. A*, **16**, 3123–3127.
16 Besling, W.F.A., Ignacimouttou, M.L., Humbert, A., Mellier, M., and Torres, J. (2004) Continuity and morphology of TaN barriers deposited by atomic layer deposition and comparison with physical vapor deposition. *Microelectron. Eng.*, **76**, 60–69.
17 Radhakrishnan, G., Robertson, R.E., Cole, R.C., and Adams, P.M. (2003) Hybrid pulsed laser deposition and Si-surface-micromachining process for integrated TiC coating in moving MEMS. *Appl. Phys. A*, **77**, 175–184.
18 Depinto, G., Dunnigan, S., and Schwechel, K. (1997) Effect of aluminum sputtering process parameters on via step coverage in micro-electronic device manufacturing. *J. Electron. Mater.*, **26**, 376–382.
19 Lan, J.K., Wang, Y.L., Chao, C.G., Lo, K.Y., and Cheng, Y.L. (2003) Effect of substrate on the step coverage of plasma-enhanced chemical-vapor deposited tetraethylorthosilicate films. *J. Vac. Sci. Technol. B*, **21**, 1224–1229.
20 Raupp, G.B. and Cale, T.S. (1989) Step coverage prediction in low-pressure chemical vapor deposition. *Chem. Mater.*, **1**, 207–214.
21 Cale, T.S., Richards, D.F., and Yang, D. (1999) Opportunities for materials modeling in microelectronics: programmed rate chemical vapor deposition. *J. Comput. Aided Mater. Des.*, **6**, 283–309.
22 Ince, G.O. and Gleason, K.K. (2010) Tunable conformality of polymer coating on high aspect ratio features. *Chem. Vapor Depos.*, **16**, 100–105.
23 Bauer, E., Mueller, A.H., Usov, I., Suvorova, N., Janicke, M.T., Waterhouse, G.I.N., Waterland, M.R., Jia, Q.X., Burrell, A.K., and McCleskey, T.M. (2008) Chemical solution route to conformal phosphor coatings on nanostructures. *Adv. Mater.*, **20**, 4704–4707.
24 Shukla, P., Minogue, E.M., McCleskey, T.M., Jia, Q.X., Lin, Y., Lu, P., and Burrell, A.K. (2006) Conformal coating of nanoscale features of microporous Anodisc™ membranes with zirconium and titanium oxides. *Chem. Commun.*, 847–849.
25 Xu, Y., Zhu, X., Dan, Y., Moon, J.H., Chen, V.W., Johnson, A.T., Perry, J.W., and Yang, S. (2008) Electrodeposition of three-dimensional titania photo crystals from holographically patterned microporous polymer templates. *Chem. Mater.*, **20**, 1816–1823.
26 Kim, J., Hong, H., Oh, K., and Lee, C. (2003) Properties including step coverage of TiN thin films prepared by atomic layer deposition. *Appl. Surf. Sci.*, **210**, 231–239.
27 Mayer, T.M., Elam, J.W., George, S.M., Kotula, P.G., and Goeke, R.S. (2003) Atomic-layer deposition of wear-resistant coatings for microelectromechanical devices. *Appl. Phys. Lett.*, **82**, 2883–2885.
28 Elam, J.W., Routkevitch, D., Mardilovich, P.P., and George, S.M. (2003) Conformal coating on ultrahigh-aspect-ratio nanopores of anodic alumina by atomic layer deposition. *Chem. Mater.*, **15**, 3507–3517.
29 Kim, H., Lee, H.B.R., and Maeng, W.J. (2009) Applications of atomic layer deposition to nanofabrication and

emerging nanodevices. *Thin Solid Films*, **517**, 2563–2580.

30 Kong, B.H., Choi, M.K., Cho, H.K., Kim, J.H., Baek, S., and Lee, J.H. (2010) Conformal coating of conductive ZnO:Al films as transparent electrodes on high aspect ratio Si microrods. *Electrochem. Solid-State Lett.*, **13**, K12–K14.

31 Knez, M., Nielsch, K., and Niinisto, L. (2007) Synthesis and surface engineering of complex nanostructures by atomic layer deposition. *Adv. Mater.*, **19**, 3425–3438.

32 Groner, M.D., Fabreguette, F.H., Elam, J.W., and George, S.M. (2004) Low temperature Al_2O_3 atomic layer deposition. *Chem. Mater.*, **16**, 639–645.

33 Liang, X.H., George, S.M., and Weimer, A.W. (2007) Synthesis of a novel porous polymer/ceramic composite material by low-temperature atomic layer deposition. *Chem. Mater.*, **19**, 5388–5394.

34 Biercuk, M.J., Monsma, D.J., Marcus, C.M., Backer, J.S., and Gordon, R.G. (2003) Low-temperature atomic-layer-deposition lift-off method for microelectronic and nanoelectronic applications. *Appl. Phys. Lett.*, **83**, 2405–2407.

35 Puurunen, R.L., Vandervorst, W., Besling, W.F.A., Richard, O., Bender, H., Conard, T., Zhao, C., Delabie, A., Caymax, M., De Gendt, S., Heyns, M., Viitanen, M.M., de Ridder, M., Brongersma, H.H., Tamminga, Y., Dao, T., de Win, T., Verheijen, M., Kaiser, M., and Tuominen, M. (2004) Island growth in the atomic layer deposition of zirconium oxide and aluminum oxide on hydrogen-terminated silicon: growth mode modeling and transmission electron microscopy. *J. Appl. Phys.*, **96**, 4878–4889.

36 Puurunen, R.L. and Vandervorst, W. (2004) Island growth as a growth mode in atomic layer deposition: a phenomenological model. *J. Appl. Phys.*, **96**, 7686–7695.

37 Johansson, A., Torndahl, T., Ottosson, L.M., Boman, M., and Carlsson, J.O. (2003) Copper nanoparticles deposited inside the pores of anodized aluminum oxide using atomic layer deposition. *Mater. Sci. Eng. C*, **23**, 823–826.

38 Gordon, R.G., Hausmann, D., Kim, E., and Shepard, J. (2003) A kinetic model for step coverage by atomic layer deposition in narrow holes or trenches. *Chem. Vapor Depos.*, **9**, 73.

39 Alberty, R.A. and Sibey, R.J. (1997) *Physical Chemistry*, 2nd edn, John Wiley & Sons, Inc., New York, pp. 607.

40 Kim, J.Y., Ahn, J.H., Kang, S.W., and Kim, J.H. (2007) Step coverage modeling of thin films in atomic layer deposition. *J. Appl. Phys.*, **101**, 073502:1–073502:17.

41 Patterson, G.N. (1971) *Introduction to the Kinetic Theory of Gas Flows*, University of Toronto Press, Toronto.

42 Perez, I., Robertson, E., Banerjee, P., Lecordier, L.H., Son, S.J., Lee, S.B., and Rubloff, G.W. (2008) TEM-based metrology for HfO_2 layers and nanotubes formed in anodic aluminum oxide nanopore structures. *Small*, **4**, 1223–1232.

43 Kim, J.Y., Kim, J.H., Ahn, J.H., Park, P.K., and Kang, S.W. (2007) Applicability of step-coverage modeling to TiO_2 thin films in atomic layer deposition. *J. Electrochem. Soc.*, **154**, H1008–H1013.

3
Precursors for ALD Processes

Matti Putkonen

3.1
Introduction

Since atomic layer deposition (ALD) is a chemical gas-phase thin film deposition method based on sequential, self-saturating surface reactions, the selection of the suitable reactants, that is, precursors, is extremely critical. Typically, two or more chemical precursors are needed, each containing different elements of the materials being deposited. Precursors are adsorbed onto the substrate surface separately. Each precursor saturates the surface forming a monolayer of material and purging between reactants excludes gas-phase reactions. Process temperatures range from room temperature to over 500 °C. The largest selection of ALD processes operates around 300 °C. ALD processes have been developed for oxides, nitrides, carbides, fluorides, certain metals, II–VI and III–V compounds, and recently also for the organic materials [1].

Development of new ALD precursors and processes has been very active over the past 35 years at several universities, R&D centers, and companies. ALD was first demonstrated for the TFEL application by ZnS deposition using elemental zinc and sulfur in mid 1970s, followed by the use of $ZnCl_2$ and H_2S [1]. After the initial period, several other precursor types have emerged. During the second half of 1980s, metal-organic precursors were also adapted for ALD [2, 3]. This opened plenty of possibilities for new ALD chemistry and processes. A large number of ALD processes have been demonstrated and published up to date, and the list is growing all the time.

In this chapter, different precursors used in ALD are discussed in detail. Generally, ALD processes can be divided according to ligand types used in the volatile precursors. The volatile metal-containing precursors typically fall into the following categories: elemental precursors, inorganic precursors, and organometallic ALD precursors. These classes are discussed in detail and some examples of chemical surface reactions are provided.

Atomic Layer Deposition of Nanostructured Materials, First Edition. Edited by Nicola Pinna and Mato Knez.

3.2
General Requirements for ALD Precursors

Although ALD resembles chemical vapor deposition (CVD) quite closely, the requirements for ALD precursors differ from other chemical gas-phase methods since all gas-phase reactions should be excluded and reactions should occur only at the surface. In the early history of ALD, CVD reactants were also used, but nowadays specific tailor-made precursors often provide better results [4, 5]. For example, the use of different exchange reactions at the surface makes it possible to create novel ways to produce challenging thin film materials other than traditional oxides and nitrides. ALD growth mode allows the use of significantly more reactive precursors than CVD. In CVD methods, a constant flow of the precursor vapor is needed to obtain a controlled process, but because ALD relies on self-limiting reactions, only a limited amount of the precursor is required during one pulse to cover the adsorption sites on the surface; the excess amount is removed during the inert gas purge or pumpdown. This makes it possible to use, in some cases, *in situ* generation of the metal precursors and thus have a fresh supply of otherwise instable precursor molecules onto the substrate surface [6, 7]. This technique has been successfully applied to produce β-diketonate compounds of alkaline earth metals (Sr, Ba) by pulsing ligand vapor over heated metal or metal hydroxide [8].

In order to avoid uncontrolled reactions, sufficient thermal stability of the precursors is needed in the gas phase as well as on the substrate surface within the deposition temperature range, which is typically 150–450 °C. Because ALD is a gas-phase process, solid and liquid precursors must be volatile under the operating temperature and pressure, and if heating is required to obtain sufficient vapor pressure, thermal stability of the precursor over a prolonged time is necessary. Stability is the key issue especially when coating porous materials or nanostructures requiring long precursor diffusion times.

There are some general requirements for ALD precursors:

- sufficient volatility at the deposition temperature;
- no self-decomposition or reaction on itself at the deposition temperature;
- precursors must adsorb or react with the surface sites;
- sufficient reactivity toward the other precursors;
- no etching of the substrate or the growing film;
- availability at a reasonable price;
- safe handling and preferably nontoxicity.

Quite often, the ALD precursors do not fulfill all of these requirements. Moreover, sensitive substrates may give additional limitations to precursor selection.

3.3
Metallic Precursors for ALD

In the ALD of inorganic materials, at least one metal-containing precursor is needed (Scheme 3.1). Recently, the ALD process has also been applied to organic materials

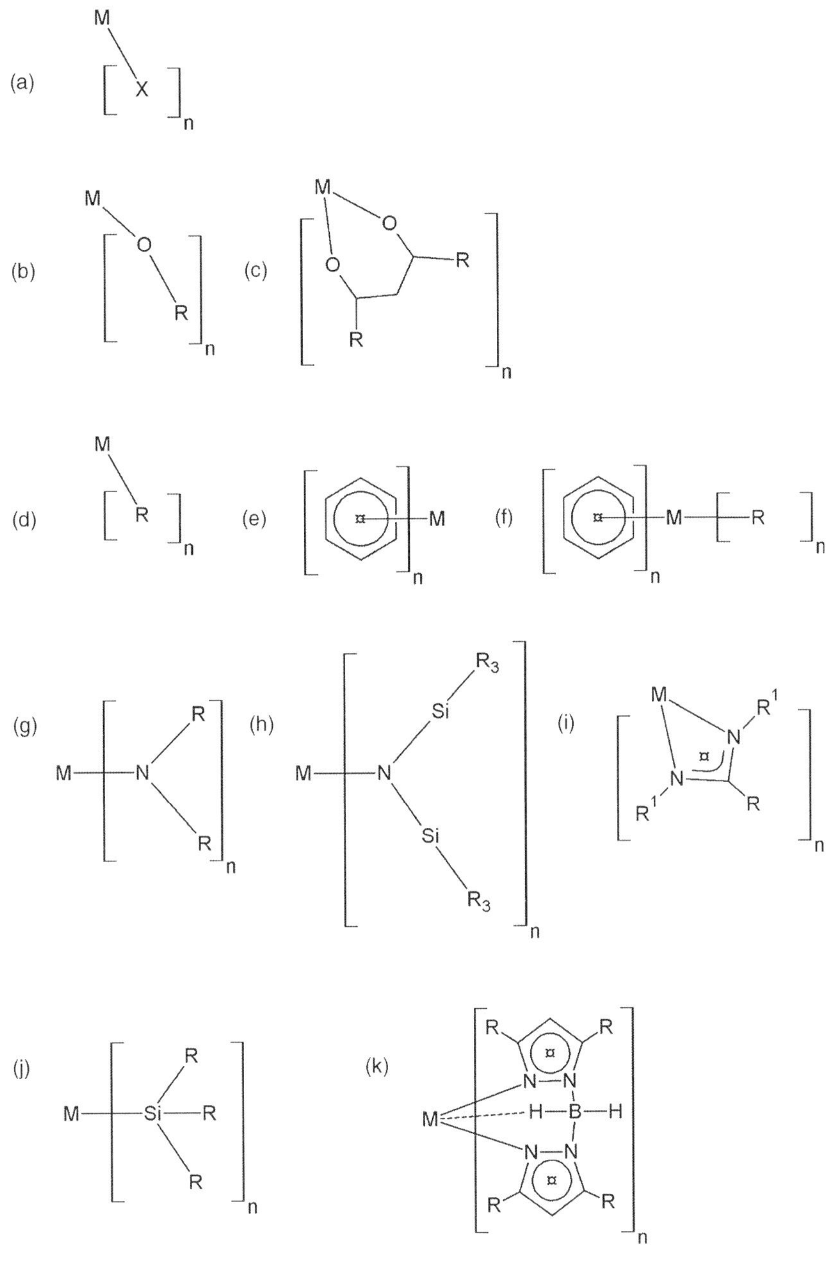

Scheme 3.1 Examples of precursors for ALD depositions. Volatile halides (a), alkoxides (b), and β-diketonates (c) are traditionally used as ALD precursors. Organometallic precursors such as metal alkyls (d) and homo- and heteroleptic cyclopentadienyl-type compounds (e and f) have often favorable properties. Nitrogen-coordinated compounds such as metal amides (g), silylamides (h), amidinates (i), and guanidinates (i, if $R = NR_2$) have also been successfully applied to ALD. More recently, alkylsilyl compounds (j) and pyrazolyl borate-type ligands (k) have been used for the demanding ALD processes.

and inorganic/organic hybrid materials, and it is now referred to as molecular layer deposition (MLD) and its features will be discussed in detail in Chapter 5.

3.3.1 Inorganic Precursors

Inorganic precursors offer low-cost alternatives for many ALD processes. In some cases, they have suffered film contamination or etching but still their thermal stability makes possible the use of relatively high deposition temperatures. Most common of these are metal halides, but other inorganic precursors such as metal nitrates, carboxylates, and isocyanates have also been tested [9–11].

3.3.1.1 Elemental Precursors

The early ALD processes utilized either volatile elements such as zinc and sulfur or halides [1]. The low volatility of pure elements limits the number of applicable processes or forces the use of very high deposition temperatures. However, deposition rates are quite high since there are no sterical hindrances in the elemental precursor.

3.3.1.2 Halides

The most common inorganic precursors are metal halides. Several metal halide precursors have been applied in ALD processes, together with water, ammonia, and hydrogen sulfide as a second reactant. However, the suitability of a particular metal halide for ALD depositions has been found to depend on the metal precursor reactivity. Metal chlorides such as $TiCl_4$ [12–14], $ZrCl_4$ [15, 16], and $HfCl_4$ [17–19] are commonly used as ALD precursors, but metal fluorides [20], bromides [21], and iodides [22] have also been successfully applied. Generally, halide precursors are quite thermally stable, so relatively high deposition temperatures can be applied. However, halide contamination of the film may cause problems at low deposition temperatures [23]. Liberation of HX (X = F, Cl, Br, I) during the deposition process may also cause problems, such as corrosion and etching.

Oxyhalides are another type of volatile inorganic precursors, which have been used in only a few studies. For instance, tungsten oxyfluoride (WO_xF_y) and H_2O have been used as precursors in the deposition of WO_3 [7], while CrO_2Cl_2 has been used in the deposition of chromium oxide [24, 25].

3.3.1.3 Oxygen-Coordinated Compounds

β-Diketonate-type compounds are perhaps the most common oxygen-coordinated precursors for ALD. These metal chelates are known for their volatility and relative stability and therefore they were originally synthesized for the separation of metals by fractional sublimation [26] and gas chromatography [27]. In the 1970s, they were utilized for the first time in CVD depositions [28]. β-Diketonates in ALD processes were first utilized in the 1980s as dopant precursors (e.g., for Tb, Ce, Eu, and Th) for ZnS [29] and other sulfide thin films [30] since halides for these dopants suffered low vapor pressures. Somewhat later, they were applied in metal oxide depositions [31], but quite high deposition temperatures are required to obtain favorable reaction with water, hydrogen peroxide, or oxygen. Another approach is to use a strong oxidizing agent, such as ozone.

Metal β-diketonates have also been studied for the deposition of metals [32–34], metal nitrides [35], and metal sulfides [36]. However, since they have metal to oxygen bond present, there might be some oxygen residues present in the deposited films. Impurity levels depend a lot on the material itself; for example, $Y(thd)_3 + H_2S$ precursor combination leads to yttrium oxysulfide [37], whereas $In(acac)_3 + H_2S$ leads to In_2S_3 [38].

Recently, metal β-diketonates have also been applied to the deposition of metal fluorides (MgF_2, CaF_2, LaF_3) using metal β-diketonates as metal precursors and solid TiF_4 or TaF_5 as a fluoride source [39]. This reaction proceeds through ligand exchange reaction between the metal β-diketonate and the metal fluoride. This reaction produces titanium or tantalum β-diketonates as volatile by-products. Another way to deposit fluoride thin films is by using fluorinated metal β-diketonates such as Ca $(hfac)_2$ or metal β-diketonates with separate fluorine source such as HF [40] or Hhfac (1,1,1,5,5,5-haxafluoroacetylacetonate) [41].

It has been observed in CVD and ALD that coordinatively unsaturated β-diketonate-type compounds may oligomerize or react with the moisture and become less volatile [6, 8]. This is observed especially with large and basic metal centers such as strontium and barium. β-Diketonate-type chelates can be protected against oligomerization and hydrolysis by adducting them with a neutral molecule [42].

With adducted precursors, volatilization behavior during heating depends on the choice of metal, ligand, and adduct and on the bond strengths between them. Adducted precursors may sublime in a single-step process, or alternatively they may dissociate during heating, releasing the neutral adduct, which leads to a two-stage volatilization process where the second step corresponds to the sublimation of the original unadducted compound [43]. ALD depositions of sulfide films have been carried out using $Ca(thd)_2(tetraen)$ or $Ce(thd)_3(phen)$, where the growth process appears to be similar to that of the unadducted precursor [44]. Similar results have also been obtained in the case of deposition process for Y_2O_3 films from $Y(thd)_3(phen)$, $Y(thd)_3(bipy)$, and O_3 as precursors [45]. It seems that, even if sublimation at one stage is observed, the adducted precursors dissociate at the deposition temperatures, producing the unadducted precursor [44].

Another interesting group of oxygen-coordinated metal precursors includes the metal alkoxides. They are usually more thermally unstable than β-diketonate compounds [46, 47], but offer lower evaporation temperatures. Titanium precursors such as $Ti(OCH_3)_4$ [48], $Ti(OCH_3CH_4)_4$ [49], and $Ti\text{-}(O^iPr)_4$ [50, 51] have been shown to be especially useful when depositing ternary strontium-containing materials to suppress the formation of strontium chloride impurities originating from the $TiCl_4$ precursor [52, 53].

Although simple alkoxide ligands can be used for some elements, metal alkoxides also offer rather straightforward possibility to modify the ligands in order to tailor the ALD process behavior and film properties [54]. For example, bulkier ligands such as *t*-butoxide [55, 56], 1-methoxy-2-methylpropan-2-ol [57], 1-dimethylamino-2-methyl-2-propanolate [58], and dimethylaminoethoxide [59] have been successfully applied for ALD deposition of metal oxides. Even SiO_2 can be deposited by using $H_2N(CH_2)_3Si(OEt)_3$ or $Si(OCH_2CH_3)_4$ as a silicon precursor [60, 61]. Recently, copper bis(dimethylamino)-2-propoxide has been applied for deposition of metallic copper films [62]. This reaction is based on the exchange reactions with diethylzinc leading to volatile $Zn(dmap)_2$ by-product and copper films.

3.3.1.4 Nitrogen-Coordinated Precursors

Another quite interesting class of ALD precursors includes nitrogen-coordinated compounds such as metal alkylamides and metal amidinates. They have been applied to ALD only recently and there are still emerging possibilities for these precursors. The biggest drawback for some of these compounds is the limited thermal stability.

The most common metal alkylamides are the ones developed for zirconium and hafnium. $M(NMe_2)_4$, $M(NMeEt)_4$, and $M(NEt_2)_4$ (M = Zr, Hf) can be used for the deposition of corresponding oxide and nitride films [63, 64]. Corresponding tantalum alkylamides show stability comparable to zirconium and hafnium, but titanium alkylamides seem to be too unstable to be used in ALD [65, 66]. Other alkylamides studied in ALD include dimethylamides of aluminum, gallium, germanium, antimony [67], and tungsten and alkylimidoalkylamides of tungsten and molybdenum [68, 69], for example. Recently, different aminosilanes have been studied for the deposition of silicon oxides and nitrides, such as bis(dimethylamino)silane [70], tris(dimethylamino)silane [71], and tetraethylsilaneamine [72].

Metal silylamides provide new opportunities for the selection of ALD materials but they are usually more thermally unstable compared to the corresponding metal alkylamides. $Ce[N(SiMe_3)_2]_3$ was applied for the first time in ALD for growing electroluminescent SrS:Ce thin films [73]. More recently, $M[N(SiMe_3)_2]_3$ complexes have been utilized in the ALD of LaO_x, PrO_x, and GdO_x as well as bismuth-containing films [74, 75]. In addition, $ZrCl_2[N(SiMe_3)_2]_2$ has been used for ZrO_2 thin films at relatively low temperatures [76]. In addition, the biggest drawback is that one possible decomposition pathway can lead to silicon impurities in the films.

Metal amidinates have been studied for the deposition of several metals and metal oxides [77]. Amidinates are structurally analogous to the carboxylates, except that the ligand is coordinating through nitrogen, making these compounds more reactive than the corresponding chelates coordinating through oxygen. For example, recently bis(*N,N′*-di-*tert*-butylacetamidinato)ruthenium [78], lanthanum tris(*N,N′*-diisopropylacetamidinate) [65], and tris(*N,N*-diisopropylformamidinato)lanthanum [66] have been used as ALD precursors. A special group of η^2-coordinated compounds includes guanidinates. For example, Gd_2O_3 films have been deposited by using $Gd(^iPrN)_2CNMe_2)_3$ as metal precursor [79]. In addition, heteroleptic guanidinate-based precursors such as $Hf(NEtMe)_2(EtMeNC(N^iPr)_2)_2$ have been used in ALD [80].

3.3.1.5 Precursors Coordinated through Other Inorganic Elements

Although most of the inorganic ALD precursors are halides and N- or O-coordinated compounds, other precursor classes are also emerging. For example, boron-coordinated precursors such as bis(tris(3,5-diethylpyrazolyl)borate)barium have been used for growing barium borate thin films [81].

Recently introduced silicon-coordinated compounds offer interesting route for complex materials. Alkylsilyl tellurium and selenium precursors make it possible to deposit $Ge_2Sb_2Te_5$ and $CuInSe_2$ [82].

3.3.2 Organometallic ALD Precursors

By IUPAC definition, organometallic compounds have bonding between at least one carbon atom and the metal atom. The chemistry of organometallic compounds has been studied for almost 200 years and the ALD relevant cyclopentadienyl compounds have been known from the beginning of the 1950s. During the past few years, the volatility and high reactivity of organometallics have been exploited in ALD. First, ALD processes exploiting organometallic precursors were studied using metal alkyls as precursors. Later on, various compounds containing cyclopentadienyl or other bulkier ligands have been successfully utilized.

3.3.2.1 Metal Alkyls

ALD processes based on organometallic precursors were first developed for the GaAs depositions by using either $(CH_3)_3Ga$ [83, 84] or $(CH_3CH_2)_3Ga$ [85] and AsH_3 as an arsenic source. Analogous indium alkyls, mainly $(CH_3)_3In$, have been used to deposit InAs [86], InP [87], and In_2O_3 [88] thin films.

Aluminum alkyls, for example, $(CH_3)_3Al$, $(CH_3)_2AlH$, and $(CH_3CH_2)_3Al$, have been used for the deposition of several III–V materials, such as AlAs [89], AlP [90], and AlN [91], using AsH_3, PH_3, and NH_3 as second precursors, respectively. The most common use of aluminum alkyls (e.g., $(CH_3)_3Al$, $(CH_3)_2AlCl$ [92], $(CH_3)_2AlH$ [93], or $(CH_3CH_2)_3Al$) is for the deposition of Al_2O_3 thin films. Al_2O_3 is probably the most frequently studied ALD-processed material since $(CH_3)_3Al$ has been used with H_2O [94], H_2O_2 [95], NO_2, [96] N_2O, [97], O_2 plasma [98], or O_3 [99] as an oxygen source.

Transparent conductive SnO_2 has been deposited using $(CH_3)_4Sn$ or $(CH_3CH_2)_4Sn$ as a metal precursor and N_2O_4 as an oxygen source [100]. In order to dope SnO_2 films, BF_3 pulses were added to the processes resulting in almost two orders of magnitude higher conductivity of the as-deposited material. In a similar manner, zinc alkyls, for example, $(CH_3)_4Zn$ and $(CH_3CH_2)_4Zn$, have been used as ALD precursors for ZnS [101], ZnSe [102], and ZnTe [103] thin films with H_2S, H_2Se, and H_2Te as chalcogen sources, respectively. In addition, they have been utilized for the deposition of ZnO thin films using H_2O as an oxygen source and have also been doped with boron [104], gallium [105], or aluminum [106] from B_2H_6, $(CH_3)_3Ga$, or $(CH_3)_3Al$, respectively. Furthermore, $(CH_3CH_2)_4Pb$ has been utilized as an ALD precursor for PbS thin films with H_2S as the second precursor [107].

3.3.2.2 Cyclopentadienyl-Type Compounds

Chemistry of cyclopentadienyl compounds is studied for almost every element [108]. The main application is their use as catalysts, but they are also suitable for use as precursors in MOCVD [109]. Although cyclopentadienyl compounds have attracted considerable interest as precursors in CVD depositions, they are sometimes too reactive [110]. However, high reactivity and thermal stability make cyclopentadienyl compounds very attractive precursors for ALD, where high reactivity can be controlled by sequential pulsing of the precursors.

For group 2 (Be, Mg, Sr, Ba) elements, several different Cp precursors are utilized. For example, $(C_5H_5)_2Mg$ has been successfully used in the ALD deposition of MgO thin films [111, 112]. Because of the larger size of the heavier alkaline earth metals, simple $(C_5H_5)_2Sr$ and $(C_5H_5)_2Ba$ compounds are not suitable for ALD. Therefore, deposition of strontium- and barium-containing films has been studied by using bulkier ligands. For example, $SrTiO_3$ and $BaTiO_3$ thin films have been deposited from $(C_5{}^iPr_3H_2)_2Sr$, $(C_5Me_5)_2Ba$, and $(C_5{}^tBu_3H_2)_2Ba$ as alkaline earth metal precursors and $Ti(OCH(CH_3)_2)_4$ and H_2O as titanium and oxygen sources, respectively [113]. In addition, SrS and BaS thin films have been deposited from $(C_5{}^iPr_3H_2)_2Sr$, $(C_5Me_5)_2Sr$, and $(C_5Me_5)_2Ba$ using H_2S as the second reactant [114].

The smallest cyclopentadienyl compounds of the series can be utilized in ALD rather straightforwardly for lanthanide elements [115], but thermal stability of the corresponding lanthanum and cerium compounds needs more precursor tailoring. Generally, bulkier ligands will protect the metal central better and decrease the tendency of thermal decomposition. Recently, $La({}^iPrCp)_3$ has been successfully utilized for La_2O_3 deposition [116]. $(C_5Me_4H)_3Ce$ has been successfully used for the doping of ALD-grown SrS thin films [117]. More recently, mixed ligand-type Cp compounds have also been applied [118].

Several volatile cyclopentadienyl-type compounds of group 4 elements have been utilized in ALD. First organometallic ALD experiments to process ZrO_2 and HfO_2 thin films were carried out using $(C_5H_5)_2MCl_2$ and $(C_5H_5)_2M(CH_3)_2$ [119, 120]. Additional heteroleptic compounds of Ti, Zr, and Hf containing alkoxy [121, 122] or N-coordinated ligands [123, 124] have been successfully used as precursors.

Only few other transition metal cyclopentadienes have been studied in ALD. For example, volatile $(C_5H_5)_2Mn$ and $(C_5MeH_4)Mn(CO)_3$ have been used as manganese sources for doping ZnS thin films to produce yellow-emitting thin film electroluminescent devices as well as for the deposition of MnO [125]. In addition, ferrocene has been utilized for the deposition of Fe_2O_3 thin films [126] and CpIn for In_2O_3 [127]. A different approach has been utilized for processing metallic nickel thin films by ALD [128]. During the first stage, NiO thin films were deposited by one ALD cycle of $(C_5H_5)_2Ni$ and H_2O. The NiO layer formed was then reduced to metallic nickel by pulsing hydrogen radicals into the reactor. Nickel and cobalt can also be deposited by three-stage deposition using an oxidizer followed by hydrogen as a reducing agent [129] or directly using a plasma-enhanced process [130].

Cyclopentadienyl precursors for noble metals have gained significant interest. Typically, oxygen or air is used as a second reactant in these processes [131]. These processes are based on the fact that O_2 does not oxidize the noble metal film, but only decomposes the organic ligand. Adsorbed oxygen atoms react with the ligands of the noble metal precursor during the metal precursor pulse [132]. For example, Ru films have been deposited from $(C_5H_5)_2Ru$ [132] or $(C_5EtH_4)_3Ru$ [133]. If longer oxidizer pulses or additional energy is used, for example, in the form of plasma, noble metal oxides (e.g., RuO_2 and PtO_2) can also be deposited [134, 135]. Deposition of platinum metal films has been studied in a similar manner using $(C_5MeH_4)PtMe_3$ and O_2 as precursors [136], but β-diketonate-type precursors can also be used [137], as shown in Section 3.3.1.3.

3.3.2.3 Other Organometallic Precursors

In addition to metal alkyls and cyclopentadienyl compounds, other types of true organometallic precursors have also been applied to ALD. For example, $(C_6H_5)_3Bi$ [138, 139] and $(C_6H_5)_4Pb$ [139] have been used for the preparation of bismuth- and lead-containing oxide thin films. These compounds are not extremely reactive and therefore ozone is needed [138, 139]. Metallic bismuth was observed as an impurity if H_2O_2 was used as an oxygen source [138].

3.4 Nonmetal Precursors for ALD

Because reactants are separated in the ALD process, it is possible to control the reactivity and the reactions between the precursors by using different precursor combinations. In addition to traditional ALD deposition mechanism that is based on metal and nonmetal precursors, more innovative reaction mechanisms have also been developed. For example, metal alkoxides have been used as oxygen sources [140]. No separate oxygen source is needed in this type of special process because the metal alkoxide serves as both metal and oxygen sources. Other reactions such as ALD mechanism based on the catalytic precursors [61, 141], ligand exchange reactions [39], or replacement reactions at the surface [142] have also been developed.

3.4.1 Reducing Agents

Sometimes a reducing precursor is needed although in many ALD processes no redox reactions occur and the oxidation state of the metal does not change. Hydrogen is perhaps the most widely used reducing agent, but metallic zinc vapor [143], different silanes [144], TMA [145], and B_2H_6 [146] have also been successfully applied. Molecular hydrogen is quite inert toward typical metal precursors and therefore either high deposition temperatures or plasma is needed to maintain ALD reactions.

3.4.2 Oxygen Sources

Typically in the case of ALD-processed oxide films, precursors attached to the surface can be transformed to the respective oxide with H_2O, H_2O_2 [147], N_2O_4 [100], N_2O [96], O_2 [126], or O_3 [148], the choice depending on the metal precursor reactivity or desired properties. Water has frequently been used as an oxygen source and indeed it readily reacts with many metal precursors. Alcohols have also been successfully applied as oxygen sources in processes involving inorganic precursors and alkoxides. For example, deposition of Al_2O_3 has been carried out with an $AlCl_3$ and ROH precursor combination, where R is H, CH_3, CH_2OHCH_2OH, t-C_4H_9OH, or n-C_4H_9OH [149].

The high thermal stability of the β-diketonate-type complexes or low reactivity of cyclopentadienyl-based precursors usually does not allow their use in ALD with mild oxygen sources. The use of H_2O, H_2O_2, N_2O, CH_3COOH, and O_2 has been explored for β-diketonate-type compounds [150, 151], but the reactivities are usually insufficient, producing either no film growth or films with a high impurity content. The use of a strong oxidizer, such as O_3 or oxygen plasma, guarantees that only a small amount of carbon is left in the film as residues and can also improve the film properties [152, 153].

3.4.3 Nitrogen Sources

Deposition of nitride thin films by ALD requires both a nitrogen source and a reducing agent in order to obtain clean surface reactions [154]. In many cases, one compound, for example, NH_3, serves as both a nitrogen source and a reducing agent. In addition, other nitrogen-containing compounds such as N_2H_4, $(CH_3)NNH_2$, $^tBuNH_2$, and CH_2CHCH_2NH have also been studied as nitrogen sources [155, 156]. Ammonia has been used for depositing several nitride materials such as TiN, Ta_3N_5, W_2N, NbN, and WCN thin films by ALD [157]. Quite often, plasma processes are utilized due to low reactivity of ammonia as well as need for lower resistivity films [158, 159].

3.4.4 Sulfur and Other Chalcogenide Precursors

For chalcogenide thin films, it is possible to use elemental S, Se, and Te as precursors provided that the other source is volatile and reactive enough [160]. ZnS deposition using elemental zinc and sulfur was the first ALD process developed. For other than elemental metal precursors, the reactivity of elemental chalcogens is often not sufficient. For halides, β-diketonates, and organometallics, simple hydrides such as H_2S, H_2Se, and H_2Te have been typically used as a second precursor [161, 162]. Due to the toxicity of these materials, *in situ* generation of H_2S has been studied by using thioacetamide decomposition [163]. Recently, different alkylsilylamines with lower toxicity have also been developed for Se and Te [82].

3.5 Conclusions

Since atomic layer deposition is a chemical gas-phase method, its success relies on the development of well-behaving ALD precursors. Although a number of good ALD processes are available, there is still room for improvement. Conventional inorganic precursors have been used since the beginning of the ALD era over some 35 years ago. To replace the classical precursors, for example, volatile halides and inorganic complexes, organometallics and nitrogen-coordinated precursors are widening the ALD process selection at the moment. Recent developments of metal processes and other "difficult" materials confirm that novel processes are constantly emerging. Nevertheless, additional research is still needed to develop special precursors and further improve process and thin film quality. In the future, more precursors and processes relying on smart reaction mechanisms are expected to appear.

References

1 Ritala, M. and Niinistö, J. (2009) Atomic layer deposition, in *Chemical Vapour Deposition: Precursors, Processes and Applications* (eds A.C. Jones and M.L. Hitchman), Royal Society of Chemistry, pp. 158–206.

2 Higashi, G.S. and Rothberg, L.J. (1985) Surface photochemical phenomena in laser chemical vapor deposition. *J. Vac. Sci. Technol. B*, **3**, 1460–1463.

3 Higashi, G.S. and Fleming, C.G. (1989) Sequential surface chemical reaction limited growth of high quality Al_2O_3 dielectrics. *Appl. Phys. Lett.*, **55** (19), 1963–1965.

4 Leskelä, M. and Ritala, M. (1999) ALD precursor chemistry: evolution and future challenges. *J. Phys. IV*, **9**, Pr8-837–PR8-852.

5 Hårsta, A. (1999) Precursor selection in halide CVD of oxides. *Chem. Vapor Depos.*, **5** (4), 191–193.

6 Soininen, P., Nykänen, E., Niinistö, L., and Leskelä, M. (2004) Atomic layer epitaxy of strontium sulfide thin films using *in situ* synthesized strontium precursors. *Chem. Vapor Depos.*, **2** (2), 69–74.

7 Tägtström, P., Mårtensson, P., Jansson, U., and Carlsson, J.-O. (1999) Atomic layer epitaxy of tungsten oxide films using oxyfluorides as metal precursors. *J. Electrochem. Soc.*, **146** (8), 3139–3143.

8 Saanila, V., Ihanus, J., Ritala, M., and Leskelä, M. (1998) Atomic layer epitaxy growth of BaS and BaS:Ce thin films from *in situ* synthesized $Ba(thd)_2$. *Chem. Vapor Depos.*, **4** (6), 227–233.

9 Badot, J.C., Ribes, S., Yousfi, E.B., Vivier, V., Pereira-Ramos, J.P., Baffier, N., and Lincot, D. (2000) Atomic layer epitaxy of vanadium oxide thin films and electrochemical behavior in presence of lithium ions. *Electrochem. Solid-State Lett.*, **3** (10), 485–488.

10 Niskanen, A., Hatanpää, T., Arstila, K., Leskelä, M., and Ritala, M. (2007) Radical-enhanced atomic layer deposition of silver thin films using phosphine-adducted silver carboxylates. *Chem. Vapor Depos.*, **13** (8), 408–413.

11 Kobayashi, K. and Okudaira, S. (1997) Preparation of ZnO films on sapphire (0001) substrates by alternate supply of zinc acetate and H_2O. *Chem. Lett.*, **6**, 511–512.

12 Ritala, M., Leskelä, M., Nykänen, E., Soininen, P., and Niinistö, L. (1993) Growth of titanium dioxide films by atomic layer epitaxy. *Thin Solid Films*, **225**, 288–295.

13 Tiznado, H. and Zaera, F. (2006) Mechanistic details of atomic layer deposition (ALD) processes for metal nitride film growth. *J. Phys. Chem. B*, **110** (27), 13491–13498.

14 Kubala, N.G., Rowlette, P.C., and Wolden, C.A. (2009) Plasma-enhanced atomic layer deposition of anatase TiO_2 using $TiCl_4$. *J. Phys. Chem. C*, **113** (37), 16307–16310.

15 Ritala, M. and Leskelä, M. (1994) Zirconium dioxide thin films deposited by ALE using zirconium tetrachloride as precursor. *Appl. Surf. Sci.*, **75** (1–4), 333–340.

16 Niinistö, J., Kukli, K., Heikkilä, M., Ritala, M., and Leskelä, M. (2009) Atomic layer deposition of high-*k* oxides of the group 4 metals for memory applications. *Adv. Eng. Mater.*, **11** (4), 223–234.

17 Ritala, M., Leskelä, M., Niinistö, L., Prohaska, T., Friedbacher, G., and Grasserbauer, M. (1994) Development of crystallinity and morphology in hafnium dioxide thin films grown by atomic layer deposition. *Thin Solid Films*, **250** (1–2), 72–80.

18 Delabie, A., Pourtois, G., Caymax, M., De Gendt, S., Ragnarsson, L.-A., Heyns, M., Fedorenko, Y., Swerts, J., and Maes, J.W. (2007) Atomic layer deposition of hafnium silicate gate dielectric layers. *J. Vac. Sci. Technol. A*, **25**, 1302–1308.

19 Nyns, L., Delabie, A., Pourtois, G., Van Elsocht, S., Vinckier, C., and De Gent, S. (2010) Study of the surface reactions in ALD hafnium aluminates. *J. Electrochem. Soc.*, **157** (1), G7–G12.

20 Scharf, T.W., Prasad, S.V., Dugger, M.T., Kotula, P.G., Goeke, R.S., and Grubbs, R.K. (2006) Growth, structure, and tribological behavior of atomic layer-deposited tungsten disulphide solid lubricant coatings with applications to MEMS. *Acta Mater.*, **54** (18), 4731–4743.

21 Putkonen, M., and Niinistö, L. (2006) Atomic layer deposition of B_2O_3 thin films at room temperature. *Thins Solid Films*, **514** (1–2), 145–149.

22 Rosental, A., Tarre, A., Gerst, A., Sundqvist, J., Hårsta, A., Aidla, A., Aarik, A., Sammelselg, V., and Uustare, T. (2003) Gas sensing properties of epitaxial SnO_2 thin films prepared by atomic layer deposition. *Sens. Actuators B*, **93** (1–3), 552–555.

23 Aarik, J., Aidla, A., Kiisler, A.-A., Uustare, T., and Sammelselg, V. (1999) Influence of substrate temperature on atomic layer growth and properties of HfO_2 thin films. *Thin Solid Films*, **340** (1–2), 110–116.

24 Tarre, A., Aarik, J., Mändar, H., Niilisk, A., Pärna, R., Rammula, R., Uustare, T., Rosental, A., and Sammelselg, V. (2008) Atomic layer deposition of Cr_2O_3 thin films: effect of crystallization on growth and properties. *Appl. Surf. Sci.*, **254** (16), 5149–5256.

25 Ezhovskii, Yu.K. and Kholkin, V.Yu. (2008) Formation and some properties of chromium oxide nanolayers on semiconductors. *Russ. Microelectron.*, **37** (6), 356–417.

26 Berg, E.W. and Acosta, J.J.C. (1968) Fractional sublimation of the β-diketone chelates of the lanthanide and related elements. *Anal. Chim. Acta*, **40**, 101–113.

27 Eisentraut, K.J. and Sievers, R.E. (1965) Volatile rare earth chelates. *J. Am. Chem. Soc.*, **87** (22), 5254–5254.

28 Ben-Dor, L., Druilhe, R., and Gibart, P. (1974) Thin films of binary metal oxides by chemical vapor deposition from organometallic chelates. *J. Cryst. Growth*, **24–25**, 172–174.

29 Tammenmaa, M., Leskelä, M., Koskinen, T., and Niinistö, L. (1986) Zinc sulphide thin films doped with rare earth ions. *J. Less-Common Met.*, **126**, 209–214.

30 Tammenmaa, M., Antson, H., Asplund, M., Hiltunen, L., Leskelä, M., and Niinistö, L. (1987) Alkaline earth sulfide thin films grown by atomic layer epitaxy. *J. Cryst. Growth*, **84** (1), 151–154.

31 Aarik, J., Aidla, A., Jaek, A., Leskelä, M., and Niinistö, L. (1994) Precursor properties of calcium β-diketonate in vapor phase atomic layer epitaxy. *Appl. Surf. Sci.*, **75** (1–4), 33–38.

32 Hämäläinen, J., Puukilainen, E., Kemell, M., Costelle, L., Ritala, M., and Leskelä, M. (2009) Atomic layer deposition of iridium thin films by consecutive oxidation and reduction steps. *Chem. Mater.*, **21** (20), 4868–4872.

33 Niskanen, A., Rahtu, A., Sajavaara, T., Arstila, K., Ritala, M., and Leskelä, M. (2005) Radical-enhanced atomic layer deposition of metallic copper thin films. *J. Electrochem. Soc.*, **152** (1), G25–G28.

34 Kim, S.K., Hoffmann-Eifert, S., and Waser, R. (2009) Growth of noble metal Ru thin films by liquid injection atomic layer deposition. *J. Phys. Chem. C*, **113** (26), 11329–11335.

35 Törndahl, T., Ottosson, M., and Carlsson, J.-O. (2006) Growth of copper(I) nitride by ALD using copper(II) hexafluoroacetylacetonate, water, and ammonia as precursors. *J. Electrochem. Soc.*, **153** (3), C146–C151.

36 Dasgupta, N.P., Lee, W., and Prinz, F.B. (2009) Atomic layer deposition of lead sulfide thin films for quantum confinement. *Chem. Mater.*, **21** (17), 3973–3978.

37 Kukli, K., Peussa, M., Johansson, L.-S., Nykänen, E., and Niinistö, L. (1999) Controlled growth of yttrium oxysulfide thin films by atomic layer deposition. *Mater. Sci. Forum*, **315–317**, 216–221.

38 Sarkar, S.K., Kim, J.Y., Goldstein, D.N., Neale, N.R., Zhu, K., Elliott, M., Frank, A.J., and George, S.M. (2010) In_2S_3 atomic layer deposition and its application as a sensitizer on TiO_2 nanotube arrays for solar energy conversion. *J. Phys. Chem. C*, **114** (17), 8032–8039.

39 Pilvi, T., Ritala, M., Leskelä, M., Bischoff, M., Kaiser, U., and Kaiser, N. (2008) Atomic layer deposition process with TiF_4 as a precursor for depositing metal fluoride thin films. *Appl. Opt.*, **47** (13), C271–C274.

40 Ylilammi, M. and Ranta-aho, T. (1994) Metal fluoride thin films prepared by atomic layer deposition. *J. Electrochem. Soc.*, **141** (5), 1278–1284.

41 Putkonen, M., Szeghalmi, A., Pippel, E., and Knez, M. (2011) Atomic layer deposition of metal fluorides through oxide chemistry, *J. Mater. Chem.*, accepted.

42 Gardiner, R., Brown, D.W., Kirlin, P.S., and Rheingold, A.L. (1991) Volatile barium β-diketonate polyether adducts. Synthesis, characterization, and metalloorganic chemical vapor deposition. *Chem. Mater.*, **3** (6), 1053–1059.

43 Gordon, R.G., Barry, S., Broomhall-Dillard, R.N.R., and Teff, D.J. (2000) Synthesis and solution decomposition kinetics of flash-vaporizable liquid barium beta-diketonates. *Adv. Mater. Opt. Electron.*, **10** (3), 201–211.

44 Hänninen, T., Mutikainen, I., Saanila, V., Ritala, M., Leskelä, M., and Hanson, J.C. (1997) $[Ca(Thd)_2(Tetraen)]$: a monomeric precursor for deposition of CaS thin films. *Chem. Mater.*, **9** (5), 1234–1240.

45 Putkonen, M., Sajavaara, T., Johansson, L.-S., and Niinistö, L. (2001) Low temperature ALE deposition of Y_2O_3 thin films from β-diketonate precursors. *Chem. Vapor Depos.*, **7** (1), 44–50.

46 Musschoot, J., Deduytsche, D., Poelman, H., Haemers, J., Van Meirhaeghe, R.L., Van den Berghe, S., and Detavernier, C. (2009) Comparison of thermal and plasma-enhanced ALD/CVD of vanadium pentoxide. *J. Electrochem. Soc.*, **156** (7), P122–P126.

47 Kim, J.C., Heo, J.S., Cho, Y.S., and Moon, S.H. (2009) Atomic layer deposition of an HfO_2 thin film using $Hf(O\text{-}iPr)_4$. *Thin Solid Films*, **517** (19), 5695–5699.

48 Pore, V., Rahtu, A., Leskelä, M., Ritala, M., Sajavaara, T., and Keinonen, J. (2004) Atomic layer deposition of photocatalytic TiO_2 thin films from titanium tetramethoxide and water. *Chem. Vapor Depos.*, **10** (3), 143–148.

49 Aarik, J., Aidla, A., Sammelselg, V., Uustare, T., Ritala, M., and Leskelä, M. (2000) Characterization of titanium dioxide atomic layer growth from titanium ethoxide and water. *Thin Solid Films*, **370** (12), 163–172.

50 Kim, S.K., Choi, G.J., Kim, J.H., and Hwang, C.S. (2008) Growth behavior of Al-doped TiO_2 thin films by atomic layer deposition. *Chem. Mater.*, **20** (11), 3723–3727.

51 Xie, Q., Musschoot, J., Deduytsche, D., Van Meirhaeghe, R.L., Detavernier, C., Van den Berghe, S., Jiang, Y.-L., Ru, G.-P., Li, B.-Z., and Qu, X.-P. (2008) Growth kinetics and crystallization behavior of TiO_2 films prepared by plasma enhanced atomic layer deposition. *J. Electrochem. Soc.*, **155** (9), H688–H692.

52 Vehkamäki, M., Hänninen, T., Ritala, M., Leskelä, M., Sajavaara, T., Rauhala, E., and Keinonen, J. (2001) Atomic layer deposition of $SrTiO_3$ thin films from a novel strontium precursor strontium bis (tri-isopropyl cyclopentadienyl). *Chem. Vapor Depos.*, **7** (2), 75–80.

53 Lee, J.H., Cho, Y.J., Min, Y.S., Kim, D., and Rhee, S.W. (2002) Plasma enhanced atomic layer deposition of $SrTiO_3$ thin films with $Sr(tmhd)_2$ and $Ti(i\text{-}OPr)_4$. *J. Vac. Sci. Technol. A*, **20** (5), 1828–1830.

54 Hatanpää, T., Vehkamäki, M., Ritala, M., and Leskelä, M. (2010) Study of bismuth alkoxides as possible precursors for ALD. *Dalton Trans.*, **39** (13), 3219–3226.

55 Bachmann, J., Jing, J., Knez, M., Barth, S., Shen, H., Mathur, S., Gösele, U., and

Nielsch, K. (2007) Ordered iron oxide nanotube arrays of controlled geometry and tunable magnetism by atomic layer deposition. *J. Am. Chem. Soc.*, **129** (31), 9554–9555.

56 Putkonen, M., Aaltonen, T., Alnes, M., Sajavaara, T., Nilsen, O., and Fjellvåg, H. (2009) Atomic layer deposition of lithium containing thin films. *J. Mater. Chem.*, **19**, 8767–8771.

57 Jones, A.C., Aspinall, H.C., Chalker, P.R., Potter, R.J., Manning, T.D., Loo, Y.F., O'Kane, R., Gaskell, J.M., and Smith, L.M. (2006) MOCVD and ALD of high-*k* dielectric oxides using alkoxide precursors. *Chem. Vapor Depos.*, **12** (2–3), 83–98.

58 Yang, T.S., Cho, W., Kim, M., An, K.-S., Chung, T.-M., Kim, C.G., and Kim, Y. (2005) Atomic layer deposition of nickel oxide films using $Ni(dmamp)_2$ and water. *J. Vac. Sci. Technol.*, **23** (4), 1238–1243.

59 Lee, J.P., Park, M.H., Chung, T.-M., Kim, Y., and Sung, M.M. (2004) Atomic layer deposition of TiO_2 thin films from Ti $(O^iPr)_2(dmae)_2$ and H_2O. *Bull. Korean Chem. Soc.*, **25** (4), 475–479.

60 Bachmann, J., Zierold, R., Chong, J.T., Hauert, R., Sturm, C., Schmidt-Grund, R., Rheinländer, B., Grundmann, M., Gösele, U., and Nielsch, K. (2008) A practical, self-catalytic, atomic layer deposition of silicon dioxide. *Angew. Chem., Int. Ed.*, **47**, 6177–6179.

61 Ferguson, J.D., Smith, E.R., Weimer, A.W., and George, S.M. (2004) Atomic layer deposition of SiO_2 at room temperature using TEOS and H_2O with NH_3 as the catalyst. *J. Electrochem. Soc.*, **151** (8), G528–G535.

62 Lee, B.H., Hwang, J.K., Nam, J.W., Lee, S.U., Kim, J.T., Koo, S.-M., Baunemann, A., Fischer, R.A., and Sung, M.M. (2009) Low-temperature atomic layer deposition of copper metal thin films: self-limiting surface reaction of copper dimethylamino-2-propoxide with diethylzinc. *Angew. Chem., Int. Ed.*, **48** (25), 4536–4539.

63 Becker, J.S., Kim, E., and Gordon, R.G. (2004) Atomic layer deposition of insulating hafnium and zirconium nitrides. *Chem. Mater.*, **16** (18), 3497–3501.

64 Kim, J.C., Cho, Y.S., and Moon, S.H. (2009) Atomic layer deposition of HfO_2 onto Si using $Hf(NMe_2)_4$. *Jpn. J. Appl. Phys.*, **48**, 066515–066521.

65 Kwon, J., Dai, M., Halls, M.D., Langereis, E., Chabal, Y.J., and Gordon, R.G. (2009) *In situ* infrared characterization during atomic layer deposition of lanthanum oxide. *J. Phys. Chem. C*, **113** (2), 654–660.

66 Lee, B., Park, T.J., Hande, S., Kim, M.J., Wallace, R.M., Kim, J., Liu, X., Yi, J.H., Li, H., Rousseau, M., Shenai, D., and Suydam, J. (2009) Electrical properties of atomic-layer-deposited La_2O_3 films using a novel La formamidinate precursor and ozone. *Microelectron. Eng.*, **86** (7–9), 1658–1661.

67 Yang, R.B., Bachmann, J., Reiche, M., Gerlach, J.W., Gösele, U., and Nielsch, K. (2009) Atomic layer deposition of antimony oxide and antimony sulfide. *Chem. Mater.*, **21** (13), 2586–2588.

68 Becker, J.S. and Gordon, R.G. (2003) Diffusion barrier properties of tungsten nitride films grown by atomic layer deposition from bis(*tert*-butylimido)bis (dimethylamido) tungsten and ammonia. *Appl. Phys. Lett.*, **82** (14), 2239–2241.

69 Miikkulainen, V., Suvanto, M., Pakkanen, T.A., Siitonen, S., Karvinen, P., Kuittinen, M., and Kisonen, H. (2008) Thin films of MoN, WN, and perfluorinated silane deposited from dimethylamido precursors as contamination resistant coatings on micro-injection mold inserts. *Surf. Coat. Technol.*, **202** (21), 5103–5109.

70 Kamiyama, S., Miura, T., and Nara, Y. (2006) Comparison between SiO_2 films deposited by atomic layer deposition with $SiH_2[N(CH_3)_2]_2$ and $SiH[N(CH_3)_2]_3$ precursors. *Thin Solid Films*, **515** (4), 1517–1521.

71 Burton, B.B., Kang, S.W., Rhee, S.W., and George, S.M. (2009) SiO_2 atomic layer deposition using tris(dimethylamino) silane and hydrogen peroxide studied by *in situ* transmission FTIR spectroscopy. *J. Phys. Chem. C*, **113** (19), 8249–8257.

72 Suzuki, I., Yanagita, K., and Dussarrat, C. (2007) Extra low-temperature SiO_2 deposition using aminosilanes. *ECS Trans.*, **3**, 119–128.

73 Rees, W.S., Jr., Just, O., and van Derveer, D.S. (1999) Molecular design of dopant precursors for atomic layer epitaxy of SrS: Ce. *J. Mater. Chem.*, **9**, 249–252.

74 He, W., Schuetz, S., Solanki, R., Belot, J., and McAndrew, J. (2004) Atomic layer deposition of lanthanum oxide films for high-kappa gate dielectrics. *Electrochem. Solid-State Lett.*, **7** (7), G131–G133.

75 Vehkamäki, V.M., Hatanpää, T., Kemell, M., Ritala, M., and Leskelä, M. (2006) Atomic layer deposition of ferroelectric bismuth titanate $Bi_4Ti_3O_{12}$ thin films. *Chem. Mater.*, **18** (16), 3883–3888.

76 Nam, W.-H. and Rhee, S.-W. (2010) Atomic layer deposition of ZrO_2 thin films using dichlorobis[bis-(trimethylsilyl)amido]zirconium and water. *Chem. Vapor Depos.*, **10** (4), 201–205.

77 Lim, B.S., Rahtu, A., Park, J.-S., and Gordon, R. (2003) Synthesis and characterization of volatile, thermally stable, reactive transition metal amidinates. *Inorg. Chem.*, **42** (24), 7951–7958.

78 Wang, H., Gordon, R.G., Alvis, R., and Ulfig, R.M. (2009) Atomic layer deposition of ruthenium thin films from an amidinate precursor. *Chem. Vapor Depos.*, **15** (10–12), 312–319.

79 Milanov, A.P., Xu, K., Laha, A., Bugiel, E., Ranjith, R., Schwendt, D., Osten, H.J., Parala, H., Fischer, R.A., and Devi, A. (2010) Growth of crystalline Gd_2O_3 thin films with a high-quality interface on Si (100) by low-temperature H_2O-assisted atomic layer deposition. *J. Am. Chem. Soc.*, **132** (1), 36–37.

80 Barreca, D., Milanov, A., Fischer, R.A., Devi, A., and Tondello, E. (2009) Hafnium oxide thin film grown by ALD: an XPS study. *Surf. Sci. Spectra*, **14** (1), 34–40.

81 Saly, M.J., Munnik, F., Baird, R.J., and Winter, C.H. (2009) Atomic layer deposition growth of BaB_2O_4 thin films from an exceptionally thermally stable tris (pyrazolyl)borate-based precursor. *Chem. Mater.*, **21** (16), 3742–3744.

82 Pore, V., Hatanpää, T., Ritala, M., and Leskelä, M. (2009) Atomic layer deposition of metal tellurides and selenides using alkylsilyl compounds of tellurium and selenium. *J. Am. Chem. Soc.*, **131** (10), 3478–3480.

83 Yokoyama, H., Tanimoto, M., Shinohara, M., and Inoue, N. (1994) Self-limiting and step-propagating nature of GaAs atomic layer epitaxy revealed by atomic force microscopy. *Appl. Surf. Sci.*, **82/83**, 158–163.

84 Reidl, J.G., Uridianyk, H.M., and Bedair, S.M. (1991) Role of trimethylgallium exposure time in carbon doping and high temperature atomic layer epitaxy of GaAs. *Appl. Phys. Lett.*, **59** (19), 2397–2399.

85 Simko, J.P., Meguro, T., Iwai, S., Ozasa, K., Aoyagi, Y., and Sugano, T. (1993) Surface photo-absorption study of the laser-assisted atomic layer epitaxial growth process of GaAs. *Thin Solid Films*, **225** (1–2), 40–46.

86 Jeong, W.G., Menu, E.P., and Dapkus, P.D. (1989) Steric hindrance effects in atomic layer epitaxy of InAs. *Appl. Phys. Lett.*, **55**, 244–246.

87 Pan, N., Carter, J., Hein, S., Howe, D., Goldman, L., Kupferberg, L., Brierley, S., and Hsieh, K.C. (1993) Atomic layer epitaxy of InP using trimethylindium and tertiarybutylphosphine. *Thin Solid Films*, **225** (1–2), 64–69.

88 Ott, A.W., Johnson, J.M., Klaus, J.W., and George, S.M. (1997) Surface chemistry of In_2O_3 deposition using $In(CH_3)_3$ and H_2O in a binary reaction sequence. *Appl. Surf. Sci.*, **112**, 205–215.

89 Hirose, S., Yoshida, A., Yamaura, M., Kano, N., and Munekata, H. (1995) Control of carbon incorporation in AlAs grown by atomic layer epitaxy using variously orientated substrates. *J. Mater. Sci.*, **11** (1), 7–10.

90 Ishii, M., Iwai, S., Ueki, T., and Aoyagi, Y. (1998) Surface reaction mechanism and morphology control in AlP atomic layer epitaxy. *Thin Solid Films*, **318**, 6–10.

91 Lee, Y.J. and Kang, S.-W. (2004) Growth of aluminum nitride thin films prepared by plasma-enhanced atomic layer deposition. *Thin Solid Films*, **446** (2), 227–231.

92 Kukli, K., Ritala, M., Leskelä, M., and Jokinen, J. (1997) Atomic layer epitaxy growth of aluminum oxide thin films from a novel $Al(CH_3)_2Cl$ precursor and H_2O. *J. Vac. Sci. Technol. A*, **15** (4), 2214–2218.

93 Huang, R. and Kitai, A.H. (1993) Preparation and characterization of thin films of MgO, Al_2O_3 and $MgAl_2O_4$ by atomic layer deposition. *J. Electron. Mater.*, **22** (2), 215–220.

94 Puurunen, R. (2005) Surface chemistry of atomic layer deposition: a case study for the trimethylaluminum/water process. *Appl. Phys. Rev.*, **97** (12), 121301-1–121301-52.

95 Fan, J.-F., Toyoda, K. (2005) (1992) Self-limiting behavior of the growth of Al_2O_3 using sequential vapor pulses of TMA and H_2O_2. *Appl. Surf. Sci.*, **60–61**, 765–769.

96 Drozd, V.E., Baraban, A.P., and Nikiforova, I.O. (1994) Electrical properties of Si–Al_2O_3 structures grown by ML-ALE. *Appl. Surf. Sci.*, **82–83**, 583–586.

97 Kumagai, H., Toyoda, K., Matsumoto, M., and Obara, M. (1993) Comparative study of Al_2O_3 optical crystalline thin films grown by vapor combinations of Al$(CH_3)_3/N_2O$ and $Al(CH_3)_3/H_2O_2$. *Jpn. J. Appl. Phys.*, **32**, 6137–6140.

98 Jeong, C.-W., Lee, J.-S., and Joo, S.-K. (2001) Plasma-assisted atomic layer growth of high-quality aluminum oxide thin films. *Jpn. J. Appl. Phys.*, **40**, 285–289.

99 Elliot, S.D., Scarel, G., Wiemer, C., Fanciulli, M., and Pavia, G. (2006) Ozone-based atomic layer deposition of alumina from TMA: growth, morphology, and reaction mechanism. *Chem. Mater.*, **18** (16), 3764–3773.

100 Drozd, V.E. and Aleskovski, V.B. (1994) Synthesis of conducting oxides by ML-ALE. *Appl. Surf. Sci.*, **82–83**, 591–594.

101 Yousfi, E.B., Asikainen, T., Pietu, V., Cowache, P., Powalla, M., and Lincot, D. (2000) Cadmium-free buffer layers deposited by atomic later epitaxy for copper indium diselenide solar cells. *Thin Solid Films*, **361–362**, 183–186.

102 Yokoyama, M., Chen, N.T., and Ueng, H.Y. (2000) Growth and characterization of ZnSe on Si by atomic layer epitaxy. *J. Cryst. Growth*, **212** (1), 97–102.

103 Wang, W.-S., Ehsani, H., and Bhat, I. (1993) Improved CdTe layers on GaAs and Si using atomic layer epitaxy. *J. Electron. Mater.*, **22** (8), 873–878.

104 Sang, B., Yamada, A., and Konagai, M. (1997) Growth of boron-doped ZnO thin films by atomic layer deposition. *Sol. Energy Mater. Sol. Cells*, **49** (1–4), 19–26.

105 Saito, K., Hiratsuka, Y., Omata, A., Makino, H., Kishimoto, S., Yamamoto, T., Horiuchi, N., and Hirayama, H. (2007) Atomic layer deposition and characterization of Ga-doped ZnO thin films. *Superlatt. Microstruct.*, **42** (1–6), 172–175.

106 Lujala, V., Skarp, J., Tammenmaa, M., and Suntola, T. (1994) Atomic layer epitaxy growth of doped zinc oxide thin films from organometals. *Appl. Surf. Sci.*, **82–83**, 34–40.

107 Yun, S.J., Kim, Y.S., and Park, S.-H.K. (2001) Fabrication of CaS:Pb blue phosphor by incorporating dimeric Pb^{2+} luminescent centers. *Appl. Phys. Lett.*, **78** (6), 721–723.

108 Long, N.J. (1998) *Metallocenes: An Introduction to Sandwich Compounds*, Blackwell Science Ltd, Oxford.

109 Weber, A., Suhr, H., Schumann, H., Köhn, R.-D. (1990) Thin yttrium and rare earth oxide films produced by plasma enhanced CVD of novel organometallic π-complexes. *Appl. Phys. A*, **51** (6), 520–525.

110 Russell, D.K. (1996) Gas-phase pyrolysis mechanisms in organometallic CVD. *Chem. Vapor Depos.*, **2** (6), 223–233.

111 Putkonen, M., Sajavaara, T., and Niinistö, L. (2000) Enhanced growth rate in atomic layer epitaxy deposition of magnesium oxide thin films. *J. Mater. Chem.*, **10**, 1857–1861.

112 Huang, R. and Kitai, A.H. (1992) Temperature-dependence of the growth orientation of atomic layer growth MgO. *Appl. Phys. Lett.*, **61** (12), 1450–1452.

113 Vehkamäki, M., Hänninen, T., Ritala, M., Leskelä, M., Sajavaara, T., Rauhala, E., and Keinonen, J. (2001) Atomic layer deposition of $SrTiO_3$ thin films from a novel strontium precursor: strontium-bis (tri-isopropyl cyclopentadienyl). *Chem. Vapor Depos.*, **7** (2), 75–80.

114 Ihanus, J., Hänninen, T., Hatanpää, T., Aaltonen, T., Mutikainen, I., Sajavaara, T., Keinonen, J., Ritala, M., and Leskelä, M. (2002) Atomic layer deposition of SrS and BaS thin films using cyclopentadienyl precursors. *Chem. Mater.*, **14** (5), 1937–1944.

115 Putkonen, M., Nieminen, M., Niinistö, J., Sajavaara, T., and Niinistö, L. (2001) Surface-controlled deposition of Sc_2O_3 thin films by atomic layer epitaxy using β-diketonate and organometallic precursors. *Chem. Mater.*, **13** (12), 4701–4707.

116 Kim, W.-S., Park, S.-K., Moon, D.-Y., Kang, B.-W., Kim, H.-D., and Park, J.-W. (2009) Characteristics of La_2O_3 thin films deposited using the ECR atomic layer deposition method. *J. Korean Phys. Soc.*, **55** (2), 590–593.

117 Lau, J.E., Peterson, G.G., Endisch, D., Barth, K., Topol, A., Kaloyeros, A.E., Tuenge, R.T., and King, C.N. (2001) *In situ* studies of the nucleation mechanisms of tris(cyclopentadienyl)cerium as cerium dopant source in SrS:Ce thin films for electroluminescent displays. *J. Electrochem. Soc.*, **148** (6), C427–C432.

118 Katamreddy, R., Stafford, N.A., Guerin, L., Feist, B., Dussarrat, C., Pallem, V., Weiland, C., and Opila, R. (2009) Atomic layer deposition of rare-earth oxide thin films for high-*k* dielectric applications. *ECS Trans.*, **19** (2), 525–536.

119 Putkonen, M. and Niinistö, L. (2001) Zirconia thin films by atomic layer epitaxy. A comparative study on the use of novel precursors with ozone. *J. Mater. Chem.*, **11**, 3141–3147.

120 Niinistö, J., Putkonen, M., Niinistö, L., Stoll, S.L., Kukli, K., Sajavaara, T., Ritala, M., and Leskelä, M. (2005) Controlled growth of HfO_2 thin films by atomic layer deposition from cyclopentadienyl-type precursor and water. *J. Mater. Chem.*, **15** (23), 2271–2275.

121 Rose, M., Niinistö, J., Michalowski, P., Gerlich, L., Wilde, L., Endler, I., and Bartha, J.W. (2009) Atomic layer deposition of titanium dioxide thin films from $Cp^*Ti(OMe)_3$ and ozone. *J. Phys. Chem. C*, **113** (52), 21825–21830.

122 Niinistö, J., Putkonen, M., Niinistö, L., Song, F., Williams, P., Heys, P.N., and Odedra, R. (2007) Atomic layer deposition of HfO_2 thin films exploiting novel cyclopentadienyl precursors at high temperatures. *Chem. Mater.*, **19** (13), 3319–3324.

123 Katamreddy, R., Wang, Z., Omarjee, V., Rao, P.V., Dussarrat, C., and Blasco, N. (2009) Advanced precursor development for Sr and Ti based oxide thin film applications. *ECS Trans.*, **25** (4), 217–230.

124 Niinistö, J., Mäntymäki, M., Kukli, K., Costelle, L., Puukilainen, E., Ritala, M., and Leskelä, M. (2010) Growth and phase stabilization of HfO_2 thin films by ALD using novel precursors. *J. Cryst. Growth*, **312** (1), 245–249.

125 Burton, B.B., Fabreguette, F.H., and George, S.M. (2009) Atomic layer deposition of MnO using bis (ethylcyclopentadienyl)manganese and H_2O. *Thin Solid Films*, **517** (19), 5658–5665.

126 Rooth, M., Johansson, A., Kukli, K., Aarik, J., Boman, M., and Hårsta, A. (2008) Atomic layer deposition of iron oxide thin films and nanotubes using ferrocene and oxygen as precursors. *Chem. Vapor Depos.*, **14** (3–4), 67–70.

127 Elam, J.W., Martinson, A.B.F., Pellin, M.J., and Hupp, J.T. (2006) Atomic layer deposition of In_2O_3 using cyclopentadienyl indium: a new synthetic route to transparent conducting oxide films. *Chem. Mater.*, **18** (15), 3571.

128 Chae, J., Park, H.-S., and Kang, S. (2002) Atomic layer deposition of nickel by the reduction of preformed nickel oxide. *Electrochem. Solid-State Lett.*, **5** (10), C64–C66.

129 Daub, M., Knez, M., Göesele, U., and Nielsch, K. (2007) Ferromagnetic nanotubes by atomic layer deposition in anodic alumina membranes. *J. Appl. Phys.*, **101** (9), 09J111–09J113.

130 Kim, J.-M., Lee, H.-B.-R., Lansalot, C., Dussarat, C., Gatineau, J., and Kim, H. (2010) Plasma-enhanced atomic layer deposition of cobalt using cyclopentadienyl isopropyl acetamidinato-cobalt as a precursor. *Jpn. J. Appl. Phys.*, **49**, 05FA10-1–05FA10-5.

131 Aaltonen, T., Ritala, M., Tung, Y.-L., Chi, Y., Arstila, K., Meinander, K., Leskelä, M. (2004) Atomic layer deposition of noble metals: exploration of the low limit of the deposition temperature. *J. Mater. Res.*, **19**, 3353–3358.

132 Aaltonen, T., Rahtu, A., Ritala, M., and Leskelä, M. (2003) Reaction mechanism studies on atomic layer deposition of ruthenium and platinum. *Electrochem. Solid-State Lett.*, **6** (9), C130–C133.

133 Park, S.-J., Kim, W.-H., Lee, H.-B.-R., Maeng, W.J., and Kim, H. (2008) Thermal and plasma enhanced atomic layer deposition ruthenium and electrical characterization as a metal electrode. *Microelectron. Eng.*, **85** (1), 39–44.

134 Kim, W.-H., Park, S.-J., Kim, D.Y., and Kim, H. (2009) Atomic layer deposition of ruthenium and ruthenium-oxide thin films by using a $Ru(EtCp)_2$ precursor and oxygen gas. *J. Korean Phys. Soc.*, **55** (1), 32–37.

135 Knoops, H.C.M., Mackus, A.J.M., Donders, M.E., van de Sanden, M.C.M., Notten, P.H.L., and Kessels, W.M.M. (2009) Remote plasma ALD of platinum and platinum oxide films. *Electrochem. Solid-State Lett.*, **12** (7), G34–G36.

136 Aaltonen, T., Ritala, M., Sajavaara, T., Keinonen, J., and Leskelä, M. (2003) Atomic layer deposition of platinum thin films. *Chem. Mater.*, **15** (9), 1924–1928.

137 Aaltonen, T., Ritala, M., and Leskelä, M. (2005) Atomic layer deposition of noble metals, in *Advanced Metallization Conference 2004 (AMC 2004)* (eds D. Erb, P. Ramm, K. Masu, and A. Osaki), Materials Research Society, pp. 663–667.

138 Schuisky, M., Kukli, K., Ritala, M., Hårsta, A., and Leskelä, M. (2000) Atomic layer chemical vapor deposition in the Bi–Ti–O system. *Chem. Vapor Depos.*, **6** (3), 139–145.

139 Harjuoja, J., Väyrynen, S., Putkonen, M., Niinistö, L., and Rauhala, E. (2006) Crystallization of bismuth titanate and bismuth silicate grown as thin films by atomic layer deposition. *J. Cryst. Growth*, **286** (2), 376–383.

140 Ritala, M., Kukli, K., Rahtu, A., Räisänen, P.I., Leskelä, M., Sajavaara, T., and Keinonen, J. (2000) Atomic layer deposition of oxide thin films with metal alkoxides as oxygen sources. *Science*, **288** (5464), 319–321.

141 Hausmann, D., Becker, J., Wang, S., and Gordon, R.G. (2002) Rapid vapor deposition of highly conformal silica nanolaminates. *Science*, **298** (5592), 402–406.

142 Putkonen, M., Sajavaara, T., Rahkila, P., Xu, L., Cheng, S., Niinistö, L., and Whitlow, H.J. (2009) Atomic layer deposition and characterization of biocompatible hydroxyapatite thin films. *Thin Solid Films*, **517** (20), 5819–5824.

143 Juppo, M., Vehkamäki, M., Ritala, M., and Leskelä, M. (1998) Deposition of molybdenum thin films by an alternate supply of $MoCl_5$ and Zn. *J. Vac. Sci. Technol. A*, **16** (5), 2845–2850.

144 Elam, J.W., Nelson, C.E., Grubbs, R.K., and George, S.M. (2001) Kinetics of the WF_6 and Si_2H_6 surface reactions during tungsten atomic layer deposition. *Surf. Sci.*, **479**, 121–135.

145 Alén, P., Juppo, M., Ritala, M., Sajavaara, T., Keinonen, J., and Leskelä, M. (2001) Atomic layer deposition of Ta(Al)N(C) thin films using trimethylaluminum as a reducing agent. *J. Electrochem. Soc.*, **148** (10), G566–G571.

146 Kim, S.-H., Kim, S.-K., Kwak, N., Sohn, H., Kim, J., Jung, S.-H., Hong, M.-R., Lee, S.H., and Collins, J. (2006) Atomic layer deposition of low-resistivity and high-density tungsten nitride thin films using B_2H_6, WF_6, and NH_3. *Electrochem. Solid-State Lett.*, **9** (3), C54–C57.

147 Kukli, K., Forsgren, K., Aarik, J., Uustare, T., Aidla, A., Niskanen, A., Ritala, M., Leskelä, M., and Hårsta, A. (2001) Atomic layer deposition of zirconium oxide from zirconium tetraiodide, water and hydrogen peroxide. *J. Cryst. Growth*, **231** (1–2), 262–272.

148 Delabie, A., Caymax, M., Gielis, S., Maes, J.W., Nyns, L., Popovici, M., Swerts, J., Tielens, H., Peeters, J., and Van Elshocht, S. (2010) Ozone-based metal oxide atomic layer deposition: impact of N_2/O_2 supply ratio in ozone generation. *Electrochem. Solid-State Lett.*, **13** (6), H176–H178.

149 Hiltunen, L., Kattelus, H., Leskelä, M., Mäkelä, M., Niinistö, L., Nykänen, E., Soininen, P., and Tiitta, M. (1991) Growth and characterization of aluminium oxide thin films deposited from various source materials by atomic layer epitaxy and chemical vapor deposition processes. *Mater. Chem. Phys.*, **28** (4), 379–388.

150 Mölsä, H., Niinistö, L., and Utriainen, M. (1994) Growth of yttrium oxide thin films from beta-diketonate precursor. *Adv. Mater. Opt. Electron.*, **4** (6), 389–400.

151 Hatanpää, T., Ihanus, J., Kansikas, J., Mutikainen, I., Ritala, M., and Leskelä, M. (1999) Properties of $[Mg_2(thd)_4]$ as a precursor for atomic layer deposition of MgO thin films and crystal structures of $[Mg_2(thd)_4]$ $[Mg(thd)_2(EtOH)_2]$. *Chem. Mater.*, **11**, 1846.

152 Liu, X., Ramanathan, S., Longdergan, A., Srivastava, A., Lee, E., Seidel, T.E., Barton, J.T., Pang, D., and Gordon, R.G. (2005) ALD of hafnium oxide thin films from tetrakis(ethylmethylamino)hafnium and ozone. *J. Electrochem. Soc.*, **152** (3), G213–G219.

153 Niinistö, J., Putkonen, M., Niinistö, L., Arstila, K., Sajavaara, T., Lu, J., Kukli, K., Ritala, M., and Leskelä, M. (2006) HfO_2 films grown by ALD using cyclopentadienyl-type precursors and H_2O or O_3 as oxygen source. *J. Electrochem. Soc.*, **153** (3), F39–F45.

154 Juppo, M., Alén, P., Ritala, M., and Leskelä, M. (2001) Trimethylaluminium as a reducing agent in the atomic layer deposition of Ti(Al)N thin films. *Chem. Vapor Depos.*, **7** (5), 211.

155 Burton, B.B., Lavoie, A.R., and George, S.M. (2008) Tantalum nitride atomic layer deposition using tris(diethylamido)(*tert*-butylimido)tantalum and hydrazine. *J. Electrochem. Soc.*, **155** (7), D508–D516.

156 Alén, P., Juppo, M., Ritala, M., Leskelä, M., Sajavaara, T., and Keinonen, J. (2002) *tert*-Butylamine and allylamine as reductive nitrogen sources in atomic layer deposition of TaN thin films. *J. Mater. Res.*, **17** (1), 107–114.

157 Kim, H. (2003) Atomic layer deposition of metal and nitride thin films: current research efforts and applications for semiconductor device processing. *J. Vac. Sci. Technol. B*, **21** (6), 2231–2261.

158 Elers, K.-E., Winkler, J., Weeks, K., and Marcus, S. (2005) $TiCl_4$ as a precursor in the TiN deposition by ALD and PEALD. *J. Electrochem. Soc.*, **152** (8), G589–G593.

159 Maeng, W.J. and Kim, H. (2006) Thermal and plasma-enhanced ALD of Ta and Ti oxide thin films from alkylamide precursors. *Electrochem. Solid-State Lett.*, **9** (6), G191–G194.

160 Guziewicz, E., Godlewski, M., Kopalko, K., Lusakowska, E., Dynowska, E., Guziewicz, M., Godlewski, M.M., and Phillips, M. (2004) Atomic layer deposition of thin films of ZnSe: structural and optical characterization. *Thin Solid Films*, **446**, 172–177.

161 Nanu, M., Reijnen, L., Meester, B., Schoonman, J., and Goossens, A. (2004) $CuInS_2$ thin films deposited by ALD. *Chem. Vapor Depos.*, **10** (1), 45–49.

162 Drozd, V.E., Nikiforova, I.O., Bogevolnov, V.B., Yafyasov, A.M., Filatova, E.O., and Papazoglou, D. (2009) ALD synthesis of SnSe layers and nanostructures. *J. Phys. D*, **42** (12), 125306–125310.

163 Bakke, J.R., King, J.S., Jung, H.J., Sinclair, R., and Bent, S.F. (2010) Atomic layer deposition of ZnS via *in situ* production of H_2S. *Thin Solid Films*. doi: 10.1016/j.tsf.2010.03.074.

4
Sol–Gel Chemistry and Atomic Layer Deposition

Guylhaine Clavel, Catherine Marichy, and Nicola Pinna

4.1
Aqueous and Nonaqueous Sol–Gel in Solution

4.1.1
Aqueous Sol–Gel

4.1.1.1 Introduction

As the name suggests, the sol–gel is a process in which an inorganic network is formed from a homogeneous solution of precursors through the formation of a sol (colloidal suspension) and then of a gel (cross-linked solid network surrounding a continuous liquid phase) by inorganic polymerizations [1, 2]. Basically, the principle of a sol–gel reaction is the formation of an inorganic polymer by hydrolysis and condensation reactions generally induced by water, with or without the presence of acid/base, the precursor being either a metal salt or a metal-organic compound. Metal alkoxides are the most convenient precursors used, reacting readily with water and known for almost all metals [3]. The characteristics, structure, and properties of a sol–gel inorganic network depend on the factors that affect the rate of hydrolysis and condensation reactions, for example, pH, temperature and time of reaction, reagent concentrations, nature of the solvent, catalyst, aging temperature and time, and drying conditions [1, 2]. By adjusting these parameters, homogeneous inorganic oxides such as bulk oxides as well as glasses, films, fibers, and particles can be produced by the sol–gel technique. Advantages such as cost effectiveness, low process temperature, molecular level homogeneity of multicomponent systems, and the possibility of easy shape processing explain the high scientific interest in sol–gel approaches [4]. Especially in the past few decades, sol–gel processes received much attention in the glass and ceramic fields due to the possibility to synthesize at low temperature materials with desirable properties of optical transparency, porosity, and chemical, mechanical, and thermal resistance [1].

4.1.1.2 Development of Sol–Gel Chemistry

Because of its mild conditions, sol–gel chemistry found an undeniable echo in the synthesis of hybrid oxides and phosphates, offering a versatile access to new hybrid

Atomic Layer Deposition of Nanostructured Materials, First Edition. Edited by Nicola Pinna and Mato Knez.

organic–inorganic materials [5–7]. Furthermore, a series of related processes that do not strictly follow the definition of sol–gel, that is, do not include the formation of a gel but make use of sol–gel chemistry, have also been developed. For example, by a careful control of sol preparation and processing, colloids or monodispersed nanoparticles of various oxides and organic–inorganic hybrids can be synthesized [8, 9]. However, the obtained particles are generally amorphous and need a subsequent calcination to be crystallized. Another intriguing approach is the "surface sol–gel process" that is, somehow, an atomic layer deposition (ALD) in solution. In this technique, an ultrathin film of metal oxide [10–16], phosphate [17, 18], or organic–inorganic hybrid [19] is grown, layer by layer, by successive dip coatings of a substrate. For example, a typical experiment consists of immersing a substrate into a metal precursor solution and then in an aqueous solution, each step being separated by a rinse in order to purge the surface of unreacted precursors [14].

While the sol–gel approach is extremely successful in the case of silica, it was found to be less attractive for the synthesis of transition metal oxides, as transition metal oxide precursors are generally much more reactive toward water preventing an accurate control over the growth of metal oxides [3, 20]. Much effort has been made to overcome this problem and achieve more controlled systems. First, the use of organic additives such as functional alcohols, carboxylic acids, or β-diketones that modify the precursor and thereby its reactivity was thoroughly studied [3, 20]. An alternative approach is to contain the hydrolysis of the metal oxide precursors by controlling the local concentration of water by chemical or physical processes. For example, the generation of water *in situ* or the slow dissolution of water from the gas phase in the solvent has been employed as strategies [21–24].

Finally, another approach to gain a better control over the formation of inorganic network is to completely avoid the used of water. The sol–gel chemistry under these nonaqueous conditions is known as nonaqueous or nonhydrolytic sol–gel (NHSG).

4.1.2
Nonaqueous Sol–Gel

4.1.2.1 Introduction

During the past few decades, the use of nonaqueous conditions has proven to be an elegant approach, eliminating the main drawbacks of aqueous sol–gel chemistry and offering advantages such as high reproducibility and better control of the composition and homogeneity of multicomponent oxides [25–29]. Following this approach, a large variety of metal oxide gels were synthesized and reported, including silica, alumina, titania, transition metal oxides, multicomponent oxides (Al/Si, Al/Ti, Si/Ti, Si/Zr, etc.), and organic–inorganic hybrids [25, 27, 29].

In the NHSG chemistry, the M–O–M bonds are formed without a hydrolysis step, the oxygen being provided by other molecular species such as the solvent (alcohol, ether, ketone) or by the metal precursor itself (alkoxide, acetate, acetylacetonate) [27, 29, 30]. It is interesting to note that, besides the number of possible precursor and/or solvent combinations, there are few different reaction pathways responsible for the metal–oxygen–metal bond formation. For instance, the formation of the metal oxide network can involve only a condensation step, occurring between ligands coordinated to two different metal centers under the elimination of organic molecules. It is possible to summarize most of these condensation reactions in only three distinct mechanisms: the alkyl halide, the ether, and ester eliminations [25, 27, 28]. It is important to note that in the case of these direct condensation reactions no intermediate formation of an OH group is involved. When the NHSG approaches engage the reaction between a metal precursor and an organic solvent, the following reaction mechanisms are possible: a ligand exchange followed by one of the condensation reactions mentioned above, a hydroxylation reaction of the metal complex leading to the formation of a hydroxyl group (nonhydrolytic hydroxylation reactions [29]) that further reacts with a metal precursor or another hydroxyl group, and finally more complex mechanisms such as Guerbet-like reactions or aldol condensations [25, 27, 28].

4.1.2.2 Synthesis of Nanomaterials

NHSG processes were extensively applied to the synthesis of oxide nanoparticles because the as-synthesized oxides are generally characterized by high crystallinity and because of the ability to control the crystal growth without the use of any additional ligand [25–28]. Taking into account that the synthesis parameters (precursor, solvent, temperature, etc.) strongly influence the size and the shape of the final materials, a large number of available precursors and organic solvents offer many possible combinations for the synthesis of a specific system [25, 27, 28]. Furthermore, as mentioned earlier, an advantage of nonhydrolytic sol–gel processes, especially in comparison to aqueous systems, is the accessibility of multimetal oxide. One reason is that, in general, the reactivity of metal oxide precursors is significantly reduced under water exclusion, making it easier to match the reactivity of different metal precursors to obtain single-phase multicomponent oxides [25, 27, 28]. As a matter of fact, nonhydrolytic sol–gel has been successfully employed for the synthesis of various binary, ternary, and multimetal oxides such as $BaTiO_3$, $InNbO_4$, $ZnGa_2O_4$, $NaNbO_3$, $NaTaO_3$, $BaZrO_3$, $SrTiO_3$, and $(Ba, Sr)TiO_3$, as well as for the synthesis of doped oxide nanoparticles [25–28, 31].

Recently, nonaqueous conditions were also successfully used for the deposition of oxide thin films using the surface sol–gel process [32] or ALD [33–35].

4.2 Sol–Gel and ALD: An Overview

ALD of metal oxides involves the reaction of a metal oxide precursor with an oxygen source. Water is the most commonly used oxygen precursor; the formation of M–O–M bonds (oxo–metal bonds) can be grouped into the two successive reactions: (i) a hydrolysis step during the water pulse forming –OH groups and oxo–metal bonds, and (ii) a condensation step in which the resulting hydroxyl groups react with the metal oxide precursors supplied by a new pulse. The formation of the oxide is therefore achieved through hydrolysis and condensation steps as in the traditional solution sol–gel process. Recently, inspired by the success of nonaqueous sol–gel, similar nonaqueous conditions were developed in ALD. The evolution and strategies developed for the deposition of metal oxide thin films by ALD show similarities with the development of the sol–gel approach. Both techniques use common precursors such as metal halides, alkoxides, and diketonates and modifications of precursors were also pursued in ALD to enhance their efficiency by tuning their reactivity/stability/volatility [36, 37]. Based on the above discussion, it would be interesting to determine and compare the chemistry responsible for the metal oxide formation in ALD and in sol–gel. Indeed, ALD could greatly benefit from the established knowledge and experience in sol–gel chemistry especially in the development of new chemical approaches and in the design of precursors.

In this section, different chemical approaches for metal oxide thin film growth by ALD will be discussed and compared with oxide formation in solution. The ALD reactions can be divided into two approaches depending on whether a hydrolytic step is involved or not. First, we will focus on the hydrolytic route; the reaction mechanism involving water or hydrogen peroxide as an oxygen source will be presented, classified according to the various possible reagents. In the second part, an overview of the reaction taking place under nonaqueous conditions will be given.

4.2.1
Metal Oxide Formation via Hydrolytic Routes

The hydrolytic sol–gel route involves the formation of hydroxyl intermediate species. The oxygen source used is, in general, water but hydrogen peroxide is also common, while metal halides, metal alkoxides, and organometallic compounds can be employed as metal sources. Depending on the metal precursor, different reactions are responsible for the metal oxide formation. In the following, for each precursor family evocated, the general mechanism observed will be illustrated through a nonexhaustive overview of *in situ* studied depositions.

4.2.1.1 Reaction with Metal Halides

Metal halides have been used in ALD as precursors for the deposition of many oxides [38]. An early *in situ* study of this reaction dealt with the deposition of titania, from titanium chloride and water, investigated by FTIR spectroscopy. A ligand exchange during both precursor pulses was observed [39]. Later on, this reaction was examined by *in situ* QMS and QCM while using D_2O. During both the $TiCl_4$ and D_2O pulses, DCl is released, confirming the results of the previously published study, that is, the formation of the metal oxide occurs via $-OH$ to $-OTiCl_x$ and $-Cl$ to $-OH$ surface exchanges [40]. Moreover, the author observed a decrease in the amount of DCl released while increasing the temperature. It had been proposed that this phenomenon is caused by dehydroxylation of the surface at high temperature, which reduced the number of active sites, by molecular adsorption of $TiCl_4$, and/or by readsorption of DCl [40]. Rahtu and Ritala investigated the mechanism of ZrO_2 deposition from $ZrCl_4$ and H_2O by *in situ* QCM and QMS [41]. They observed a reaction mechanism similar to that for titania. However, they quantified that, at high temperature, only 1 equiv of HCl was released during the metal precursor pulse to form Zr–O bond and 3 equiv of HCl was released during the water pulse, while at moderate temperature (300–325 °C) 2 equiv of HCl was released during each half-sequence. A similar behavior was observed for HfO_2 deposition with hafnium halide precursor [41]. Another example of reaction with metal halide is the deposition of SnO_2 using $SnCl_4$ and H_2O or H_2O_2. The reaction with hydrogen peroxide should be thermodynamically more favorable than that with water as shown by the calculated enthalpies of the overall reactions (−34.6 and 4.5 kcal/mol in the cases of H_2O_2 and water, respectively) [42], and effectively, a decrease in the deposition temperature and an increase in the growth per cycle (GPC) by changing the oxygen source from water to hydrogen peroxide was demonstrated [42].

The deposition of other metal oxides was also studied by *in situ* techniques. Generally, the reaction pathway observed consists of the hydrolysis of the surface by the oxygen source with formation of hydrogen halide as a by-product (Eq. (4.1)). The resulting OH-terminated surface then reacts with the incoming metal halide to form the M–O–M bond under the elimination of hydrogen halide (Eq. (4.2)). Moreover, depending on the temperature, phenomena of dehydroxylation of the surface and/or readsorption of the by-product, among other reactions, have been observed [43].

$$\equiv \mathrm{OMX}_x + x\mathrm{H_2O} \rightarrow \; \equiv \mathrm{OM(OH)}_x + x\mathrm{HX} \tag{4.1}$$

$$\equiv \mathrm{OH} + \mathrm{MX}_x \rightarrow \; \equiv \mathrm{OMX}_{x-1} + \mathrm{HX} \tag{4.2}$$

4.2.1.2 **Reaction with Metal Alkoxide**

In solution, one of the most widely used sol–gel approaches to form metal oxide is the reaction between metal alkoxide and water that releases the corresponding alcohol as a by-product. In ALD, the related reaction can be represented by the following two half-equations (Eqs. (4.3) and (4.4)):

$$\equiv \mathrm{OM(OR)}_x + x\mathrm{H_2O} \rightarrow \; \equiv \mathrm{OM(OH)}_x + x\mathrm{ROH} \tag{4.3}$$

$$\equiv \mathrm{OH} + \mathrm{M(OR)}_x \rightarrow \; \equiv \mathrm{OM(OR)}_{x-1} + \mathrm{ROH} \tag{4.4}$$

Various metal oxide depositions via this approach have been investigated. For example, tantalum oxide deposition from $Ta(OC_2H_5)_5$ and water was studied by *in situ* QCM. The two possible mechanisms considered were either growth via molecular adsorption of the metal precursor or growth via the formation of intermediate hydroxyl species by exchange reactions. The mass gain and surface mass exchange ratio, determined by QCM monitoring, pointed out an exchange reaction process as the main reaction path [44]. Aarik *et al.* studied the reaction mechanism of the titanium ethoxide/water process [45], as well as titanium isopropoxide/water and titanium isopropoxide/H_2O_2 processes [46], by *in situ* QCM. For all the precursor combinations tested, the surface species changed from isopropoxide or ethoxide to hydroxyl species during the oxygen precursor pulses. However, the GPC appeared to be much lower with titanium ethoxide than with the isopropoxide. In fact, the authors observed that during the ethoxide precursor pulse, less than one ligand is released, the remaining ethanol being liberated during the water pulse. The resulting steric hindrance explained the low GPC found with this precursor [45]. Comparing the titanium isopropoxide/water and titanium isopropoxide/hydrogen peroxide processes, it appeared that a higher GPC has been obtained at low temperature using H_2O_2 as a result of its greater ability to eliminate alkoxide ligands. At higher temperatures, the growth rate was independent of the oxygen precursor.

Basically, the main reaction pathway observed during depositions from metal alkoxide and water or hydrogen peroxide consists of hydrolysis of the alkoxide surface species leading to an –OH-terminated surface by the release of alcohol. The hydroxyl groups then react with the gas-phase metal alkoxide to form the M–O–M bond under alcohol elimination. Moreover, phenomena of dehydroxylation of the surface at high temperature can occur. Due to the decrease in OH surface groups, only one ligand is in general released, during the metal pulse [45, 47].

4.2.1.3 Reaction with Organometallic Compounds

Trimethylaluminum (TMA) is probably the most widely employed organometallic compound in ALD, and its use with water for alumina deposition has been intensively studied [38]. The two half-sequences of this process are exchange reactions releasing methane as a by-product (Eqs. (4.5) and (4.6)), as confirmed by *in situ* FTIR, QCM, and QMS experiments [38, 48].

$$\equiv \mathrm{OAl(CH_3)}x + x\mathrm{H_2O} \rightarrow \equiv \mathrm{OAl(OH)}_x + x\mathrm{CH_4} \quad (4.5)$$

$$\equiv \mathrm{OH} + \mathrm{Al(CH_3)_3} \rightarrow \equiv \mathrm{OAl(CH_3)_2} + \mathrm{CH_4} \quad (4.6)$$

Furthermore, it has also been proposed that dissociation reactions occur during both water and TMA pulses [38]. The formation of Al_2O_3 from TMA and H_2O, being thermodynamically favorable, can be achieved at temperature as low as 33 °C, even if in this case the GPC is low because of the slow surface reaction kinetics [49]. A decrease in the GPC was also observed at high temperature due to the diminution of the hydroxyl group concentration on the surface caused by dehydroxylation [48]. Besides alumina, zinc and indium oxides have also been deposited from a metal alkyl/water process (diethylzinc [50] and trimethylindium [51], respectively). The reaction mechanisms found for these two processes were similar to that involving TMA. However, in the case of In_2O_3 deposition, rough oxide films with a low GPC have been obtained resulting probably from low hydroxyl coverage of the surface [51]. Other organometallic precursors than metal alkyl were also used to deposit metal oxide, such as cyclopentadienyl-based metal compounds. As an example, Niinisto *et al.* have investigated the ALD deposition of ZrO_2 from $Cp_2Zr(CH_3)_2$ and H_2O by *in situ* QMS [52]. The by-products released during both pulses were mostly methane and cyclopentadiene. Due to its higher reactivity, it appeared that the methyl ligand first reacted on the hydroxylated surface. In conclusion, the main reaction pathway observed during metal oxide deposition from organometallic compound and water is a ligand exchange reaction producing the corresponding alkane/alkene as by-products.

4.2.2
Metal Oxide Deposition under Nonaqueous Conditions

During the past decade, nonaqueous conditions were attempted to be adapted to ALD [33, 35]. Different approaches were studied, including the use of alkoxides, alcohols, or carboxylic acids as oxygen sources. In the following, the elucidated or proposed reaction mechanisms under nonaqueous conditions will be grouped and discussed in view of the three main condensation mechanisms discussed previously in Section 4.1.2.1, that is, the halide, the ether, and the ester condensations. Some other approaches such as depositions involving the use of organometallic compounds will be introduced as well.

4.2.2.1 Alkyl Halide Elimination

In solution, the alkyl halide elimination, that is, the condensation between metal chlorides and metal alkoxides (Eq. (4.7)) or alcohols, was the first NHSG reaction introduced for the formation of metal oxide gels [29] and nanoparticles [26, 28], and it is probably the most studied one [25, 27, 29].

$$\equiv \mathrm{M}-\mathrm{X}+\equiv \mathrm{M}-\mathrm{OR}\rightarrow \equiv \mathrm{M}-\mathrm{O}-\mathrm{M}\equiv +\mathrm{RX} \quad (4.7)$$

In ALD, the first example of the NHSG process and first mechanistic study found in the literature is attributable to Brei *et al.* [53]. In 1993, they proved that titanium silicate growth from $TiCl_4$ and $Si(OEt)_4$ takes place via an alkyl halide elimination condensation mechanism by monitoring the growth by IR and analyzing the by-products of the reaction by mass spectrometry. Later on, Rahtu and Ritala [54] studied the reaction mechanism for the deposition of $Zr_xTi_yO_z$ from $ZrCl_4$ and $Ti(O^iPr)_4$. They also found out that the principal reaction responsible for the metal oxide formation is an alkyl halide elimination. However, in their study the deposition temperature was slightly too high (300 °C) to safely exclude thermal decomposition of $Ti(O^iPr)_4$. Ritala *et al.* [35] showed that this approach can be applied for the deposition of various metal oxide and silicate thin films. Up to now, Al_2O_3 [35], Ta_2O_5 [35], Hf, Ti, and Zr aluminate [35], $Ti_xHf_yO_z$ [35], $Ti_xZr_yO_z$ [35, 54], and Hf [55], Zr [35, 56] and Ti [53, 57] silicate were deposited from their corresponding metal alkoxides and halides.

The approaches using a metal halide with an alcohol or an ether are expected, in solution, to proceed via an alkyl halide elimination condensation. Indeed, an alcoholysis/etherolysis takes place forming an alkoxy group on the metal center that can, further on, condense with a halide ligand as in Eq. (4.7). However, the formation of hydroxyl groups is a competitive pathway especially if electron-donor substituent groups are present in the alkyl radical of the alcohol (e.g., tertiary and benzylic alcohols). Indeed, contrarily to primary alcohols, they preferentially react with metal halides forming OH groups under the elimination of alkyl halides instead of forming alkoxy groups under the elimination of HCl [29]. This approach was first applied in ALD at the beginning of the 1990s for the formation of alumina from $AlCl_3$ and different alcohols [58, 59]. More recently, Evans *et al.* [60] showed that titania can be grown from $TiCl_4$ and *tert*-butyl alcohol or diisopropyl ether. Even though they did not discuss it, the idea behind these experiments arises directly from similar reactions performed in solution. In fact, the use of *tert*-butyl alcohol involves preferentially the formation of hydroxyl groups while the use of the diisopropyl ether can only lead to the formation of alkoxy groups [61]. However, the authors concluded that their approaches do not bring clear advantages over traditional water-assisted depositions [60].

4.2.2.2 Ether Elimination

Another nonaqueous condensation step, in solution, is the ether elimination (Eq. (4.8)). For example, the direct condensation of alkoxides was found to proceed via an ether elimination for multivalent early transition metals and leads to oxoclusters [62]. Mixed metal oxides can also be directly produced if a basic alkoxide is used [63].

$$\equiv \mathrm{M-OR} + \equiv \mathrm{M-OR'} \rightarrow \equiv \mathrm{M-O-M} \equiv + \mathrm{ROR'} \tag{4.8}$$

An equivalent approach that makes use of the same chemical principle consists of the reaction of a metal alkoxide with an alcohol. This route was extensively used in solution for the formation of metal oxides [27, 28].

In ALD, the growth of metal oxides using only one metal alkoxide source was already tested [34, 35]. Naturally, no deposition was observed to take place at temperatures below the thermal self-decomposition of the metal precursors. There are only few reports on the reaction of alkoxides and alcohols, and no mechanistic studies have been carried out [58]. The deposition of titania using $Ti(O^iPr)_4$ and methanol or ethanol has been already tested (Knez, M. and Pinna, N., unpublished results.). However, under similar experimental conditions, water leads to a larger growth per cycle than in the case of alcohols.

4.2.2.3 Ester Elimination

The third possible route involves the condensation between metal carboxylates and alkoxides under the elimination of esters (Eq. (4.9)).

$$\equiv \mathrm{M-OR} + \equiv \mathrm{M-OOCR'} \rightarrow \equiv \mathrm{M-O-M} \equiv + \mathrm{R'COOR} \tag{4.9}$$

In solution, this reaction has been used for the formation of clusters, silicates, and metal and mixed metal oxides [27, 29, 64–66]. The mechanism of the ester elimination is similar to acid-catalyzed transesterification, where the electropositive metal center acts as a Lewis acid by coordinating to the carbonyl oxygen atom [65]. The transfer of the alkoxide ligand to the carbonyl carbon leading to the formation of the ester was demonstrated by labeling experiments on the oxygen of the alkoxy groups by Caruso *et al.* [65]. Metal oxides have also been prepared, in solution, by reacting metal alkoxides and carboxylic acids for the production of titania clusters and nanoparticles, for example [22, 67]. In those cases, the metal precursor is hydrolyzed by water generated *in situ* by the reaction between the carboxylic acid and the alcohol resulting from ligand exchanges. However, a competitive mechanism in which direct esterification takes place in the coordination sphere of the metal was also proposed but could not be verified in liquid phase [67, 68]. More recently, another approach based on the reaction between a metal acetate and an alcohol was used for the formation of zinc oxide [69, 70] and indium oxide [71].

One application of this NHSG route to ALD has been recently reported [34]. Indeed, it was demonstrated that the deposition of titania and hafnia from simple metal alkoxides and acetic acid reaches acceptable GPC at temperatures as low as 50 °C, and a self-limiting growth was observed through a large temperature range (e.g., the ALD window for TiO_2 is 150–200 °C). The proposed reaction mechanism for this process, supported by GC–MS and kinetic studies, involved an ester elimination condensation. During the carboxylic acid pulse, the alkoxy ligand at the surface is replaced by a carboxylate group under the elimination of alcohol, while during the metal alkoxide pulse, a reaction between the carboxylate surface species and the metal alkoxides occurs under the elimination of an ester. In this experiment, no water was formed, therefore proving a direct condensation mechanism between alkoxide and acetate species. Later on, this process was also applied to the deposition of V_2O_4 [72] and SnO_2 [73].

4.2.2.4 **Other Reactions**

Additional nonaqueous routes, for example, the reaction of organometallic compounds with alcohols or metal alkoxides, can be considered. In solution, these routes have not been extensively studied, as organometallic compounds are generally very reactive and therefore would not permit the control of gel formation or particle growth. In ALD, however, their high reactivity could be an advantage and depositions of alumina [35, 74] and aluminosilicates [75] were investigated. It was suggested for the TMA/aluminum triisopropoxide process that the reaction involved the elimination of *tert*-butane [35]. The ALD from TMA and isopropyl alcohol should proceed by a similar chemistry [74]. Indeed, the reaction of a metal alkyl and an alcohol usually leads to the formation of alkoxy species by ligand exchange [62]. Jeon *et al.* [74] made a direct comparison of the alumina deposition using this combination of precursors or water as an oxygen source. In contrast to the traditional approach, no oxidation of silicon was observed under nonaqueous conditions. Finally, inspired by NHSG, other reactions can be tested in ALD. For example, ketones and aldehydes were applied in solution for the synthesis of metal oxide nanoparticles [27]. In particular, titania nanoparticles were synthesized in various ketones and aldehydes via an aldol condensation mechanism [76]. Although a similar NHSG approach was not yet reported for ALD, preliminary tests for the deposition of titania from $Ti(O^iPr)_4$ and acetone show a GPC at 200 °C slightly lower than in the case of water (Knez, M. and Pinna, N., unpublished results.). However, so far no mechanistic studies were performed.

4.2.3
Concluding Remarks

In this section, an overview of some relevant reactions taking place in ALD was given. However, it is important to mention that there exist several alternative ALD processes that were not taken into account here, for example, those using strong oxidizing agents such as ozone or radical oxygen as an oxygen source. From the few mechanistic studies found in the literature on these oxygen sources, it seems that condensation reactions similar to those reported above are engaged. Indeed, depending on the metal precursor used and the flow rate of O_3, ozone leads to the formation of OH-terminated surfaces [77, 78] or to the formation of carbonate and formate species in addition to the hydroxyl groups [79–82] or to alkoxide species [83]. For example, according to Goldstein *et al.*, a large number of formate surface species were formed by the reaction between trimethylaluminum and O_3 [79]. Similar findings were established for reactions between titanium tetraisopropoxide and ozone or O_2 plasma by Rai and Agarwal [84]. These surface species are the active sites for the next precursor pulse reaction, so the expected mechanisms are different from those for the hydrolyzed surface [79, 82, 83, 85]. Unfortunately, the reactions responsible for the oxide growth from these species are still not clearly identified [80, 81]. On the other hand, the reaction of ozone can produce water as a by-product leading to a hydroxyl-terminated surface [86, 87]. Thus, in those cases, the metal precursor half-reactions are similar to those of processes employing H_2O or H_2O_2 sources.

In the metal oxide ALD deposition from water or hydrogen peroxide, the main reaction pathway consists of the hydrolysis of surface species leading to the formation of $-$OH surface groups. The resulting OH-terminated surface then reacts with the incoming gas-phase metal precursor species to form the M$-$O$-$M bond. Therefore, in the hydrolytic ALD approaches, the mechanisms involved in the M$-$O$-$M bond formation are proven to be similar to the pathways present in solution, that is, hydrolysis and condensation. Even though the number of mechanistic studies of NHSG conditions applied to ALD is still limited, it seems evident that, here again, similar reactions are accountable for the metal oxide formation in solution and in ALD. Nevertheless, depending on the temperature, the contribution of dehydroxylation phenomena of the surface and/or readsorption of the by-product, as well as the metal precursor self-decomposition, should be taken into account while analyzing reaction mechanisms.

Concerning the metal oxide film formation by NHSG, the approaches did not, up to now, demonstrate significant advantages compared to water-assisted processes. Although one of the foreseen advantages offered by this approach was the possibility to obtain sharp silicon–metal oxide interfaces in the case of metal oxides grown on silicon, it was so far only demonstrated for alumina depositions [35, 74]. An interesting aspect of NHSG approaches is the possibility to deposit metal oxide films without intermediate hydroxyl group formation. Furthermore, in the case of the direct condensation, some benefits can be listed. Notably, the growth per cycle can be twice as large as compared to traditional depositions since both reactants can contain a metal center. Moreover, during the reaction between alkoxy and halide precursors, the formation of mineral acids, which could subsequently damage the freshly formed oxide film, is avoided contrary to the reaction of metal halides with water or alcohols.

4.3 Mechanistic and *In Situ* Studies

In view of what was discussed above, it emerges that similar chemistry is responsible for the metal oxide formation in ALD and in solution even though some differences exist. Therefore, ALD could greatly benefit from the established knowledge and experience of sol–gel chemistry. In order to highlight the underlying reaction mechanisms in both techniques, a thorough comparison of particular cases of metal oxide formation will be given. Initially, we will describe, after a brief reminder of the silica formation in solution, some pathways already developed in ALD. Later on, a detailed comparison of the underlying reactions taking place in sol–gel chemistry and in ALD for the most widely studied nonaqueous approach, namely, the alkyl halide elimination, will be made. However, before entering the discussion of these mechanisms in detail, some principal differences between reactions performed in solution and in ALD as well as differences in the study of mechanisms have to be pointed out.

4.3.1 General Considerations

4.3.1.1 Differences between Sol–Gel in Solution and in ALD

There are some notable differences between oxide formation in solution and in ALD. First of all, the nature of the reactions is obviously different. In sol–gel, the reactions occurring in solution are step growth condensation polymerizations. The precursors/monomers and already formed dimers and oligomers condense to give larger colloidal species under the elimination of simple molecules. In ALD, the metal oxide growth is based on subsequent self-terminating heterogeneous reactions between the surface species and the monomers coming from the gas phase. The latter thereby undergo irreversible chemisorption. Another important difference concerns the timescales of these two processes, the typical pulse length in ALD being a few seconds compared to some hours/days for sol–gel. Indeed, to better control the metal oxide formation in a sol–gel process, it is important to curb hydrolysis and/or condensation reactions. On the contrary, in ALD, fast reactions are sought after in order to obtain short deposition times. For example, nonhydrolytic conditions were introduced in sol–gel chemistry because in these cases the reactions are slower. Although this suggests that nonhydrolytic conditions are not favorable for ALD of metal oxides, they bring some significant benefits as listed above. Similarly, the modification of precursors in both techniques is not pursuing the same purpose due to the very different expected requirements for sol–gel and ALD precursors in terms of reactivity, volatility, thermal stability, and solubility [20, 36, 88].

4.3.1.2 Reaction Mechanism Study

Reaction mechanisms are challenging to investigate in ALD due to the fact that the products are in the gas phase and mixed with a large amount of unreacted precursors at each step. Consequently, the quantity of the chemical species available for analysis is very small and highly diluted by the reactants. Dedicated setups have to be conceived in order to detect the relevant species *in situ* [47, 78, 89]. The majority of the experiments devoted to the elucidation of the reaction mechanisms applied Fourier transform infrared spectroscopy to examine surface adsorbates or the quartz crystal microbalance (QCM) technique for the detection of the mass variation of the sample. This latter provides direct information about the kinetics of the reactions and the absorption/desorption phenomena. Direct identification of the reaction by-products is possible via mass spectrometry studies performed at the outlet of the ALD reactor. So far, only a few reactions have been examined by this technique. Moreover, in order to obtain a better understanding of the reaction mechanisms, the various experimental conditions must be carefully controlled. Especially, valuable data can be obtained if the reactions are investigated under self-limiting growth [38]. Up to date, a limited number of studies discussed the possible reaction mechanisms, and unfortunately, some of these depositions were investigated under nonideal conditions, that is, not in the regime of a self-limiting growth.

However, it is important to keep in mind that in ALD the reactions are naturally separated due to the alternating introduction of the two reactants. In contrast, in solution, solvolysis, exchange reactions, and condensations occur simultaneously between the precursors and intermediate species, making it difficult to investigate their respective mechanisms. Therefore, due to the divided steps, the *in situ* study of mechanisms in ALD is an attractive approach to better understand reactions performed in solution during oxide growth or in sol–gel chemistry and that might also be interesting in surface chemistry such as in heterogeneous catalysis, for example. Indeed, the surface science approach, that is, growth processes and study of interfaces at the molecular level, is based on model systems that are usually clean single-crystal surfaces with preset crystallographic orientation in an ultrahigh vacuum (UHV) environment. However, one may wonder the extent to which these model systems are able to mimic the chemistry of processes operated under real reaction conditions, under ambient conditions, and on very complex systems that are reactions at the liquid/solid interface (not well-defined chemical composition or geometrical structure, contaminated surfaces, etc.) [90–92]. ALD is a process that can be operated at any pressure from UHV to atmospheric pressure, being just limited by the precursor physical properties. Furthermore, the reactions are not performed on perfect surfaces as in typical UHV experiments but on a wide range of substrates such as absorbed species, thin films, high aspect ratio structures, particles, and organic, amorphous, or crystalline materials. In addition, the chemistry of ALD is continually enriched by new reactions due, in part, to the development of molecular layer deposition, catalyzed depositions, and atmospheric pressure ALD [93]. All in all, ALD might be regarded as a tool to study reactions occurring in solution.

4.3.2 Comparison of Selected Reactions

4.3.2.1 Case of the Silica Deposition

The ALD deposition of conformal and homogeneous film of silica is very challenging. Even if its sol–gel chemistry is very well known, only a few articles describe its deposition.

The sol–gel synthesis, in solution, of silica is principally based on silicon alkoxide precursors. Indeed, silicon alkoxides appear to be poorly reactive toward hydrolysis due to the weak electronegativity of the silicon atom that permits very slow

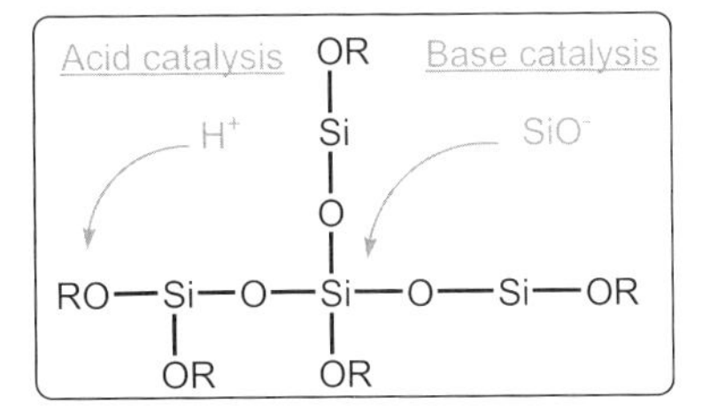

Figure 4.1 Scheme of acid/base catalysis during silica formation.

polymerizations [2]. As a consequence, without any catalyst, the gelation occurs within several days after addition of water. Both hydrolysis and condensation involve S_N2 nucleophilic substitution reactions (Figure 4.1) that can be catalyzed by acids, bases, and nucleophile compounds [2]. The acid catalysis promotes the loss of the alkoxy group by its protonation. The reaction is then favored at terminal Si and leads mainly to the formation of linear polymeric gels [1]. The basic catalysis promotes the condensation step, HO^- and SiO^- having a high nucleophilic character. Here, the reactions at central Si are favored leading to branched polymer products [1]. Bases are also used to obtain particles; the well-known Stöber method, for example, used NH_3 as a catalyst for the synthesis of well-calibrated silica particles [94].

In aqueous sol–gel, few processes are based on silicon halide precursors due to their high reactivity toward water and the formation of hydrogen halides as by-products of the reaction [95]. In nonaqueous sol–gel, the reaction of metal halides with metal alkoxides or alcohols by alkyl halide elimination was applied for the formation of silica [96] and various silicates [97]. However, the condensation of Si–Cl/Si–OR is very slow in the absence of an additional Lewis acid. For the formation of silicates, this requirement is not a problem because a large majority of transition metal salts present some Lewis acidity. In the case of pure silica, it was demonstrated that the addition of a very small amount of such a promoter leads to a rapid gelation. However, the exact catalytic mechanism is not yet known due to the complexity of the medium and the fact that the promoters are included in the final product [96].

The use of silicon halide has been investigated in ALD by George *et al.* [98, 99] who established a complete, conformal, and self-limited growth at 600 K with a GPC of 1.1 Å per cycle by using $SiCl_4$ and water. The chemical stability of the silanol group and the undissociative adsorption of $SiCl_4$ lead to a direct substitution of the hydroxyl groups by the silicon chloride species. However, a high deposition temperature was required. In fact, like in solution, a catalyst is mandatory for the deposition of silica and novel gas surface catalytic reactions were developed in ALD. For example, pyridine introduced at the beginning of each precursor sequence has permitted the formation of silica at room temperature from silicon chloride and water [100]. Indeed, pyridine by interacting strongly with surface functional groups and gas-phase reactants accelerated the metal oxide growth leading to a high GPC without contamination of the as-deposited film. This route was also investigated by using ammonia as a catalyst. Although a similar mechanism to the one involving pyridine

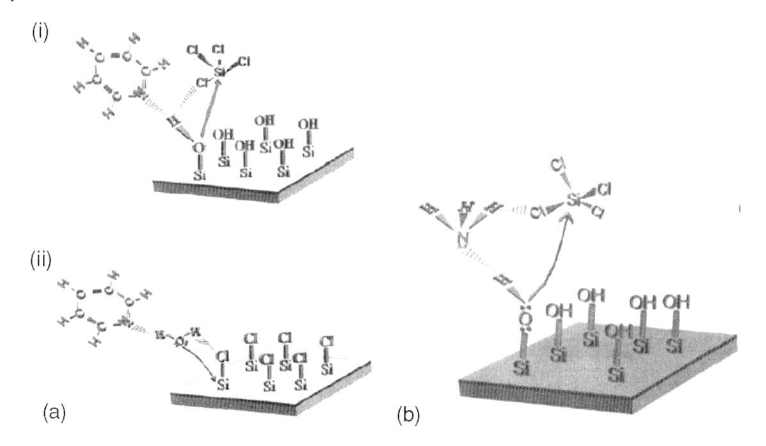

Figure 4.2 (a) Mechanisms proposed by Klaus *et al.* [100] for the pyridine catalysis of SiO_2 deposition during (i) the $SiCl_4$ and (ii) the H_2O pulses. (b) Proposed mechanism for Lewis catalysis (NH_3) of SiO_2 during the $SiCl_4$ pulse. Reproduced with permission from Ref. [100, 101].

was observed, a stronger catalytic effect was noticed due to the formation of a six-membered ring transition state, more favorable than the four-membered one of pyridine (Figure 4.2). However, ammonia can react with the released HCl and form ammonium salts [101].

As mentioned above, in solution, the silica synthesis is performed mainly from silicon alkoxides. In ALD, tetraethoxysilane (TEOS) and water have been employed for the deposition of silica, NH_3 being the catalyst for both half-sequences. This approach has an advantage that no side reaction between the alcohol by-product and the catalyst occurs and that alcohols are not corrosive as hydrogen halides. The mechanism of the reaction, studied by *in situ* FTIR spectroscopy, revealed a catalytic pathway similar to the one observed in the case of halide precursors. Without catalyst, only minimal changes were observed in the FTIR spectra after each pulse, due to the inertia of silicon alkoxide under these conditions [102]. Later on, Bachmann *et al.* developed a self-catalytic ALD using 3-aminopropyltriethoxysilane, which is both the Si precursor and the catalyst. The amino group permits the catalysis of the hydrolysis and condensation of the alkoxy ligands [103]. However, the use of a third precursor, ozone, is necessary in order to convert the remaining amino groups into OH species. In solution, alkyl or aminoalkyl groups that are inert toward hydrolysis are mostly employed to obtain either hybrid or porous materials [6, 7].

The alkyl halide elimination reaction approach has also been tested in ALD for the formation of silicate thin films. However, pure silica could not be obtained by this approach [35] because the condensation has to be catalyzed, like in solution [96]. The only examples available make use of alkoxysilanols catalyzed by a Lewis acid previously deposited. Indeed, He *et al.* [104] showed that the formation of "silica-rich nonstoichiometric silicates" from $(^tBuO)_3SiOH$ can occur only when Lewis acids such as $AlCl_3$, $ZrCl_4$, and $HfCl_4$ were first absorbed. Furthermore, the proposed catalytic phenomenon responsible for the GPC well above the monolayer is limited to silanol precursors since the presence of a silanol group is a prerequisite for the thin film growth [75]. The ability of the Si precursor to continue to diffuse down to the surface and to insert into the catalyst metal–oxygen bond by a concerted mechanism leads to rapid oxide formation, while the self-limited character of this pathway is due to the β-elimination and cross-linking that gradually prevent the access to the catalyst [75, 105].

4.3.2.2 **Case of Alkyl Halide Condensation**

Comparing the different nonhydrolytic approaches applied to ALD, it appears that the reaction of metal alkoxides with carboxylic acids demonstrated the possibility of a low-temperature deposition and self-limiting growth, while metal oxide depositions based on alkyl halide and ether elimination require high deposition temperatures.

Indeed, from the data reported in the literature for reactions between halide and alkoxide precursors in the ALD process, it is clear that deposition temperatures are relatively high (300–500 °C) and definitely higher than those required in solution. For instance, the formation of oxide network in solution can occur at room temperature [29] and is extremely fast at higher temperature [106, 107]. A closer look at the reaction mechanism about the same approaches in liquid phase provides insights and explains the different behavior observed. Some studies have pointed out that, in solution, the condensation step needs to be promoted by a metal center acting as a Lewis acid [96, 108]. For example, it was proposed that the condensation between $Ti(O^iPr)_4$ and $TiCl_4$ leading to the Ti–O–Ti bond formation follows a concerted mechanism mediated by an alkoxohalide species formed *in situ* by ligand exchange reactions [108]. In this case, $TiCl_3O^iPr$ molecules present in the reaction mixture are the key species for the first condensation reaction, converting alkoxo bridges to oxo bridges by the activation of the Ti–Cl bonds [29, 108]. The different behavior observed in the metal oxide formation via alkyl halide elimination might be explained by these findings. In solution, the condensation reaction does not take place directly between metal alkoxides and metal halides but between intermediate species, formed immediately due to a rapid ligand exchange reaction (e.g., alkoxohalide, oxoalkoxohalide, etc.) [97, 109], either because they are more reactive or because they act as a catalyst. Thus, the reaction kinetics strictly depend on the concentration of these key species and hence on the initial chloride/alkoxide concentration [109]. The importance of the presence of these intermediate species was further demonstrated by the observation that an induction period is generally needed to initiate the condensation. Obviously, this reaction cannot be directly transposed to ALD because the ratio of reactants is difficult to control and the majority of experiments are performed using flow-type reactors without long residence time. Instead, it is expected that the same chemical approach would require higher temperatures in ALD compared to the liquid phase, and indeed, this approach is reported to work efficiently only for temperatures around 300 °C and above. However, it is not yet clear if this fact is due to kinetic or thermodynamic reasons. Moreover, at these elevated temperatures, alkoxy groups at the surface (mainly secondary and tertiary alkyl groups) start to decompose leaving hydroxyl groups. These groups open an alternative and energetically favorable condensation pathway for the metal halides provided in the subsequent pulse. For example, Rahtu and Ritala [54] observed from mass spectrometric detection, besides the alkyl halide, an alkene produced via the well-known beta-elimination reaction of alkoxides [36] leaving hydroxyl groups at the surface. Although they demonstrated that this was only a slow side reaction, the formation of some hydroxyl groups being available for reaction with the next metal pulse certainly influences the thin film formation.

4.3.3 Conclusions

Sol–gel technology is a well-developed domain and was employed for decades for the synthesis of various oxides, phosphates, hybrids, and composites. Besides the complexity of the reaction mixtures and the numerous factors influencing the reactions, substantial endeavors were dedicated to the study of the molecular reaction mechanisms and the thermodynamics and kinetics of sol–gel systems in order to control and understand the inorganic network growth. Inorganic polymerization (hydrolysis and condensation reactions), aqueous chemistry of metal cations, chemistry of metal alkoxides, and physical chemistry (phenomena of aggregation, gelation, and drying) are keys to the sol–gel science. Hence, the knowledge gained by the sol–gel community is broad, encompassing diverse fields of chemistry and physics.

In this chapter, it was demonstrated that ALD makes use of sol–gel chemistry and can benefit from the chemical concepts built during the past 40 years for the formation of oxide network. Conversely, the ALD community has at hand the tools and the knowledge needed to solve problems related to reaction mechanisms, catalytic activities, and M–O–M bond formation. The latter is otherwise difficult or even impossible to solve, depending on the phase in which the reaction takes place. In fact, due to the alternate introduction of the precursors, ALD may be considered as a model system for mechanistic studies and might help to answer open questions in some distinctive sol–gel reactions. ALD can also be viewed as a tool operating at the frontier between surface science (i.e., studies performed in UHV on model systems such as single crystals) and industrially relevant heterogeneous reaction conditions (i.e., performed at atmospheric pressure or above). Therefore, ALD might experimentally bridge the apparent gap between the two fields if *in situ* analytical techniques such as FTIR, XPS, NEXAFS, and mass spectrometry are applied. Furthermore, the use of *in situ* techniques in ALD permits to better control and understand thin film deposition and to study chemical phenomena that are difficult to be investigated in solution or under conventional working conditions.

References

1 Hench, L.L. and West, J.K. (1990) The sol–gel process. *Chem. Rev.*, **90**, 33–72.

2 Livage, J. and Sanchez, C. (1992) Sol–gel chemistry. *J. Non-Cryst. Solids*, **145**, 11–19.

3 Livage, J., Henry, M., and Sanchez, C. (1988) Sol–gel chemistry of transition metal oxides. *Prog. Solid State Chem.*, **18**, 259–341.

4 Zarzycki, J. (1997) Past and present of sol–gel science and technology. *J. Sol-Gel Sci. Technol.*, **8**, 17–22.

5 Livage, J. (1999) Sol–gel synthesis of hybrid materials. *Bull. Mater. Sci.*, **22**, 201–205.

6 Sanchez, C., Julian, B., Belleville, P., and Popall, M. (2005) Applications of hybrid organic–inorganic nanocomposites. *J. Mater. Chem.*, **15**, 3559–3592.

7 Wen, J. and Wilkes, G.L. (1996) Organic/inorganic hybrid network materials by the sol–gel approach. *Chem. Mater.*, **8**, 1667–1681.

8 Cushing, B.L., Kolesnichenko, V.L., and O'Connor, C.J. (2004) Recent advances in the liquid-phase syntheses of inorganic nanoparticles. *Chem. Rev.*, **104**, 3893–3946.

9 Livage, J., Henry, M., Jolivet, J.P., and Sanchez, C. (1990) Chemical synthesis of fine powders. *MRS Bull.*, **XV**, 18–25.

10 Aoki, Y., Kunitake, T., and Nakao, A. (2005) Sol–gel fabrication of dielectric HfO_2 nano-films: formation of uniform, void-free layers and their superior electrical properties. *Chem. Mater.*, **17**, 450–458.

11 Foong, T.R.B., Shen, Y., Hu, X., and Sellinger, A. (2010) Template-directed liquid ALD growth of TiO_2 nanotube arrays: properties and potential in photovoltaic devices. *Adv. Funct. Mater.*, **20**, 1390–1396.

12 Freiman, G., Barboux, P., Perrière, J., and Giannakopoulos, K. (2009) Layer by layer deposition of zirconium oxide onto silicon. *Thin Solid Films*, **517**, 2670–2674.

13 Huang, J. and Kunitake, T. (2003) Nano-precision replication of natural cellulosic substances by metal oxides. *J. Am. Chem. Soc.*, **125**, 11834–11835.

14 Ichinose, I., Senzu, H., and Kunitake, T. (1997) A surface sol–gel process of TiO_2 and other metal oxide films with molecular precision. *Chem. Mater.*, **9**, 1296–1298.

15 Kovtyukhova, N.I., Mallouk, T.E., and Mayer, T.S. (2003) Templated surface sol–gel synthesis of SiO_2 nanotubes and SiO_2-insulated metal nanowires. *Adv. Mater.*, **15**, 780–785.

16 Yan, W., Mahurin, S., Overbury, S., and Dai, S. (2006) Nanoengineering catalyst supports via layer-by-layer surface functionalization. *Top. Catal.*, **39**, 199–212.

17 Freiman, G., Barboux, P., Perrière, J., and Giannakopoulos, K. (2007) Sequential grafting of dielectric phosphates onto silicon oxide. *Chem. Mater.*, **19**, 5862–5867.

18 Zhang, J., Ma, Z., Jiao, J., Yin, H., Yan, W., Hagaman, E.W., Yu, J., and Dai, S. (2009) Layer-by-layer grafting of titanium phosphate onto mesoporous silica SBA-15 surfaces: synthesis, characterization, and applications. *Langmuir*, **25**, 12541–12549.

19 Ichinose, I., Kawakami, T., and Kunitake, T. (1998) Alternate molecular layers of metal oxides and hydroxyl polymers prepared by the surface sol–gel process. *Adv. Mater.*, **10**, 535–539.

20 Sanchez, C., Livage, J., Henry, M., and Babonneau, F. (1988) Chemical modification of alkoxide precursors. *J. Non-Cryst. Solids*, **100**, 65–76.

21 Corriu, R.J.P. and Leclercq, D. (1996) Recent developments of molecular chemistry for sol–gel processes. *Angew. Chem., Int. Ed. Engl.*, **35**, 1420–1436.

22 Ivanda, M., Music, S., Popovic, S., and Gotic, M. (1999) XRD, Raman and FT-IR spectroscopic observations of nanosized TiO_2 synthesized by the sol–gel method based on an esterification reaction. *J. Mol. Struct.*, **480–481**, 645–649.

23 Kominami, H., Kohno, M., Takada, Y., Inoue, M., Inui, T., and Kera, Y. (1999) Hydrolysis of titanium alkoxide in organic solvent at high temperatures: a new synthetic method for nanosized, thermally stable titanium(IV) oxide. *Ind. Eng. Chem. Res.*, **38**, 3925–3931.

24 Monge, M., Kahn, M.L., Maisonnat, A., and Chaudret, B. (2003) Room-temperature organometallic synthesis of soluble and crystalline ZnO nanoparticles of controlled size and shape. *Angew. Chem., Int. Ed.*, **42**, 5321–5324.

25 Mutin, P.H. and Vioux, A. (2009) Nonhydrolytic processing of oxide-based materials: simple routes to control homogeneity, morphology, and nanostructure. *Chem. Mater.*, **21**, 582–596.

26 Niederberger, M. (2007) Nonaqueous sol–gel routes to metal oxide nanoparticles. *Acc. Chem. Res.*, **40**, 793–800.
27 Niederberger, M. and Pinna, N. (2009) *Metal Oxide Nanoparticles in Organic Solvents: Synthesis, Formation, Assembly and Application*, Springer.
28 Pinna, N. and Niederberger, M. (2008) Surfactant-free nonaqueous synthesis of metal oxide nanostructures. *Angew. Chem., Int. Ed.*, **47**, 5292–5304.
29 Vioux, A. (1997) Nonhydrolytic sol–gel routes to oxides. *Chem. Mater.*, **9**, 2292–2299.
30 Pinna, N., Garnweitner, G., Antonietti, M., and Niederberger, M. (2005) A general nonaqueous route to binary metal oxide nanocrystals involving a C–C bond cleavage. *J. Am. Chem. Soc.*, **127**, 5608–5612.
31 Pinna, N., Karmaoui, M., and Willinger, M.-G. (2011) The "benzyl alcohol route": an elegant approach towards doped and multimetal oxide nanocrystals. *J. Sol-Gel Sci. Technol.*, **57**, 323–329.
32 Yan, W., Mahurin, S.M., Overbury, S.H., and Dai, S. (2005) Nonhydrolytic layer-by-layer surface sol–gel modification of powdered mesoporous silica materials with TiO_2. *Chem. Mater.*, **17**, 1923–1925.
33 Clavel, G., Rauwel, E., Willinger, M.G., and Pinna, N. (2009) Non-aqueous sol–gel routes applied to atomic layer deposition of oxides. *J. Mater. Chem.*, **19**, 454–462.
34 Rauwel, E., Clavel, G., Willinger, M.-G., Rauwel, P., and Pinna, N. (2008) Non-aqueous routes to metal oxide thin films by atomic layer deposition. *Angew. Chem., Int. Ed.*, **47**, 3592–3595.
35 Ritala, M., Kukli, K., Rahtu, A., Raisanen, P.I., Leskelä, M., Sajavaara, T., and Keinonen, J. (2000) Atomic layer deposition of oxide thin films with metal alkoxides as oxygen sources. *Science*, **288**, 319–321.
36 Jones, A.C., Aspinall, H.C., Chalker, P.R., Potter, R.J., Manning, T.D., Loo, Y.F., O'Kane, R., Gaskell, J.M., and Smith, L.M. (2006) MOCVD and ALD of high-κ dielectric oxides using alkoxide precursors. *Chem. Vapor Depos.*, **12**, 83–98.
37 Leskelä, M. and Ritala, M. (2002) Atomic layer deposition (ALD): from precursors to thin film structures. *Thin Solid Films*, **409**, 138–146.
38 Puurunen, R.L. (2005) Surface chemistry of atomic layer deposition: a case study for the trimethylaluminum/water process. *J. Appl. Phys.*, **97**, 121301–121352.
39 Lakomaa, E.L., Haukka, S., and Suntola, T. (1992) Atomic layer growth of TiO_2 on silica. *Appl. Surf. Sci.*, **60–61**, 742–748.
40 Matero, R., Rahtu, A., and Ritala, M. (2001) *In situ* quadrupole mass spectrometry and quartz crystal microbalance studies on the atomic layer deposition of titanium dioxide from titanium tetrachloride and water. *Chem. Mater.*, **13**, 4506–4511.
41 Rahtu, A. and Ritala, M. (2002) Reaction mechanism studies on the zirconium chloride–water atomic layer deposition process. *J. Mater. Chem.*, **12**, 1484–1489.
42 Du, X., Du, Y., and George, S.M. (2005) *In situ* examination of tin oxide atomic layer deposition using quartz crystal microbalance and Fourier transform infrared techniques. *J. Vac. Sci. Technol. A*, **23**, 581–588.
43 Puurunen, R.L. (2005) Formation of metal oxide particles in atomic layer deposition during the chemisorption of metal chlorides: a review. *Chem. Vapor Depos.*, **11**, 79–90.
44 Kukli, K., Aarik, J., Aidla, A., Siimon, H., Ritala, M., and Leskelä, M. (1997) *In situ* study of atomic layer epitaxy growth of tantalum oxide thin films from Ta $(OC_2H_5)_5$ and H_2O. *Appl. Surf. Sci.*, **112**, 236–242.
45 Aarik, J., Aidla, A., Sammelselg, V., Uustare, T., Ritala, M., and Leskelä, M. (2000) Characterization of titanium dioxide atomic layer growth from titanium ethoxide and water. *Thin Solid Films*, **370**, 163–172.
46 Aarik, J., Aidla, A., Uustare, T., Ritala, M., and Leskelä, M. (2000) Titanium isopropoxide as a precursor for atomic layer deposition: characterization of titanium dioxide growth process. *Appl. Surf. Sci.*, **161**, 385–395.

47 Rahtu, A. and Ritala, M. (2002) Reaction mechanism studies on titanium isopropoxide–water atomic layer deposition process. *Chem. Vapor Depos.*, **8**, 21–28.

48 Juppo, M., Rahtu, A., Ritala, M., and Leskelä, M. (2000) *In situ* mass spectrometry study on surface reactions in atomic layer deposition of Al_2O_3 thin films from trimethylaluminum and water. *Langmuir*, **16**, 4034–4039.

49 Groner, M.D., Fabreguette, F.H., Elam, J.W., and George, S.M. (2004) Low-temperature Al_2O_3 atomic layer deposition. *Chem. Mater.*, **16**, 639–645.

50 Yousfi, E.B., Fouache, J., and Lincot, D. (2000) Study of atomic layer epitaxy of zinc oxide by *in-situ* quartz crystal microgravimetry. *Appl. Surf. Sci.*, **153**, 223–234.

51 Ott, A.W., Johnson, J.M., Klaus, J.W., and George, S.M. (1997) Surface chemistry of In_2O_3 deposition using $In(CH_3)_3$ and H_2O in a binary reaction sequence. *Appl. Surf. Sci.*, **112**, 205–215.

52 Niinisto, J., Rahtu, A., Putkonen, M., Ritala, M., Leskelä, M., and Niinisto, L. (2005) *In situ* quadrupole mass spectrometry study of atomic-layer deposition of ZrO_2 using $Cp_2Zr(CH_3)_2$ and water. *Langmuir*, **21**, 7321–7325.

53 Brei, V.V., Kaspersky, V.A., and Gulyanitskaya, N.U. (1993) Synthesis and study of boron phosphate and titanium silicate compounds on silica surface. *React. Kinet. Catal. Lett.*, **50**, 415–421.

54 Rahtu, A. and Ritala, M. (2002) Reaction mechanism studies on the atomic layer deposition of $Zr_xTi_yO_z$ using the novel metal halide–metal alkoxide approach. *Langmuir*, **18**, 10046–10048.

55 Kukli, K., Ritala, M., Leskelä, M., Sajavaara, T., Keinonen, J., Hegde, R.I., Gilmer, D.C., and Tobin, P.J. (2004) Properties of oxide film atomic layer deposited from tetraethoxy silane, hafnium halides, and water. *J. Electrochem. Soc.*, **151**, F98–F104.

56 Kim, W.-K., Kang, S.-W., and Rhee, S.-W. (2003) Atomic layer deposition of zirconium silicate films using zirconium tetra-*tert*-butoxide and silicon tetrachloride. *J. Vac. Sci. Technol. A*, **21**, L16–L18.

57 Prince, K.E., Evans, P.J., Triani, G., Zhang, Z., and Bartlett, J. (2006) Characterisation of alumina–silica films deposited by ALD. *Surf. Interface Anal.*, **38**, 1692–1695.

58 Hiltunen, L., Kattelus, H., Leskelä, M., Makela, M., Niinisto, L., Nykanen, E., Soininen, P., and Tiitta, M. (1991) Growth and characterization of aluminium oxide thin films deposited from various source materials by atomic layer epitaxy and chemical vapor deposition processes. *Mater. Chem. Phys.*, **28**, 379–388.

59 Tiitta, M., Nykanen, E., Soininen, P., Niinisto, L., Leskelä, M., and Lappalainen, R. (1998) Preparation and characterization of phosphorus-doped aluminum oxide thin films. *Mater. Res. Bull.*, **33**, 1315–1323.

60 Evans, P.J., Mutin, P.H., Triani, G., Prince, K.E., and Bartlett, J.R. (2006) Characterisation of metal oxide films deposited by non-hydrolytic ALD. *Surf. Interface Anal.*, **38**, 740–743.

61 Arnal, P., Corriu, R.J.P., Leclercq, D., Mutin, P.H., and Vioux, A. (1996) Preparation of anatase, brookite and rutile at low temperature by non-hydrolytic sol–gel methods. *J. Mater. Chem.*, **6**, 1925–1932.

62 Bradley, D.C., Mehrota, R.C., Rothwell, I.P., and Singh, A. (2001) *Alkoxo and Aryloxo Derivatives of Metals*, Academic Press.

63 Kessler, V.G. (2003) Molecular structure design and synthetic approaches to the heterometallic alkoxide complexes (soft chemistry approach to inorganic materials by the eyes of a crystallographer). *Chem. Commun.*, 1213–1222.

64 Caruso, J., Hampden-Smith, M.J., Rheingold, A.L., and Yap, G. (1995) Ester elimination versus ligand exchange: the role of the solvent in tin–oxo cluster-building reactions. *J. Chem. Soc., Chem. Commun.*, 157–158.

65 Caruso, J., Roger, C., Schwertfeger, F., Hampden-Smith, M.J., Rheingold, A.L., and Yap, G. (1995) Solvent-dependent

ester elimination and ligand exchange reactions between trimethylsilyl acetate and tin(IV) tetra-*tert*-butoxide. *Inorg. Chem.*, **34**, 449–453.

66 Chandler, C.D., Roger, C., and Hampden-Smith, M.J. (1993) Chemical aspects of solution routes to perovskite-phase mixed-metal oxides from metal-organic precursors. *Chem. Rev.*, **93**, 1205–1241.

67 Doeuff, S., Henry, M., and Sanchez, C. (1990) Sol–gel synthesis and characterization of titanium oxo-acetate polymers. *Mater. Res. Bull.*, **25**, 1519–1529.

68 Cozzoli, P.D., Kornowski, A., and Weller, H. (2003) Low-temperature synthesis of soluble and processable organic-capped anatase TiO_2 nanorods. *J. Am. Chem. Soc.*, **125**, 14539–14548.

69 Clavel, G., Willinger, M.G., Zitoun, D., and Pinna, N. (2007) Solvent dependent shape and magnetic properties of doped ZnO nanostructures. *Adv. Funct. Mater.*, **17**, 3159–3169.

70 Joo, J., Kwon, S.G., Yu, J.H., and Hyeon, T. (2005) Synthesis of ZnO nanocrystals with cone, hexagonal cone, and rod shapes via non-hydrolytic ester elimination sol–gel reactions. *Adv. Mater.*, **17**, 1873–1877.

71 Narayanaswamy, A., Xu, H., Pradhan, N., Kim, M., and Peng, X. (2006) Formation of nearly monodisperse In_2O_3 nanodots and oriented-attached nanoflowers: hydrolysis and alcoholysis vs pyrolysis. *J. Am. Chem. Soc.*, **128**, 10310–10319.

72 Willinger, M.G., Neri, G., Rauwel, E., Bonavita, A., Micali, G., and Pinna, N. (2008) Vanadium oxide sensing layer grown on carbon nanotubes by a new atomic layer deposition process. *Nano Lett.*, **8**, 4201–4204.

73 Marichy, C., Donato, N., Willinger, M.-G., Latino, M., Karpinsky, D., Yu, S.-H., Neri, G., and Pinna, N. (2011) Tin dioxide sensing layer grown on tubular nanostructures by a non-aqueous atomic layer deposition process. *Adv. Funct. Mater.*, **21**, 658–666.

74 Jeon, W.-S., Yang, S., Lee, C.-S., and Kang, S.-W. (2002) Atomic layer deposition of Al_2O_3 thin films using trimethylaluminum and isopropyl alcohol. *J. Electrochem. Soc.*, **149**, C306–C310.

75 Hausmann, D., Becker, J., Wang, S.L., and Gordon, R.G. (2002) Rapid vapor deposition of highly conformal silica nanolaminates. *Science*, **298**, 402–406.

76 Garnweitner, G., Antonietti, M., and Niederberger, M. (2005) Nonaqueous synthesis of crystalline anatase nanoparticles in simple ketones and aldehydes as oxygen-supplying agents. *Chem. Commun.*, 397–399.

77 Elam, J.W., Martinson, A.B.F., Pellin, M.J., and Hupp, J.T. (2006) Atomic layer deposition of In_2O_3 using cyclopentadienyl indium: a new synthetic route to transparent conducting oxide films. *Chem. Mater.*, **18**, 3571–3578.

78 Langereis, E., Keijmel, J., van de Sanden, M.C.M., and Kessels, W.M.M. (2008) Surface chemistry of plasma-assisted atomic layer deposition of Al_2O_3 studied by infrared spectroscopy. *Appl. Phys. Lett.*, **92**, 231904–231903.

79 Goldstein, D.N., McCormick, J.A., and George, S.M. (2008) Al_2O_3 atomic layer deposition with trimethylaluminum and ozone studied by *in situ* transmission FTIR spectroscopy and quadrupole mass spectrometry. *J. Phys. Chem. C*, **112**, 19530–19539.

80 Kwon, J., Dai, M., Halls, M.D., and Chabal, Y.J. (2008) Detection of a formate surface intermediate in the atomic layer deposition of high-κ dielectrics using ozone. *Chem. Mater.*, **20**, 3248–3250.

81 Rai, V.R., Vandalon, V., and Agarwal, S. (2010) Surface reaction mechanisms during ozone and oxygen plasma assisted atomic layer deposition of aluminum oxide. *Langmuir*, **26**, 13732–13735.

82 Wang, Y., Dai, M., Ho, M.-T., Wielunski, L.S., and Chabal, Y.J. (2007) Infrared characterization of hafnium oxide grown by atomic layer deposition using ozone as the oxygen precursor. *Appl. Phys. Lett.*, **90**, 022906–022903.

83 Kwon, J., Dai, M., Halls, M.D., and Chabal, Y.J. (2010) Suppression of substrate oxidation during ozone based atomic layer deposition of Al_2O_3: effect of ozone flow rate. *Appl. Phys. Lett.*, **97**, 162903–162903.

84 Rai, V.R. and Agarwal, S. (2009) Surface reaction mechanisms during plasma-assisted atomic layer deposition of titanium dioxide. *J. Phys. Chem. C*, **113**, 12962–12965.

85 Lee, H.J., Park, M.H., Min, Y.-S., Clavel, G., Pinna, N., and Hwang, C.S. (2010) Unusual growth behavior of atomic layer deposited $PbTiO_3$ thin films using water and ozone as oxygen sources and their combination. *J. Phys. Chem. C*, **114**, 12736–12741.

86 Knapas, K. and Ritala, M. (2008) *In situ* reaction mechanism studies on atomic layer deposition of ZrO_2 from $(CpMe)_2Zr(OMe)Me$ and water or ozone. *Chem. Mater.*, **20**, 5698–5705.

87 Rose, M., Niinisto, J., Endler, I., Bartha, J.W., Kucher, P., and Ritala, M. (2010) *In situ* reaction mechanism studies on ozone-based atomic layer deposition of Al_2O_3 and HfO_2. *ACS Appl. Mater. Interfaces*, **2**, 347–350.

88 O'Brien, P. and Sullivan, A. (2004) From molecules to materials: materials discussion 7. *J. Mater. Chem.*, **14**, E11–E15.

89 Wang, Y., Ho, M.T., Goncharova, L.V., Wielunski, L.S., Rivillon-Amy, S., Chabal, Y.J., Gustafsson, T., Moumen, N., and Boleslawski, M. (2007) Characterization of ultra-thin hafnium oxide films grown on silicon by atomic layer deposition using tetrakis(ethylmethyl-amino) hafnium and water precursors. *Chem. Mater.*, **19**, 3127–3138.

90 Bent, B.E. (1996) Mimicking aspects of heterogeneous catalysis: generating, isolating, and reacting proposed surface intermediates on single crystals in vacuum. *Chem. Rev.*, **96**, 1361–1390.

91 Diebold, U. (2003) The surface science of titanium dioxide. *Surf. Sci. Rep.*, **48**, 53–229.

92 Ma, Z. and Zaera, F. (2006) Organic chemistry on solid surfaces. *Surf. Sci. Rep.*, **61**, 229–281.

93 George, S.M. (2009) Atomic layer deposition: an overview. *Chem. Rev.*, **110**, 111–131.

94 Stöber, W., Fink, A., and Bohn, E. (1968) Controlled growth of monodisperse silica spheres in the micron size range. *J. Colloid Interface Sci.*, **26**, 62–69.

95 Heley, J.R., Jackson, D., and James, P.F. (1997) The production of ultrafine silica powders from silicon tetrachloride: control of the primary particle size. *J. Sol-Gel Sci. Technol.*, **8**, 177–181.

96 Bourget, L., Corriu, R.J.P., Leclercq, D., Mutin, P.H., and Vioux, A. (1998) Non-hydrolytic sol–gel routes to silica. *J. Non-Cryst. Solids*, **242**, 81–91.

97 Lafond, V., Mutin, P.H., and Vioux, A. (2002) Non-hydrolytic sol–gel routes based on alkyl halide elimination: toward better mixed oxide catalysts and new supports. Application to the preparation of a SiO_2–TiO_2 epoxidation catalyst. *J. Mol. Catal. A*, **182–183**, 81–88.

98 George, S.M., Sneh, O., Dillon, A.C., Wise, M.L., Ott, A.W., Okada, L.A., and Way, J.D. (1994) Atomic layer controlled deposition of SiO_2 and Al_2O_3 using ABAB... binary reaction sequence chemistry. *Appl. Surf. Sci.*, **82–83**, 460–467.

99 Sneh, O., Wise, M.L., Ott, A.W., Okada, L.A., and George, S.M. (1995) Atomic layer growth of SiO_2 on Si(100) using $SiCl_4$ and H_2O in a binary reaction sequence. *Surf. Sci.*, **334**, 135–152.

100 Klaus, J.W., Sneh, O. and George, S.M. (1997) Growth of SiO_2 at room temperature with the use of catalyzed sequential half-reactions. *Science*, **278**, 1934–1936.

101 Klaus, J.W. and George, S.M. (2000) Atomic layer deposition of SiO_2 at room temperature using NH_3-catalyzed sequential surface reactions. *Surf. Sci.*, **447**, 81–90.

102 Ferguson, J.D., Smith, E.R., Weimer, A.W., and George, S.M. (2004) ALD of SiO_2 at room temperature using TEOS and H_2O with NH_3 as the catalyst. *J. Electrochem. Soc.*, **151**, G528–G535.

103 Bachmann, J., Zierold, R., Chong Yuen, T., Hauert, R., Sturm, C., Schmidt-Grund, R., Rheinländer, B., Grundmann, M., Gösele, U., and Nielsch, K. (2008) A practical, self-catalytic, atomic layer deposition of silicon dioxide. *Angew. Chem., Int. Ed.*, **47**, 6177–6179.

104 He, W., Solanki, R., Conley, J.F., and Ono, Y. (2003) Pulsed deposition of silicate films. *J. Appl. Phys.*, **94**, 3657–3659.

105 Burton, B.B., Boleslawski, M.P., Desombre, A.T., and George, S.M. (2008) Rapid SiO_2 atomic layer deposition using tris(*tert*-pentoxy)silanol. *Chem. Mater.*, **20**, 7031–7043.

106 Joo, J., Yu, T., Kim, Y.W., Park, H.M., Wu, F., Zhang, J.Z., and Hyeon, T. (2003) Multigram scale synthesis and characterization of monodisperse tetragonal zirconia nanocrystals. *J. Am. Chem. Soc.*, **125**, 6553–6557.

107 Trentler, T.J., Denler, T.E., Bertone, J.F., Agrawal, A., and Colvin, V.L. (1999) Synthesis of TiO_2 nanocrystals by nonhydrolytic solution-based reactions. *J. Am. Chem. Soc.*, **121**, 1613–1614.

108 Arnal, P., Corriu, R.J.P., Leclercq, D., Mutin, P.H., and Vioux, A. (1997) A solution chemistry study of nonhydrolytic sol–gel routes to titania. *Chem. Mater.*, **9**, 694–698.

109 Tang, J., Fabbri, J., Robinson, R.D., Zhu, Y., Herman, I.P., Steigerwald, M.L., and Brus, L.E. (2004) Solid-solution nanoparticles: use of a nonhydrolytic sol–gel synthesis to prepare HfO_2 and $Hf_xZr_{1-x}O_2$ nanocrystals. *Chem. Mater.*, **16**, 1336–1342.

5
Molecular Layer Deposition of Hybrid Organic–Inorganic Films

Steven M. George, Byunghoon Yoon, Robert A. Hall, Aziz I. Abdulagatov, Zachary M. Gibbs, Younghee Lee, Dragos Seghete, and Byoung H. Lee

5.1
Introduction

The field of atomic layer deposition (ALD) has grown dramatically in the past 5–10 years [1]. ALD process technologies have met the needs of the semiconductor industry. The virtues of ALD have also been introduced to many new areas. Moreover, the types of materials that can be grown using ALD-inspired processes have expanded dramatically with the introduction of organic precursors that can be incorporated in the growing film. Since a molecular fragment can be added to the film, this area is no longer "atomic" layer deposition. Instead, the definition of "molecular" layer deposition can be used to describe this new area [2]. A schematic illustrating the sequential, self-limiting growth leading to molecular layer deposition (MLD) is shown in Figure 5.1 [3].

The original definition of MLD described the sequential, self-limiting chemistry used for the growth of a polyimide, an all-organic polymer [4]. A number of Japanese groups worked in this area of organic MLD in the 1990s [4–7]. The various MLD systems were based on condensation polymerization reactions. This early work was an outgrowth of the field of vapor deposition polymerization within polymer science [8]. There was no apparent communication between these early all-organic MLD investigators and the atomic layer epitaxy (ALE) community at that time. The reemergence of all-organic MLD occurred later within the ALD community with the demonstration of the MLD of polyimide [9] and polyamide [3, 10]. Other MLD systems were also demonstrated including polyurea [11] and polyurethane [12]. We note that the MLD of polyimides and polyamides had been observed earlier by Japanese groups in the 1990s [4, 6].

Another development occurred with the realization that organic precursors used for all-organic MLD could be mixed with the inorganic precursors for ALD to define new hybrid organic–inorganic materials [13, 14]. These hybrid materials were nearly without precedence in either the organic or inorganic chemistry literature. The expanded basis set introduced by these hybrid materials has changed the prospects

Atomic Layer Deposition of Nanostructured Materials, First Edition. Edited by Nicola Pinna and Mato Knez.

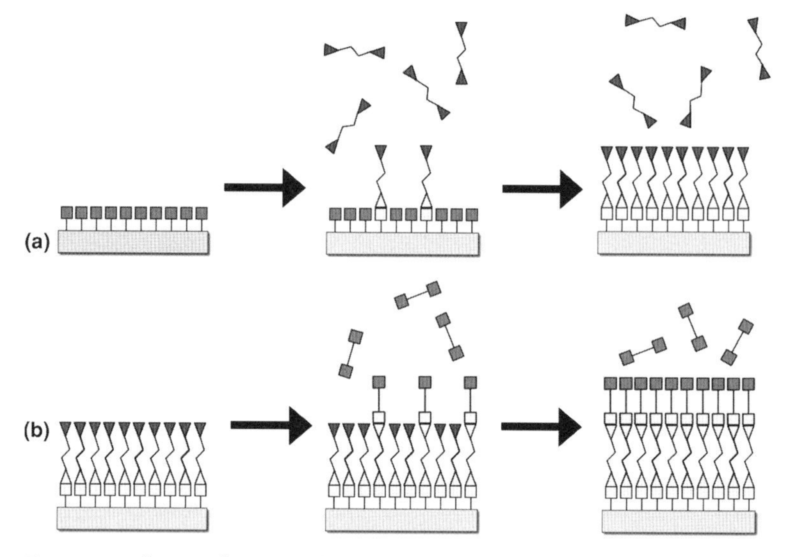

Figure 5.1 Schematic illustrating ideal sequential, self-limiting reactions for MLD growth using two homobifunctional reactants.

for materials development using ALD and MLD. The number of inorganic precursors available for ALD can be limiting. Inorganic precursor synthesis is often the rate-limiting step in ALD materials development. In contrast, a vast number of organic precursors are available from organic chemistry. The sheer magnitude of organic precursors leads to an overwhelming variety of possibilities for hybrid organic–inorganic MLD.

The broad palette of organic precursor choices is reduced severely by two factors. One factor is the vapor pressure of the organic precursor. For a useful MLD process, a vapor pressure $>$100 mTorr is preferable. The second factor is the thermal decomposition temperature of the organic precursor. The vapor pressure can be increased by raising the temperature. However, the vapor pressure is limited by the thermal decomposition temperature. Consequently, only limited vapor pressures can be obtained before the organic precursor is no longer stable. These two factors end up ruling out a great number of organic precursors. The remaining organic precursors are generally small organic molecules with a limited number of chemical functional groups.

Despite these limitations, a number of hybrid organic–inorganic materials have been developed recently using MLD techniques [2, 13–19]. These systems have begun to define the wide range of materials that can be deposited using MLD. The possibility to mix and match organic and inorganic precursors and their relative incorporation in the film will also lead to a wide spectrum of film properties. The mechanical properties will be able to be tuned by controlling the organic and inorganic relative proportions. Either organic–inorganic alloys or nanolaminates can be produced because the ALD and MLD cycles can be defined at the atomic or molecular fragment level to control film composition.

This chapter will first review some of the general issues concerning the MLD of hybrid organic–inorganic films. The first several MLD systems that have been demonstrated will then be described to illustrate the current state of the art. Some new systems will then be introduced to show the diversity of chemistries that can be employed to grow various hybrid organic–inorganic films. Additional systems will be suggested that may offer a range of thin film properties. Finally, speculations will be offered for the future prospects for the MLD of hybrid organic–inorganic materials.

5.2
General Issues for MLD of Hybrid Organic–Inorganic Films

The robustness of MLD growth of hybrid organic–inorganic films is expected from the heats of reaction of the underlying surface reactions. The reactions yielding hybrid organic–inorganic films are very similar to some of the most favorable ALD reactions such as Al_2O_3 ALD [20–22]. One of the first examined hybrid organic–inorganic materials grown using MLD was an "alucone" based on the reaction between trimethylaluminum (TMA) and ethylene glycol (EG) [13]. The EG molecule, $HO{-}CH_2{-}CH_2{-}OH$, contains two hydroxyls groups and is very analogous to H_2O as a reactant in Al_2O_3 ALD. The difference is that the $-CH_2-CH_2-$ molecular fragment is introduced into the hybrid organic–inorganic film. A schematic showing the growth of the alucone based on TMA and EG is displayed in Figure 5.2 [13].

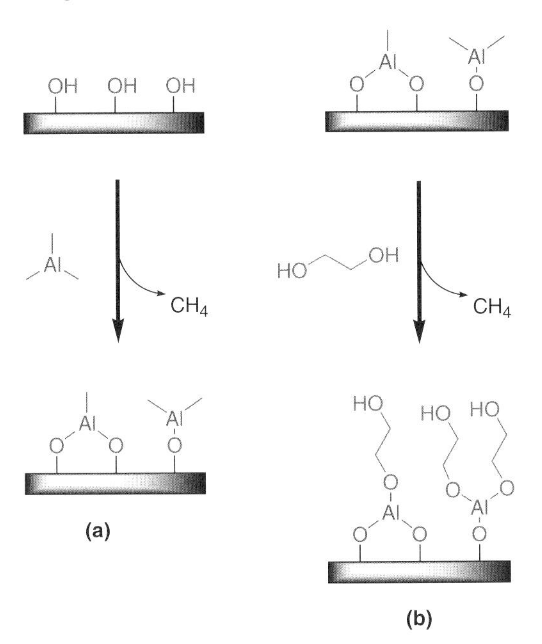

Figure 5.2 Schematic depicting two-step AB alucone MLD growth using TMA and EG. TMA is exposed to a hydroxylated surface and produces a surface covered with $-AlCH_3$ species. The subsequent EG exposure produces a surface covered with $-OCH_2CH_2OH$ species.

EG is one of the many organic diols that can be used together with TMA for alucone film growth. Other possible diols include 1,4-butanol and hydroquinone. One possible difficulty with diols is that they are homobifunctional precursors and can react twice with the $AlCH_3^*$ surface species [2, 10]. These “double reactions” lead to a loss of reactive surface sites and could produce a decreasing growth per cycle during MLD. The problem of double reactions may be minimized using polyols to assure that a hydroxyl group will be available for the subsequent TMA exposure. This strategy will be discussed in Section 5.5 for the MLD of the alucone based on TMA and glycerol that is a common triol.

Alternatively, a heterobifunctional precursor, such as ethanolamine (EA), $HO{-}CH_2{-}CH_2{-}NH_2$, could be employed that shows preferential reactivity between its hydroxyl group and the $AlCH_3^*$ surface species [19]. This preference leaves an amine ($-NH_2$) group available for the subsequent surface reaction. Likewise, ring-opening reactions can be employed that will react and then express a new functional group when the ring is opened [2, 19]. The ring-opening reaction also has the advantage of containing the functional group in a hidden form. The hidden functionality leads to higher vapor pressures and shorter purge times compared to precursors that have the same exposed functionality. The ABC MLD system discussed in Section 5.4 will illustrate the use of both heterobifunctional and ring-opening precursors.

In addition to the reactions of diols, triol, and polyols with TMA, other chemical functional groups will react with TMA including carboxylic acids, aldehydes, carbonyls, and isocyanate groups (Derk, A.R., Zimmerman, P., and Musgrave, C.B. (2010) Unpublished results.). These precursors can all be used together with TMA to define sequential, self-limiting MLD procedures. The high reactivity of TMA facilitates a huge number of possible MLD reactions. The heats of reaction of the underlying surface reactions are all very exothermic.

In addition to TMA, other inorganic ALD precursors can also be matched with various organic precursors to define other classes of hybrid organic–inorganic materials. For example, diethylzinc (DEZ) can react with diols to produce “zincone” MLD films [17, 18]. $TiCl_4$ can react with diols to produce “titanicone” MLD films (Abdulagatov, A.I. and George, S.M. (2010) Unpublished results.). Zirconium tetra-*t*-butoxide can react with diols to produce “zircone” MLD films (Lee, B.H. and George, S.M. (2010) Unpublished results.). A vast number of other systems can be defined using other inorganic precursors [2]. These different inorganic precursors can all have reactivities with a large number of different organic precursors. The possibilities are vast and may yield films with a number of functional and tunable properties.

One complication that has been noticed for the MLD of hybrid organic–inorganic films is the diffusion of the inorganic precursor, for example, TMA, into the underlying MLD film. The hybrid organic–inorganic film is similar to an organic polymer and contains much more free volume than typical inorganic materials. The polymeric network of the MLD film allows small precursor molecules such as TMA to diffuse into the polymer [13, 23]. This precursor diffusion leads to a chemical vapor deposition (CVD) component of the MLD growth [24]. The TMA can react with incoming precursors and add to the deposition that results from the purely surface reactions.

The CVD contribution to the MLD growth can be considerable as illustrated during ABC MLD [24]. This CVD leads to a dependence of the MLD on precursor exposure and purge time [24]. This behavior is nonideal and will lead to problems for the conformality of MLD in high aspect ratio structures. However, the precursor diffusion during MLD also has unexpected benefits. The precursor diffusion adds new reactive sites to the growing MLD film that helps to compensate for the loss of surface sites resulting from double reactions, steric hindrance, or inefficient reactions. These new reactive sites allow many MLD processes to display linear growth that may not be possible without the precursor diffusion continually adding new reactive sites.

5.3 MLD Using Trimethylaluminum and Ethylene Glycol in an AB Process

As mentioned in Section 5.2, one of the first examples of the MLD of hybrid organic–inorganic films was the growth of an aluminum alkoxide film based on the reaction of TMA and EG [13]. In general, the two-step MLD reaction between a metal alkyl, such as TMA, and a diol, such as EG, can be written as follows [13]:

$$\text{(A)}\quad \mathrm{SR'OH^* + MR_x \rightarrow SR'O - MR_{x-1}{}^* + RH} \qquad (5.1)$$

$$\text{(B)}\quad \mathrm{SMR^* + HOR'OH \rightarrow SM - OR'OH^* + RH} \qquad (5.2)$$

The asterisks indicate the surface species and S denotes the substrate with the reaction products from the previous reactions. In reaction A, the reaction stops when all the $SR'OH^*$ species have completely reacted to produce $SR'O{-}MR_{x-1}{}^*$ species. In reaction B, the reaction stops when all the SMR^* species have completely reacted to produce $SM{-}OR'OH^*$ species. The sequential and self-limiting reactions of TMA and EG ideally yield a polymeric film described by $(Al{-}(O{-}CH_2{-}CH_2{-}O)_x)_n$. This new polymer is an alucone [25] and can be called poly(aluminum ethylene glycol). A schematic illustrating the growth of poly(aluminum ethylene glycol) is shown in Figure 5.2 [13].

Previous studies have demonstrated that alucone MLD using TMA and EG is very efficient [13]. *Ex situ* X-ray reflectivity (XRR) investigations have shown that the MLD growth rate is temperature dependent [13]. The growth rate decreased from 4.0 Å per TMA/EG cycle at 85 °C to 0.4 Å per TMA/EG cycle at 175 °C. Figure 5.3 shows that alucone MLD was also linear versus the number of TMA/EG cycles [13]. In addition, XRR analysis determined that the density of these alucone films was independent of the deposition temperature and constant at ~1.5 g/cm^3 [13]. This density is low and much more characteristic of an organic polymer than an inorganic solid.

Quartz crystal microbalance (QCM) measurements also revealed the linearity of alucone MLD growth versus TMA and EG exposures. Figure 5.4 displays QCM results that reveal a mass gain of 18 ng/cm^2 per TMA/EG cycle at 135 °C [13]. The QCM results also show a large mass increase during the TMA exposures that

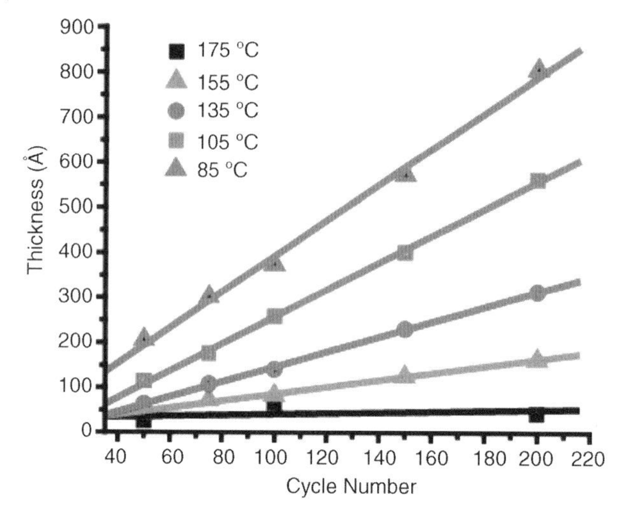

Figure 5.3 AB alucone MLD film thickness measured using XRR analysis versus number of AB reaction cycles for a variety of substrate temperatures.

subsequently decays immediately after the TMA exposure. This mass transient is consistent with TMA diffusion into and out of the AB alucone MLD film [23]. The TMA diffusion into and out of the MLD film also helps explain the temperature dependence of the MLD growth. Less TMA will diffuse into the MLD film and more TMA will diffuse out of the film at higher temperatures. Consequently, higher temperatures will lower the growth per cycle.

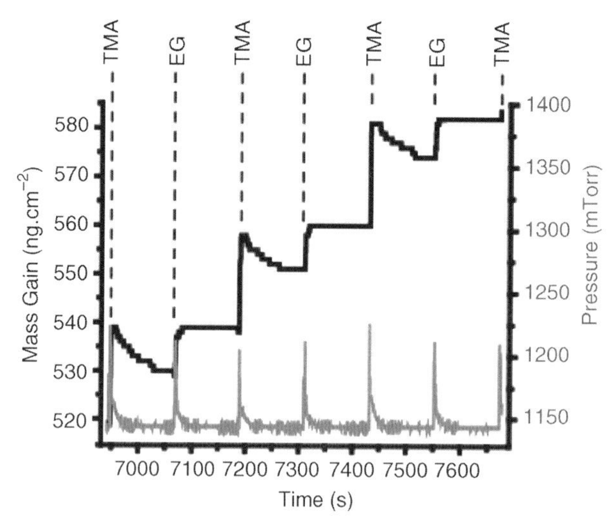

Figure 5.4 Mass gain from QCM measurements and corresponding reactor pressure during three TMA and EG reaction cycles in the steady-state regime for AB alucone MLD at 135 °C.

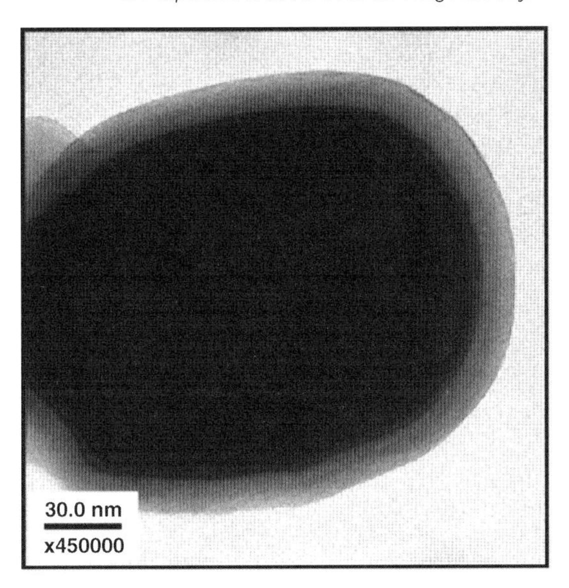

Figure 5.5 TEM image of a $BaTiO_3$ particle that was coated at 135 °C with 40 AB cycles of Al_2O_3 ALD and then 50 AB cycles of AB alucone MLD using TMA and EG.

The surface reactions during MLD with TMA and EG also displayed self-limiting behavior as evidenced by the infrared absorbance of the surface species that are removed or added during each reaction [13]. The AB alucone MLD films were also fairly stable in air and displayed a contraction of ~22% over the first 3 days that the films were exposed to air. After this contraction, the films were very stable. The AB alucone films also were extremely smooth and conformal when deposited on nanoparticles. Figure 5.5 shows the TEM image of a $BaTiO_3$ particle that was coated with 40 AB cycles of Al_2O_3 ALD and then 50 AB cycles of AB alucone MLD at 135 °C [13]. The quality of the overlying MLD film is equivalent to the underlying ALD film.

5.4 Expansion to an ABC Process Using Heterobifunctional and Ring-Opening Precursors

The MLD of hybrid organic–inorganic materials can be performed without homobifunctional precursors and their "double reactions" using different types of precursors [2]. One three-step ABC MLD process that can be accomplished without using homobifunctional precursors is based on (1) TMA, a homomultifunctional inorganic precursor; (2) EA, a heterobifunctional organic reactant; and (3) maleic anhydride (MA), a ring-opening organic reactant [19]. The proposed surface reactions during the ABC alucone growth are [19]

$$\text{(A)}\quad S-COOH^{*}+Al(CH_3)_3 \rightarrow S-COO-Al(CH_3)_2^{\ *}+CH_4 \qquad (5.3)$$

$$(B)\quad S-AlCH_3{}^* + HO(CH_2)_2 - NH_2 \rightarrow S - Al - O(CH_2)_2NH_2{}^* + CH_4 \tag{5.4}$$

$$(C)\quad S-NH_2{}^* + C_4H_2O_3(MA) \rightarrow S - NH - C(O)CHCHCOOH^* \tag{5.5}$$

This surface reaction mechanism is illustrated in Figure 5.6 [19].

In this ABC reaction sequence, TMA reacts with carboxylic groups in reaction A given by Eq. (5.3) to form $AlCH_3{}^*$ species. Subsequently, the $AlCH_3{}^*$ species react preferentially with the hydroxyl end of the EA reactant to form $Al-OCH_2CH_2NH_2{}^*$ surface species in reaction B given by Eq. (5.4). MA then reacts with amine-terminated surface functional groups to reform carboxylic groups through a ring-opening reaction in reaction C given by Eq. (5.5). The three-step reaction sequence is repeated by exposure to TMA, EA, and MA to grow the ABC MLD film.

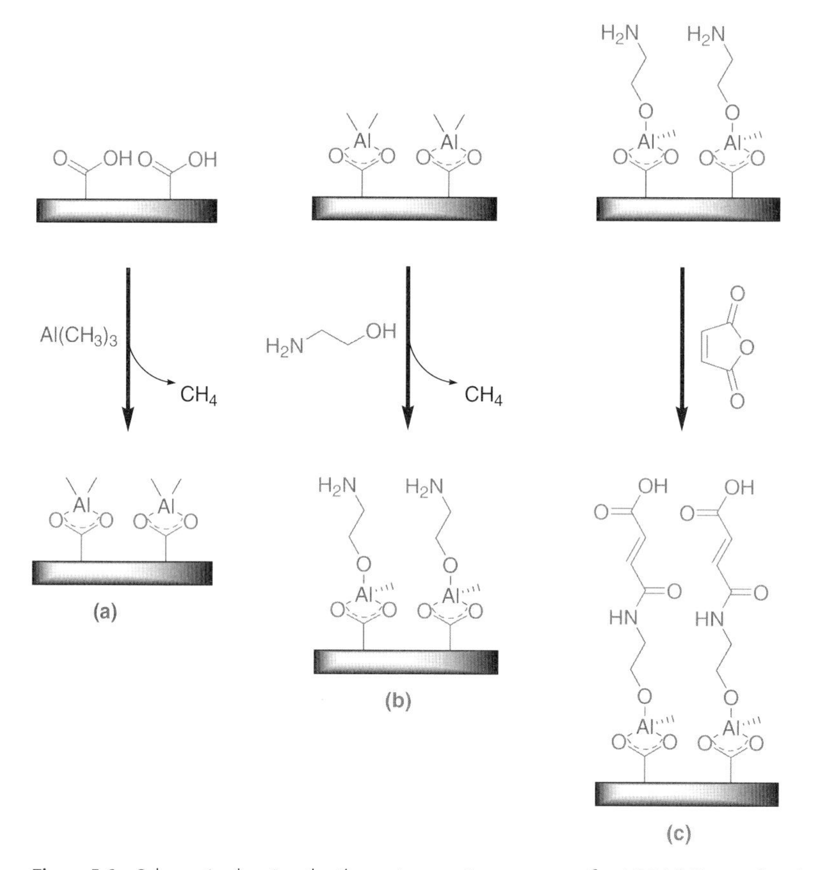

Figure 5.6 Schematic showing the three-step reaction sequence for ABC MLD growth using (a) TMA, (b) EA, and (c) MA.

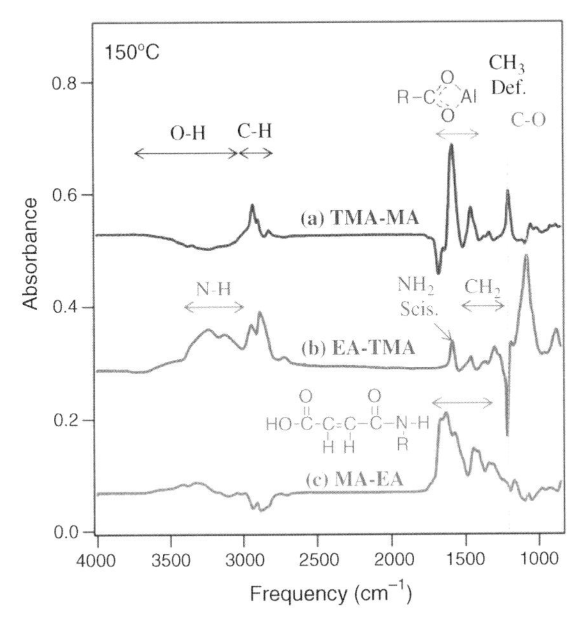

Figure 5.7 FTIR difference spectra after TMA, EA, and MA exposures during ABC MLD at 150 °C: (a) TMA–MA; (b) EA–TMA; and (c) MA–EA. The FTIR difference spectra after each reactant exposure are referenced with respect to the previous reactant exposure.

The change of surface species during each reaction in the three-step ABC reaction sequence can be monitored by FTIR difference spectra. In the observed FTIR difference spectra, the added surface species appear as positive absorbance features and the removed surface species appear as negative absorbance features. Figure 5.7 displays the observed FTIR difference spectra after each surface reaction at 150 °C [19]. The TMA–MA difference spectrum in Figure 5.7a is consistent with the reaction of TMA with the carboxylic acid species produced from the previous MA reaction to form aluminum carboxylate species. The EA–TMA difference spectrum in Figure 5.7b is in agreement with the reaction of EA with $AlCH_3^*$ species from the previous TMA reaction to form alkoxyamine species. The features for the MA–EA difference spectrum in Figure 5.7c are consistent with the ring-opening reaction of MA with the alkoxyamine surface species from the previous EA reaction to form new carboxylic acid species.

The ABC MLD also displays linear growth as evidenced by the QCM measurements. Figure 5.8 shows QCM results in the steady-state growth region for deposition at 90 °C [24]. The TMA exposure causes a very large mass gain of ~2500 ng/cm^2 per ABC cycle. This large mass gain may indicate the diffusion of a substantial quantity of TMA into the ABC MLD film. After the TMA exposure, there is a subsequent mass loss that is consistent with the diffusion of TMA out of the ABC MLD film. The subsequent EA exposure also produces a mass gain that is much larger than the mass

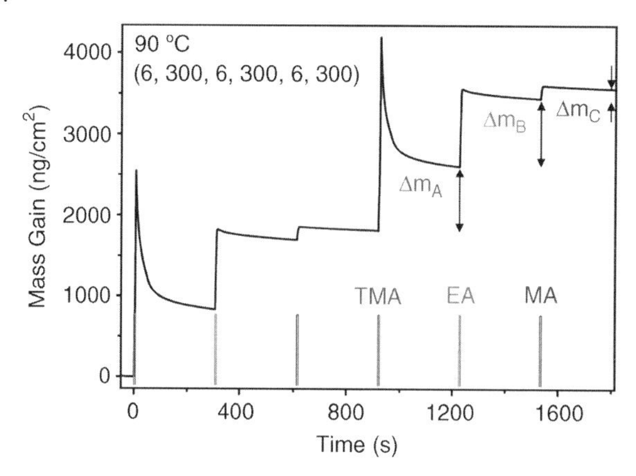

Figure 5.8 Mass gain from QCM measurements for two cycles of ABC MLD film growth in the linear growth region at 90 °C. The pulse sequence was TMA 6 s, N_2 purge 300 s, EA 6 s, N_2 purge 300 s, MA 6 s, and N_2 purge 300 s.

gain expected for only a surface reaction. This large mass gain may be produced by the reaction of EA with both $AlCH_3^*$ surface species and molecular TMA in the ABC MLD film. The MA exposure then produces a mass gain that is much smaller than either the TMA or EA exposures. However, this mass gain is consistent with a surface reaction between MA and the surface amine species [24].

The ABC MLD system yields large growth rates that argue against any "double" reactions that may limit the MLD growth rate. In fact, the growth rates are so large that they cannot be explained based on only surface reactions. TMA diffusion into the ABC MLD film is needed to justify the large growth rates. The diffusion of TMA into and out of the ABC film was measured experimentally and then fitted using a numerical model [24]. The mass gain profiles could be fitted well assuming Fickian diffusion of TMA into and out of the ABC MLD films. These results illustrate the importance of TMA diffusion in the ABC MLD process. TMA diffusion may be important because of the small size of the TMA precursor. Other larger aluminum precursors, such as triethylaluminum and triisobutylaluminum, could be used instead of TMA to determine the importance of TMA size for the large effect of TMA diffusion.

The importance of TMA diffusion into and out of the ABC film can also be verified by measuring the mass gain per cycle versus TMA purge time. Longer TMA purge times should allow more TMA to diffuse out of the ABC film and reduce the film growth rate. QCM experiments were able to confirm the dependence of the mass gain on the purge time. Figure 5.9 shows QCM experiments performed using purge times of 90 and 300 s [24]. The shorter purge time of 90 s yields a large mass gain of 3980 ng/cm^2 per cycle. The longer purge time of 300 s yields a much lower mass gain of 1950 ng/cm^2 per cycle. The larger mass gains are attributed to less time for TMA diffusion out of the ABC film.

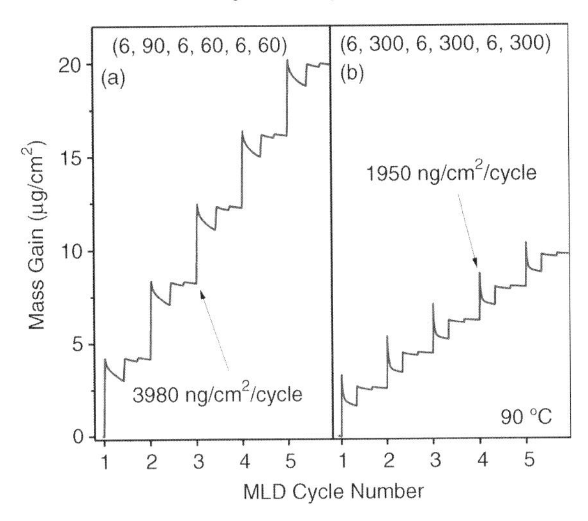

Figure 5.9 Mass gain from QCM measurements for ABC MLD film growth in the linear growth region at 90 °C for purge times of 90 s (a) and 300 s (b) with dosing times of 6 s for all the reactants.

5.5
Use of a Homotrifunctional Precursor to Promote Cross-Linking in an AB Process

The AB alucone MLD system using TMA and EG displayed very efficient reactions. However, this MLD system may also suffer from double reactions because EG is a homobifunctional precursor. This MLD system also displayed some film contraction over the first several days after this film was exposed to air. In addition, tensile strain measurements of an AB alucone film grown using TMA and EG with a thickness of 100 nm had a low critical tensile strain of 0.69% [26]. This low critical tensile strain may result from the low amount of cross-linking in the MLD film. These problems with the TMA + EG MLD system led to the exploration of the TMA + glycerol system. Glycerol provides an additional hydroxyl group for reaction and should increase the cross-linking between the chains in the deposited MLD film.

The proposed reaction sequence using TMA and glycerol (GL) is displayed in Figure 5.10 (Hall, R.A., Yoon, B., and George, S.M. (2010) Unpublished results.). GL is a sugar alcohol that has three hydroxyl groups on its three-carbon chain. In contrast, EG has only two hydroxyl groups on its two-carbon chain as shown in Figure 5.2. Although GL has only one extra hydroxyl group, the addition of the extra carbon and hydroxyl group severely reduces the GL vapor pressure. The vapor pressure of EG is 0.10 Torr at 20 °C. In contrast, the GL vapor pressure is only 3.5×10^{-9} Torr at 20 °C. Consequently, the GL was heated to 140 °C to obtain a higher vapor pressure of 1.5 Torr for the MLD growth studies.

The FTIR difference spectra revealed that the TMA and GL reactions are efficient and proceed to near completion. Figure 5.11 shows the FTIR difference spectra for glycerol–TMA and TMA–glycerol (Hall, R.A., Yoon, B., and George, S.M. (2010) Unpublished results.). The FTIR spectra show the "flipping" of the O–H stretching

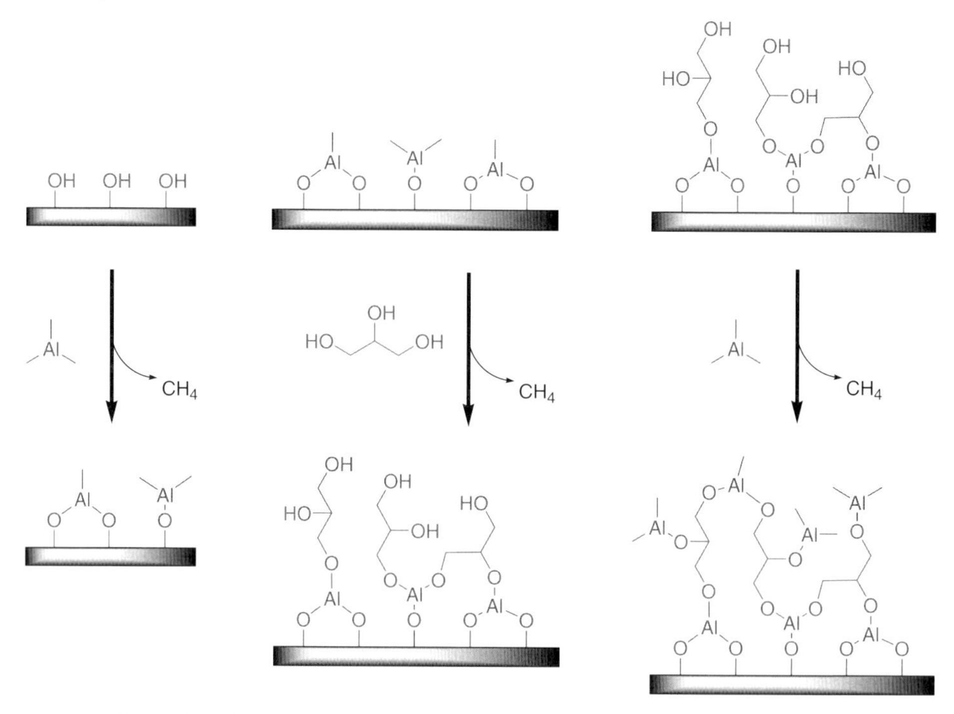

Figure 5.10 Schematic depicting two-step AB alucone MLD growth using TMA and GL.

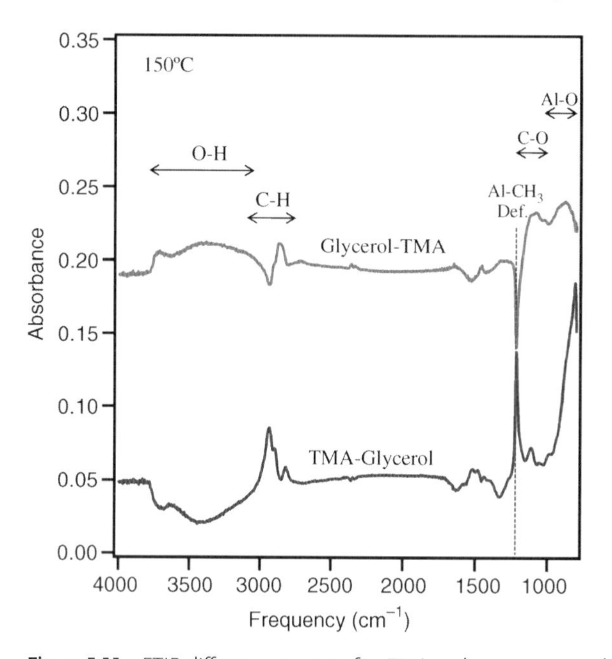

Figure 5.11 FTIR difference spectra after TMA and GL exposures during AB alucone MLD at 150 °C. The FTIR difference spectra are referenced with respect to the previous reactant exposure.

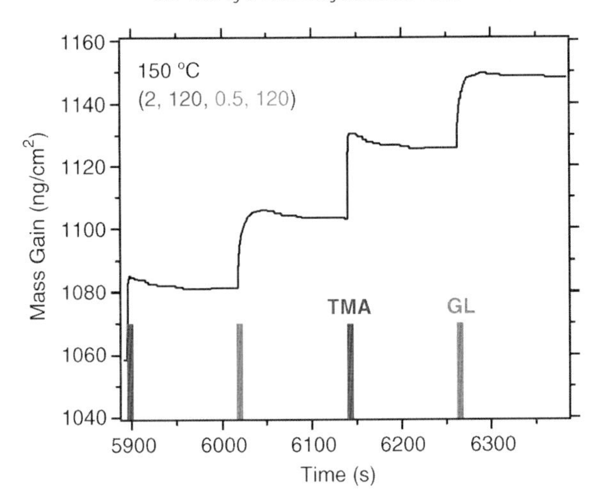

Figure 5.12 Mass gain from QCM measurements for two cycles of AB alucone MLD film growth with TMA and GL in the linear growth region at 150 °C. The pulse sequence was TMA 2 s, N_2 purge 120 s, GL 0.5 s, and N_2 purge 120 s.

vibrations at higher frequencies with each TMA and GL exposure. This flipping between positive absorbance for one reactant and then a mirror image negative absorbance for the second reactant is consistent with repetitive self-limiting reactions. There is also a flipping of the strong $AlCH_3$ deformation mode at lower frequencies that is consistent with the addition and subtraction of the $AlCH_3^*$ surface species. Other FTIR experimental measurements are consistent with self-limiting TMA and GL reactions.

The TMA + GL reaction can also be characterized using QCM studies. The QCM analysis revealed linear MLD growth with an average mass gain of 41.5 ng/cm^2 per cycle at 150 °C. This mass gain of 41.5 ng/cm^2 per cycle is equivalent to a growth rate of 2.5 Å per cycle. Figure 5.12 displays QCM results for two TMA + GL cycles at 150 °C (Hall, R.A., Yoon, B., and George, S.M. (2010) Unpublished results.). The QCM shows that a mass gain is observed during the TMA exposure. Likewise, a small mass loss is observed after the TMA exposure. This behavior suggests that some TMA is diffusing out of the MLD film after the TMA exposure. A similar mass gain is observed during the GL exposure. The slight mass loss after the GL exposure may indicate that some GL diffuses out of the MLD film or that some GL may be desorbing from the surface of the MLD film. Given the low vapor pressure of GL, the mass loss is more likely explained by GL desorption.

The TMA + GL system also shows a growth rate that is much less dependent on temperature than the growth rate for TMA + EG. XRR analysis was employed to study the film thickness after various numbers of MLD cycles at temperatures of 150, 170, and 190 °C. These XRR results are shown in Figure 5.13 (Hall, R.A., Yoon, B., and George, S.M. (2010) Unpublished results.). The film thicknesses are very similar for all three temperatures and are consistent with a growth rate of 2.0–2.3 Å per cycle. The growth rate of 2.3 Å per cycle at 150 °C is in reasonable agreement with the QCM

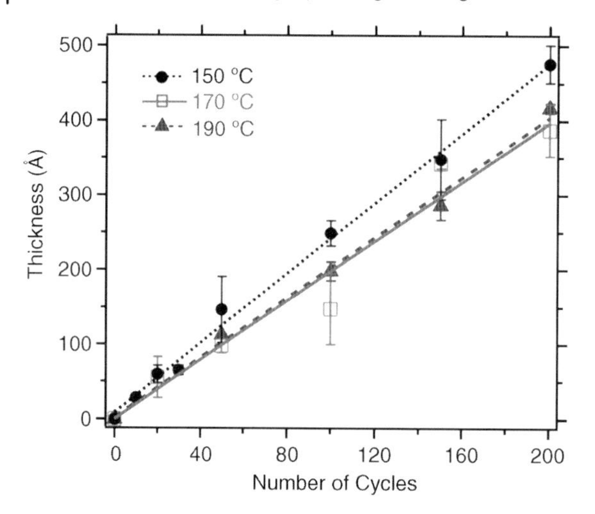

Figure 5.13 Thickness of AB alucone MLD films grown using TMA and GL measured using XRR analysis versus number of AB reaction cycles. Results are shown for growth temperatures of 150, 170, and 190 °C.

measurement of 2.5 Å per cycle at 150 °C under similar reaction conditions. These more constant growth rates versus temperature compared to TMA + EG suggest that TMA diffusion may be less of a factor for the TMA + GL MLD system because of the more extensive cross-linking between the growing chains.

The XRR analysis of the TMA + GL MLD films also indicated that the film thickness was nearly constant versus time after exposure to ambient conditions (Hall, R.A., Yoon, B., and George, S.M. (2010) Unpublished results.). The MLD films grown using TMA + GL were not observed to contract like the MLD films grown using TMA + EG. This higher film stability may also indicate that there is more cross-linking in these MLD films that increases their stability. Mechanical testing should reveal whether the increased cross-linking for the TMA + GL system also enhances its critical tensile strain for cracking relative to the TMA + EG system.

5.6
Use of a Heterobifunctional Precursor in an ABC Process

The ABC system with TMA + EA + MA utilized EA as a heterobifunctional precursor and MA as a ring-opening precursor to avoid using homobifunctional precursors that have problems with "double reactions." The hydroxyl end of EA reacts preferentially with $AlCH_3{}^*$ surface species. The remaining amine end of EA is then available to react with MA and leads to a ring-opening reaction that reveals a carboxylic acid group. This carboxylic acid group is then able to react with TMA and the ABC sequence can be repeated to grow the ABC MLD film.

4-Carboxyphenyl isothiocyanate (CI) is another heterobifunctional precursor that may be able to define an AB or ABC process with TMA. The carboxylic acid group is

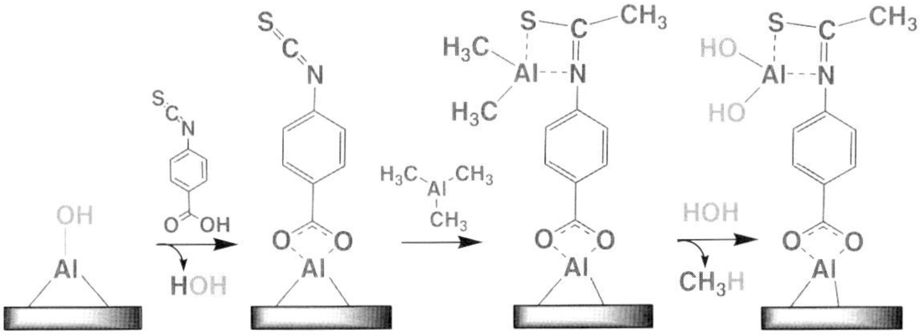

Figure 5.14 Schematic showing the proposed three-step reaction sequence for ABC MLD growth using TMA, H_2O, and CI.

known to react with TMA as a result of the previous studies of the ABC system with TMA + EA + MA. The isothiocyanate group is also predicted to react with TMA (Derk, A.R., Zimmerman, P., and Musgrave, C.B. (2010) Unpublished results.). However, density functional theory (DFT) calculations indicate that the reaction between TMA and carboxylic groups is much more favored relative to the reaction between TMA and isocyanate groups (Derk, A.R., Zimmerman, P., and Musgrave, C.B. (2010) Unpublished results.). Consequently, the relative reactivities may be different enough to avoid double reactions and obtain constant growth versus number of MLD cycles.

Figure 5.14 shows CI in an ABC process defined by CI + TMA + H_2O (Gibbs, Z. M., Yoon, B., and George, S.M. (2010) Unpublished results.). In this ABC process, the carboxylic group of CI reacts preferentially with surface $AlOH^*$ groups to form an aluminum carboxylate species. The remaining CI molecular fragment covers the surface with isothiocyanate groups. TMA then reacts with the isothiocyanate group via a $-CH_3$ transfer mechanism (Derk, A.R., Zimmerman, P., and Musgrave, C.B. (2010) Unpublished results.). Subsequently, the $AlCH_3^*$ surface species react with H_2O to form a surface covered with $AlOH^*$ species that can then react with CI again to repeat the ABC process. Because TMA can react with the isothiocyanate group, CI may also be able to define an AB process with TMA.

QCM results for the CI + TMA + H_2O reaction at 135 °C are displayed in Figure 5.15 (Gibbs, Z.M., Yoon, B., and George, S.M. (2010) Unpublished results.). The TMA exposure leads to a pronounced mass gain. The subsequent H_2O exposure produces very little mass gain as expected from the transition from $AlCH_3^*$ to $AlOH^*$ surface species. The reaction between CI and the surface $AlOH^*$ species then leads to a larger mass gain. This reaction takes a number of CI micropulses for the reaction to reach completion. Figure 5.15 shows the results for 25 individual CI micropulses performed in a (10, 5) sequence, where 10 s is the CI exposure time and 5 s is the purge time after each individual CI micropulse. Many CI micropulses are required because of the low vapor pressure of the CI precursor.

Figure 5.16 displays the FTIR difference spectra for the TMA + H_2O + CI reaction sequence at 150 °C (Gibbs, Z.M., Yoon, B., and George, S.M. (2010) Unpublished results.). The results for the TMA reaction given by TMA–CI show a loss of absorbance corresponding with O–H stretching vibrations and the N=C=S

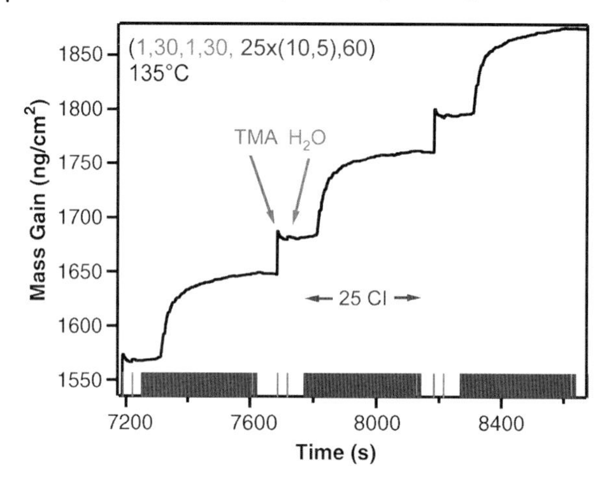

Figure 5.15 Mass gain from QCM measurements for two cycles of ABC MLD film growth in the linear growth region at 135 °C. The pulse sequence was TMA 1 s, N_2 purge 30 s, H_2O 1 s, N_2 purge 30 s, 25× (CI 10 s, N_2 purge 5 s), and N_2 purge 60 s. The multiple CI pulsing was required because of the low vapor pressure of CI.

isothiocyanate group at ~2050 cm^{-1}. There is also a gain of absorbance for Al–carboxylate and $AlCH_3{}^*$ species as expected for the reaction of TMA with carboxylic acid groups. The results for the H_2O reaction given by H_2O–TMA show a gain of absorbance corresponding with the O–H stretching vibrations and a loss of absorbance for $AlCH_3$ species. These changes are expected when H_2O reacts with $AlCH_3{}^*$ surface species.

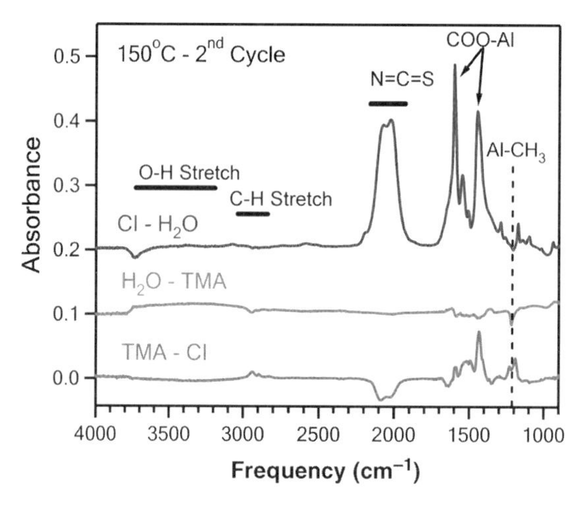

Figure 5.16 FTIR difference spectra after TMA, H_2O, and CI exposures during ABC MLD at 150 °C during the second ABC cycle on ZrO_2 nanoparticles. The FTIR difference spectra are referenced with respect to the previous reactant exposure.

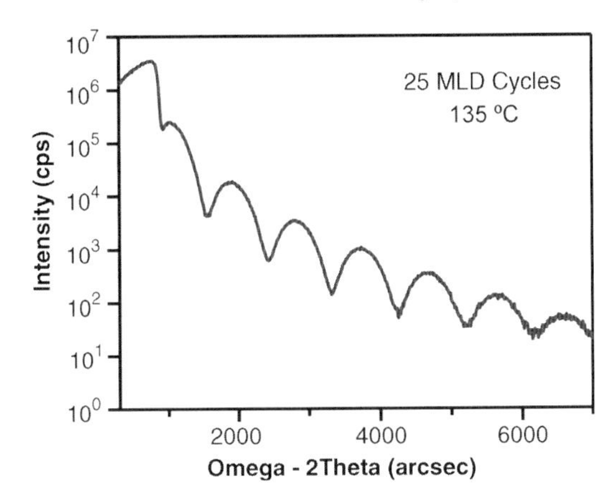

Figure 5.17 XRR scan of an ABC MLD film grown using 25 cycles of TMA, H_2O, and CI. The fits of this XRR scan yield a film thickness of 170 Å.

More pronounced changes in the FTIR difference spectra are observed for the CI reaction with the hydroxylated surface. The CI reaction leads to a large gain in the absorbance for the isothiocyanate group at ~2050 cm^{-1} and the Al–carboxylate at 1444 and 1602 cm^{-1} (Gibbs, Z.M., Yoon, B., and George, S.M. (2010) Unpublished results.). The appearance of these large absorbances argues that the carboxylic acid end of the CI molecule reacts with the hydroxylated surface, binds through an Al–carboxylate linkage, and yields free isothiocyanate groups. The FTIR difference spectra also reveal that the CI leads to the loss of some O–H stretching vibrations as expected by the reaction of the carboxylic acid with surface hydroxyl groups. The gain in the absorbance for the isothiocyanate group in the CI–H_2O difference spectrum is larger than the loss of absorbance for the isothiocyanate group in TMA–CI difference spectrum. This difference indicates that either the TMA does not react with all the isothiocyanate groups during the second MLD cycle or the $(CH_3)_2Al-(S-C{=}N)$ complex has a similar vibrational frequency and absorption intensity.

XRR analysis was used to study MLD films grown using the TMA + H_2O + CI process. Figure 5.17 shows an XRR scan after 25 MLD cycles of TMA + H_2O + CI on a silicon wafer at 135 °C (Gibbs, Z.M., Yoon, B., and George, S.M. (2010) Unpublished results.). This XRR scan is consistent with a film thickness of 170 Å. The density of the MLD film is determined to be 1.55 g/cm^3. The film thickness of 170 Å after 25 MLD cycles is consistent with a growth rate of 6.8 Å per cycle for this TMA + H_2O + CI process. This growth rate is close to the growth rate of 7.5 Å per cycle obtained from the QCM measurements assuming the density obtained from the XRR analysis.

5.7 MLD of Hybrid Alumina–Siloxane Films Using an ABCD Process

Polydimethylsiloxane (PDMS) is one of the most important organic–inorganic polymers and contains $[-Si(CH_3)_2-O]_n$ chains. The strength and flexibility of the

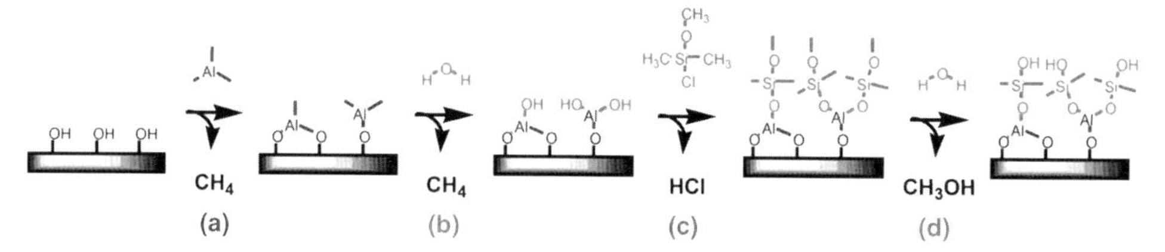

Figure 5.18 Schematic showing the four-step reaction sequence for ABCD MLD growth of an alumina–siloxane film using TMA, H_2O, DMMCS, and H_2O.

Si–O bonds and bond angles give PDMS desirable thermal and mechanical properties [27, 28]. PDMS MLD would be extremely useful for the growth of flexible and compliant thin films. However, initial attempts at PDMS MLD revealed that the growth rate became negligible after approximately 15 MLD cycles. These attempts were made using the sequential dosing of water with homobifunctional silane molecules such as bis(dimethylamino)dimethylsilane and 1,3-dichlorotetramethyldisiloxane or heterobifunctional silane molecules such as dimethylmethoxychlorosilane (DMMCS). The lack of growth after approximately 15 MLD cycles was attributed to the competing desorption of cyclic siloxanes such as hexamethylcyclotrisiloxane (D3) or decamethylcyclopentasiloxane (D5) from the PDMS film [29, 30].

To prevent the desorption of cyclic siloxanes, a new approach was pursued where DMMCS and H_2O were used together with TMA in an ABCD process defined by TMA/H_2O/DMMCS/H_2O. A schematic of this reaction sequence is given in Figure 5.18 (Abdulagatov, A.I., Goldstein, D.N., Yoon, B. and George, S.M. (2010) Unpublished results.). This reaction sequence introduces the $-Si(CH_3)_2-O-$ linkage into the growing film. The addition of TMA adds –Al–O– subunits into the growing chain and prevents the competing desorption of cyclic siloxanes. The TMA could be introduced during every reaction cycle. The TMA could also be introduced less frequently to grow longer $[-Si(CH_3)_2-O-]_n$ chains before inserting the –Al–O– subunit. Initial work has explored the ABCD process to demonstrate the growth of alumina–siloxane hybrid organic–inorganic films.

Figure 5.19 presents QCM results for the growth of the alumina–siloxane MLD film at 200 °C (Abdulagatov, A.I., Goldstein, D.N., Yoon, B. and George, S.M. (2010) Unpublished results.). The MLD growth is linear with a mass gain of ∼21 ng/cm^2 per cycle. Each reactant was dosed twice using 1 s exposures and 30 s purge times. The TMA reaction yielded a mass gain of ∼15 ng/cm^2 per cycle. The DMMCS reaction yielded a mass gain of ∼6 ng/cm^2 per cycle. The H_2O exposures were necessary but did not produce noticeable mass changes. The film growth at 200 °C was also examined using XRR analysis. The XRR measurements confirmed linear growth at 200 °C with a growth rate of 0.9 Å per cycle (Abdulagatov, A.I., Goldstein, D.N., Yoon, B. and George, S.M. (2010) Unpublished results.). Using the density of 2.3 g/cm^3 for the alumina–siloxane MLD films, the mass gain of ∼21 ng/cm^2 per cycle is consistent with a growth rate of 0.9 Å per cycle.

The growth of the alumina–siloxane MLD film was also explored using FTIR spectroscopy. Figure 5.20 shows the FTIR difference spectra after each reactant

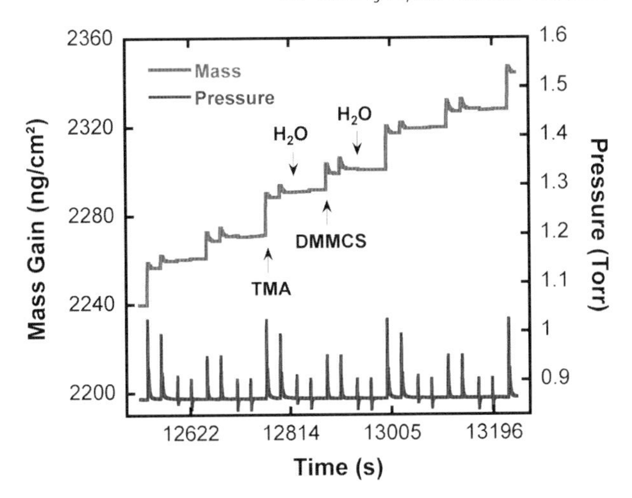

Figure 5.19 Mass gain from QCM measurements for three cycles of ABCD MLD film growth in the linear growth region at 200 °C. Each reactant was dosed twice using 1 s exposures and 30 s purge times.

exposure in the TMA/H_2O/DMMCS/H_2O sequence at 180 °C (Abdulagatov, A.I., Goldstein, D.N., Yoon, B. and George, S.M. (2010) Unpublished results.). As illustrated by the TMA–H_2O difference spectra, the TMA exposure leads to the loss

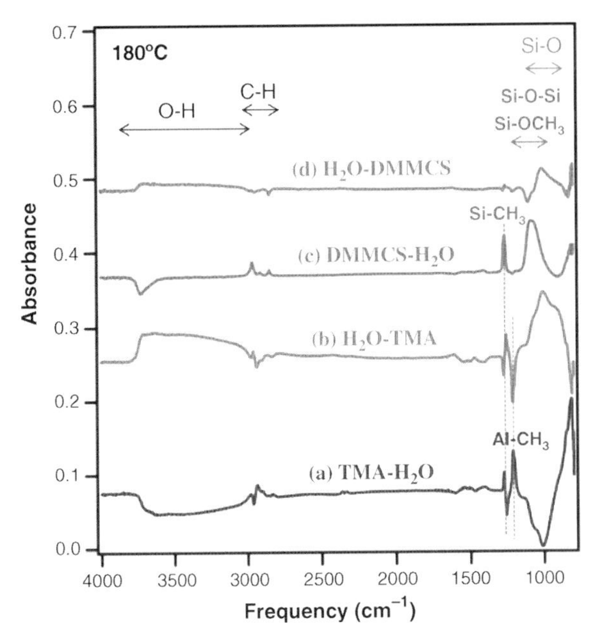

Figure 5.20 FTIR difference spectra after TMA, H_2O, DMMCS, and H_2O exposures during ABCD MLD at 180 °C: (a) TMA–H_2O; (b) H_2O–TMA; (c) DMMCS–H_2O; and (d) H_2O–DMMCS. The FTIR difference spectra are referenced with respect to the previous reactant exposure.

of absorbance for O–H stretching vibrations and gain of absorbance for the C–H stretching vibrations and $AlCH_3$ deformation modes. The H_2O–TMA difference spectrum shows that the H_2O exposure removes absorbance for the C–H stretching vibrations and $AlCH_3$ deformation modes and adds absorbance for the O–H stretching vibrations. These absorbance changes are very similar to the absorbance changes observed during Al_2O_3 ALD [31].

The DMMCS exposure then adds absorbance for Si–OCH_3 and Si–CH_3 vibrational features and removes absorbance for isolated O–H stretching vibrations. These changes are expected for the reaction of the chlorosilane precursor with surface hydroxyl groups. The subsequent H_2O exposure adds absorbance for O–H stretching vibrations and removes absorbance for Si–OCH_3 vibrational features. This difference spectrum is consistent with the replacement of the $SiOCH_3$ group by SiOH during the H_2O exposure. Additional FTIR experiments confirmed that the DMMCS reaction and the H_2O reaction after the DMMCS reaction were both self-limiting (Abdulagatov, A.I., Goldstein, D.N., Yoon, B. and George, S.M. (2010) Unpublished results.).

Compositional analysis using XPS showed that the alumina–siloxane MLD films grown at 200 °C had an average atomic concentration of 45% oxygen, 19% carbon, 5% silicon, and 32% aluminum (Abdulagatov, A.I., Goldstein, D.N., Yoon, B. and George, S.M. (2010) Unpublished results.). The low atomic concentration of silicon indicates that the chlorosilane reaction with the hydroxylated surface is not very efficient. However, the chlorine level was below the detection limit. In contrast, films deposited at <180 °C resulted in films with an average atomic concentration of ~2% chlorine. The presence of chlorine impurities suggests that some of the silicon methoxy groups begin to react with the hydroxylated surface at lower temperatures.

There are many silanes and silanols that can react with TMA and H_2O in ABC or ABCD processes. Silanols can react directly after TMA exposures and avoid the need for two H_2O exposures. One possible reaction sequence utilizes diisopropyl-iso-propoxy-silane (DIPS) together with TMA and H_2O in a DIPS/H_2O/TMA sequence. A schematic of this reaction sequence is presented in Figure 5.21 (Abdulagatov, A.I., Goldstein, D.N., and George, S.M. (2010) Unpublished results.). This ABC process has been demonstrated and yields a mass gain of ~22 ng/cm^2 per cycle and a growth rate of ~1.3 Å per cycle at 150 °C.

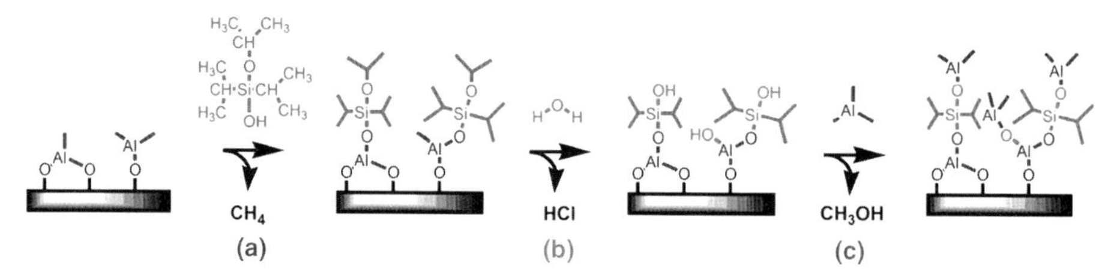

Figure 5.21 Schematic depicting the three-step reaction sequence for ABC MLD growth of an alumina–siloxane film using DIPS, H_2O, and TMA.

5.8
Future Prospects for MLD of Hybrid Organic–Inorganic Films

The use of various organic and inorganic precursors offers a nearly limitless set of combinations for the MLD of hybrid organic–inorganic films. Many of these combinations can be used to fabricate films with specific functional properties. One example of a functional hybrid organic–inorganic film is an MLD film grown using TMA and triethylenediamine (TED). TMA is a Lewis acid and TED is a Lewis base. An exposure sequence of TMA and TED can be used to grow an MLD film with unreacted $AlCH_3$ species remaining in the film (Yoon, B. and George, S.M. (2010) Unpublished results.). A schematic of this reaction sequence is given in Figure 5.22 (Yoon, B. and George, S.M. (2010) Unpublished results.). These $AlCH_3$ species can react with H_2O and serve as a H_2O getter. The H_2O getters may be useful as interlayers in multilayer gas diffusion barrier films.

Conductive hybrid organic–inorganic films may also be useful for flexible displays. ZnO ALD films are known to have a low resistivity of $\sim 1 \times 10^{-2}\,\Omega\,\text{cm}$ [32]. ZnO ALD films are grown using DEZ and H_2O [33]. Hybrid organic–inorganic MLD films can

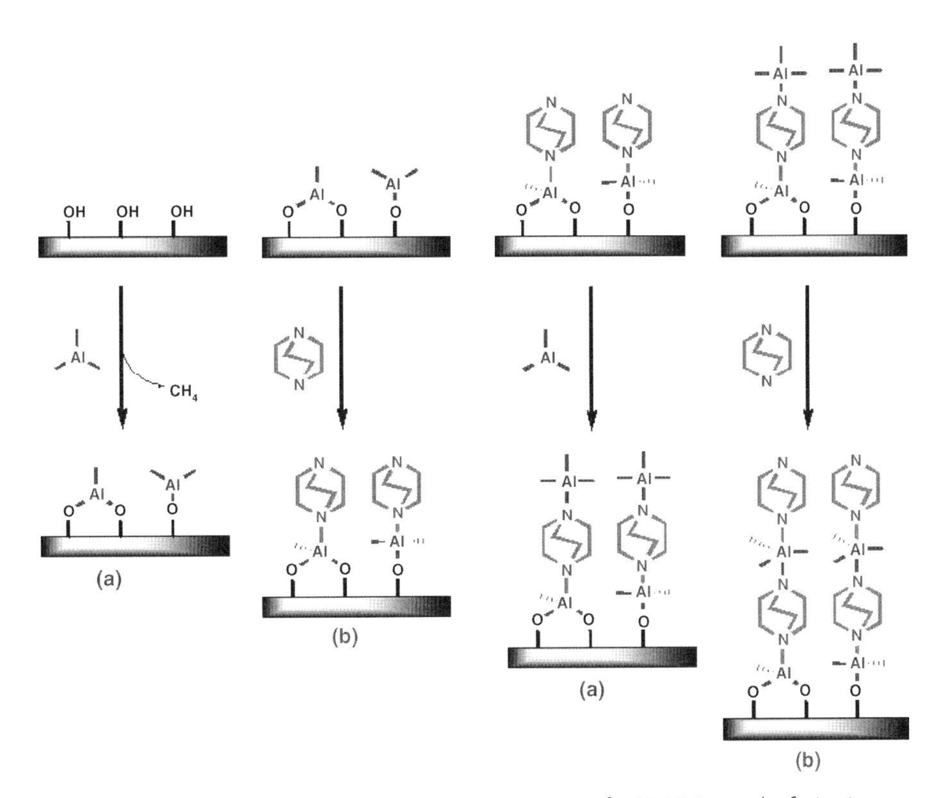

Figure 5.22 Schematic depicting the two-step reaction sequence for AB MLD growth of a Lewis acid–Lewis base film using TMA and TED.

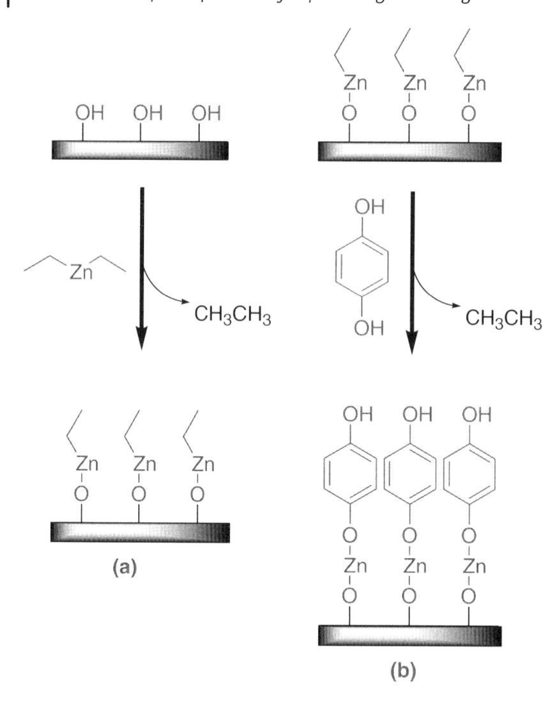

Figure 5.23 Schematic showing the two-step reaction sequence for AB zincone MLD growth using DEZ and HQ.

be grown using DEZ and EG as mentioned earlier in Section 5.2 and are called "zincones" [17, 18]. Although the zincone MLD film based on DEZ and EG does not display significant conductivity, zincone alloy films based on DEZ, hydroquinone (HQ) and H_2O have displayed conductivity similar to ZnO ALD films [34]. The schematic showing the surface chemistry for zincone MLD using DEZ and HQ is given in Figure 5.23 [34]. If these conducting MLD films display sufficient toughness, they may be useful for flexible displays and may be candidates to replace indium tin oxide (ITO).

The hybrid organic–inorganic MLD films have a low density that approaches the low densities of organic polymers. In contrast, inorganic ALD films have a much higher density. Mixtures of hybrid organic–inorganic MLD layers with ALD layers can be used to obtain films with a density that varies from the low density of the pure MLD film to the high density for the inorganic ALD film (Lee, B.H. and George, S.M. (2010) Unpublished results.). As an example, the densities of hybrid Al_2O_3 ALD:AB alucone MLD films are shown in Figure 5.24 (Lee, B.H. and George, S.M. (2010) Unpublished results.). Al_2O_3 ALD was grown using TMA and H_2O [20, 21]. AB alucone MLD was grown using TMA and EG [13]. The density was varied by changing the relative number of ALD and MLD cycles during the alloy growth.

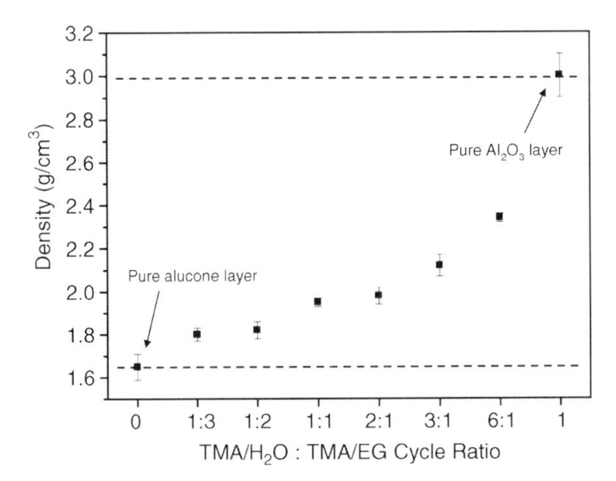

Figure 5.24 Density of alloys of Al_2O_3 and AB alucone using TMA and EG from XRR analysis. The alloys were prepared using different numbers of TMA/H_2O and TMA/EG cycles. For example, the 3 : 1 ratio sample was prepared using repetitive sequences of three cycles of TMA/H_2O and then one cycle of TMA/EG.

Figure 5.24 indicates that the density can be varied widely with changing organic–inorganic film composition. Other properties that depend on density will also change accordingly. For example, mechanical properties such as the elastic modulus and stiffness should be tunable [35]. Optical and electrical properties such as refractive index and dielectric constant should also vary with the composition of the alloy film [36]. In general, films with a variety of tunable properties should be possible by changing the ratio of ALD and MLD cycles used to grow the alloy film.

Most of the MLD systems reviewed in this chapter have been based on AB, ABC, or ABCD processes using TMA. As mentioned earlier, other organometallic and organic precursors are also possible. Hybrid organic–inorganic films based on zinc are possible using DEZ [17, 18]. Other hybrid organic–inorganic systems based on zirconium and titanium are possible using $Zr(O\text{-}t\text{-}Bu)_4$ and $TiCl_4$, respectively (Abdulagatov, A.I. and George, S.M. (2010) Unpublished results.; Lee, B.H. and George, S.M. (2010) Unpublished results.). Many other organometallic precursors can also be used to define other hybrid organic–inorganic MLD polymers. For example, metal alkyls based on magnesium and manganese are available as Mg$(EtCp)_2$ and $Mn(EtCp)_2$. These metal alkyls are expected to react with diols or carboxylic acids to define new MLD systems [37, 38].

The possibilities for the MLD of hybrid organic–inorganic films are virtually unlimited given all the metals on the periodic table and organic compounds available from organic chemistry. The challenge over the next few years will be to determine the hybrid organic–inorganic films that may be grown easily and that may display useful properties. The tunable mechanical, optical, dielectric, conductive, and chemical properties of the hybrid organic–inorganic films should be valuable for a wide range of applications.

Acknowledgments

This research was funded by the National Science Foundation (NSF), the Air Force Office of Scientific Research (AFOSR), the Defense Advanced Research Program Agency (DARPA), and DuPont Central Research and Development.

References

1 George, S.M. (2010) *Chem. Rev.*, **110**, 111.
2 George, S.M., Yoon, B., and Dameron, A.A. (2009) *Acc. Chem. Res.*, **42**, 498.
3 Du, Y. and George, S.M. (2007) *J. Phys. Chem. C*, **111**, 8509.
4 Yoshimura, T., Tatsuura, S., and Sotoyama, W. (1991) *Appl. Phys. Lett.*, **59**, 482.
5 Kubono, A., Yuasa, N., Shao, H.L., Umemoto, S., and Okui, N. (1996) *Thin Solid Films*, **289**, 107.
6 Shao, H.I., Umemoto, S., Kikutani, T., and Okui, N. (1997) *Polymer*, **38**, 459.
7 Yoshimura, T., Tatsuura, S., Sotoyama, W., Matsuura, A., and Hayano, T. (1992) *Appl. Phys. Lett.*, **60**, 268.
8 Kubono, A. and Okui, N. (1994) *Prog. Polym. Sci.*, **19**, 389.
9 Putkonen, M., Harjuoja, J., Sajavaara, T., and Niinisto, L. (2007) *J. Mater. Chem.*, **17**, 664.
10 Adamcyzk, N.M., Dameron, A.A., and George, S.M. (2008) *Langmuir*, **24**, 2081.
11 Kim, A., Filler, M.A., Kim, S., and Bent, S.F. (2005) *J. Am. Chem. Soc.*, **127**, 6123.
12 Lee, J.S., Lee, Y.J., Tae, E.L., Park, Y.S., and Yoon, K.B. (2003) *Science*, **301**, 818.
13 Dameron, A.A., Seghete, D., Burton, B.B., Davidson, S.D., Cavanagh, A.S., Bertand, J.A., and George, S.M. (2008) *Chem. Mater.*, **20**, 3315.
14 Nilsen, O., Klepper, K.B., Nielson, H.O., and Fjellvag, H. (2008) *ECS Trans.*, **16**, 3.
15 Lee, B.H., Im, K.K., Lee, K.H., Im, S., and Sung, M.M. (2009) *Thin Solid Films*, **517**, 4056.
16 Lee, B.H., Ryu, M.K., Choi, S.Y., Lee, K.H., Im, S., and Sung, M.M. (2007) *J. Am. Chem. Soc.*, **129**, 16034.
17 Peng, Q., Gong, B., VanGundy, R.M., and Parsons, G.N. (2009) *Chem. Mater.*, **21**, 820.
18 Yoon, B., O'Patchen, J.L., Seghete, D., Cavanagh, A.S., and George, S.M. (2009) *Chem. Vapor Depos.*, **15**, 112.
19 Yoon, B., Seghete, D., Cavanagh, A.S., and George, S.M. (2009) *Chem. Mater.*, **21**, 5365.
20 Dillon, A.C., Ott, A.W., Way, J.D., and George, S.M. (1995) *Surf. Sci.*, **322**, 230.
21 Ott, A.W., Klaus, J.W., Johnson, J.M., and George, S.M. (1997) *Thin Solid Films*, **292**, 135.
22 Puurunen, R.L. (2005) *J. Appl. Phys.*, **97**, 121301.
23 Wilson, C.A., Grubbs, R.K., and George, S.M. (2005) *Chem. Mater.*, **17**, 5625.
24 Seghete, D., Hall, R.A., Yoon, B., and George, S.M. (2010) *Langmuir*, **26**, 19045.
25 McMahon, C.N., Alemany, L., Callender, R.L., Bott, S.G., and Barron, A.R. (1999) *Chem. Mater.*, **11**, 3181.
26 Miller, D.C., Foster, R.R., Zhang, Y., Jen, S.H., Bertrand, J.A., Lu, Z., Seghete, D., O'Patchen, J.L., Yang, R., Lee, Y.C., George, S.M., and Dunn, M.L. (2009) *J. Appl. Phys.*, **105**, 093527.
27 Li, Z.L., Brokken-Zijp, J.C.M., and de With, G. (2004) *Polymer*, **45**, 5403.
28 Lotters, J.C., Olthuis, W., Veltink, P.H., and Bergveld, P. (1997) *J. Micromech. Microeng.*, **7**, 145.
29 Camino, G., Lomakin, S.M., and Lageard, M. (2002) *Polymer*, **43**, 2011.
30 Grassie, N. and Macfarlane, I.G. (1978) *Eur. Polym. J.*, **14**, 875.
31 Ferguson, J.D., Weimer, A.W., and George, S.M. (2000) *Thin Solid Films*, **371**, 95.

32 Elam, J.W., Routkevitch, D., and George, S.M. (2003) *J. Electrochem. Soc.*, **150**, G339.
33 Elam, J.W. and George, S.M. (2003) *Chem. Mater.*, **15**, 1020.
34 Yoon, B., Lee, Y., Derk, A., Musgrave, C.B., and George, S.M. (2011) *ECS Transactions*, **33** (27), 191.
35 Salmi, L.D., Puukilainen, E., Vehkamaki, M., Heikkila, M., and Ritala, M. (2009) *Chem. Vapor Depos.*, **15**, 221.
36 Zaitsu, S., Jitsuno, T., Nakatsuka, M., Yamanaka, T., and Motokoshi, S. (2002) *Appl. Phys. Lett.*, **80**, 2442.
37 Burton, B.B., Fabreguette, F.H., and George, S.M. (2009) *Thin Solid Films*, **517**, 5658.
38 Burton, B.B., Goldstein, D.N., and George, S.M. (2009) *J. Phys. Chem. C*, **113**, 1939.

6
Low-Temperature Atomic Layer Deposition

Jens Meyer and Thomas Riedl

6.1
Introduction

The vast majority of atomic layer deposition (ALD) processes occur at substrate/reactor temperatures exceeding 100 °C. However, coating thermally fragile substrates, for example, polymers, biomolecules, or sensitive electronic devices, demands ALD processes that bring about the characteristic benefits of ALD (homogeneity, conformity, density) along with a reliable processing window well below 100 °C. To date, there are a very limited number of precursor systems that in principle allow for ALD processes at low temperatures [1–3]. Among them, the deposition of Al_2O_3 via trimethylaluminum (TMA) and water has become a model precursor system for low-temperature ALD (LT-ALD) [2, 4]. The relatively high vapor pressure of TMA and its high reactivity are the keys that allow for LT-ALD. The exploration of the TMA/water system for applications as flexible protective coatings on polymer substrates against erosion by atomic oxygen or as barrier layers to suppress the permeation of molecular species (e.g., water, oxygen) has started only very recently [5–9]. Within a short time period, a plethora of studies on the deposition of ALD layers on polymeric substrates and fibrous structures has been reported [10, 11]. Among the many applications, the surface functionalization of nanostructured polymer templates for surface-enhanced Raman spectroscopy [12] or catalytic surfaces [13] are a few tantalizing examples that require a LT-ALD process. Recently, the marriage of the beneficial properties of cellulose paper and the ability of ALD to coat complex surfaces has led to the preparation of functional substrates that allow for novel or substantially improved applications. For example, the individual fibers of cellulose paper have been coated by functional ALD layers to obtain large-area bioadhesive surfaces [14] or photocatalytic composites [15].

The prototyping of biological nanostructures [16] holds the promise for complex yet easy to prepare functional elements. In a very recent report, the modification of biological structures such as spider silk by ALD infiltration has been shown to yield a novel hybrid material with outstanding mechanical properties [17]. Ultimately, the highly efficient encapsulation of moisture/air-sensitive optoelectronic (organic)

Atomic Layer Deposition of Nanostructured Materials, First Edition. Edited by Nicola Pinna and Mato Knez.

devices [7, 8, 18, 19] or the preparation of protective coatings for sensor systems or MEMS for operation under harsh/corrosive ambient conditions (*in vivo* monitoring) [20] relies on processes that deliver dense layers without deterioration of the device due to excessive thermal impact. It is plain to imagine that the requirement of LT-ALD touches not only the limits of precursor chemistry but also those of processing technology. Processes that can cope with a reduced vapor pressure at low temperatures and avoid the pileup of reactants are mandatory in order to rule out unwanted gas-phase reactions and parasitic CVD. At the same time, in order to be a serious technology for production, the processing times in LT-ALD must not exceed those of conventional ALD processes. Rather, for high-throughput manufacturing, novel ALD processing strategies have to be developed that can operate at low temperatures and at coating rates in $nm/(s\ m^2)$ on the same order as that of other coating techniques.

6.2 Challenges of LT-ALD

As detailed in the previous chapters, ALD is seen as the CVD – analogue in which the precursors are fed sequentially into a dedicated reaction chamber. Thereby, the film growth exclusively proceeds via self-limiting surface reactions. The reaction between reactants A and B occurs efficiently if the reaction enthalpy ΔH of the two partners is sufficiently negative. There exist some well-established A/B systems that, owing to the reactivity of the species involved, allow for thermal ALD processes at reactor temperatures well below 100 °C (Table 6.1). Among them are TMA/H_2O [2, 4], DEZ/H_2O [21, 22], and TDMAZr/H_2O [7, 23]. Aside from the mere reaction chemistry of a particular ALD process, which can be categorized according to the reaction enthalpy, there are some issues at low process temperatures brought about by the particular design of the reaction chamber and the mode of running a specific ALD process (see Section 6.4). Generally, for lower deposition temperatures, aside from the reactivity of the precursors, their vapor pressure is the predominant limitation. For ALD processes involving H_2O as a reactant, long purge times (>100 s) at temperatures below 100 °C have been found necessary in order to avoid a pileup of H_2O molecules in the reaction chamber [2]. Similarly, long purging times at reactor temperatures below 100 °C were found necessary for a class of rather volatile metal amide precursors used to prepare ZrO_2 and HfO_2 layers [23] as shown in Figure 6.1. In the case of incomplete removal of precursors from the reactor volume, the onset of parasitic CVD is evidenced by a substantial variation in the growth rates, lack of homogeneity, and overall deteriorated film properties (lower density, increased roughness, etc.) [8, 23]. On the other hand, purging times of 100 s to guarantee a "clean, CVD-free" ALD process may lead to overall processing times for a 100 nm thick metal oxide layer on the order of 50 h. In the following, the particular properties of Al_2O_3 LT-ALD are discussed and concepts are introduced to overcome certain limitations.

Groner *et al.* showed in a detailed study how the temperature affects the film growth and film properties, which is briefly reviewed here [2]. It was found that the

Table 6.1 List of low-temperature (RT–100 °C) ALD processes.

Material	Precursor 1	Precursor 2	Temperature	References
Al_2O_3	TMA	H_2O	33 °C 45 °C	[2] [41]
Al_2O_3	TMA	O_3	RT	[42]
Al_2O_3	TMA	O_2 plasma	RT	[43]
B_2O_3	BBr_3	H_2O	RT	[3]
CdS	$Cd(CH_3)_2$	H_2S	RT	[44]
HfO_2	$Hf[N(Me_2)]_4$	H_2O	90 °C	[45]
Pd	$Pd(hfac)_2$	H_2	80 °C	[36]
Pd	$Pd(hfac)_2$	H_2 plasma	80 °C	[36]
Pt	$MeCpPtMe_3$	O_2 plasma + H_2	100 °C	[39]
PtO_2	$MeCpPtMe_3$	O_2 plasma	100 °C	[39]
SiO_2	$Si(NCO)_4$	H_2O	RT	[46]
SiO_2	$SiCl_4$	H_2O	RT[a]	[28]
SnO_2	TDMASn	H_2O_2	50 °C	[47]
SnO_x	$C_{12}H_{26}N_2Sn$	H_2O_2	50 °C	[48]
Ta_2O_5	$TaCl_5$	H_2O	80 °C	[49]
Ta_2O_5	$Ta[N(CH_3)_2]_5$	O_2 plasma	100 °C	[50]
Ta	$TaCl_5$	H plasma	RT	[40]
Ti	$TiCl_4$	H plasma	RT	[51]
TiO_2	$Ti[OCH(CH_3)]_4$	H_2O	35 °C	[52]
TiO_2	$TiCl_4$	H_2O	100 °C	[53]
V_2O_5	$VO(OC_3H_9)_3$	O_2	90 °C	[54]
ZnO	$Zn(CH_2CH_3)_2$	H_2O	60 °C	[55]
ZnO	$Zn(CH_2CH_3)_2$	H_2O_2	RT	[31]
ZrO_2	$Zr(N(CH_3)_2)_4)_2$	H_2O	80 °C	[7]

a) When using a pyridine catalyst.

growth rate decreased with decreasing temperature due to slower reaction kinetics. The maximum growth rate of 1.33 Å per cycle was obtained at temperatures between 100 and 125 °C. Exposure time and nitrogen purge times between precursor doses increase with decreasing temperature as a result of the reduced activation energy of the reaction and reduced vapor pressure of the reactants. Especially, the purging after the H_2O dosing takes longer because of slower desorption of H_2O from the reactor walls at low temperature. Also, the film density of Al_2O_3 is significantly lower compared to high-temperature depositions. The typical values of the density of amorphous Al_2O_3 are in the range of 3.5–3.7 g/cm^3. At 177 °C the average density decreases to 3.0 g/cm^3 and at 33 °C values around 2.5 g/cm^3 have been reported. According to this, the refractive index also decreases slightly with decreasing growth temperature, which was attributed to a higher impurity level at lower temperatures [24]. In addition, the relative dielectric constant ε shows a similar temperature-dependent characteristic. In the case of Al_2O_3 grown at 250 °C, the relative dielectric constant is about $\varepsilon \approx 9$. At lowered processing temperatures, a dielectric constant of $\varepsilon \approx 7.9$ was found, again demonstrating a reduced film density [7]. Figure 6.2a shows the structural properties of the Al_2O_3 layer depicted by transmission electron

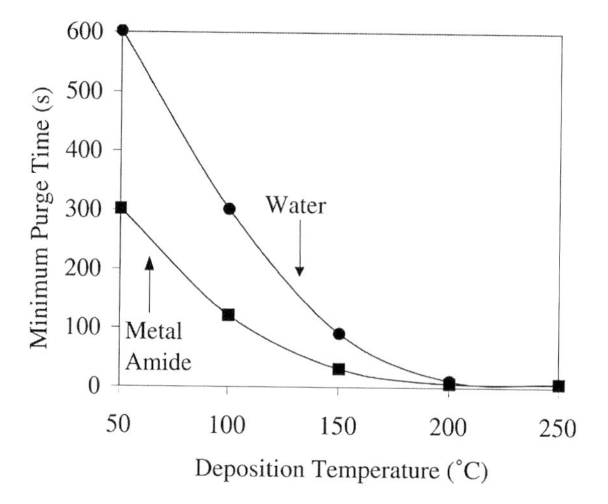

Figure 6.1 Purge times required to achieve 0.096 nm per cycle thickness for zirconium oxide films (0.093 nm for the hafnium oxide) at low deposition temperatures. The purge time to reach the saturated thickness per cycle increases drastically with decreasing temperature for the metal precursors (identical tendency for both metal amides) and even longer purge times are required for the H_2O precursor [23].

microscopy (TEM). Defects can be identified as tiny voids leading to more sponge-like morphology. Dillon *et al.* have shown by FTIR spectroscopy that there is an accumulation of Al–OH species due to incomplete precursor reaction in films prepared with a TMA/H_2O ALD process at growth temperatures below 450 K (177 °C) [25]. It is likely that small amounts of Al–OH species are piled up in the Al_2O_3 film at LT-ALD as in the film produced with an 80 °C process shown in Figure 6.2a. As a result, the Al_2O_3 films grow with a reduced packing density that consequently leads to a higher areal density of permeation channels for gaseous species and therefore affects their suitability, for example, as gas diffusion barrier layers (see Section 6.5). In contrast to neat Al_2O_3 layers, it has been reported that no such voids are found in nanolaminates consisting of alternating Al_2O_3 and ZrO_2 layers, as shown in Figure 6.2b and c, prepared at the same reactor temperature

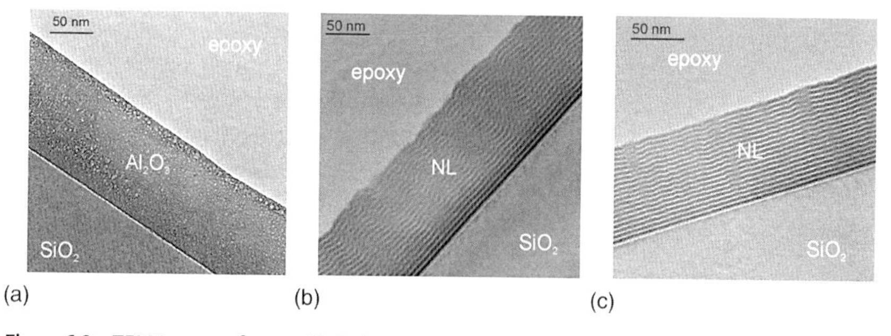

Figure 6.2 TEM images of a neat Al_2O_3 layer (a) and Al_2O_3/ZrO_2 nanolaminates prepared at 80 °C with a N_2 purge time of 5 s (a, b) and 20 s (c).

(80 °C) [7]. The reason for the absence of voids in this case is still under investigation. A possible explanation may be linked to the Zr precursor tetrakis(dimethylamido) zirconium(IV) (TDMA(Zr)) used in this case. This chemical has a lower vapor pressure than TMA. Therefore, residues of TDMA(Zr) may remain in the reactor during the entire process [23] and may function as a scavenger for H_2O. Thus, a pileup of H_2O in the layers is suppressed. The assumption of the TDMA(Zr) precursor staying in the reactor means, on the other hand, that the process takes place at the edge of the ALD window and approaches the regime of chemical vapor deposition (CVD). This is partially reflected in Figure 6.2b, where the TEM images show that the nanolaminate structure is perfect with sharp and well-defined interfaces between the sublayers only during the initial stages of the deposition process. With increasing number of sublayers, the nanolaminate evolves in a wavy structure. The origin of this change in the growth is not clarified yet. It is believed that the onset of parasitic CVD is the reason for this effect [26, 27]. A strong hint to this assumption is given by an experiment where the purging time between precursor doses is increased from 5 to 20 s. In this case, the nearly ideal nanolaminate structure can be preserved as shown in Figure 6.2c. In addition, the film thickness decreases by around 10% when 20 s purge time is used indicating again the parasitic CVD toward short purge times.

In general, with decreasing process temperature the process moves toward the edge of the ALD window. As a consequence, the purge time between the precursor pulses has to be extended. Longer purging times, however, directly lead to overall longer processing times. Groner *et al.* increased the purge time for a 33 °C Al_2O_3 film process from 5 s at 177 °C by a factor of 36 to 180 s [2]. It is important to note that even though the film properties are altered in LT-ALD, it is still possible to grow high-quality films at very low temperatures with ALD typical features. For example, at 33 °C growth temperature a very low leakage current of Al_2O_3 $<10^{-7}$ nA/cm^2 at a 5 V bias and a catastrophic breakdown field of 3.7 MV/cm was measured [2]. Moreover, very efficient gas diffusion barriers can be realized by LT-ALD as discussed in Section 6.5. In some low-temperature processes, mixing of ALD and CVD can be tolerated without loss of functionality. Certainly, the possibility of a trade-off between processing times and "purity" of the ALD process has to be judged for each application. Nevertheless, transferring the potential of ALD toward novel areas that demand low-temperature processing brings about a tantalizing amount of opportunities. Only a tiny selection of those will be presented below.

6.3 Materials and Processes

Highly reactive precursors with a strongly exothermic reaction are a key factor for efficient low-temperature ALD processes. There exist a variety of precursors whose specific reaction mechanisms show a favorable enthalpy in an ALD process for the preparation of materials such as Al_2O_3, TiO_2, SiO_2, and ZnO. The use of TMA and H_2O provides an ALD window down to 33 °C with good film properties [2]. The reaction of titanium tetraisopropyl oxide and H_2O is also substantially exothermic, allowing for a reduced processing temperature of 35 °C [16]. It has been found that a pyridine catalyst lowers the deposition temperature of SiO_2 ALD with $SiCl_4$ and H_2O to RT [28]. Therefore, these exemplary precursor systems would not *a priori* require some advanced ALD processes that are assisted by direct or remote plasmas to promote reactivity of the precursors. Nevertheless, great care has to be taken with the plain statement that a certain combination of A/B precursors supports a thermal ALD process at low temperatures without judging the layer properties for a particular application. For example, it is well known that DEZ/H_2O forms ZnO layers at temperatures well below 80 °C. Even Al-doped ZnO layers in the form of nanolaminates have been demonstrated and analyzed. However, while Al:ZnO layers prepared at 150 °C show a specific conductivity of 750 S/cm, this value drops by more than two orders of magnitude to 3 S/cm when the same nominal structure with same precursor system is deposited at 80 °C [29]. This decay in conductivity renders the resulting layers virtually useless for application as transparent conducting electrodes in large-area optoelectronics. A similar degradation of conductivity toward low processing temperatures has been reported by several groups for ZnO layers prepared by thermal ALD [22, 30]. Employing a stronger oxidizing species (H_2O_2), an optimum was obtained at 100 °C and LT-ALD as low as RT has been achieved [31]. A further partial workaround is the use of ozone or oxygen plasma (PEALD) to further improve the film quality and thus allow LT-ALD for a multiplicity of materials [32]. In a recent paper, PEALD growth of ZnO layers in the temperature range between 25 and 120 °C was reported [33]. Therein, a similar trend of reduced conductivity with lower reactor temperatures has been evidenced. Notably, residues of carbonyl and hydroxyl species were found in the layers at temperatures below 65 °C [34]. Obviously, PEALD has the potential to widen the temperature window for ALD growth of functional materials such as TCOs. We will not further address PEALD here, because it will be extensively treated in Chapter 7. A serve side effect of PEALD is the induced damage of sensitive materials such as polymer or biological substrates. Another drawback of PEALD is related to the deposition of surfaces with a high aspect ratio. Since the excited species produced by the plasma have only a limited lifetime before they become deexcited, surface reactions are hampered at surfaces that are not easily accessible by the plasma species.

The development of metal ALD systems that may be used at low temperatures is still in its early stages. This is in part due to the lack of appropriate reducing agents and proper surface functionalization to create the necessary sticking sites for the chemisorption of precursors. Thus, only a small selection of precursors exists so far that allows the deposition of metals at low temperatures. Pd is one of the few metals that can be deposited using LT-ALD, both with and without plasma on arbitrary substrates. Senkevich *et al.* reported on a method based on activating the surface with a tetrasulfide self-assembled monolayer at 200 °C. Once this seed layer is deposited, the growth temperature can be reduced to 80 °C by sequential pulsing using palladium(II) hexafluoroacetylacetonate ($Pd^{II}(hfac)_2$) and H_2 [35]. A direct film growth of Pd and Ta on non-noble metal surfaces such as Si and SiO_2 was demonstrated by the use of plasma ALD [36, 37]. More recently, Eyck *et al.* reported on Pd thermal ALD at 80 °C on air-exposed tantalum and silicon using molecular hydrogen and $Pd^{II}(hfac)_2$ as the precursor [38]. Good film properties of Pt (19 μΩ cm at 22 nm thickness) have been achieved with an additional H_2 gas exposure step in the deposition cycle during O_2 plasma ALD at 100 °C [39]. Ta is another LT-ALD system that facilities the deposition down to RT with a $TaCl_5$ precursor and atomic hydrogen as the reactive species in a plasma-enhanced process [40]. Table 6.1 summarizes materials, precursors, and temperatures in the region from RT to 100 °C that have been demonstrated for thermal ALD as well as PEALD systems.

6.4 Toward Novel LT-ALD Processes

Generally, one can categorize ALD processes by the pumping strategy and use of a carrier gas during and in between the precursor dosing. In one hypothetical limit, where no carrier gas is used, the residence time of the precursors in the reaction chamber can be very long. To avoid parasitic CVD, the pump cycles between precursor doses may become quite substantial, especially at low reactor temperatures. This could ultimately lead to unacceptable processing times particularly when aiming at high-throughput manufacturing of moisture barriers for organic electronics. Therefore, an inert purge gas is typically used in a viscous flow-type reactor, where the base pressure is on the order of 1 mbar [56]. Owing to a more efficient removal of residual precursors and reaction by-products, the time interval between precursor doses can be significantly lowered. Nevertheless, conventional ALD reactor concepts will reach limits in terms of short processing times at low temperatures.

In a very recent approach presented by Kodak, the established concept of reactor-based ALD has been overcome and an ALD process has been introduced that takes place at atmospheric pressure without a reaction chamber [57, 58]. In Kodak's process, a specially designed deposition head is used, in which each reactant is confined to controlled regions along the substrate surface. Reactants are spatially separated by outlets of an inert gas to avoid mixing of precursor gases between the channels (Figure 6.3). Additional exhaust slots between the gas inlets guarantee an efficient removal of unwanted excess material and reaction products. In this setup, the distance d between the deposition head and the substrate is critical in order to adjust the pressure gradient between inlet and exhaust. For too large d, gas-phase reactions of the precursors take place and the ALD regime is abandoned. Ideally, d is on the order of 20–30 μm [58]. When the deposition head and substrate remain stationary with respect to each other, different areas on the substrate surface are exposed to a single precursor species only. As a result, no deposition occurs. When the substrate moves in a direction perpendicular to the gas delivery slots in the ALD head, the surface of the substrate will undergo a sequential exposure to the reactant gases, very similar to that known in traditional chamber-based ALD. In a linear setup with a single pass of the substrate relative to the ALD head, the layer thickness or number of A/B cycles would be given by the number of the precursor delivery slots in the ALD head. For thicker layers, a long ALD head is required. Alternatively, the substrate could pass by the ALD head multiple times either in oscillatory fashion or in a conveyor-like setup. In this case, a compact

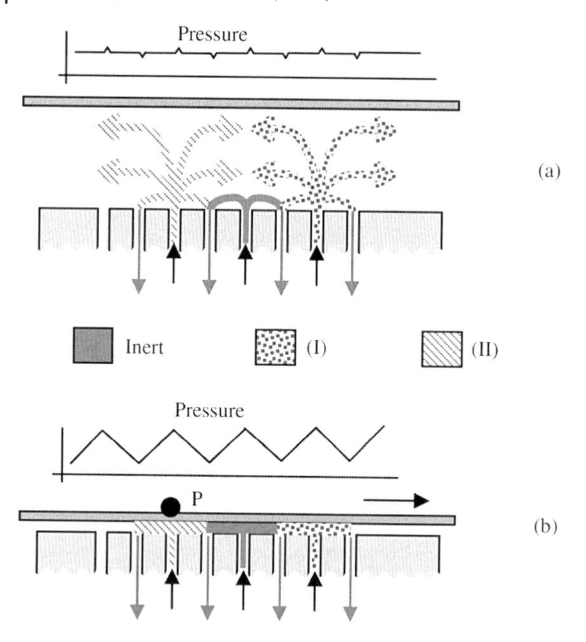

Figure 6.3 (Kodak ALD head) Side view of the spatial ALD coating head showing gas inlet slots (upward arrows) and gas exhaust slots (downward arrows). The precursors are denoted by (I) and (II), respectively. (a) A relatively large separation between the coating head and the substrate, leading to low pressure and significant gas mixing. (b) A smaller separation between the substrate and the head resulting in large pressure fields between channels [58].

ALD head could be used. With a tiny ALD head, even spatially selective ALD or ALD printing could be achieved. Aiming at low-temperature ALD, this approach appears favorable as the previously discussed pileup of precursors in a reaction chamber is not possible here. Nevertheless, potential limitations of this atmospheric ALD process toward low temperatures have not yet been fully explored.

As of yet, this process has been used to prepare ZnO-based thin film transistors (TFTs) with impressive operational stability and performance. The saturation field effect mobility of the TFTs prepared at a deposition temperature of 200 °C was higher than 10 $cm^2/(V\,s)$, an order of magnitude better than that of conventional amorphous silicon TFTs [59]. High-speed ring oscillator circuits based on these TFTs have also been reported [60]. The TFT characteristics are quite comparable to those of other metal oxide TFTs prepared by vacuum processes, such as sputtering or pulsed laser deposition [61]. Thus, atmospheric ALD seeds the prospect to prepare high-performance large-area electronics at ambient air and relatively low substrate temperatures. Concepts that harvest the sequential exposure of a surface with spatial rather than temporal separation of the precursors have been disclosed before. Notably, in 2007 General Electric and Planar Systems Inc. have filed patents on roll-to-roll ALD processes, where a flexible substrate is passed through two separate chambers each filled with one of the two ALD precursors [62, 63]. The precursor filled areas are separated by a region of inert (purge) gas. With these approaches, ALD is on the verge of becoming a module in a roll-to-roll production environment. Thereby, the outstanding properties of ALD deposited layers, such as thin film barriers, will be available for continuous manufacturing.

6.5 Thin Film Gas Diffusion Barriers

Gas diffusion barriers are needed for various applications ranging from food packing to flat panel displays or solar cells grown on plastic substrates. Since ALD has the advantage of allowing for the deposition of very dense films at very low temperatures ($<100\,^\circ C$), it is a promising technique to realize high-quality gas diffusion barriers even on thermally very fragile substrates. It has been demonstrated that Al_2O_3 films grown by LT-ALD provide effective gas diffusion barriers with water vapor transmission rates (WVTRs) below 10^{-5} g/(m^2 day) at room temperature [9]. With multilayer structures consisting of alternating metal oxide/polymer layers or metal oxide/metal oxide layers, even an improvement in terms of WVTR and pinhole density was reported [64]. For Al_2O_3/ZrO_2 nanolaminates grown at 80 °C, as shown schematically in Figure 6.4, a very low WVTR of 4.7×10^{-5} g/(m^2 day) has been measured in a climate cabinet at 70 °C and 70% relative humidity [7]. The activation energy E_a of the WVTR through the nanolaminate was found to be 92 kJ/mol. Thus, the WVTR for the nanolaminate barrier at room temperature was estimated to be on the order of 5×10^{-7} g/(m^2 day). In a further study, it was investigated how the variation of the nanolaminate thickness affects the WVTR. As depicted in Figure 6.4, the WVTR does not substantially depend on the layer thickness as long as the thickness does not fall below a critical value of about 40 nm. Below 40 nm, an abrupt increase in

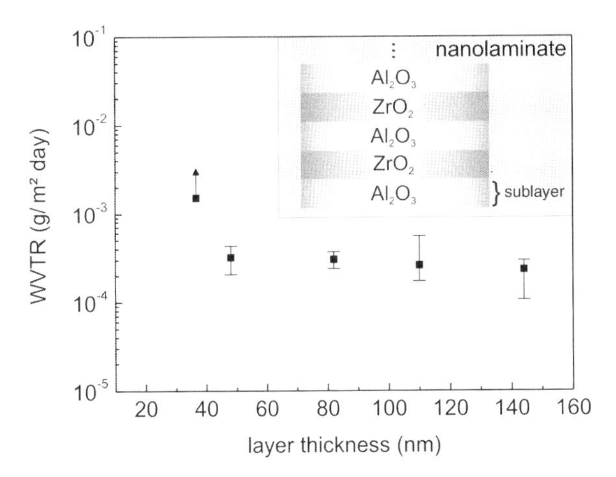

Figure 6.4 WVTR versus Al_2O_3/ZrO_2 nanolaminate thickness, grown at 80 °C. WVTR measured at 80 °C and 80% relative humidity. For thicknesses less than 37 nm, a significantly higher failure rate was observed. *Inset*: Scheme of a nanolaminate structure.

the WVTR was observed under environmental conditions of 80 °C and 80% relative humidity. This can be attributed to defect-dominated gas diffusion, caused by pinholes and/or microcracks, with decreasing layer thickness of the barrier [65]. Previous reports on Al_2O_3 or SiO_2 barriers have demonstrated a similar threshold-like characteristic for the WVTR versus barrier thickness [9, 65]. On the other hand, the leveling off of the WVTR at $\sim 3 \times 10^{-4}$ g/(m^2 day) toward thicker barriers is substantially consistent with gas permeability through pinhole defects in barrier films [65].

In addition to a low WVTR, a reliable gas diffusion barrier must bring about the ability to form homogeneous films without statistical local defects or paths of elevated water/oxygen permeation on large areas. The probability of statistical defects can be derived simply by the experiment shown in Figure 6.5, where an array of 100 individual Ca pads (150 μm in diameter) has been encapsulated with four types of barrier coatings in order to characterize their pinhole density. The encapsulated Ca arrays have been photographed after 16, 40, and 90 h under storage in controlled environmental conditions of 50 °C and 80% relative humidity. Many (70%) of the initially metallic Ca pads in the case of the neat Al_2O_3 encapsulation turned transparent due to corrosion, demonstrating the high density of defects in the Al_2O_3 barrier coating. The neat ZrO_2 as well as 95 nm Al_2O_3 + 5 nm ZrO_2 already provide a substantial barrier; however, a significant pinhole density can still be observed. Very remarkably, if the Ca sensors are covered by the Al_2O_3/ZrO_2 nanolaminate, the failure rate of the Ca pads stored for 160 h under the same conditions as above is only about 3%. The origin of the reliable and efficient nanolaminate gas diffusion barrier was attributed to essentially three key factors: (i) A higher packing density due to suppressed crystallization [7, 66]. (ii) It was found that ZrO_2 efficiently hampers the

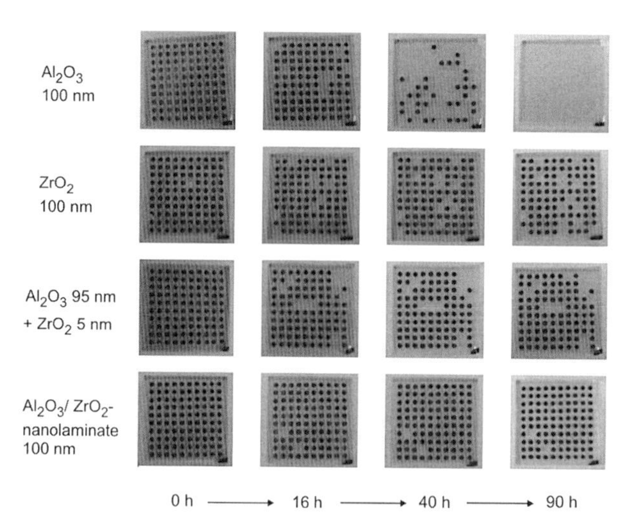

Figure 6.5 Array of 100 Ca pads encapsulated by 100 nm Al_2O_3, 100 nm ZrO_2, 95 nm Al_2O_3 + 5 nm ZrO_2, and 100 nm Al_2O_3/ZrO_2 nanolaminate. The photographs were taken at the beginning and after 16, 40, and 90 h under environmental conditions of 50 °C and 80% relative humidity.

corrosion of the neat Al_2O_3 by water, a major source of deterioration of the barrier functionality. Previously, Dameron *et al.* demonstrated that SiO_2 can protect an Al_2O_3 layer from water corrosion [5]. (iii) An aluminate phase was identified by X-ray photoemission spectroscopy at the Al_2O_3/ZrO_2 sublayer interfaces, which leads to stabilization and densification of the NL barrier coating [67].

So far, most work on inorganic barrier layers prepared by ALD has been done on rigid substrates. On the other hand, in many visions the success of organic electronics is associated with the prospect of flexible applications, like roll-up displays and solar cells [68]. There are some reports that demonstrate efficient inorganic ALD barriers on flexible substrates such as polyethylene naphthalate (PEN) [6]. On the other hand, studies of the mechanical properties of these barrier layers are limited. Especially, crack formation upon repeated bending, which would severely compromise the applicability of these gas permeation barriers for flexible applications, is typically not studied in great detail. Only very recently, the mechanical robustness of ALD layers on flexible substrates has been assessed [69]. Even though a substantial critical strain of 5% for the onset of crack formation was found for 5 nm thin alumina layers, the critical strain significantly decreased to only 0.88% for 125 nm thick layers. In the same paper, multilayers of Al_2O_3 and alucone [70], an inorganic/organic hybrid, were also proposed to increase the overall mechanical strength of the layer system. The idea behind this approach is the mechanical decoupling of the neighboring inorganic sublayers. Thereby, the internal stress of the layer system is expected to decrease significantly. Even though this approach did not quite fulfill the intention, as the multilayer structures typically showed a smaller critical strain compared to that of the single layers, this concept points to a promising strategy for next-generation flexible barrier coatings – the combination of ALD and molecular layer deposition (MLD). Recently, MLD has been considered as a technique to create organic, polymer-like films with a similar control and conformity as known from ALD [71]. An in-depth treatment of MLD is presented in Chapter 5. MLD growth with bifunctional reactants has been demonstrated for a number of "passive" polymers [72, 73], such as polyamide, polyurethane, and so on. Owing to their increased toughness compared to that of alucone, these compounds may now prove extremely useful for durable and flexible organic/inorganic hybrid systems applied as protective coatings for organic electronic devices and beyond. Work in this field is still in its infancy and substantial progress is to be expected in the near future.

6.6 Encapsulation of Organic Electronics

6.6.1 Encapsulation of Organic Light Emitting Diodes

Organic electronics such as organic light emitting devices (OLEDs) and organic photovoltaic cells (OPVs) state a particularly demanding application. Without encapsulation, a rapid degradation and a severely limited lifetime of the devices have been reported [74]. For devices with a serious long-term stability, a barrier is required with a WVTR $< 10^{-6}$ g/(m^2 day). However, a direct exposure of the functional organic materials to ALD precursors can cause strong degradation effects of the devices. Chang *et al.* have found that TMA attacks the double bonds in the conjugated OLED polymer poly[1-methoxy-4-(2′-ethyl-hexyloxy)-2,5-phenylenevinylene] (MEH-PPV) [75]. Typically, in organic electronic devices a metallic top electrode is used that acts as a precursor diffusion barrier [8]. Thus, LT-ALD can be applied to realize dense and leakproof encapsulation layers so that even OLEDs and OPVs, consisting of highly sensitive and reactive materials (e.g., low work function cathode materials) in the device stack, can be encapsulated. Effective OLED encapsulation has been reported for Al_2O_3 layers prepared by ALD [19, 76]. The use of Al_2O_3/ZrO_2 nanolaminate structures as discussed above led to further improvement of the encapsulation. Figure 6.6 shows the *L–I–V* characteristics between Al_2O_3 and Al_2O_3/ZrO_2 nanolaminate encapsulated samples compared to a reference device encapsulated by a conventional glass lid with getter. The OLED structure is based on a state-of-the-art p-i-n OLED with a phosphorescent emitter system. The encapsulation process in this study was carried out at 80 °C as discussed above. For an OLED without encapsulation, a rapid decay of the luminance to 50% of its starting value of 1000 cd/m^2 is found and a characteristic lifetime $t_{50} = 350$ h is derived. The active area at this stage already shows significant nonemitting, black areas, which indicate severe degradation of the device [77]. For the encapsulated OLEDs, the lifetime test has been stopped after 1000 h of operation and a model based on a stretched exponential decay can be fitted to derive an estimate for t_{50} [78]. For the glass lid encapsulated device, $t_{50} = 55\,000$ h can be extracted. The devices sealed by ALD reach 8700 and 22 000 h for the Al_2O_3 and ZrO_2/Al_2O_3 encapsulating layers, respectively [8]. In view of the uncertainty associated with the extrapolation of a 1000 h measurement to a timescale of 10 000 h, it can be stated that the NL encapsulation allows for a substantial lifetime close to that of the reference OLEDs. It is essential to note that the active OLED area in case of the device encapsulated with the nanolaminate is essentially free from black spots (not shown here). This is in favorable agreement with the vast absence of statistical defects in this encapsulation system.

As opposed to conventional OLEDs where the light is emitted via the transparent glass substrate, in top-emitting devices the light is emitted via the top electrode. Top-emitting OLEDs are favorable for active matrix OLED displays, where the opaque Si-based backplane electronics positioned under each pixel substantially compromises the amount of light coupled out for bottom-emitting structures. Recently, transparent thin film transistors based on metal oxide semiconductors have evolved as a powerful replacement for a-Si backplane technology, which would allow for efficient driver electronics that do not interfere with the light emission of bottom-emitting OLEDs [79]. At the same time, entirely see-through active matrix OLED displays become feasible [61, 80, 81]. In view of these prospects, the encapsulation of top-emitting or entirely transparent OLEDs has to be considered. As opposed to the previous case, where only the barrier functionality of the encapsulation layer was required, in this case also the optical properties of the thin-film encapsulation are of paramount importance. Ideally,

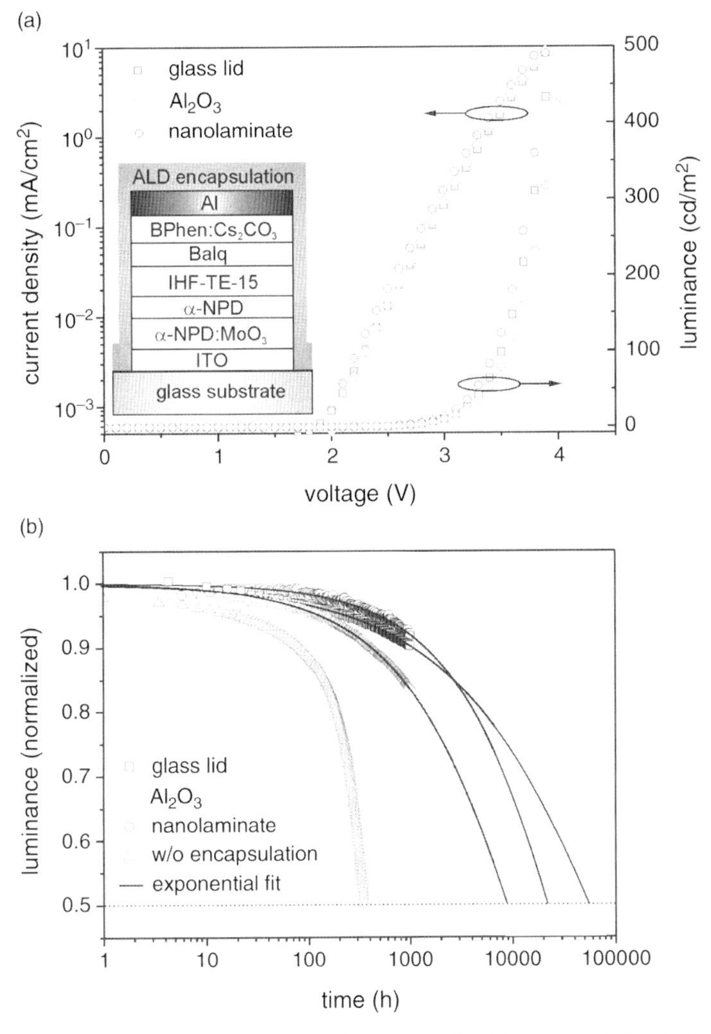

Figure 6.6 (a) *L–I–V* characteristics of encapsulated OLEDs measured directly after fabrication. (b) Normalized luminance versus time characteristics of neat and encapsulated OLEDs. Starting luminance is 1000 cd/m^2. *Inset*: Layer sequence of an OLED structure with ALD encapsulation.

the dielectric properties of the barrier layer can be tuned to improve the light extraction from the top-emitting OLED. Figure 6.7 shows the study of a top-emitting OLED with an encapsulation layer on top. The layer sequence is based on a state-of-the-art device with the peculiarity of a 15 nm thin aluminum top cathode that allows for some out-coupling of light. The bottom electrode consists of 100 nm Ag that is reflective. *I–V* characteristics of the fresh and encapsulated OLEDs are shown in Figure 6.7a. No substantial change due to the ALD process is encountered. A closer look reveals some increased current density in the encapsulated devices. A similar behavior is observed for OLEDs treated at elevated temperatures (like those used in an ALD process),

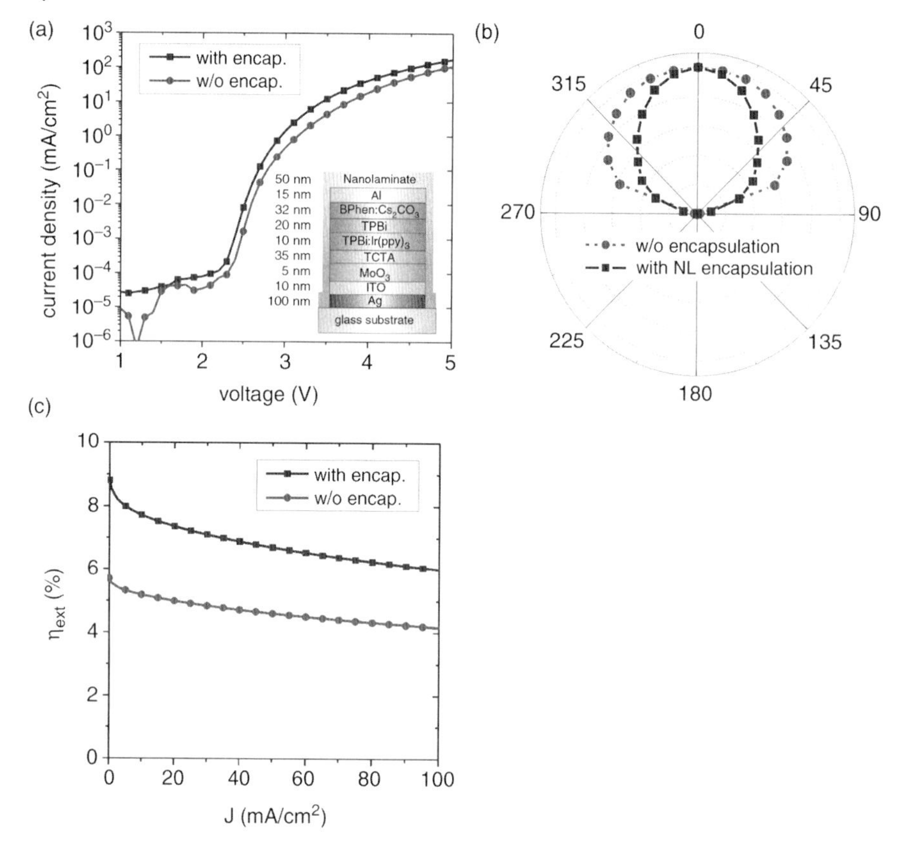

Figure 6.7 (a) *I–V* characteristics of the encapsulated OLED versus neat devices directly after preparation. *Inset*: Layer sequence of a top-emitting OLED structure with a 50 nm thick nanolaminate barrier layer on top. (b) Angle-resolved light emission from the OLEDs. (c) External quantum efficiency η_{ext} versus current density for the neat and encapsulated top emitters.

leading to slightly increased current densities. The quantum efficiency, on the other hand, does not change. Note that the layer sequence of the unencapsulated devices has been optimized for maximum light extraction. A maximum current efficiency of 27.5 cd/A (at 100 cd/m^2) has been determined. Upon careful design of the optical properties of the encapsulation layer (thickness and composition), the amount of light extraction can be further increased. Specifically, in the present case, a ZrO_2/Al_2O_3 nanolaminate (20/20 subcycles) structure with a thickness of 50 nm has been used. Under 0° observation angle, the current efficiency almost doubles and reaches a value of 45 cd/A. It is well known that the capping layer may substantially alter the angular dependence of the top emitter. This is also observed in an angle-resolved light emission study as shown in Figure 6.7b. The radiation pattern narrows toward the forward (0°) direction. One could speculate that this changed emission pattern is the sole reason for the increased efficiency found under 0° observation. While this wouldstill be a substantial improvement for some display applications, for lighting the amount of photons extracted over the entire half-space has to be considered. To this end, the OLEDs have been studied in an integrating sphere that allows determining the external quantum efficiency of both OLED structures. While for the optimized OLED without encapsulation layer an external efficiency η_{ext} on the order of 4–6% is derived, the OLED with tuned encapsulation layer shows an approximately 40% increased external efficiency (Figure 6.7c). In both devices, η_{ext} is limited by the semitransparent 15 nm thick aluminum top electrode. Substantially increased external efficiencies are expected for metal oxide-based top electrodes (e.g., ITO or Al:ZnO).

6.6.2
Encapsulation of Organic Solar Cells

Organic solar cells (OSCs) based on polymers and small molecules have seen a tremendous increase in interest during the past few years. Some significant progress in this field seeded the prospect for a cost-effective and easy to fabricate photovoltaic technology – typical advantages claimed for organic (opto)electronic devices. Cell efficiencies of up to 5–7.4% have recently been reported [82]. In a very recent press release, Mitsubishi Chemical has announced an efficiency as high as 9.2%, without disclosing details about the materials used and the device structure of their solar cells. Considering their intrinsic physical properties such as lightweight, mechanical flexibility, and semitransparency, these devices may open up new opportunities for applications of novel photovoltaic cells. However, in order to meet the demands of the market, not only must this new technology demonstrate reasonable cell efficiencies at a concomitantly low production cost, but organic solar cells must also provide lifetimes of at least 10 000 h and even more than 10 years if architectural applications are envisaged. As discussed in the case of OLEDs, organic materials are by nature more susceptible to chemical degradation from oxygen and water than inorganic materials. A plethora of studies has been carried out and they show that the stability/degradation mechanisms are indeed rather complicated and are certainly still not fully understood. To date, it is not clear whether organic solar cells are as demanding in terms of encapsulation as OLEDs are. Nevertheless, there are functional elements in both types of devices, for example, low work function cathodes, which show an intrinsic susceptibility toward corrosion by water and oxygen. Early experiments have clearly evidenced the requirement for an encapsulation of organic solar cells [83].

Typical organic solar cells are based on a layer sequence with the *anode* side adjacent to the typically transparent substrate. There may, however, be serious reasons associated with the internal morphology of the organic donor–acceptor blend system to flip the cell layout to an inverted structure, where the bottom electrode is the cathode (see Figure 6.8a) [84]. In the following example, the lifetime of a series of these inverted organic solar cells based on a bulk heterojunction (BHJ) of region-regular poly(3-hexylthiophene) (P3HT):(6,6)-phenyl-C_{61}-butyric acid methyl ester (PCBM) has been studied with a simple variation of the electron selective layer. Specifically, interlayers of cesium carbonate and titanium oxide have been used and their degradation has been studied for both as-prepared and encapsulated devices.

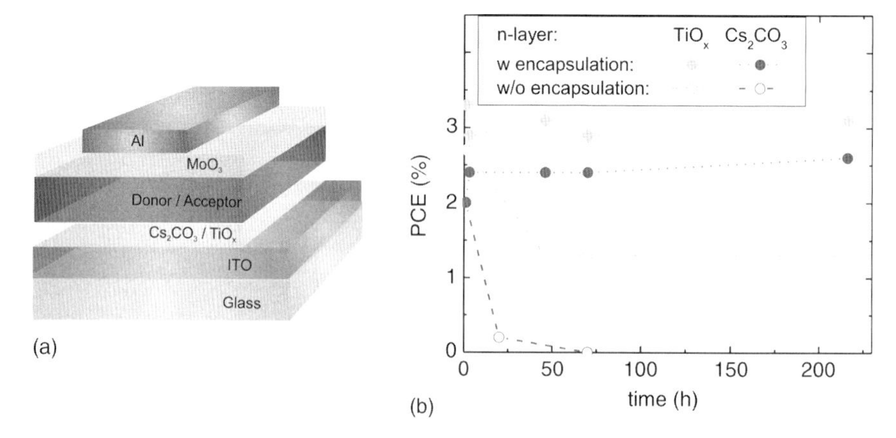

Figure 6.8 Inverted cell. (a) Layer sequence of an inverted BHJ solar cell with varied electron selective interlayers (Cs_2CO_3, TiO_x). (b) PCE of the cells with and without encapsulation.

For the encapsulation, a 60 nm thick Al_2O_3/ZrO_2 nanolaminate structure prepared by ALD, as discussed above, has been applied on top of the cells at 80 °C. The typical power conversion efficiency (PCE) of the cells in this study is about 3% (Figure 6.8b), a typical value for P3HT:PCBM BHJ structures. As expected, the unencapsulated cells show a rapid decay of the power conversion efficiency. This is especially evident in cells that contain the hygroscopic Cs_2CO_3 interlayers, which degrade within the first 20 h of operation. Even though TiO_x interlayers appear to be more stable in this cell structure, they still lose efficiency within 50 h. On the contrary, the cells encapsulated with the ALD moisture barriers do not show any change in PCE within the time range of this experiment (~200 h). Even though, extended lifetime studies are still under way, it is already clear that these ALD nanolaminate encapsulation layers prove extremely beneficial to improve the operational lifetime of organic solar cells by forming efficient barriers for oxygen and moisture. In a previous report, neat Al_2O_3 barriers prepared at 100 °C have been used on top of pentacene and C_{60} organic solar cells [18]. The Al_2O_3 barrier in this work was 200 nm thick and increased the operational lifetime of the cells from 10 h in the unencapsulated case to 6000 h. This remarkable result is somewhat weakened by the substantial thickness of the barrier requiring a 20 h long ALD process. To overcome these limitations, hybrid barrier structures have been proposed, which comprise layers prepared by "fast" CVD or PECVD processes and very thin ALD layers. For example, in a very recent report such a hybrid encapsulation structure has been used for organic solar cells. Here, a 100 nm thick layer of SiO_x or SiN_x (prepared by fast PECVD) is followed by a thin Al_2O_3 layer (10–50 nm) prepared by ALD. The stack is completed by a 1 μm thick cap of parylene, prepared by chemical vapor deposition [85]. The low WVTR of 2×10^{-5} g/(cm^2 day) allowed the passivated solar cells to operate for more than 5800 h without any degradation of the PCE. In this approach, the thin ALD layer serves as a passivation of potential pinholes formed in the SiO_x or SiN_x PECVD layers, which by themselves have been shown to exhibit orders of magnitude larger WVTRs [6].

In order to meet the requirements of large-area and high-throughput manufacturing, ALD processes have to be adapted and novel ALD concepts, such as those discussed in Section 6.4, have to be further developed.

6.7 Conclusions

While still in its infancy, it is clear that low-temperature ALD is about to pave the way toward exciting novel applications harvesting the unique advantages of conventional ALD. Specifically, the preparation of ultrahigh and reliable gas diffusion barriers on thermally fragile polymer substrates or optoelectronic devices is expected to play a key role to reach the required operational lifetimes of organic light emitting diodes and organic solar cells. A further tantalizing field is the replication or functionalization of biological nanostructures to create novel artificial materials with outstanding properties. To fulfill the demands of processing temperatures below 100 °C, sufficiently volatile and reactive precursor systems beyond those already available are required. Especially, the ALD of singular materials like several metals is far from thoroughly developed, to date. Aside from the precursor chemistry, some substantial progress on the equipment and processing side is mandatory in the near future. Novel approaches of running an ALD process in a continuous workflow (e.g., roll-to-roll manufacturing), under ambient conditions, and at competitive throughput are only some of the targets to be tackled. It is very fascinating to contribute to this rapid development and to witness the respectable dynamics in this field.

Acknowledgments

The dedicated work of our collaborators and colleagues at the Institute of High Frequency Technology (TU Braunschweig) requires particular acknowledgment. Moreover, the fruitful cooperation with our partners from academia and industry is highly appreciated. Special thanks go to the German Research Foundation (DFG) and the German Federal Ministry of Education and Research (BMBF) for the generous funding of our research. In addition, J.M. thanks the DFG for generous support within the postdoctoral fellowship program.

References

1 Knez, M., Niesch, K., and Niinisto, L. (2007) Synthesis and surface engineering of complex nanostructures by atomic layer deposition. *Adv. Mater.*, **19** (21), 3425–3438.

2 Groner, M.D., Fabreguette, F.H., Elam, J.W., and George, S.M. (2004) Low-temperature Al_2O_3 atomic layer deposition. *Chem. Mater.*, **16** (4), 639–645.

3 Putkonen, M. and Niinisto, L. (2006) Atomic layer deposition of B_2O_3 thin films at room temperature. *Thin Solid Films*, **514** (1–2), 145–149.
4 Puurunen, R.L. (2005) Surface chemistry of atomic layer deposition: a case study for the trimethylaluminum/water process. *J. Appl. Phys.*, **97** (12), 121301.
5 Dameron, A.A., Davidson, S.D., Burton, B.B., Carcia, P.F., McLean, R.S., and George, S.M. (2008) Gas diffusion barriers on polymers using multilayers fabricated by Al_2O_3 and rapid SiO_2 atomic layer deposition. *J. Phys. Chem. C*, **112** (12), 4573–4580.
6 Carcia, P.F., McLean, R.S., Groner, M.D., Dameron, A.A., and George, S.M. (2009) Gas diffusion ultrabarriers on polymer substrates using Al_2O_3 atomic layer deposition and SiN plasma-enhanced chemical vapor deposition. *J. Appl. Phys.*, **106** (2), 023533.
7 Meyer, J., Görrn, P., Bertram, F., Hamwi, S., Winkler, T., Johannes, H.H., Weimann, T., Hinze, P., Riedl, T., and Kowalsky, W. (2009) Al_2O_3/ZrO_2 nanolaminates as ultrahigh gas-diffusion barriers: a strategy for reliable encapsulation of organic electronics. *Adv. Mater.*, **21** (18), 1845–1849.
8 Meyer, J., Schneidenbach, D., Winkler, T., Hamwi, S., Weimann, T., Hinze, P., Ammermann, S., Johannes, H.H., Riedl, T., and Kowalsky, W. (2009) Reliable thin film encapsulation for organic light emitting diodes grown by low-temperature atomic layer deposition. *Appl. Phys. Lett.*, **94** (23), 233305.
9 Groner, M.D., George, S.M., McLean, R.S., and Carcia, P.F. (2006) Gas diffusion barriers on polymers using Al_2O_3 atomic layer deposition. *Appl. Phys. Lett.*, **88** (5), 051907.
10 Jur, J.S., Spagnola, J.C., Lee, K., Gong, B., Peng, Q., and Parsons, G.N. (2010) Temperature-dependent subsurface growth during atomic layer deposition on polypropylene and cellulose fibers. *Langmuir*, **26** (11), 8239–8244.
11 Spagnola, J.C., Gong, B., Arvidson, S.A., Jur, J.S., Khan, S.A., and Parsons, G.N. (2010) Surface and sub-surface reactions during low temperature aluminium oxide atomic layer deposition on fiber-forming polymers. *J. Mater. Chem.*, **20** (20), 4213–4222.
12 Zhang, X.Y., Zhao, J., Whitney, A.V., Elam, J.W., and Van Duyne, R.P. (2006) Ultrastable substrates for surface-enhanced Raman spectroscopy: Al_2O_3 overlayers fabricated by atomic layer deposition yield improved anthrax biomarker detection. *J. Am. Chem. Soc.*, **128** (31), 10304–10309.
13 King, J.S., Wittstock, A., Biener, J., Kucheyev, S.O., Wang, Y.M., Baumann, T.F., Giri, S.K., Hamza, A.V., Baeumer, M., and Bent, S.F. (2008) Ultralow loading Pt nanocatalysts prepared by atomic layer deposition on carbon aerogels. *Nano Lett.*, **8** (8), 2405–2409.
14 Hyde, G.K., McCullen, S.D., Jeon, S., Stewart, S.M., Jeon, H., Loboa, E.G., and Parsons, G.N. (2009) Atomic layer deposition and biocompatibility of titanium nitride nano-coatings on cellulose fiber substrates. *Biomed. Mater.*, **4** (2), 025001.
15 Kemell, M., Pore, V., Ritala, M., Leskelä, M., and Linden, M. (2005) Atomic layer deposition in nanometer-level replication of cellulosic substances and preparation of photocatalytic TiO_2/cellulose composites. *J. Am. Chem. Soc.*, **127** (41), 14178–14179.
16 Knez, M., Kadri, A., Wege, C., Gösele, U., Jeske, H., and Nielsch, K. (2006) Atomic layer deposition on biological macromolecules: metal oxide coating of tobacco mosaic virus and ferritin. *Nano Lett.*, **6** (6), 1172–1177.
17 Lee, S.M., Pippel, E., Gösele, U., Dresbach, C., Qin, Y., Chandran, C.V., Brauniger, T., Hause, G., and Knez, M. (2009) Greatly increased toughness of infiltrated spider silk. *Science*, **324** (5926), 488–492.
18 Potscavage, W.J., Yoo, S., Domercq, B., and Kippelen, B. (2007) Encapsulation of pentacene/C-60 organic solar cells with Al_2O_3 deposited by atomic layer deposition. *Appl. Phys. Lett.*, **90** (25), 253511.
19 Park, S.H.K., Oh, J., Hwang, C.S., Lee, J.I., Yang, Y.S., and Chu, H.Y. (2005) Ultrathin film encapsulation of an OLED by ALD. *Electrochem. Solid-State Lett.*, **8** (2), H21–H23.
20 Hoivik, N.D., Elam, J.W., Linderman, R.J., Bright, V.M., George, S.M., and Lee, Y.C.

(2003) Atomic layer deposited protective coatings for micro-electromechanical systems. *Sens. Actuators A*, **103** (1–2), 100–108.

21 Lee, S.M., Grass, G., Kim, G.M., Dresbach, C., Zhang, L.B., Gösele, U., and Knez, M. (2009) Low-temperature ZnO atomic layer deposition on biotemplates: flexible photocatalytic ZnO structures from eggshell membranes. *Phys. Chem. Chem. Phys.*, **11** (19), 3608–3614.

22 Lujala, V., Skarp, J., Tammenmaa, M., and Suntola, T. (1994) Atomic layer epitaxy growth of doped zinc oxide thin films from organometals. *Appl. Surf. Sci.*, **82–83**, 34–40.

23 Hausmann, D.M., Kim, E., Becker, J., and Gordon, R.G. (2002) Atomic layer deposition of hafnium and zirconium oxides using metal amide precursors. *Chem. Mater.*, **14** (10), 4350–4358.

24 Kukli, K., Ritala, M., Leskelä, M., and Jokinen, J. (1997) Atomic layer epitaxy growth of aluminum oxide thin films from a novel $Al(CH_3)_2Cl$ precursor and H_2O. *J. Vac. Sci. Technol. A*, **15**, 2214–2218.

25 Dillon, A.C., Ott, A.W., Way, J.D., and George, S.M. (1995) Surface chemistry of Al_2O_3 deposition using $Al(CH_3)_3$ and H_2O in a binary reaction sequence. *Surf. Sci.*, **322**, 230–242.

26 Fabreguette, F.H., Wind, R.A., and George, S.M. (2006) Ultrahigh X-ray reflectivity from W/Al_2O_3 multilayers fabricated using atomic layer deposition. *Appl. Phys. Lett.*, **88** (1), 013116.

27 Szeghalmi, A., Senz, S., Bretschneider, M., Gösele, U., and Knez, M. (2009) All dielectric hard X-ray mirror by atomic layer deposition. *Appl. Phys. Lett.*, **94** (13), 133111.

28 Klaus, J.W., Sneh, O., and George, S.M. (1997) Growth of SiO_2 at room temperature with the use of catalyzed sequential half-reactions. *Science*, **278**, 1934–1936.

29 Meyer, J., Görrn, P., Hamwi, S., Johannes, H.H., Riedl, T., and Kowalsky, W. (2008) Indium-free transparent organic light emitting diodes with Al doped ZnO electrodes grown by atomic layer and pulsed laser deposition. *Appl. Phys. Lett.*, **93** (7), 073308.

30 Sang, B., Yamada, A., and Konagai, M. (1997) Growth of boron-doped ZnO thin films by atomic layer deposition. *Sol. Energy Mater. Sol. Cells*, **49** (1–4), 19–26.

31 King, D.M., Liang, X., Li, P., and Weimer, A.W. (2008) Low-temperature atomic layer deposition of ZnO films on particles in a fluidized bed reactor. *Thin Solid Films*, **516**, 8517–8523.

32 Lim, J.W. and Yun, S.Y. (2004) Electrical properties of alumina films by plasma-enhanced atomic layer deposition. *Electrochem. Solid-State Lett.*, **7**, F45–F48.

33 Rowlette, P.C., Allen, C.G., Bromley, O.B., Dubetz, A.E., and Wolden, C.A. (2009) Plasma-enhanced atomic layer deposition of semiconductor grade ZnO using dimethyl zinc. *Chem. Vapor Depos.*, **15** (1–3), 15–20.

34 Rowlette, P.C., Allen, C.G., Bromley, O.B., and Wolden, C.A. (2009) Self-limiting deposition of semiconducting ZnO by pulsed plasma-enhanced chemical vapor deposition. *J. Vac. Sci. Technol. A*, **27** (4), 761–766.

35 Senkevich, J.J., Tang, F., Rogers, D., Drotar, J., Jezewski, C., Lanford, W.A., Wang, G.-C., and Lu, T.-M. (2003) Substrate-independent palladium atomic layer deposition. *Chem. Vapor Depos.*, **9**, 258–264.

36 Eyck, G.A.T., Senkevich, J.J., Tang, F., Liu, D., Pimanpang, S., Karaback, T., Wang, G.-C., Lu, T.-M., Jezewski, C., and Lanford, W.A. (2005) Plasma-assisted atomic layer deposition of palladium. *Chem. Vapor Depos.*, **11**, 60–66.

37 Kim, H. and Rossnagel, S.M. (2003) Plasma-enhanced atomic layer deposition of tantalum thin films: the growth and film properties. *Thin Solid Films*, **441**, 311–316.

38 Eyck, G.A.T., Pimanpang, S., Bakhru, H., Lu, T.-M., and Wang, G.-C. (2006) Atomic layer deposition of Pd on an oxidized metal substrate. *Chem. Vapor Depos.*, **12**, 290–294.

39 Knoops, H.C.M., Mackus, A.J.M., Donders, M.E., van de Sanden, M.C.M., Notten, P.H.L., and Kessels, W.M.M. (2009) Remote plasma ALD of platinum and platinum oxide films. *Electrochem. Solid-State Lett.*, **12**, G34–G36.

40 Rossnagel, S.M., Sherman, A., and Turner, F. (2000) Plasma-enhanced atomic layer

deposition of Ta and Ti for interconnect diffusion barriers. *J. Vac. Sci. Technol. A*, **18**, 2016–2020.

41 Peng, Q., Sun, X.-Y., Spagnola, J.C., Hyde, G.K., Spontak, R.J., and Parsons, G.N. (2007) Atomic layer deposition on electrospun polymer fibers as a direct route to Al_2O_3 microtubes with precise wall thickness control. *Nano Lett.*, **7**, 719–722.

42 Kim, S.K., Lee, S.W., Hwang, C.S., Min, Y.-S., Won, J.Y., and Jeong, J. (2006) Low temperature ($<100\,^{\circ}C$) deposition of aluminum oxide thin films by ALD with O_3 as oxidant. *J. Electrochem. Soc.*, **153**, F69–F76.

43 Langereis, E., Creatore, M., Heil, S.B.S., van de Sanden, M.C.M., and Kessels, W.M.M. (2006) Plasma-assisted atomic layer deposition of Al_2O_3 moisture permeation barriers on polymers. *Appl. Phys. Lett.*, **89**, 081915.

44 Luo, Y., Slater, D., Han, M., Moryl, J., and Osgood, R.M.J. (1997) Low-temperature, chemically driven atomic-layer epitaxy: *in situ* monitored growth of CdS/ZnSe (100). *Appl. Phys. Lett.*, **71**, 3799–3801.

45 Lu, Y., Bangsaruntip, S., Wang, X., Zhang, L., Nishi, Y., and Dai, H. (2006) DNA functionalization of carbon nanotubes for ultrathin atomic layer deposition of high κ dielectrics for nanotube transistors with 60 mV/decade switching. *J. Am. Chem. Soc.*, **128**, 3518–3519.

46 Gasser, W., Uchida, Y., and Matsumura, M. (1994) Quasi-monolayer deposition of silicon dioxide. *Thin Solid Films*, **250**, 213–218.

47 Elam, J.W., Baker, D.A., Hryn, A.J., Martinson, A.B.F., Pellin, M.J., and Hupp, J.T. (2008) Atomic layer deposition of tin oxide films using tetrakis(dimethylamino) tin. *J. Vac. Sci. Technol. A*, **26** (2), 244–252.

48 Heo, J., Hock, A., and Gordon, R.G. (2010) Low temperature atomic layer deposition of tin oxide. *Chem. Mater.*, **22** (17), 4964–4973.

49 Kukli, K., Aarik, J., Aidla, A., Kohan, O., Uustare, T., and Sammelselg, V. (1995) Properties of tantalum oxide thin films grown by atomic layer deposition. *Thin Solid Films*, **260**, 135–142.

50 Heil, S.B.S., Roozeboom, F., van de Sanden, M.C.M., and Kessels, W.M.M. (2008) Plasma-assisted atomic layer deposition of Ta_2O_5 from alkylamide precursor and remote O_2 plasma. *J. Vac. Sci. Technol. A*, **26**, 472–480.

51 Kim, H. and Rossnagel, S.M. (2002) Growth kinetics and initial stage growth during plasma-enhanced Ti atomic layer deposition. *J. Vac. Sci. Technol. A*, **20**, 802–808.

52 Knez, M., Kadri, A., Wege, C., Gösele, U., Jeske, H., and Nielsch, K. (2006) Atomic layer deposition on biological macromolecules: metal oxide coating of tobacco mosaic virus and ferritin. *Nano Lett.*, **6**, 1172–1177.

53 Aarik, J., Aidla, A., Uustare, T., and Sammelselg, V. (1995) Morphology and structure of TiO_2 thin films grown by atomic layer deposition. *J. Cryst. Growth*, **148**, 268–275.

54 Keranen, J., Guimon, C., Iiskola, E., Auroux, A., and Niinisto, L. (2003) Surface-controlled gas-phase deposition and characterization of highly dispersed vanadia on silica. *J. Phys. Chem. B*, **107**, 10773–10784.

55 Guziewicz, E., Kowalik, I.A., Godlewski, M., Kopalko, K., Osinniy, V., Wójcik, A., Yatsunenko, S., Łusakowska, E., Paszkowicz, W., and Guziewicz, M. (2008) Extremely low temperature growth of ZnO by atomic layer deposition. *J. Appl. Phys.*, **103**, 033515.

56 Elam, J.W., Groner, M.D., and George, S.M. (2002) Viscous flow reactor with quartz crystal microbalance for thin film growth by atomic layer deposition. *Rev. Sci. Instrum.*, **73** (8), 2981–2987.

57 Levy, D.H., Freeman, D., Nelson, S.F., Cowdery-Corvan, P.J. and Irving, L.M. (2008) Stable ZnO thin film transistors by fast open air atomic layer deposition. *Appl. Phys. Lett.*, **92** (19), 192101.

58 Levy, D.H., Nelson, S. F. and Freeman, D. (2009) Oxide electronics by spatial atomic layer deposition. *J. Display Technol.*, **5** (12), 484–494.

59 Chen, C.Y. and Kanicki, J. (1996) High field-effect-mobility a-Si:H TFT based on high deposition-rate PECVD materials. *IEEE Electron Device Lett.*, **17** (9), 437–439.

60 Sun, J., Mourey, D.A., Zhao, D.L., Park, S.K., Nelson, S.F., Levy, D.H., Freeman, D., Cowdery-Corvan, P., Tutt, L., and Jackson, T.N. (2008) ZnO thin-film

transistor ring oscillators with 31-ns propagation delay. *IEEE Electron Device Lett.*, **29** (7), 721–723.

61 Görrn, P., Sander, M., Meyer, J., Kroger, M., Becker, E., Johannes, H.H., Kowalsky, W., and Riedl, T. (2006) Towards see-through displays: fully transparent thin-film transistors driving transparent organic light-emitting diodes. *Adv. Mater.*, **18** (6), 738–741.

62 Erlat, A.G., Breitung, E.M., and Heller, C.M.A. (2007) Systems and methods for roll-to-roll atomic layer deposition on continuously fed objects. Proceedings of US2007281089A1.

63 Dickey, E.R. and Barrow, W.A. (2007) Atomic layer deposition system and method for coating flexible substrates. Proceedings of US2007224348 (A1).

64 Charton, C., Schiller, N., Fahland, M., Holländer, A., Wedel, A., and Noller, K. (2006) Development of high barrier films on flexible polymer substrates. *Thin Solid Films*, **502**, 99–103.

65 Sobrinho, A.S.d.S., Czeremuszkin, G., Latréche, M., and Wertheimer, M.R. (2000) Defect–permeation correlation for ultrathin transparent barrier coatings on polymers. *J. Vac. Sci. Technol. A*, **18**, 149–157.

66 Keneshea, F.J. and Douglass, D.L. (1971) The diffusion of oxygen in zirconia as a function of oxygen pressure. *Oxid. Met.*, **3**, 1–14.

67 Meyer, J., Schmidt, H., Kowalsky, W., Riedl, T., and Kahn, A. (2010) The origin of low water vapor transmission rates through Al_2O_3/ZrO_2 nanolaminate gas-diffusion barriers grown by atomic layer deposition. *Appl. Phys. Lett.*, **96**, 243308.

68 Nathan, A. and Chalamala, B.R. (2005) Special issue on flexible electronics technology. Part 1. Systems and applications. *Proc. IEEE*, **93** (7), 1235–1238.

69 Miller, D.C., Foster, R.R., Zhang, Y.D., Jen, S.H., Bertrand, J.A., Lu, Z.X., Seghete, D., O'Patchen, J.L., Yang, R.G., Lee, Y.C., George, S.M., and Dunn, M.L. (2009) The mechanical robustness of atomic-layer- and molecular-layer-deposited coatings on polymer substrates. *J. Appl. Phys.*, **105** (9), 093527.

70 Dameron, A.A., Seghete, D., Burton, B.B., Davidson, S.D., Cavanagh, A.S., Bertrand, J.A., and George, S.M. (2008) Molecular layer deposition of alucone polymer films using trimethylaluminum and ethylene glycol. *Chem. Mater.*, **20** (10), 3315–3326.

71 George, S.M., Yoon, B., and Dameron, A.A. (2009) Surface chemistry for molecular layer deposition of organic and hybrid organic–inorganic polymers. *Acc. Chem. Res.*, **42** (4), 498–508.

72 Du, Y. and George, S.M. (2007) Molecular layer deposition of nylon 66 films examined using *in situ* FTIR spectroscopy. *J. Phys. Chem. C*, **111** (24), 8509–8517.

73 Lee, J.S., Lee, Y.J., Tae, E.L., Park, Y.S., and Yoon, K.B. (2003) Synthesis of zeolite as ordered multicrystal arrays. *Science*, **301** (5634), 818–821.

74 Burrows, P.E., Bulovic, V., Forrest, S.R., Sapochak, L.S., McCarty, D.M., and Thompson, M.E. (1994) Reliability and degradation of organic light emitting devices. *Appl. Phys. Lett.*, **65**, 2922–2924.

75 Chang, C.-Y., Tsai, F.-Y., Jhuo, S.-J., and Chen, M.-J. (2008) Enhanced OLED performance upon photolithographic patterning by using an atomic-layer-deposited buffer layer. *Org. Electron.*, **9**, 667–672.

76 Ghosh, A.P., Gerenser, L.J., Jarman, C.M., and Fornalik, J.E. (2005) Thin-film encapsulation of organic light-emitting devices. *Appl. Phys. Lett.*, **86**, 223503–223501.

77 Schaer, M., Nuesch, F., Berner, D., Leo, W., and Zuppiroli, L. (2001) Water vapor and oxygen degradation mechanisms in organic light emitting diodes. *Adv. Funct. Mater.*, **11** (2), 116–121.

78 Féry, C., Racine, B., Vaufrey, D., Doyeux, H., and Cinà, S. (2005) Physical mechanism responsible for the stretched exponential decay behavior of aging organic light-emitting diodes. *Appl. Phys. Lett.*, **87**, 213502–213501.

79 Görrn, P., Holzer, P., Riedl, T., Kowalsky, W., Wang, J., Weimann, T., Hinze, P., and Kipp, S. (2007) Stability of transparent zinc tin oxide transistors under bias stress. *Appl. Phys. Lett.*, **90** (6), 063502.

80 Meyer, J., Winkler, T., Hamwi, S., Schmale, S., Johannes, H.H., Weimann, T., Hinze, P., Kowlasky, W., and Riedl, T. (2008) Transparent inverted organic light-

emitting diodes with a tungsten oxide buffer layer. *Adv. Mater.*, **20** (20), 3839–3843.

81 Riedl, T., Görrn, P., and Kowalsky, W. (2009) Transparent electronics for see-through AMOLED displays. *J. Display Technol.*, **5** (12), 501–508.

82 Liang, Y., Xu, Y., Xia, J., Tsai, S.-T., Wu, Y., Li, G., Ray, C., and Yu, L. (2010) For the bright future: bulk heterojunction polymer solar cells with power conversion efficiency of 7.4%. *Adv. Mater.*, **22** (20), E135–E138.

83 Dennler, G., Lungenschmied, C., Neugebauer, H., Sariciftci, N.S., Latreche, M., Czeremuszkin, G., and Wertheimer, M.R. (2006) A new encapsulation solution for flexible organic solar cells. *Thin Solid Films*, **511**, 349–353.

84 Li, X., Yang, H., Li, C.S., Xu, L.M., Zhang, Z.G., and Kim, D.H. (2008) Effects of additives on the morphologies of thin titania films from self-assembly of a block copolymer. *Polymer*, **49** (5), 1376–1384.

85 Kim, N., Potscavage, W.J., Domercq, B., Kippelen, B., and Graham, S. (2009) A hybrid encapsulation method for organic electronics. *Appl. Phys. Lett.*, **94** (16), 163308.

7
Plasma Atomic Layer Deposition

Erwin Kessels, Harald Profijt, Stephen Potts, and Richard van de Sanden

7.1
Introduction

Plasma ALD (atomic layer deposition) is an energy-enhanced ALD method that is rapidly gaining in popularity. In plasma ALD, also referred to as plasma-enhanced ALD (PEALD), plasma-assisted ALD, and in some cases radical-enhanced ALD, the surface is exposed to the species generated by a plasma during the coreactant step. See Figure 7.1 for a schematic illustration of a plasma ALD cycle. Typical plasmas used are those generated in O_2, N_2, and H_2 reactant gases or combinations thereof. Such plasmas can replace ligand exchange reactions typical of H_2O or NH_3, and they can be employed to deposit metal oxides, metal nitrides, and metal films. Moreover, plasmas generated in H_2O and NH_3 have been reported, for which there can be a combination of plasma and thermal ALD surface reactions taking place at the same time.

Plasma ALD offers several merits for the deposition of ultrathin films over other techniques, such as chemical vapor deposition (CVD) or physical vapor deposition (PVD). The high reactivity of the plasma species on the deposition surface during the plasma ALD process allows for more freedom in processing conditions and for a wider range of material properties. These ideas will be addressed in detail later in this chapter and are the primary reason why the interest in plasma ALD has increased rapidly in recent years. This interest has also been catalyzed by the many new applications of ALD that are emerging in and outside the semiconductor industry: the key driver of the field of ALD. Several non-semiconductor applications have set new requirements for the ALD parameter space, which cannot always be satisfied easily by a pure thermal ALD process.

The increasing popularity of plasma ALD is manifested by the increasing number of recent publications about the topic (see Figure 7.2) and the large set of thin film materials that have been synthesized by the method (see Figure 7.3). Such is the interest and demand in the field that the number of ALD equipment manufacturers providing dedicated plasma ALD tools has increased significantly. Currently (status 2010), the companies ASM (*Emerald* (2005) and *Stellar* (2006)) [1], Oxford

Atomic Layer Deposition of Nanostructured Materials, First Edition. Edited by Nicola Pinna and Mato Knez.

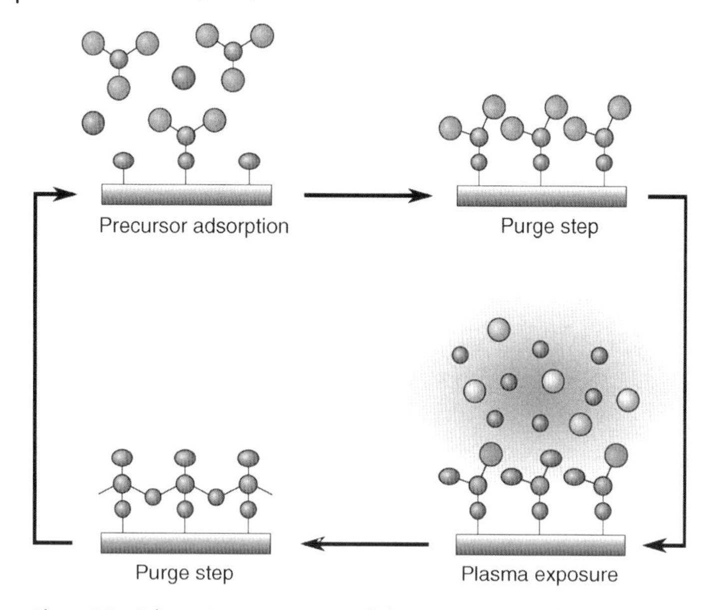

Figure 7.1 Schematic representation of plasma ALD. During the coreactant step of the cycle, the surface is exposed to species generated by a plasma.

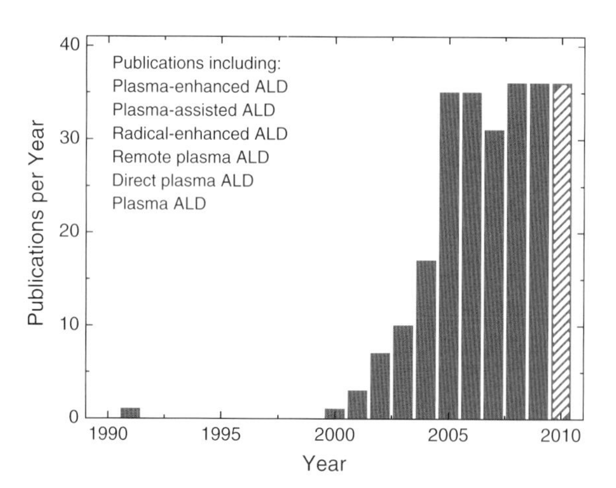

Figure 7.2 Number of publications per year on the subject of plasma ALD, between 1990 and 2010 (up to May 2010). The search was run in published abstracts using *Web of Science*®. The search terms included "plasma-enhanced ALD," "plasma-assisted ALD," "radical-enhanced ALD," "remote plasma ALD," "direct plasma ALD," and "plasma ALD." Although not found using these search terms, the first report of plasma ALD by De Keijser and Van Opdorp (Philips Research Laboratories, Eindhoven), published in 1991, is also included.

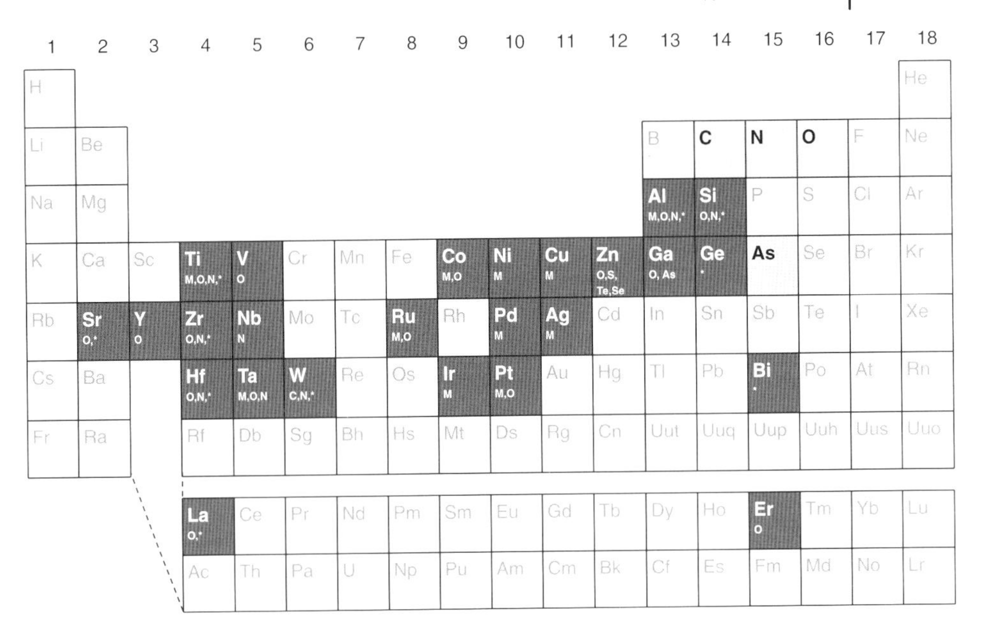

Figure 7.3 Periodic table showing the elements (dark background) that have been part of films synthesized by plasma ALD. "M" means that the pure elements were deposited; compounds with oxygen, nitrogen, and other elements are indicated by their symbols ("O", "N", etc.). Asterisks denote ternary or other compounds.

Instruments (*FlexAL* (2006) and *OpAL* (2008)) [2], Beneq (*TFS 200* (2009)) [3], and Cambridge Nanotech (*Fiji* (2009)) [4] provide tools specifically dedicated to plasma ALD.

The first reported case of plasma ALD came to light in 1991, when De Keijser and Van Opdorp of the Philips Research Laboratories in Eindhoven, the Netherlands, published a paper on atomic layer epitaxy (ALE) of GaAs using H radicals [5]. The hydrogen radicals were generated in a remote microwave-induced plasma and transported to the deposition surface through a quartz tube (see Figure 7.4). The atomic hydrogen was used to drive the surface reactions after $Ga(CH_3)_3$ and AsH_3 pulsing at substrate temperatures below 500 °C, which is close to the onset temperature for the thermal decomposition of $Ga(CH_3)_3$. Subsequently, the method remained unexplored until the end 1990s, when the semiconductor industry became interested in ALD. Sherman filed a patent on the method in 1996 [6], after which Rossnagel *et al.* reported on plasma ALD of Ta and Ti metal films in 2000 [7]. In the latter case, the anticipated application of the technique was the deposition of Cu diffusion barriers in interconnect technology, a field already very familiar with the merits and robustness of plasma-assisted processes through the broadly applied PVD technique of sputtering. Afterward, the number of materials, processes, and applications of plasma ALD has diversified and grown rapidly.

In this chapter, different plasma ALD configurations will be described and the merits offered by the technique will be discussed. These merits will also be illustrated for several materials and applications on the basis of results that have recently been

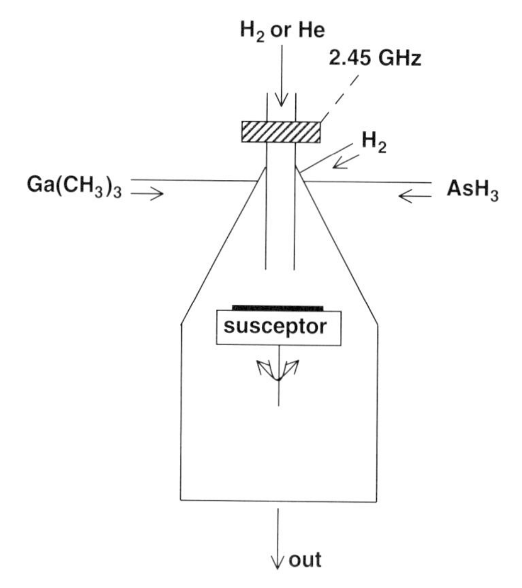

Figure 7.4 Reactor layout as used in the first plasma ALD experiments (Philips Research Laboratories, Eindhoven) reported in the literature. A H_2 plasma was generated by means of a remote microwave-induced plasma source in a quartz tube. The H radicals assisted in the ALE process of GaAs. Reprinted with permission from Ref. [5]. Copyright 1991, the American Institute of Physics.

reported. Also, the challenges that plasma ALD faces will be addressed, both in terms of limitations inherent to the use of a plasma-based process and in terms of the scale-up of the technique for industrial applications. However, in the next section, first some plasma basics will be briefly presented for the typical plasma operating conditions employed during plasma ALD. Finally, it should be emphasized that this chapter is not intended to give a complete overview of all the reports and developments related to plasma ALD over the past decade, but merely a comprehensive introduction to the method.

7.2 Plasma Basics

A plasma is a collection of free charged particles, among other gas-phase species, and is, on average, electrically neutral. This so-called quasineutrality means that at macroscopic length scales (typically >1 mm) the electron density is equal to the ion density, i.e. $n_e = n_i$, under the assumption that the negative ions can be neglected. In most plasma configurations, the free electrons are heated by electric fields to an average electron temperature, T_e, of typically 3×10^4 K (approximately 3 eV), while the gas temperature remains low ($T_g = 300$–500 K). This results in nonequilibrium conditions, caused by the relatively low gas pressure that is typically employed in processing plasmas, which, therefore, belong to the class of the so-called "cold" plasmas [8, 9]. Considering a Maxwellian or Druyvesteyn-like energy distribution, the electrons in the high-energy tail are able to ionize and dissociate the reactant gas through electron-impact collisions. This leads to the formation of ions and reactive atomic and molecular neutrals, typically referred to as "plasma radicals." Subsequently, these ions and radicals can undergo additional gas-phase reactions and they can induce surface reactions when they arrive at deposition or reactor surfaces.

Although the charged particles play a central role in sustaining the plasma, the fractional ionization or "ionization degree" of processing plasmas is very low, typically within the range of 10^{-6} to 10^{-3}. This means that the flux of electrons and ions to the deposition surface is much lower than the flux of the plasma radicals. In many cases, the surface chemistry is therefore ruled by the interaction of the plasma radicals with the surface species. However, the energy of the ions, E_{ion}, arriving at the surface can be much higher than the ion or electron temperature, as ions are accelerated within a plasma sheath at the boundary between the plasma and the substrate. For typical processing plasmas, E_{ion} is < 50 eV, although it can also be significantly lower and higher depending on the plasma operating conditions. The `plasma sheath is a thin, positive space-charge region, which develops because the electron thermal velocity is much higher than the ion thermal velocity. To make the net current to the substrate zero, an electric field develops between the plasma and the substrate, which retards the electrons and accelerates the ions. Therefore, an electropositive plasma is (time-averaged) always at a positive potential relative to any surface in contact with it. In the rudimentary case of a floating substrate, the difference between the plasma potential, V_p, and the substrate potential, V_f, is generally given by

$$V_p - V_f = \frac{T_e}{2e} + \frac{T_e}{2e} \ln\left(\frac{m_i}{2\pi m_e}\right),$$

where T_e is the electron energy in eV, and m_e and m_i are the electron and ion mass, respectively. This means that $V_p - V_f$ is typically a few multiples of T_e. However, depending on the plasma and substrate stage configuration (symmetry or asymmetry of the electrodes, grounding or biasing of electrode/substrate stage, etc.), the potential difference across a plasma sheath can also be as high as a few tens or hundreds of volts. The energy gained by the ions in the plasma sheath, and consequently whether "ion bombardment" can take place or not, also depends on the collisional mean free path of the ions and the thickness of the plasma sheath. At relatively low pressures, the ion mean free path is larger than the plasma sheath thickness, such that the ions can be accelerated over the full potential difference (i.e., $E_{ion} = e(V_p - V_f)$). An example of a distribution of ion energies for ions arriving at a substrate for an O_2 plasma under such conditions is given in Figure 7.5. At higher pressures, however, the ions collide with neutrals in the space-charge region (collisional plasma sheath) and the net energy gained by the ions can be much smaller as a result. Note also that the ions in the plasma sheath are accelerated in the direction perpendicular to the surface. This means that the flux of the ions to the surface is anisotropic with the ions having an angle of incidence around the normal to the surface.

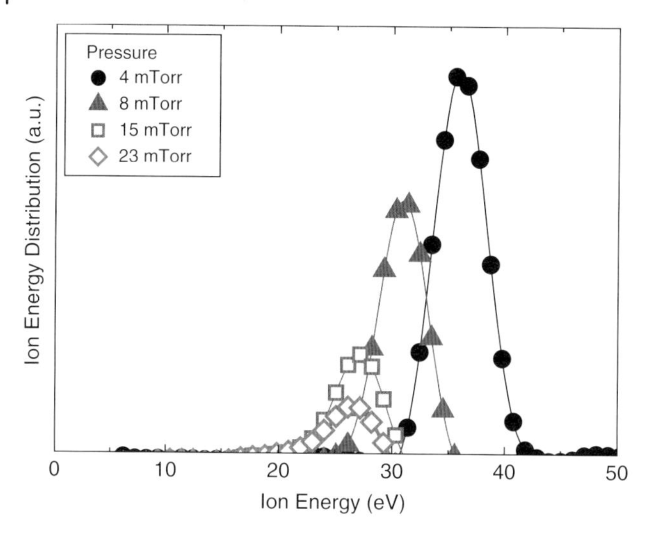

Figure 7.5 Ion energy distribution as measured by a retarding field energy analyzer (RFEA) in an O_2 plasma (plasma power: 100 W) used for remote plasma ALD. The RFEA was positioned at the substrate stage and various gas pressures were employed [10].

The key properties of the plasma step, applied during the synthesis of thin film materials by plasma ALD, are as follows:

1) The reactive species are created in the gas phase, which means that a relatively high reactivity can be provided to the deposition surface (almost) independently of the substrate conditions. The reactivity of the plasma can also be "selective" by tuning its properties and composition by carefully choosing the plasma operating conditions (gases, flows, power, pressure, etc.).
2) Typically, the plasma supplies a relatively low heat flux to the surface, despite its high reactivity. The reason is that, for cold plasmas, only the electrons are heated significantly and not the other gas-phase species.
3) Through ion bombardment, additional energy can be provided to the deposition surface [11]. This energy is locally dissipated by the surface species and can enhance surface reaction rates and processes such as surface diffusion. Possible ion–surface interactions are depicted in Figure 7.6 for typical ranges of ion energy and ion flux toward the substrate, corresponding to various plasma-assisted techniques. Moreover, the presence and level of ion bombardment can be controlled through the plasma operating conditions (mainly the gas pressure) as well as by the choice of plasma configuration and substrate (stage) conditions.

These key properties can be summarized by the phrase: plasmas can deliver a high, diverse but selective reactivity to a surface without heat, and can therefore access a parameter space in materials processing, which is unattainable with strictly chemical methods [8, 9].

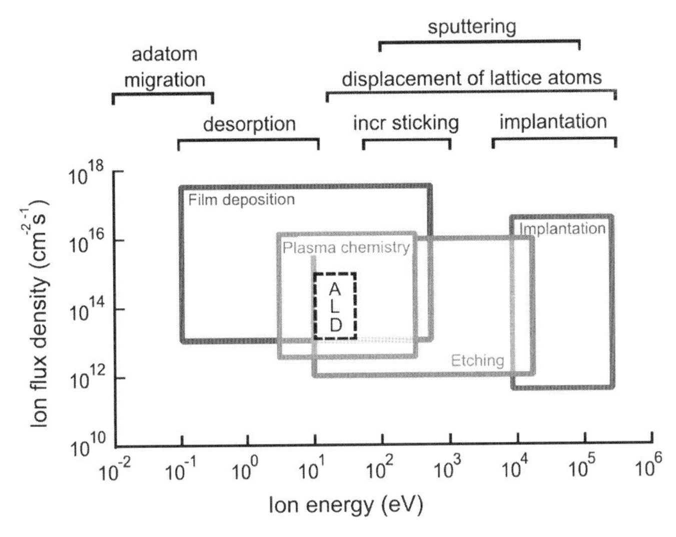

Figure 7.6 Ion–surface interactions during plasma processes displayed versus ion flux and ion energy. The typical operating windows for plasma ALD and other plasma-based processes are indicated. Reprinted with permission from Ref. [11]. Copyright 1984, the American Vacuum Society.

Other key differences between plasma ALD and thermal ALD include:

- Electron impact collisions, as well as other reactions, which lead to the excitation of atoms and molecules. This excitation can be electronic for atoms and electronic, vibrational, and rotational for molecules. When electronically excited states return to the ground state, they emit their energy as electromagnetic radiation, which can be measured using optical emission spectroscopy (OES). This excitation process accounts for the ultraviolet to visible emission by the plasma as shown in the OES spectrum of an O_2 plasma in Figure 7.7 [10], for example. The emission in the visible region gives the plasma its characteristic color and, therefore, its spectral fingerprint can be easily used to extract information about the chemical and physical processes occurring both within the plasma and at the surface. Measuring the visible emission of the plasma also provides many opportunities for plasma ALD in terms of process monitoring and optimization [12]. The emission in the ultraviolet can, however, also be sufficiently energetic to influence and induce (unfavorable) processes at surfaces or within thin films.
- The fact that the reactant species created from the reactant gas during the plasma step are mainly radicals. Apart from the ALD surface reactions, these radicals can also undergo additional reactions at the surface, even at saturated surface sites. For example, atoms can recombine on wall (and deposition) surfaces to form nonreactive molecules that desorb back into the plasma. The probability of such recombination reactions, the so-called surface recombination probability, r, can be as small as 10^{-6} and as high as 1 (see Table 7.1). The value of r has a direct impact on the density of the radicals in the plasma as it defines the surface loss term for

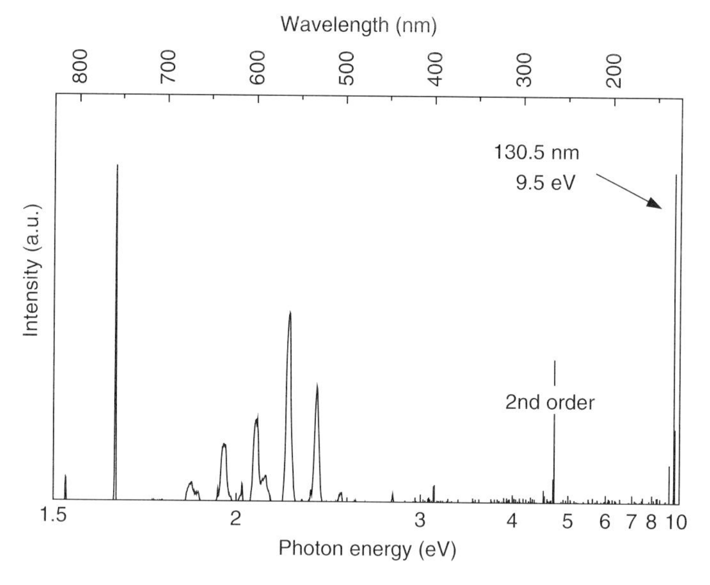

Figure 7.7 Optical emission spectrum of plasma radiation in an O_2 plasma, as used for plasma ALD (operating pressure: 8 mTorr; plasma power: 100 W). The emission in the (vacuum) ultraviolet region was measured by means of a VUV monochromator and the emission in the visible region by a simple spectrometer [10].

Table 7.1 Overview of recombination loss probabilities, *r*, for H, N, and O radicals on various materials [51].

Radical	Material	*r*
H	SiO_2	0.00004 ± 0.00003
N	SiO_2	0.0003 ± 0.0002
O	SiO_2	0.0002 ± 0.0001
H	Ti	0.35
	Al	0.29
	Ni	0.20 ± 0.09
	Cu	0.14
	Au	0.15 ± 0.05
	Pd	0.07 ± 0.015
	Pt	0.03
O	Al_2O_3	0.0021
	ZnO	0.00044
	Fe_2O_3	0.0052
	Co_3O_4	0.0049
	NiO	0.0089
	CuO	0.043

Errors are indicated where available. The values of *r* for SiO_2, H on metals, and O on oxides are taken from Refs [13–15], respectively.

Table 7.2 Densities of plasma species in an O_2 plasma, as typically used in plasma ALD processes.

Pressure (mTorr)	O_2 (cm^{-3})	O (cm^{-3})	O_2^* (cm^{-3})	O^* (cm^{-3})	O_2^+ (cm^{-3})	O^+ (cm^{-3})	O^- (cm^{-3})	n_e (cm^{-3})	T_e (eV)	E_{ion} (eV)
10	3×10^{14}	7×10^{13}	4×10^{13}	4×10^{12}	5×10^{10}	4×10^{10}	2×10^{10}	7×10^{10}	2.8	15.3
100	3×10^{15}	1×10^{14}	3×10^{14}	5×10^{10}	4×10^{10}	1×10^{9}	3×10^{10}	2×10^{10}	2.1	10.8

Data are presented for two different pressures, and the electron temperature, T_e, and energy, E_{ion}, of ions accelerated to the (grounded) substrate are also given. The data have been compiled from the modeling results described in Ref. [16] for an inductively coupled plasma operated at a source power of 500 W. The excited species O^* and O_2^* correspond to the lowest metastable states being O (^{1}D) and O_2 ($a^1\Delta_g$), respectively.

the radicals. Moreover, a relatively high r can also significantly reduce the flux of radicals in trenches or other high aspect ratio structures on the substrate, for which the radicals have to undergo multiple wall collisions to reach deep inside the structures.

- The presence of a multitude of gas-phase and surface species, which makes it not possible to identify single reactant species solely responsible for the surface reactions. For example, when admixing two reactant gases in the plasma, new molecules (and related radicals) can be formed through gas-phase or surface recombination reactions. Furthermore, volatile products from the ALD reactions can be excited, ionized, and dissociated by the plasma when leaving the surface. All of these species can contribute to the ALD surface chemistry adding to its complexity.

To illustrate which species are typically present in a plasma as well as their density, the ion energy, and other plasma parameters, an overview is given in Table 7.2 for an O_2 plasma [16]. Data are given for two operating pressures for an inductively coupled plasma (ICP), as typically employed for remote plasma ALD described in the next section.

7.3 Plasma ALD Configurations

Several equipment configurations exist for assisting an ALD process by means of a plasma step [17]. In the first configuration, a plasma generator is fitted to a thermal ALD reactor, see Figure 7.8a. Examples of such plasma sources are microwave surfatron systems [18] and the radio frequency-driven *R* Evolution* system of MKS Instruments, [19] which are also commonly used for plasma-based reactor cleaning. Due to technical constraints on existing ALD reactors, plasma generation typically takes place at a relatively far distance from ALD reaction zone. Consequently, the plasma species have to flow through the reactor tubing between the plasma source and reaction chamber. This results in many surface collisions, where ions and

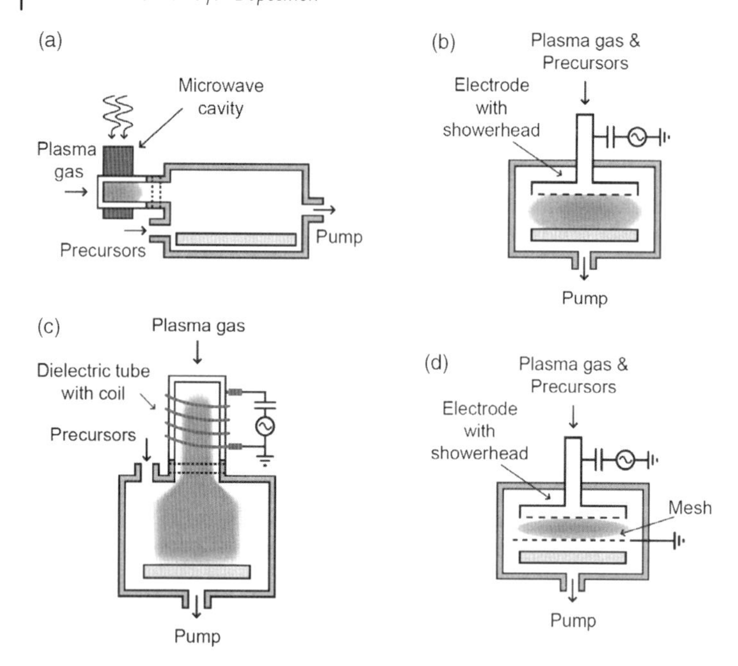

Figure 7.8 Various reactor configurations for plasma ALD: (a) radical-enhanced ALD; (b) direct plasma ALD; (c) remote plasma ALD; and (d) direct plasma reactor with mesh. The reactor layouts and plasma sources shown serve only as examples.

electrons are lost before reaching the substrate due to their recombination at surfaces. The method is, therefore, typically referred to as radical-enhanced ALD. The many surface collisions of the plasma species can, however, also significantly reduce the flux of radicals arriving at the substrate. This is especially prominent when the choice of the inner surface of the tubing material is not harmonized with the plasma radicals to reduce surface recombination. For example, H radicals have a relatively low surface recombination probability on quartz surfaces but a very high recombination probability on most metals (see Table 7.1). In the case of metallic surfaces, very long radical exposure times might be necessary to reach saturation of the reactant step in the ALD cycle.

The second configuration stems directly from the field of plasma-enhanced chemical vapor deposition (PECVD). In this case, a capacitively coupled plasma is generated at radio frequency (typically 13.56 MHz) between two parallel electrodes in the so-called RF parallel plate or RF diode reactor (see Figure 7.8b). In this case, typically one electrode is powered while the other is grounded. Generally, the substrate is positioned on the grounded electrode and the gases are introduced into the reactor either through a showerhead in the powered electrode (showerhead-type) or from the side of the electrodes (flow-type). As such, this configuration of ALD reactors is referred to as "direct plasma ALD", because the wafer is directly positioned at one of the electrodes that contributes to plasma generation. The ALD reactors provided by ASM (*Emerald* and *Stellar*) [1] and Beneq (*TFS 200*) [3], for example, can be classified as this type. Typical operating pressures used during the plasma step in direct plasma ALD are on the order of 1 Torr, although these could be < 100 mTorr for an RF parallel plate reactor. During direct plasma ALD, the fluxes of plasma radicals and ions toward the deposition surface can be very high, as the plasma species are created in very close proximity of the substrate surface. In principle, this enables short plasma exposure steps, something that is also facilitated by the relatively simple reactor layout. On the other hand, the voltage over the plasma sheath can be quite significant such that the energy of the ions arriving on the substrate can be substantial, depending on the operating pressure. The ion energy can be reduced by working at high pressures, but this can compromise on the plasma uniformity. In addition, the emission of high energy photons arriving at the deposition surface can be significant, possibly leading to plasma damage, for example, by ultraviolet photons.

A third configuration for plasma ALD equipment can be classified as "remote plasma ALD". In this case, as its name implies, the plasma source is located remotely from the substrate stage such that substrate is not involved in the generation of the plasma species (see Figure 7.8c). This configuration can be distinguished from radical-enhanced ALD by the fact that the plasma is still present above the deposition surface, i.e. the electron and ion densities have not decreased to zero. The "downstream" plasma can be of the afterglow type (where the local T_e is too low to be ionizing) or it can still be active (ionizing). The flux of the radicals toward the substrate can therefore be much higher than that for radical-enhanced ALD. Moreover, under these circumstances, the plasma and substrate conditions can be varied (relatively) independently of each other, something that is not the case for direct plasma ALD. For example, in direct plasma ALD a change in substrate temperature affects the gas temperature and consequently the density of gas-phase species and the generation of plasma species. The remote nature of the remote plasma ALD configuration therefore allows for more control of the plasma's composition and properties compared to direct plasma ALD. The plasma properties can be optimized relatively easily by tuning the operating conditions of the plasma source and the downstream conditions at the position of the substrate. This holds specifically for the presence of ion bombardment and the influence of plasma radiation. A variety of plasma sources can be employed for remote plasma ALD, including microwave plasmas [18], electron cyclotron resonance (ECR) plasmas [20], and RF-driven inductively coupled plasmas [7]. The last type is the most popular and it has proven itself extensively as a plasma source in plasma etching. In the Oxford Instruments *FlexAL* and *OpAL* reactors [2], a multiturn cylindrical coil surrounds an alumina discharge tube, while in the Cambridge Nanotech *Fiji* system [4] the inductor surrounds a quartz tube. The operating pressure for these ICP systems can be as high as 1 Torr, but, in many cases, pressures < 100 mTorr are common. For such low pressures, which are atypical for (thermal) ALD, the reactors are equipped with turbomolecular pumps. An alternative approach is provided by Beneq in their *TFS 200* system [3]. Although this is, in principle, a direct ALD reactor employing a capacitively coupled plasma, a grid can be placed between the two electrodes creating the so-called triode configuration (see Figure 7.8d). This enables confinement of the plasma between the driven electrode and the grid such that the deposition surface is not in contact with the active plasma. In essence, this leads to a lower ion energy and flux towards the deposition surface but also to a somewhat reduced radical density, despite the substrate still being subject to plasma radiation.

7.4 Merits of Plasma ALD

The use of plasma ALD for ultrathin film synthesis has several potential advantages, in addition to the benefits provided by the ALD technique itself. These merits can be useful for specific applications of ALD-synthesized thin films:

1) **Improved material properties**: There have been reports that for some materials and applications, plasma ALD affords better material properties than thermal ALD in terms of, for example, film density, impurity content, and electronic properties. In most cases, these improved material properties are a result of the high reactivity provided by the plasma, which will be addressed in more detail below. However, more specifically, this improvement can often be attributed to kinetically driven, selective ALD surface reactions, for example, the abstraction of surface halogen atoms by H radicals and several ion-assisted surface reactions, as illustrated in Figure 7.6.
2) **Deposition at reduced substrate temperatures**: As high reactivity is delivered to the deposition surface by the plasma species, less thermal energy is required at the substrate to drive the ALD surface chemistry. This means that it is possible to deposit films with equivalent material properties at lower substrate temperatures than for thermal ALD. The reactivity delivered by the plasma species is not only provided by reactive plasma radicals but is also determined by the kinetic energy of the ions accelerated in the plasma sheath, the surface-recombination energy of the ions and other species, and the energy flux caused by the plasma radiation.
3) **Increased choice of precursors and materials**: The fact that reactive plasma radicals are delivered to the deposition surface allows for the use of precursors with relatively high thermal and chemical stabilities. A classic example is the ALD of metal oxides from β-diketonate precursors, such as those with thd ligands (thd $= \eta^2$-$[(H_3C)_3CC{=}O]_2CH$; 2,2,6,6,-tetramethyl-3,5-heptanedionato). Such precursors require more reactive coreactants as they show no or low reactivity with H_2O (in essence, they do not readily undergo hydrolysis reactions). Although the use of O_3 might be a solution in such cases, plasma ALD processes can give even more reactivity. The plasma-assisted method can also yield solutions when depositing non-oxidic materials such as metal nitrides and metals. In addition to a wider choice of precursors, plasma ALD processes also enable the deposition of more material systems, for example, of the elemental metals Ti and Ta. For these metals, no (thermal ALD) reactant is

available to reduce the precursor [7, 21]. Furthermore, plasma ALD allows for a wider choice of substrate materials to be used, particularly those which are temperature sensitive.

4) **Good control of stoichiometry and film composition**: Nonthermally driven reactions can be induced at the deposition surface due to the nonequilibrium conditions in the plasma, which enables better control of the ALD surface chemistry and of the species incorporated into the film. Therefore, the use of a plasma provides additional variables with which to tune the stoichiometry and composition of the films. These include the operating pressure, plasma power, plasma exposure time, and the admixing of additional gases to the plasma. It is, for example, relatively straightforward to incorporate low concentrations of N atoms into oxide thin films by the addition of N_2 to a plasma generated in O_2. Hot electrons in the high-energy tail of the electron energy distribution are able to dissociate the strongly bonded N_2 molecules by electron-impact collisions. Such controlled doping of thin film materials is difficult to achieve with strictly thermally driven ALD reactions.
5) **Increased growth rate**: Cases exist in which the plasma species create a higher density of reactive surface sites due to the high reactivity of the plasma. Consequently, this can lead to higher growth per cycle values. Moreover, the plasma can be switched on and off very rapidly, which enables fast pulsing of the plasma reactant species and reduced purge times. The latter is especially important for the ALD of metal oxides at low temperatures (room temperature to 150 °C), where purging of H_2O, in the case of thermal ALD, requires excessively long purge times and therefore long cycle times [22]. Shorter cycle times have a significant impact on the net growth rate (thickness of material deposited within a set time) and throughput of an ALD reactor. The high plasma reactivity is also beneficial in particular cases where the nucleation delay is shorter for plasma ALD than for the equivalent thermal ALD process. This aspect also contributes to an increased throughput of ALD reactors.
6) **More processing versatility in general**: The availability of a plasma source on an ALD reactor allows for several other *in situ* treatments of the deposition surface, deposited films, and reactor walls. Plasmas can be used for *in situ* substrate modification (e.g. oxidation by an O_2 plasma and nitridation by NH_3 or N_2 plasmas), substrate cleaning and pretreatment (e.g. by a H_2 plasma), post-deposition treatments (e.g. film nitridation), reactor wall conditioning, and reactor wall cleaning. For example, a layer of TiN covering the walls of the reactor can be easily removed by running a F-based plasma such as one generated in NF_3 or SF_6 [17].

The aforementioned merits of plasma ALD can be illustrated by several results that have been obtained for various material systems in the recent years.

A classic example is the remote plasma ALD of Al_2O_3 from $Al(CH_3)_3$ and an O_2 plasma carried out in two ALD reactors, which differ in operating pressure by a factor of ~10 [23]. For both reactors, the process and material properties were compared directly with those for thermal ALD from $Al(CH_3)_3$ and H_2O. The results in Figure 7.9

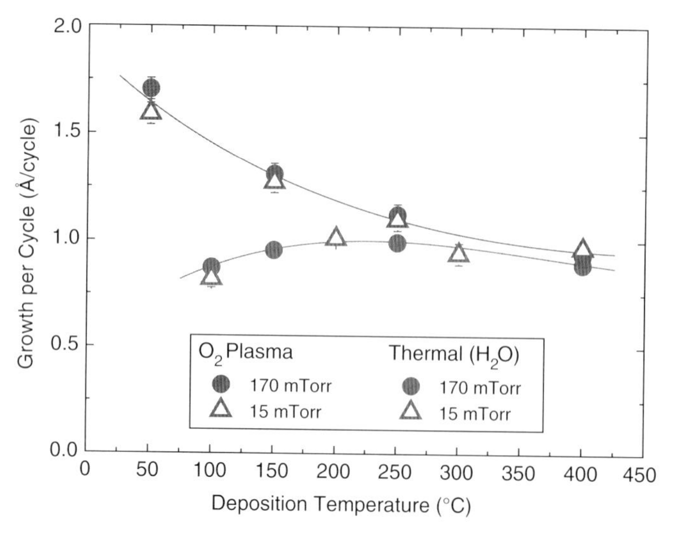

Figure 7.9 Growth per cycle of Al_2O_3 films as a function of the substrate temperature. The films were deposited by plasma ALD (O_2 plasma) and thermal ALD (H_2O). Two different ALD reactors were used, one operating at 15 mTorr and the other at 170 mTorr [23, 34].

show that the Al_2O_3 films can be deposited by plasma ALD at substrate temperatures between 25 and 400 °C, with the growth per cycle (GPC) showing no dependence on the plasma operating pressure. It can also be observed that the GPC increases proportionally with decreasing substrate temperature. Although this effect can partly be attributed to a slightly reduced mass density of the films for temperatures below 150 °C, the change in GPC is predominantly caused by an increased incorporation of Al atoms into the film at lower substrate temperatures [24]. This Al incorporation is related to a higher surface density of hydroxyl groups at lower substrate temperatures. This has been confirmed by *in situ* infrared spectroscopy, which revealed that hydroxyl groups on Al_2O_3 surfaces are the dominant adsorption sites for $Al(CH_3)_3$ during plasma ALD [25]. For the thermal ALD process of $Al(CH_3)$ and H_2O, the GPC values are comparable but slightly lower than those for plasma ALD at temperatures >200 °C. This implies that the GPC for thermal ALD is slightly lower than that for plasma ALD at these temperatures. However, it is at temperatures <200 °C that the main difference is observed, because here the GPC of thermal ALD drops significantly below that for plasma ALD. The thermal energy at these temperatures is insufficient to allow for a full reaction with the surface and so, in this case, high hydroxyl surface density cannot be created. Additionally for the thermal ALD process, the purge after the H_2O dose needs to be increased at low substrate temperatures to avoid parasitic CVD reactions. For temperatures >200 °C, it was reported that a purge time of <5 s was sufficient, whereas at 58 °C, 30 s is required to completely remove residual water. At room temperature, Al_2O_3 films can still be synthesized by thermal ALD, but the purge time required becomes impractically long (e.g. 180 s at 33 °C [22]). For plasma ALD, the purge time can be kept relatively short, even at room temperature, in essence, equal to the time used at higher temperatures or only slightly extended [22]. The structural properties of Al_2O_3 thin films are highly comparable for both thermal and plasma ALD using $Al(CH_3)_3$ as a precursor [23]. At lower substrate temperatures the film quality degrades slightly, as evidenced by a reduced mass density, an increased O/Al ratio, and a higher H concentration [23, 26]. Furthermore, small traces of C can be found in the films deposited at room temperature. For plasma ALD, the C atoms are incorporated as CO_x groups, the density of which can be reduced by increasing the plasma exposure time [25]. Despite the somewhat lower material quality of the Al_2O_3 at substrate temperatures $<150\,^{\circ}C$, these films are considered very attractive for applications that require dense, amorphous, and conformal Al_2O_3 films deposited at low temperatures. One interesting application is the encapsulation of polymeric and/or organic devices that require barriers against H_2O and O_2 permeation [27]. For example, it was found that Al_2O_3 can provide very low water vapor transmission rates for the encapsulation of organic LEDs while also significantly reducing the pinhole density [28–32]. The main advantage of plasma ALD in this case is the fact that short cycle times can be maintained over the full temperature range, including room temperature. Moreover, the striking result was reported that, for Al_2O_3 films deposited by remote plasma ALD, the best barrier performance was obtained at room temperature [30]. Another interesting application of low-temperature ALD-synthesized Al_2O_3 is the protection of high-precision and high-purity metal parts against corrosion. For this application, ultrathin, dense, and defect-free coatings are desired, which need to be deposited at reduced substrate temperatures so as to maintain the mechanical properties of the substrates and to reduce the possibility of premature surface oxidation [33].

Plasma ALD of TiO_2 films from metal-organic precursors and an O_2 plasma is another case for which the merits of the plasma-based process can clearly be illustrated. As shown in Figure 7.10, TiO_2 has also been deposited at low substrate temperatures by this method, using $Ti(O^iPr)_4$, $Ti(Cp^{Me})(O^iPr)_3$, and $Ti(Cp^*)(OMe)_3$ ($Cp = \eta^5\text{-}C_5H_5$, cyclopentadienyl; $Cp^{Me} = \eta^5\text{-}C_5H_4CH_3$, monomethylcyclopentadienyl; $Cp^* = \eta^5\text{-}C_5(CH_3)_5$, pentamethylcyclopentadienyl) as precursors. TiO_2 films have been deposited at substrate temperatures as low as 25, 50, and $100\,^{\circ}C$, respectively [24]. Moreover, the GPC values were hardly affected by the substrate temperature over the full temperature range (up to $300\,^{\circ}C$). For thermal ALD, with either H_2O or O_3, the situation is quite different [35, 36]. Firstly, the GPC increases with substrate temperature, which indicates that the surface reactions are rate-limited by thermal activation. This holds not only for $Ti(O^iPr)_4$ with both H_2O and O_3 (see Figure 7.10), but also for $Ti(Cp^*)(OMe)_3$ with O_3 (not shown) [37]. Furthermore, for all these precursors, the GPC for the thermal process is lower than that for plasma ALD (except for the case of $Ti(O^iPr)_4$ with O_3 at $300\,^{\circ}C$, where the high GPC has been attributed to thermal decomposition of the precursor [35]). This clearly demonstrates that an O_2 plasma is not only more reactive than H_2O but also more effective than O_3. In addition, O_2 plasmas (and O_3) allow for the use of Cp-based Ti precursors, which show almost no reactivity against H_2O in ALD processes [37]. This exemplifies the fact that a wider range of precursors can be employed for use

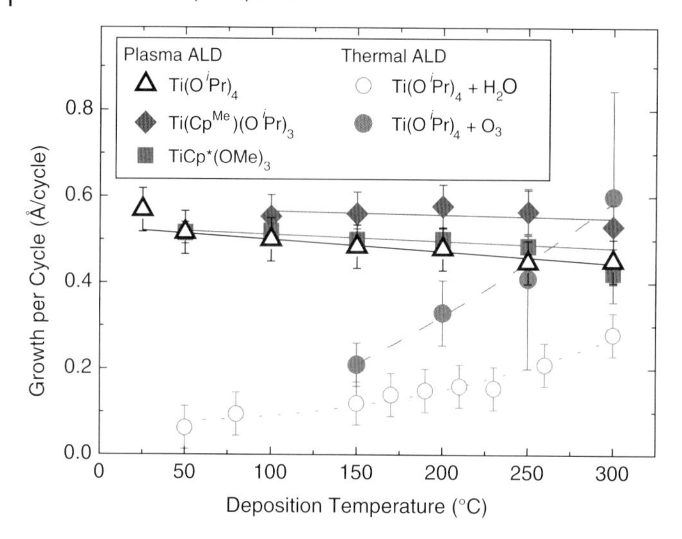

Figure 7.10 Growth per cycle of TiO_2 films as a function of the substrate temperature. Plasma ALD was carried out using $Ti(O^iPr)_4$, $Ti(Cp^{Me})(O^iPr)_3$, and $Ti(Cp^*)(OMe)_3$ as precursors in combination with an O_2 plasma [24]. Data for thermal ALD with H_2O [26] and O_3 [35] using the widely employed $Ti(O^iPr)_4$ precursor are given for comparison.

with plasma ALD. These relatively stable Cp-based precursors are currently of considerable interest for the DRAM industry, where the use of $SrTiO_3$ is being considered as an ultrahigh-k dielectric in upcoming technology nodes, because $SrTiO_3$ can be deposited by the combination of TiO_2 and SrO ALD cycles into the so-called supercycles [38].

The fact that plasma ALD can lead to improved material properties has also clearly been demonstrated by results on metallic TiN films [39, 40]. The combination of $TiCl_4$ with a H_2–N_2 plasma yielded thin TiN films with an excellent resistivity and low impurity levels, greatly surpassing the material quality achieved with the standard thermal process employing $TiCl_4$ and NH_3 [39, 40]. Moreover, in terms of acceptable material quality, thermal ALD of TiN is limited to the substrate temperature range of 300–400 °C, while the plasma ALD process can yield fair material properties down to temperatures as low as 100 °C as shown in Figure 7.11. For thicker films, resistivity values as low as 72 μΩ cm (400 °C) and 209 μΩ cm (100 °C) have been reported, which could be attributed to Cl concentrations of 0.1 and 2.1 at.%, respectively, at these temperatures [40]. This is quite a promising achievement, since TiN has numerous applications, such as an electrode material in MIM (metal-insulator-metal) capacitors and as a metal gate in CMOS (complementary metal-oxide-semiconductor) devices.

The ability of plasma ALD to synthesize materials that cannot (or hardly) be deposited by thermal ALD is exemplified by the case of TaN_x, which has applications in gate stacks as a metal gate material. Thermal ALD processes using either $TaCl_5$ or metal-organic precursors in combination with NH_3 always yield the highly resistive Ta_3N_5 phase because the reducing power of NH_3 is insufficient to reduce the Ta oxidation state from +5 to +3 [41]. However, for plasma ALD processes employing

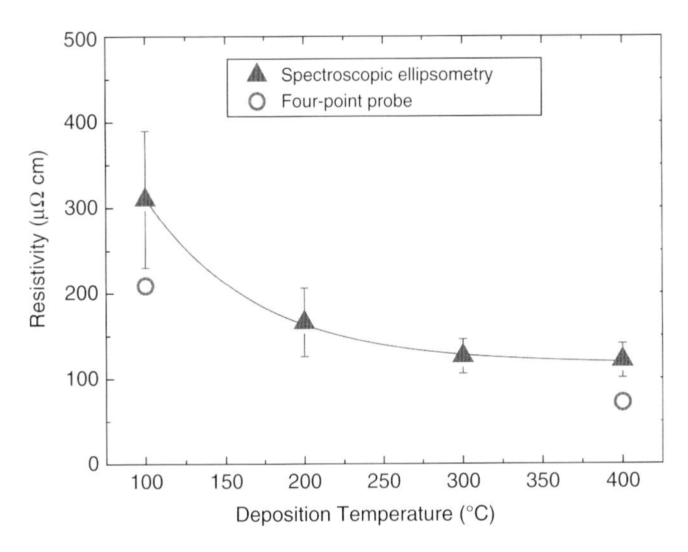

Figure 7.11 Resistivity (at room temperature) of TiN films obtained at 100–400 °C. The films were deposited by plasma ALD using $TiCl_4$ in combination with a H_2–N_2 plasma. The resistivity was determined by *in situ* spectroscopic ellipsometry (thin films, ~10 nm) and four-point probe measurements (thicker films, >45 nm) [40].

metal-organic precursors, it is possible to deposit conductive cubic TaN films using a H_2 plasma [42, 43]. This is illustrated in Figure 7.12, where the resistivity of TaN films, deposited using $Ta[N(CH_3)_2]_5$ as a precursor, as a function of the H_2 plasma exposure time is shown. It is clear that TaN films with resistivity values as low as 380 μΩ cm could be obtained for long plasma exposure times. In addition, the aforementioned highly resistive Ta_3N_5 films can also be synthesized by plasma ALD using the same $Ta[N(CH_3)_2]_5$ precursor, either by admixing a small concentration of N_2 into the H_2 plasma or by operating the plasma in NH_3 [43].

For the final example, the ALD of metal films is considered, in particular, the ALD of Pt. Noble metals, with applications as electrodes and in catalysis, are among the exceptional cases of elemental metals for which thermal ALD is relatively straightforward [44]. Pt can, for example, be deposited by thermal ALD using $Pt(Cp^{Me})Me_3$ as a precursor combined with O_2 gas [45]. However, in this case, as well as many other noble metal processes, a relatively long nucleation delay can occur on oxides and other starting surfaces, and is especially prominent when lower operating pressures are employed. This nucleation delay can be almost completely eliminated in the case of plasma ALD with an O_2 plasma instead of O_2 gas [46], as shown in Figure 7.13. This makes the ALD process and thickness control much more reliable, while also reducing the consumption of an expensive Pt precursor. Furthermore, when a (seed) layer of Pt has been deposited by plasma ALD, the thermal process can be continued without disruption in the thickness increment per cycle (see Figure 7.13). Another interesting feature of the plasma ALD process is that PtO_2 films can also be deposited from the same precursor and an O_2 plasma, the only difference (compared to plasma ALD of Pt) being the plasma exposure time. Pt can be deposited using short plasma

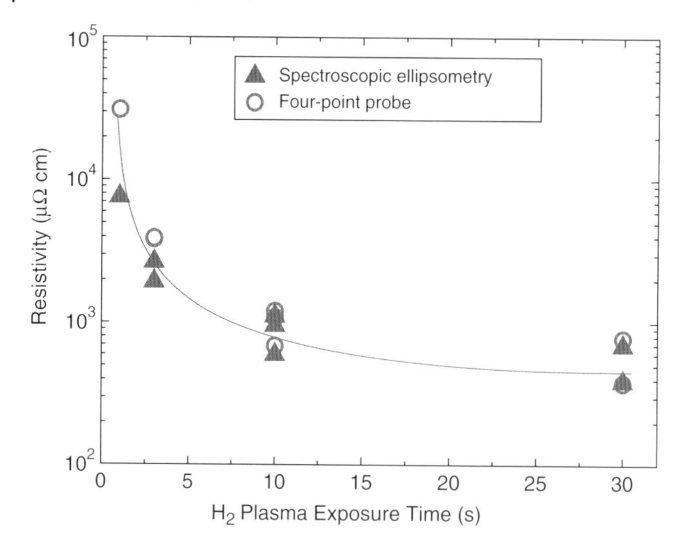

Figure 7.12 Resistivity (at room temperature) of TaN films as a function of H_2 plasma exposure time [43]. The data were obtained by *in situ* spectroscopic ellipsometry and four-point probe measurements. Reprinted with permission from Ref. [43]. Copyright 2007, the American Institute of Physics.

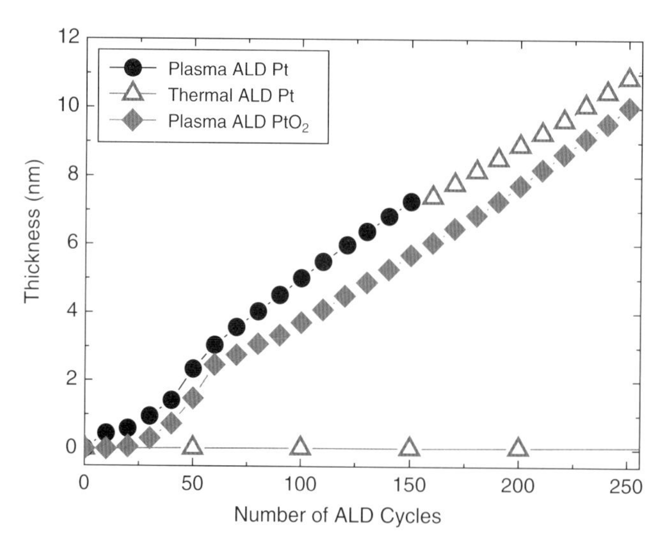

Figure 7.13 Thickness evolution of Pt and PtO_2 films deposited on an Al_2O_3 substrate [46]. The precursor was $Pt(Cp^{Me})Me_3$ and O_2 gas or an O_2 plasma was used as the oxidant. After 150 cycles, the plasma ALD process was stopped and film growth was continued using thermal ALD. The plasma exposure time was 0.5 s for Pt and 5 s for PtO_2. Reproduced from Ref. [46] with permission of ECS – the Electrochemical Society.

exposure times (~0.5 s), whereas relatively long plasma exposure times (~5 s) yield PtO_2. The GPC for PtO_2 is very similar to that for Pt (see Figure 7.13) even though the number of Pt atoms per cycle is lower. So far, the only thermal ALD process for PtO_x that has been reported was based on a combination of $Pt(acac)_2$ (acac = η^2-$(H_3CC{=}O)_2CH$, acetylacetonate) and O_3, and was restricted to the small temperature window of 120–130 °C [47]. Other examples of noble metal deposition include the ALD of Ru using $Ru(Cp)(CO)_2(Et)$ with either O_2 gas or an O_2 plasma. A similar nucleation delay was observed for the thermal process that was, again, reduced significantly by the use of a plasma in the coreactant step. This was, however, achieved at the expense of the surface roughness, which was higher when the plasma was used [48].

7.5 Challenges for Plasma ALD

Although plasma ALD can offer several benefits over thermal ALD for selected applications, the method also faces several challenges when compared to its thermal counterpart.

One limitation of plasma ALD that is often highlighted is the *reduced conformality or step coverage* that can be achieved on nonplanar substrates. These can be substrates with surface structures having high aspect ratios (e.g., wafers with trenches or vias) or substrate materials with very high surface areas (e.g., porous materials and powders). Thermal ALD is known to be the method of choice for depositing conformal thin films on such substrates, as film growth is self-limiting and independent of the precursor flux when the conditions are such that saturation of the ALD (half-) reactions is reached. The difference between plasma and thermal ALD, in this respect, is that plasma ALD involves reactive species that not only undergo ALD reactions (as in thermal ALD) but can also react on saturated surface sites. As mentioned in Section 7.2, radicals from the plasma can react with other radicals and species residing at the surface, forming nonreactive molecules that desorb back into the plasma. In the case of high aspect ratio structures or porous materials, the radicals have to undergo several surface collisions in order to reach deep inside the surface features, which significantly reduces the local flux of the radicals due to the surface recombination. This has only recently been addressed qualitatively by simulations [49] and experiments [50]. In many other reports, the impact of surface recombination of plasma radicals is generalized and the poor conformality of plasma ALD films is often claimed. However, the conformality achieved by plasma ALD under certain conditions depends strongly on the value of the recombination probability, r, which itself depends on (a) the type of radicals responsible for film growth in a certain plasma ALD process and (b) the material being deposited (see Table 7.1). The influence of the value of r can be illustrated by the results obtained from simple two-dimensional Monte Carlo simulations, in which the evolution of the deposition profile within a trench of high aspect ratio was studied as a function of the incoming radical flux [49–51]. This was achieved by considering only the plasma exposure step during a single ALD cycle. The information on the deposition profile at the sidewall of the trench was derived from the number of ALD growth events as a function of the sidewall position. For $r > 0$, it was found that the deposition profiles showed a minimum thickness at the region just above the bottom of the trench (see Figure 7.14a) and this minimum was more pronounced for larger values of r. The deposition profiles were clearly distinguishable from the cases in which the

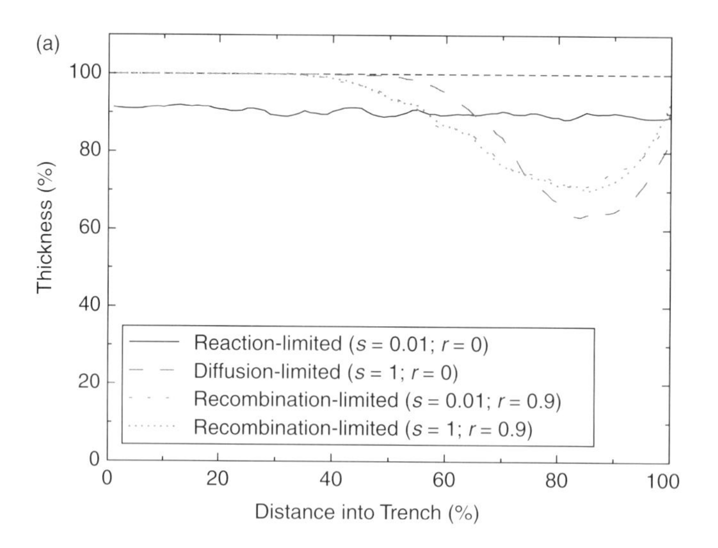

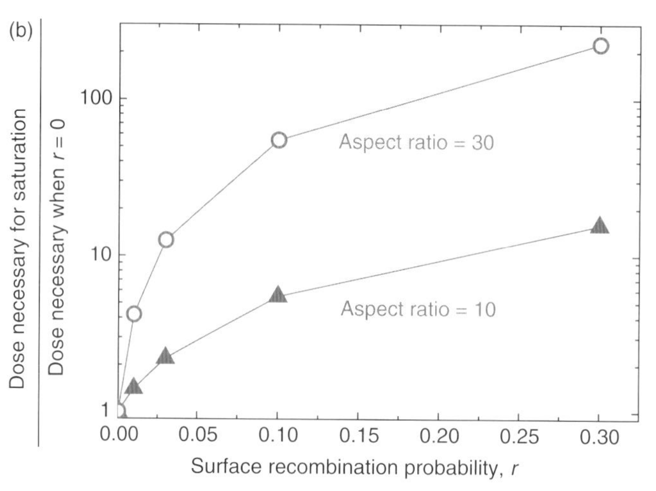

Figure 7.14 Results from Monte Carlo simulations to investigating the influence of surface recombination of radicals during plasma ALD [51]. (a) Equivalent thickness profile in a trench of aspect ratio 10 for different deposition regimes, obtained for various combinations of values for the sticking probability, s, and surface recombination probability, r. The positions within the trench labeled 0 and 100% correspond to the trench opening and trench bottom, respectively. Note that both recombination-limited cases show almost perfect overlap. (b) The dose required to reach saturation in trenches with aspect ratios of 10 and 30 for nonzero values of r. This dose is normalized to the dose required to reach saturation in these trenches when $r = 0$. For the simulations, $s = 0.01$ was assumed.

conformality was reaction-limited or diffusion-limited [52] (see Figure 7.14a). Therefore, in addition to these limiting cases (which are also observed for thermal ALD), a new recombination-limited regime was distinguished for the case of plasma ALD [51]. In order to achieve conformal deposition while in this recombination-limited regime, a longer plasma exposure time has to be employed. Compared to the case where $r = 0$, the dose required for saturation increases considerably when increasing the value of r, especially for high aspect ratios. The latter is illustrated in Figure 7.14b for aspect ratios of 10 and 30 and for r values ranging from 0.01 to 0.3. The "sticking" probability of the radicals on unreacted surface sites was assumed to be 0.01. From the figure it is clear that for the smaller r values and reasonable aspect ratios (≤ 10), conformal deposition can still be achieved relatively easily by increasing the plasma exposure time by, for example, up to 10 times longer than required for a planar substrate. However, for very high aspect ratios (e.g., ≥ 30), plasma exposure times are necessary that are practically unfeasible. Consequently, because most current applications have reasonable aspect ratios, it can be concluded that in many practical cases conformal deposition in high aspect ratio structures can certainly be achieved by plasma ALD. This is also confirmed by the results obtained for several metal oxides [26, 53–58], where the r values of oxygen radicals are known to be relatively low. For other cases, such as metals, the r values of radicals are relatively high. This makes it very difficult or even impossible to achieve a reasonable conformality, especially for aspect ratios >10. Finally, it should be noted that thermal ALD processes can also be subject to surface recombination losses of reactant species. For example, for some materials, O_3 also has considerable (nonzero) r values, which generally increase with surface temperature [59]. In this case, conformal film growth might also be recombination-limited [51].

A second limitation of plasma ALD that can have an important effect on certain applications can be described as *plasma-induced damage*. During the plasma step of the ALD cycle, the deposition surface is exposed to a multitude of reactive species from the plasma that can also induce undesired surface reactions, including oxidation and nitridation of the top surface layers of the substrate. In the case of (enhanced) surface oxidation, the interaction of oxygen atoms with the substrate during the initial ALD cycles can result in a (thicker) interfacial oxide, for example, when metal oxides are deposited on silicon [60]. Nitridation can take place when using N_2- or NH_3-based plasmas to deposit metal nitrides [61]. Note that nitridation is often also employed on purpose, for example, to increase the stability or the relative permittivity (k-value) of oxides. Plasma-induced damage can also manifest itself by the formation of defects inside the material or at the surface onto which the film is deposited. The bombardment of the substrate by energetic particles, such as ions accelerated in the plasma sheath, can lead to bond breaking, displacement of atoms in the surface region and charge accumulation on dielectric layers. Such ion bombardment effects are particularly important during plasma-activated processes, such as reactive ion etching, where the substrate is negatively biased to give the incoming ions kinetic energies of up to several hundreds of eV [8, 9]. During plasma ALD, the ion energies are typically much lower due to grounding of the substrate stage and/or the high pressures employed (i.e. when the plasma sheath is collisional), meaning that they are typically below the damage threshold (e.g. the displacement threshold) and cannot cause significant defect creation. However, there is still a significant possibility that ion bombardment may be of influence because the performance of most semiconductor devices, in which the ultrathin ALD-synthesized films are employed, is strongly affected by their interface properties. Obviously, such an influence is more likely when the flux of ions is considerable, such as in direct plasma ALD. Another often neglected, but in many cases even more important, mechanism is defect creation by plasma radiation. As mentioned earlier, a significant amount of UV radiation is formed in plasmas, the energy of which can reach up to 10 eV per photon. Oxides such as SiO_2 as well as the deposited metal oxides have band gap values smaller than 9 eV, meaning that vacuum ultraviolet (VUV) radiation can create defects, for example, by photoemission or photoinjection of charge. Plasma-induced defect creation has been studied in great detail for plasma processing of SiO_2-based gate stacks [62–65]. The influence of VUV exposure has recently also been studied for the high-k oxide HfO_2 [66–68]; however, it has not really been highlighted for the synthesis process of metal oxides by plasma ALD itself. Note that, apart from the fact that the metal oxide film might itself be affected, the interfacial SiO_x layer, which is typically present between the metal oxide and the Si substrate, can also be directly affected. From experiments (see Figure 7.15), it has recently been

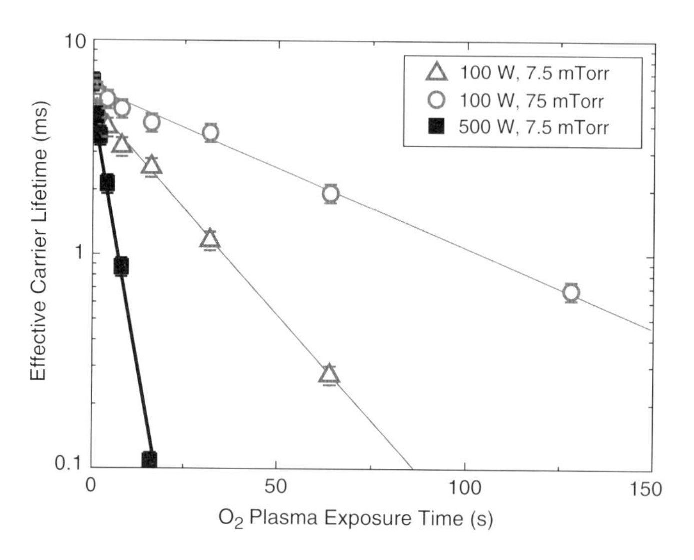

Figure 7.15 Experiments proving that VUV radiation from the plasma affects the surface passivation of crystalline Si by Al_2O_3 when deposited by plasma ALD [11, 34]. Al_2O_3 films (30 nm in thickness) were deposited by thermal ALD at both sides of double-side-polished Si wafers. After annealing for 10 min at 400 °C, the wafers were exposed to an O_2 plasma for various exposure times, t_p. The effective charge carrier lifetimes degraded for increasing exposure times at a rate proportional to increasing VUV radiation present in the plasma. After degradation, the lifetimes could be restored again by annealing. This case reflects the case of surface passivation by plasma ALD, where as-deposited Al_2O_3 by plasma ALD yielded negligible passivation, while after annealing the surface passivation was excellent. On the other hand, Al_2O_3 deposited by thermal ALD already yielded a fair level of passivation in the as-deposited state.

demonstrated that the surface passivation of crystalline Si by as-deposited Al_2O_3 is strongly affected by defects generated by VUV photons from the O_2 plasma (photon energy of 9.5 eV, see Figure 7.7) [34]. This induced a very high interface defect density ($D_{it} \approx 10^{12}\,eV^{-1}\,cm^{-2}$ at the midgap) such that extremely high surface recombination velocities of 10^7 cm/s were obtained [34]. Most of these surface defects can, however, be removed by annealing at 400 °C (D_{it} reduced to $\sim 10^{11}\,eV^{-1}$ cm^{-2}), such that excellent surface recombination velocities, as low as 1 cm/s, can be achieved on low-resistivity float-zone Si. These surface recombination velocities correspond to a surface passivation quality not yet paralleled by any other material [23]. On the other hand, as-deposited Al_2O_3 synthesized by thermal ALD affords reasonably low surface recombination velocities, as the films are not exposed to VUV radiation from a plasma (despite this fact, the level of surface passivation after annealing is not as good as for plasma ALD-synthesized Al_2O_3) [34]. Furthermore, it has been demonstrated that the level of surface passivation by as-deposited Al_2O_3 synthesized by plasma ALD can be increased by decreasing the length of the plasma exposure step, in this case, by using slightly subsaturated ALD conditions.

The aforementioned challenges for plasma ALD are inherent to the plasma–surface interaction that takes place during the plasma step. To assess their possible impact on the potential applications of the plasma ALD method, a deeper investigation of the underlying surface processes, including the plasma–surface interaction, is required in order to establish whether they lead to problems for the specific applications considered. Another class of challenges is related to the development of equipment and the *industrial scale-up of the process*. The use of plasma steps during ALD cycles requires more complex equipment in comparison with thermal ALD. Unless this additional complexity is compensated for by the significant advantages of plasma ALD, thermal ALD will be the method of choice. For research and development (R&D), the consequences for plasma ALD equipment are relatively minor, as R&D equipment is required to directly compare the results obtained by plasma and thermal ALD. This is evidenced by the fact that several combined plasma and thermal ALD tools have recently been introduced onto the market. However, for industrial scale-up the situation is different. Here, plasma ALD will only be adopted when it provides major benefits over thermal ALD and when alternative deposition techniques are not available for the specific applications envisioned. The latter situation is currently prominent in the field of crystalline silicon photovoltaics, where, at the moment, only thermal ALD is considered by equipment manufacturers for the scale-up of Al_2O_3 deposition [3, 69, 70], despite the fact that plasma ALD can provide significant benefits over thermal ALD in this case [34].

7.6 Concluding Remarks and Outlook

With the growing need for high-quality ultrathin and conformal films in and outside the semiconductor industry, the number of applications of ALD will grow substantially in the next decade. As a consequence, the requirements on process conditions and material properties will increase and diversify, requiring new experimental approaches and a variety of ALD equipment configurations. As plasma ALD can provide some unique merits over the thermal ALD method, it is therefore expected that the interest in this method will also keep growing considerably. This increase in interest is already clear because the number of ALD equipment manufacturers providing dedicated plasma ALD tools has increased significantly in the past few years. The demand for plasma ALD equipment from industrial R&D labs has, in particular, appeared to be high. It is likely that this is fueled by the fact that industrial laboratories are particularly focused on equipment that provides a high degree of flexibility in combination with the robust nature of the equipment and processes. In this respect, plasma-based techniques have been well accepted in thin film and device manufacturing. Nevertheless, plasma ALD also faces several challenges. In order to address the question whether and when aspects inherent to the plasma-based process will provide principal limitations for certain applications, a deeper understanding of film growth by plasma ALD is required. Therefore, more insight into the underlying surface reactions and the role of the plasma–surface interaction needs to be obtained. This is quite a challenging task but, consequently, also an appealing one, considering the complexity of plasma processes. Also of vital importance is that plasma ALD equipment for high-volume manufacturing will be developed and implemented. It would be best to focus initially on applications for which no alternative deposition techniques exist, or on applications for which the merits of plasma ALD are substantial and critical. Once established, the method will certainly find more applications and plasma ALD will complement existing thin film manufacturing techniques.

Acknowledgments

The current and past ALD team members of the Eindhoven University are thanked for their contribution to the measurements and the many fruitful discussions. The authors acknowledge the Dutch Technology Foundation STW for their financial support of this work through several projects.

References

1 Website ASM international, http://www.asm.com, August 1, 2010.
2 Website Oxford Instruments, http://www.oxford-instruments.com, August 1, 2010.
3 Website Beneq, http://www.beneq.com, August 1, 2010.
4 Website Cambridge Nanotech, http://www.cambridgenanotech.com, August 1, 2010.
5 De Keijser, M. and Van Opdorp, C. (1991) Atomic layer epitaxy of gallium arsenide with the use of atomic hydrogen. *Appl. Phys. Lett.*, **58**(11), 1187–1189.
6 Sherman, A. (1999) Sequential chemical vapor deposition. US Patent 5,916,365.
7 Rossnagel, S.M., Sherman, A., and Turner, F. (2000) Plasma-enhanced atomic layer deposition of Ta and Ti for interconnect diffusion barriers. *J. Vac. Sci. Technol. B*, **18**(4), 2016–2020.

8 Grill, A. (1994) *Cold Plasmas in Materials Fabrication: From Fundamentals to Applications*, Wiley–IEEE Press, New York.

9 Lieberman, M.A. and Lichtenberg, A.J. (2005) *Principles of Plasma Discharges and Materials Processing*, John Wiley & Sons, Inc., New York.

10 Profijt, H.B., Kudlacek, P., van de Sanden, M.C.M., and Kessels, W.M.M. (2011) Ion and photon–surface interaction during remote plasma ALD. *J. Electrochem. Soc.*, **158**, G88–G91.

11 Takagi, T. (1984) Ion–surface interactions during thin-film deposition. *J. Vac. Sci. Technol. A*, **2**(2), 382–388.

12 Mackus, A.J.M., Heil, S.B.S., Langereis, E., Knoops, H.C.M., van de Sanden, M.C.M., and Kessels, W.M.M. (2010) Optical emission spectroscopy as a tool for studying, optimizing, and monitoring plasma-assisted atomic layer deposition processes. *J. Vac. Sci. Technol. A*, **28** (1), 77–87.

13 Wood, B.J. and Wise, H. (1961) Kinetics of hydrogen atom recombination on surfaces. *J. Appl. Chem.*, **65**(11), 1976.

14 Greaves, J.C. and Linnett, J.W. (1959) Recombination of atoms at surfaces. 5. Oxygen atoms at oxide surfaces. *Trans. Faraday Soc.*, **55**(8), 1346–1354.

15 Kim, Y.C. and Boudart, M. (1991) Recombination of O, N, and H-atoms on silica: kinetics and mechanism. *Langmuir*, **7**(12), 2999–3005.

16 Gudmundsson, J.T., Kouznetsov, I.G., Patel, K.K., and Lieberman, M.A. (2001) Electronegativity of low-pressure high-density oxygen discharges. *J. Phys. D*, **34** (7), 1100–1109.

17 Heil, S.B.S., van Hemmen, J.L., Hodson, C.J., Singh, N., Klootwijk, J.H., Roozeboom, F., van de Sanden, M.C.M., and Kessels, W.M.M. (2007) Deposition of TiN and HfO_2 in a commercial 200mm remote plasma atomic layer deposition reactor. *J. Vac. Sci. Technol. A*, **25** (5), 1357–1366.

18 Niskanen, A., Rahtu, A., Sajavaara, T., Arstila, K., Ritala, M., and Leskelä, M. (2005) Radical-enhanced atomic layer deposition of metallic copper thin films. *J. Electrochem. Soc.*, **152**(1), G25–G28.

19 Website MKS Instruments, http://www.mksinst.com, August 1, 2010).

20 Lee, E.J., Ko, M.G., Kim, B.Y., Park, S.K., Kim, H.D., and Park, J.W. (2006) Lanthanum-oxide thin films deposited by plasma-enhanced atomic layer deposition. *J. Korean Phys. Soc.*, **49**(3), 1243–1246.

21 Kim, H. and Rossnagel, S.M. (2002) Growth kinetics and initial stage growth during plasma-enhanced Ti atomic layer deposition. *J. Vac. Sci. Technol. A*, **20** (3), 802–808.

22 Groner, M.D., Fabreguette, F.H., Elam, J.W., and George, S.M. (2004) Low-temperature Al_2O_3 atomic layer deposition. *Chem. Mater.*, **16**(4), 639.

23 Dingemans, G., van de Sanden, M.C.M., and Kessels, W.M.M. (2010) Influence of the deposition temperature on the c-Si surface passivation by Al_2O_3 films synthesized by ALD and PECVD. *Electrochem. Solid-State Lett.*, **13** (3), H76–H79.

24 Potts, S.E., Keunig, W., Langereis, E., Dingemans, G., van de Sanden, M.C.M., and Kessels, W.M.M. (2010) Low temperature plasma-enhanced atomic layer deposition of metal oxide thin films. *J. Electrochem. Soc.*, **157**(7), P66–P74.

25 Langereis, E., Keijmel, J., van de Sanden, M.C.M., and Kessels, W.M.M. (2008) Surface chemistry of plasma-assisted atomic layer deposition of Al_2O_3 studied by infrared spectroscopy. *Appl. Phys. Lett.*, **92**(23), 231904.

26 Van Hemmen, J.L., Heil, S.B.S., Klootwijk, J.H., Roozeboom, F., Hodson, C.J., van de Sanden, M.C.M., and Kessels, W.M.M. (2007) Plasma and thermal ALD of Al_2O_3 in a commercial 200mm ALD reactor. *J. Electrochem. Soc.*, **154** (7), G165–G169.

27 Chatham, H. (1996) Oxygen diffusion barrier properties of transparent oxide coatings on polymeric substrates. *Surf. Coat. Technol.*, **78**(1–3), 1–9.

28 Carcia, P.F., Mclean, R.S., Reilly, M.H., Groner, M.D., and George, S.M. (2006) Ca test of Al_2O_3 gas diffusion barriers grown by atomic layer deposition on polymers. *Appl. Phys. Lett.*, **89**(3), 031915.

29 Ghosh, A.P., Gerenser, L.J., Jarman, C.M., and Fornalik, J.E. (2005) Thin-film encapsulation of organic light-emitting devices. *Appl. Phys. Lett.*, **86**(22), 223503.

30 Langereis, E., Creatore, M., Heil, S.B.S., van de Sanden, M.C.M., and Kessels,

W.M.M. (2006) Plasma-assisted atomic layer deposition of Al_2O_3 moisture permeation barriers on polymers. *Appl. Phys. Lett.*, **89**(8), 081915.

31 Park, S.H.K., Oh, J., Hwang, C.S., Lee, J.I., Yang, Y.S., and Chu, H.Y. (2005) Ultrathin film encapsulation of an OLED by ALD. *Electrochem. Solid-State Lett.*, **8** (2), H21–H23.

32 Yun, S.J., Ko, Y.W., and Lim, J.W. (2004) Passivation of organic light-emitting diodes with aluminum oxide thin films grown by plasma-enhanced atomic layer deposition. *Appl. Phys. Lett.*, **85** (21), 4896–4898.

33 Matero, R., Ritala, M., Leskelä, M., Salo, T., Aromaa, J., and Forsen, O. (1999) Atomic layer deposited thin films for corrosion protection. *J. Phys. IV*, **9**(P8), 493–499.

34 Dingemans, G., Terlinden, N.M., Pierreux, D., Profijt, H.B., van de Sanden, M.C.M., and Kessels, W.M.M. (2011) Chemical and field-effect passivation of Si by thermal and plasma atomic layer deposited Al_2O_3, submitted *Electrochem. Solid-State Lett.*, **14**(1), H1–H4.

35 Williams, P., Kingsley, A., Leese, T., Otsuka, Y., and Uotani, K. (2008) 8th International Conference on Atomic Layer Deposition, Bruges, Belgium, February 7, 2008, Book of Abstracts.

36 Xie, Q., Jiang, Y.L., Detavernier, C., Deduytsche, D., Van Meirhaeghe, R.L., Ru, G.P., Li, B.Z., and Qu, X.P. (2007) Atomic layer deposition of TiO_2 from tetrakis-dimethyl-amido titanium or Ti isopropoxide precursors and H_2O. *J. Appl. Phys.*, **102**(8)

37 Katamreddy, R., Wang, Z., Omarjee, V., Venkateswara Raoa, P., Dussarrat, C., and Blasco, C. (2009) Advanced precursor development for Sr and Ti based oxide thin film applications. *ECS Trans.*, **25** (4), 217–230.

38 Kittl, J.A., Opsomer, K., Popovici, M., Menou, N., Kaczer, B., Wang, X.P., Adelmann, C., Pawlak, M.A., Tomida, K., Rothschild, A., Govoreanu, B., Degraeve, R., Schaekers, M., Zahid, M., Delabie, A., Meersschaut, J., Polspoel, W., Clima, S., Pourtois, G., Knaepen, W., Detavernier, C., Afanas'ev, V.V., Blomberg, T., Pierreux, D., Swerts, J., Fischer, P., Maes, J.W., Manger, D., Vandervorst, W., Conard, T., Franquet, A., Favia, P., Bender, H., Brijs, B., Van Elshocht, S., Jurczak, M., Van Houdt, J., and Wouters, D.J. (2009) High-*k* dielectrics for future generation memory devices (Invited Paper). *Microelectron. Eng.*, **86**(7–9), 1789–1795.

39 Elers, K.E., Winkler, J., Weeks, K., and Marcus, S. (2005) $TiCl_4$ as a precursor in the TiN deposition by ALD and PEALD. *J. Electrochem. Soc.*, **152**(8), G589–G593.

40 Heil, S.B.S., Langereis, E., Roozeboom, F., van de Sanden, M.C.M., and Kessels, W.M.M. (2006) Low-temperature deposition of TiN by plasma-assisted atomic layer deposition. *J. Electrochem. Soc.*, **153**(11), G956–G965.

41 Ritala, M., Kalsi, P., Riihela, D., Kukli, K., Leskelä, M., and Jokinen, J. (1999) Controlled growth of TaN, Ta_3N_5, and TaO_xN_y thin films by atomic layer deposition. *Chem. Mater.*, **11** (7), 1712–1718.

42 Kim, H., Detavernier, C., Van der Straten, O., Rossnagel, S.M., Kellock, A.J., and Park, D.G. (2005) Robust TaN_x diffusion barrier for Cu-interconnect technology with subnanometer thickness by metal-organic plasma-enhanced atomic layer deposition. *J. Appl. Phys.*, **98**(1), 014308.

43 Langereis, E., Knoops, H.C.M., Mackus, A.J.M., Roozeboom, F., van de Sanden, M.C.M., and Kessels, W.M.M. (2007) Synthesis and *in situ* characterization of low-resistivity TaN_x films by remote plasma atomic layer deposition. *J. Appl. Phys*, **102**(8), 083517.

44 Aaltonen, T. (2005) Atomic layer deposition of noble metal thin films. PhD thesis, University of Helsinki, Finland.

45 Aaltonen, T., Ritala, M., Sajavaara, T., Keinonen, J., and Leskelä, M. (2003) Atomic layer deposition of platinum thin films. *Chem. Mater.*, **15**(9), 1924–1928.

46 Knoops, H.C.M., Mackus, A.J.M., Donders, M.E., van de Sanden, M.C.M., Notten, P.H.L., and Kessels, W.M.M. (2009) Remote plasma ALD of platinum and platinum oxide films. *Electrochem. Solid-State Lett.*, **12**(7), G34–G36.

47 Hämäläinen, J., Munnik, F., Ritala, M., and Leskelä, M. (2008) Atomic layer deposition of platinum oxide and metallic platinum thin films from $Pt(acac)_2$ and ozone. *Chem. Mater.*, **20**, 6840.

48 Leick, N., Verkuijlen, R.O.F., Langereis, E., Rushworth, S., Roozeboom, F., van de

Sanden, M.C.M., and Kessels, W.M.M. (2011) Thermal and plasma-assisted ALD of Ru from $CpRu(CO)_2Et$ and O_2, submitted *J. Vac. Sci. Technol.*, A **29**(2), 021016.
49 Kessels, W.M.M. (2006) Opportunities and challenges of plasma-enhanced ALD. Baltic Conference on Atomic Layer Deposition, Oslo, Norway, June 19, 2006, Book of Abstracts.
50 Dendooven, J., Deduytsche, D., Musschoot, J., Vanmeirhaeghe, R.L., and Detavernier, C. (2009) Modeling the conformality of atomic layer deposition: the effect of sticking probability. *J. Electrochem. Soc.*, **156**(4), 63–67.
51 Knoops, H.C.M., Langereis, E., van de Sanden, M.C.M., and Kessels, W.M.M. (2010) Conformality of plasma-assisted ALD: physical processes and modeling. *J. Electrochem. Soc.*, **157**(12), G241–G249
52 Elam, J.W., Routkevitch, D., Mardilovich, P.P., and George, S.M. (2003) Conformal coating on ultrahigh-aspect-ratio nanopores of anodic alumina by atomic layer deposition. *Chem. Mater.*, **15** (18), 3507–3517.
53 Kim, J.Y., Kim, J.H., Ahn, J.H., Park, P.K., and Kang, S.W. (2007) Applicability of step-coverage modeling to TiO_2 thin films in atomic layer deposition. *J. Electrochem. Soc.*, **154**(12), H1008–H1013.
54 Kubala, N.G., Rowlette, P.C., and Wolden, C.A. (2009) Plasma-enhanced atomic layer deposition of anatase TiO_2 using $TiCl_4$. *J. Phys. Chem. C*, **113**(37), 16307–16310.
55 Kwon, O.S., Kim, S.K., Cho, M., Hwang, C.S., and Jeong, J. (2005) Chemically conformal ALD of $SrTiO_3$ thin films using conventional metallorganic precursors. *J. Electrochem. Soc.*, **152**(4), C229–C236.
56 Lee, J.H., Cho, Y.J., Min, Y.S., Kim, D., and Rhee, S.W. (2002) Plasma enhanced atomic layer deposition of $SrTiO_3$ thin films with $Sr(tmhd)_2$ and $Ti(i\text{-}OPr)_4$. *J. Vac. Sci. Technol. A*, **20**(5), 1828–1830.
57 Lee, W.J., You, I.K., Ryu, S.O., Yu, B.G., Cho, K.I., Yoon, S.G., and Lee, C.S. (2001) $SrTa_2O_6$ thin films deposited by plasma-enhanced atomic layer deposition. *Jpn. J. Appl. Phys. Part 1*, **40**(12), 6941–6944.
58 Niskanen, A., Kreissig, U., Leskelä, M., and Ritala, M. (2007) Radical enhanced atomic layer deposition of tantalum oxide. *Chem. Mater.*, **19**(9), 2316–2320.
59 Oyama, S.T. (2000) Chemical and catalytic properties of ozone. *Catal. Rev. Sci. Eng.*, **42**(3), 279–322.
60 Ha, S.C., Choi, E., Kim, S.H., and Roh, J.S. (2005) Influence of oxidant source on the property of atomic layer deposited Al_2O_3 on hydrogen-terminated Si substrate. *Thin Solid Films*, **476**(2), 252–257.
61 Langereis, E., Heil, S.B.S., van de Sanden, M.C.M., and Kessels, W.M.M. (2006) *In situ* spectroscopic ellipsometry study on the growth of ultrathin TiN films by plasma-assisted atomic layer deposition. *J. Appl. Phys.*, **100**(2), 023534.
62 Cismaru, C. and Shohet, J.L. (1999) Plasma vacuum ultraviolet emission in an electron cyclotron resonance etcher. *Appl. Phys. Lett.*, **74**(18), 2599–2601.
63 Druijf, K.G., Denijs, J.M.M., Vanderdrift, E., Granneman, E.H.A., and Balk, P. (1994) Nature of defects in the Si–SiO_2 system generated by vacuum-ultraviolet irradiation. *Appl. Phys. Lett.*, **65**(3), 347–349.
64 Hughes, H.L. and Benedetto, J.M. (2003) Radiation effects and hardening of MOS technology: devices and circuits. *IEEE Trans. Nucl. Sci.*, **50**(3), 500–521.
65 Samukawa, S., Ishikawa, Y., Kumagai, S., and Okigawa, N. (2001) On-wafer monitoring of vacuum-ultraviolet radiation damage in high-density plasma processes. *Jpn. J. Appl. Phys. Part 2*, **40** (12B), L1346–L1348.
66 Lauer, J.L., Shohet, J.L., and Nishi, Y. (2009) Effect of thermal annealing on charge exchange between oxygen interstitial defects within HfO_2 and oxygen-deficient silicon centers within the SiO_2/Si interface. *Appl. Phys. Lett.*, **94**(16), 162907.
67 Ren, H., Cheng, S.L., Nishi, Y., and Shohet, J.L. (2010) Effects of vacuum ultraviolet and ultraviolet irradiation on ultrathin hafnium-oxide dielectric layers on (100)Si as measured with electron-spin resonance. *Appl. Phys. Lett.*, **96** (19), 192904.
68 Upadhyaya, G.S. and Shohet, J.L. (2007) Comparison of the vacuum-ultraviolet radiation response of HfO_2/SiO_2/Si dielectric stacks with SiO_2/Si. *Appl. Phys. Lett.*, **90**(7), 0729094.
69 Website Levitech, http://www.levitech.nl, August 1, 2010.
70 Website TNO, http://www.tno.nl, August 1, 2010.

Part Two
Nanostructures by ALD

Atomic Layer Deposition of Nanostructured Materials, First Edition. Edited by Nicola Pinna and Mato Knez.

8
Atomic Layer Deposition for Microelectronic Applications

Cheol Seong Hwang

8.1
Introduction

The enormous improvement in computing technology has been one of the most important impetuses that has driven the development of modern civilization over the past 50 years. Improving computer technology has been achieved by the appropriate collaboration (or competition) between design technology, such as recent multicore architecture of microprocessors, and highly advanced process technology for chip fabrication. While design technology will further evolve for more energy-saving and faster architectures, the old paradigm for chip fabrication appears to be coming under increasing scrutiny. The old paradigm for the chip fabrication can be well represented by the famous Moore's law, which has been accomplished through strict scaling technology in both memory and logic chips. As the design rule of modern semiconductor chips has already decreased to ~20 nm, the economical fabrication of chips becomes extremely challenging due to the huge expense for fine patterning and related process complexities. Therefore, it is natural that an alternative way to "just shrink" the device be pursued. It can be duly stated that the alternative way in the fabrication technology is to seek new functional materials and new film process technology. Hopefully, combining these should provide a way to solve the above problems, that is, a new functional material processed by an innovative process technology for new innovative device structure.

Atomic layer deposition (ALD) appears to be a very timely and crucial process technique for microelectronics from this point of view, even though it was originally suggested to achieve uniformly rather thick film (~μm) deposition over a very large area for other applications [1]. This is more evident as the necessary film thickness range reaches the several tens (or even less) of nm scale for highly scaled semiconductor devices. The properties of thin films grown by conventional technologies, such as chemical vapor deposition (CVD) and sputtering, tend to deviate rather severely from their bulk values in this thickness range. In addition, the smaller feature size of highly scaled semiconductor devices naturally calls for the fabrication of extreme three-dimensional structures. Thin films should be deposited conformally over these 3D structures with atomic scale uniformity in thickness, composition, and electrical/

Atomic Layer Deposition of Nanostructured Materials, First Edition. Edited by Nicola Pinna and Mato Knez.

chemical properties. ALD is possibly the best known thin film growth technique for these purposes.

Making the ever-shrinking feature size, where "pitch" is a more appropriate term in memory, whereas feature size generally represents the smallest dimension in logic devices, has always been the biggest hurdle to overcome at each technology node. Even with the very recent immersion lithographic technology combined with the 193 nm ArF excimer laser exposure tool, making patterns with a feature size <30 nm relies on supplementary processes. These include deposition/etching back for making a smaller space or contact hole, and a double patterning technique for denser and smaller patterns. For these resolution enhancement technologies, several sacrificial layers that are mostly compatible with the photoresist (PR) are necessary. Finely patterned PR maintains its fidelity only up to $\sim$200 °C. Therefore, the generally low processing temperature of ALD is well suited to processes where thin films are deposited on top or sides of PR patterns. Several ALD materials are used as the sacrificial hard masks for dry etching processes, which mean that they must have reasonable density and chemical stability with very tight thickness control. These requirements are another challenge for ALD films, which require some further developments in the precursors and process conditions.

3D stacking trends in semiconductor chip fabrication also offer other chances as well as challenges to ALD. In contrast to multichip package, the wafer level, such as through-silicon via (TSV) technology, or device level 3D stacking technologies require a higher growth rate and higher conformality of the ALD layers.

This chapter reviews the necessity of various ALD layers, which will eventually function in the final products (called active layers) or will be removed during the subsequent processes (called sacrificial layers) in memory and logic microprocessors. Sometimes these requirements are cross-linked so that it is not easy to discriminate them. Therefore, the application fields for memory and logic devices are classified, and the active and sacrificial layers are described in each field.

8.2 ALD Layers for Memory Devices

Various semiconductor memories are classified according to the mass production level, and the application of ALD in each subcategory is described. This is a useful method for understanding the status of ALD in the semiconductor industry, even though such a classification appears to be somewhat unusual.

8.2.1 Mass Production Level Memories

This includes the dynamic random access memory (DRAM), flash memory, and phase change RAM (PCRAM). DRAM is volatile memory, whereas flash memory and PCRAM are nonvolatile memories. DRAM has been considered to have reached its scaling limit at each generation for a quite long time, but this has not occurred yet.

It may not be also the case in the near future due to the continuing materials and technology innovations as well as the need for higher speed and density from computer systems. The overall performance of a computer system depends not only on the microprocessor speed but also on the speed and density of the main memory, which is DRAM. Sluggish development of other emerging memories has made the importance of DRAM even more evident in modern high-performance computers.

High-density flash memory becomes very popular as mobile appliances flourish in everyday life. Due to its unique chain (or string) structure of memory cells, NAND-type flash memory has increased its packing density and commercial use dramatically over the past decade, and may encroach on the share of hard disks. However, flash memory is not expected to replace DRAM due to its slow writing speed and limited number of rewrite and inefficient utilization of memory. PCRAM appears very attractive considering its enormous evolution into the semi-mass production level during the past 5 years. Such fast evolution has not been observed in any other types of semiconductor memory. It has random write/read accessibility (i.e., high-speed operation) as well as a very small cell size. Although it is currently at only the 0.5–1 Gb density level and has limited write/read numbers ($<10^{10}$) compared to the practically unlimited number of DRAM, it is expected to undergo rapid development over the next decade. This is basically because the performance of PCRAM will improve with a smaller cell size, which is certainly not expected from charge storage-type DRAM and flash memories.

8.2.1.1 Dynamic Random Access Memory

Each DRAM cell is composed of one select transistor and one capacitor. There are also periphery circuits, such as sense amplifiers, decoders, encoders, and input/output interfaces, which drive the memory cells. The role of the transistor is to select a certain cell to be written or read out at the cross-point, where both the bit and word lines are activated. When the word line is deactivated, it works as a closed switch that keeps the stored charge in the drain junction-cell capacitor. These two operations require a very contradictory function of the transistor: fully open during writing and reading but fully closed during data retention. This can be represented by a high on/off current (I_{on}/I_{off}) ratio (as high as 10^8), which normally requires a high threshold voltage ($V_{th} \sim 0.7$–0.8 V) and an operation voltage. This high V_{th} is not compatible with the lower drive voltage of modern DRAMs ($V_{dd} < \sim 2$ V) because the overvoltage that can be written to the capacitor, which is approximately proportional to $V_{dd} - V_{th}$, decreases rapidly with decreasing V_{dd}. Therefore, innovations in select transistors and cell capacitors are necessary to maintain the function of DRAM.

As the requirement for faster on–off action in a DRAM select transistor is less severe than logic transistors, the deleterious short channel effects by the drain voltage have been overcome by adopting a 3D transistor structure, such as recessed channel array transistor (RCAT) [2]. In the RCAT structure, the channel length has been increased by etching the Si channel (see Figure 8.1) to form a trench shape. The physically longer channel length decreases the drain-induced barrier lowering and local electric field enhancement effect by the lower channel doping concentration. The latter is also important for achieving a lower junction leakage current. It should

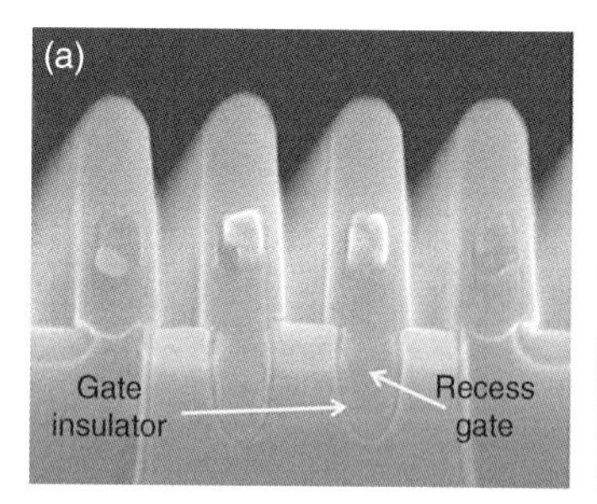

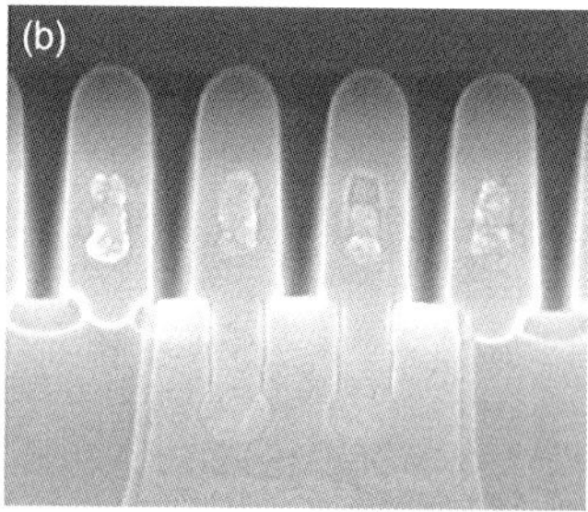

Figure 8.1 (a) SEM image of the fabricated RCAT DRAM cell [2]. (b) SEM image of the fabricated S-RCAT DRAM cell [2].

be noted that this strategy cannot be applied to the logic transistors because this will certainly reduce the on current and the operation speed. In that case, a higher gate capacitance is the only option to overcome the drain-related side effects (see Section 8.3.1). Due to this physically longer dimension, it is not expected that high-*k* gate dielectrics will be used in the select transistor of DRAM at this moment. However, the nonuniform growth of gate SiO_2 on the various crystallographic planes of an etched Si channel surface (see Figure 8.1) is becoming more problematic as the gate insulator scales. ALD of thin SiO_2 with extremely high quality might be necessary in future DRAMs as the gate insulator of the select transistor for conformal formation of the gate dielectric layer. This will also be the case for the vertical type 3D flash memory, as shown in Section 8.2.1.2. However, the requirements for the extremely high quality of the gate dielectric layer may impose several large hurdles to overcome for this type of application of ALD SiO_2. As the select transistor is n-type, metal gate engineering is also less problematic than the complementary MISFETs (CMISFETs) of logic devices. If metal gate technology is necessary, the highly developed TiN ALD process can be well adopted with some tuning for work function control. Metal-organic (MO) ALD would be beneficial compared to $TiCl_4$-based ALD on account of its higher flexibility of composition control, which will allow more room for work function control.

On the other hand, the transistors in the periphery area require a higher operation speed so that high-*k* gate dielectric and related technology are necessary. As the transistors in the periphery area are composed of n- and p-type transistors, similar complications as that of high-*k* logic transistors exist. This will be described in detail in Section 8.3.1. The required equivalent oxide thickness ($EOT = t_{phy} \times 3.9/k$, where t_{phy} is the physical thickness of the film) is slightly less stringent than logic devices. There are many reports of high-*k* gate dielectrics and related integration issues. Recent reviews by Kuesters *et al.* and Niinistö *et al.* summarize these issues quite well [3, 4]. In addition, it might be necessary to develop additional mask sets and integration schemes because the select transistors in the memory cell area do not adopt high-*k* dielectric materials in DRAM.

Dense word line (gate line) and bit line patterning is another key technology for DRAM fabrication. Double patterning technology (DPT) is now indispensable in this area, which will be discussed in detail in Section 8.2.1.2. For DPT, ALD is an indispensable technology.

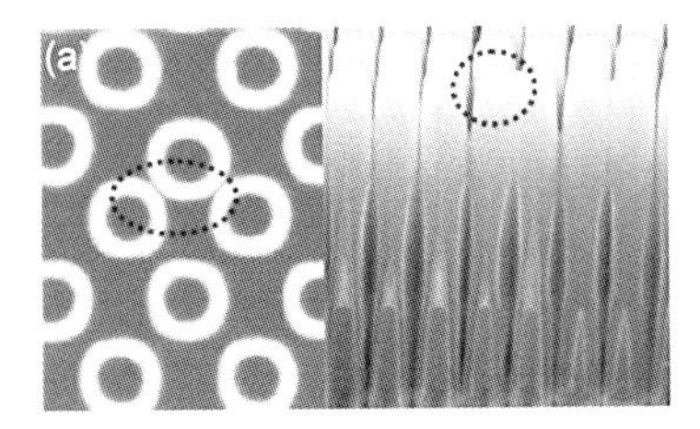

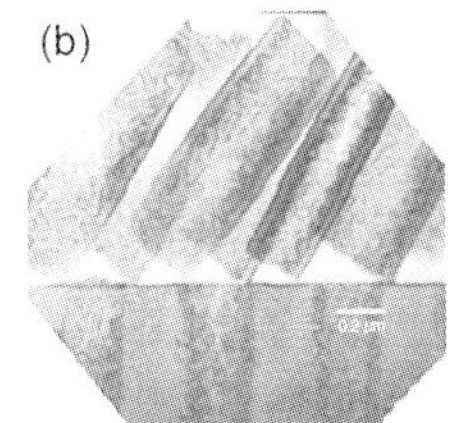

Figure 8.2 (a) SEM image of storage node leaning [2]. (b) TEM image of broken storage node (poly-Si storage node).

Another important application of ALD is certainly capacitors. As discussed above, the electrical and structural environment for capacitors in highly scaled DRAM is becoming harsher, which does not allow sufficient design space for maintaining the performance with smaller cell capacitance. The sluggish improvement in the performance of the sense amplifier requires an almost constant cell capacitance of ~25 fF irrespective of the technology node. The smaller footprint of the shrunken cell capacitor has a negligible effect on the cell capacitance because >95% of the surface area of the cell capacitor is provided by the sidewall area of the bottom electrode. The larger influence of the smaller design rule comes from the weaker mechanical strength of the tall storage node (bottom electrode) with a smaller footprint. This generally results in the leaning or even breakage of the storage node during capacitor fabrication (see Figure 8.2 [2]). Therefore, the storage node height should be decreased as the design rule shrinks further. The shift in the bottom electrode material from heavily doped polycrystalline Si (poly-Si) to TiN has improved the mechanical strength of the capacitor structure greatly and decreased the detrimental interfacial low-*k* dielectric (mostly SiO_2) effect. Figure 8.3a and b shows the sustainable height that is reported in ITRS roadmap [5] and necessary heights, respectively, of the storage node to achieve a cell capacitance of 25 fF for the various EOT values of the capacitor dielectric as a function of the design rule <50 nm. Here, the structure was assumed to be a simple box type not a cylinder type. From these figures, it can be understood that an EOT value of 0.3 nm is essential for the DRAM with a design rule of 20 nm. Another critical factor is the low thickness of the dielectric as

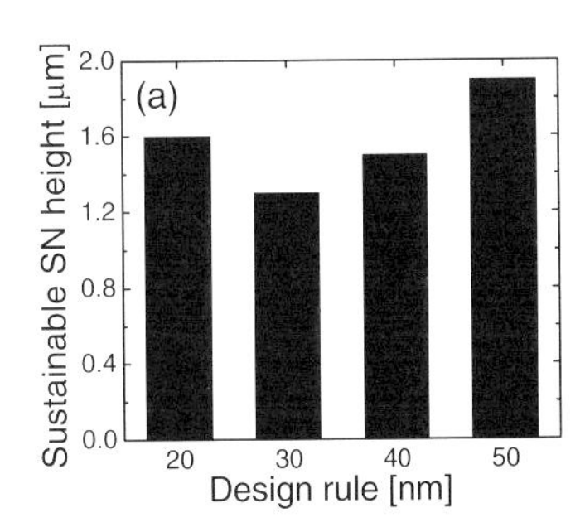

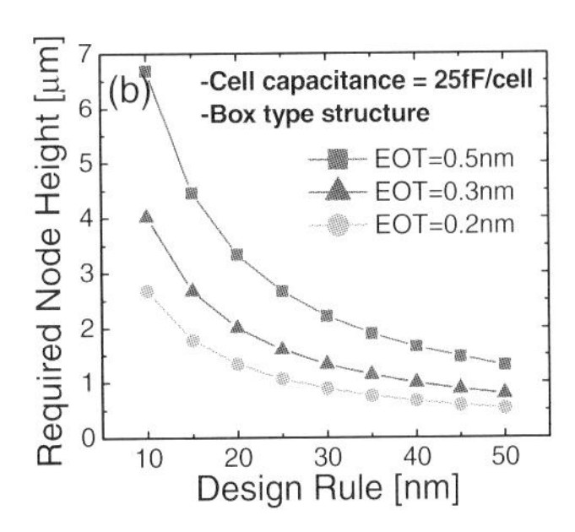

Figure 8.3 (a) Sustainable storage node height that is reported in ITRS 2009 [5]. (b) Required storage node height to achieve a cell capacitance of 25 fF for the various EOTs of the capacitor dielectric as a function of the design rule.

well as the top electrode to fit them into the narrow gap between the neighboring storage nodes. For the design rule of 20 nm (pitch 40 nm), only 6–8 nm thick films (dielectric and top electrode) are allowed. Considering the height of the storage node, this corresponds to an aspect ratio of 80 and 200 for the capacitor dielectric and top electrode, respectively. Therefore, it is evident that all the dielectric and electrode layers should be deposited by ALD.

Achieving such a challenging EOT value definitely requires a very high dielectric constant material and a metal electrode that does not form low-k interfacial layers. The most recent approach for this can be found from the trilayered structure of tetragonal (or cubic) ZrO_2 ($k \sim 40$)/amorphous Al_2O_3 ($k \sim 9$)/tetragonal (or cubic) ZrO_2 with TiN electrodes (called the ZAZ capacitor) [6, 7]. The use of TiN as the electrode instead of conventional heavily doped poly-Si also contributed largely to the improvement in DRAM capacitors owing to its higher electrical conductivity (negligible carrier depletion thickness) and no formation of low interfacial SiO_2. All these engineering efforts have allowed the production of DRAM down to ~45–50 nm technology node. In these DRAMs, a minimum EOT of ~0.7–0.8 nm and capacitor height of ~2 μm with a cylindrical structure are used. This has been well summarized by Kuesters *et al.* [3]. Niinistö *et al.* [4] summarized the recent trends of ALD of these high-k materials, including the newly highlighted rutile structure TiO_2, for the dielectric and electrode for such an extreme geometry [8]. Rutile structure TiO_2 films on Ru and Ru/TiN electrodes have a dielectric constant as high as 80–130 [8, 9]. This is due to the local lattice match between the rutile TiO_2 and RuO_2 formed between the TiO_2 and Ru layers by the *in situ* or *ex situ* oxidation of the Ru electrode. Al-doped TiO_2 films showed a minimum achievable EOT as small as 0.48 nm [10].

Along with these group IV binary oxides (and their slight modification by doping), there have been extensive studies on perovskite dielectrics, most typically $SrTiO_3$ (STO), which may offer a much higher k value [11]. However, perovskite oxides have two serious drawbacks; first, their k value decreases severely with decreasing physical thickness in the range of interest. This has been dealt with extensively both experimentally and theoretically [11–14]. The second is the much more complicated processes compared to binary oxides due to their multication composition. The optimum k value is acquired only in a very narrow stoichiometric range [15, 16]. In addition, the TiN electrode appears to be incompatible with these very high-k materials due to (local) lattice mismatch, chemical interactions (interfacial oxidation and reaction), and low work function. This is also the case for rutile TiO_2. Therefore, there is a strong need for better electrodes. Ru or RuO_2 appears to be the most promising material at this moment [17, 18]. Another concern for the ALD process of perovskite STO dielectric and metal Ru/RuO_2 electrodes is the low growth rate. Recent reports on the use of cyclopentadienyl-based group II precursors for STO films have largely mitigated this problem [19, 20]. The use of RuO_4 as the Ru precursor appears to provide a promising way to meet the mass production level growth rate and film quality [21]. Growth of the RuO_2 layer by ALD is another challenge as this oxide is easily reduced during the precursor injection step. Oxidized RuO_2 has inferior properties to ALD RuO_2 due to the rougher surface morphology, necessitating the development of an ALD process for this material.

The fabrication of DRAM capacitors depends heavily on the development of bottom and top electrodes composed of noble metals (or noble metal oxides) and higher-*k* dielectric layers. Unfortunately, this is a very challenging area of ALD in terms of the precursor chemistry, process control, and mass productivity (high cost and low growth rate). Recent reviews by the authors discussed this issue in detail [20]. These are the applications of ALD layers as the active materials.

There is another important application of ALD SiO_2 layers for fabricating the present DRAM capacitors using the ALD SiO_2 layer as a sacrificial layer. Figure 8.4 shows the process flow for the fabrication of ZAZ-type capacitors with a cylindrical cell structure using a TiN bottom electrode. After depositing the TiN layer on the capacitor hole pattern (called mold oxide, Figure 8.4a), it needs to be separated to form each storage node (Figure 8.4c). This can be achieved by either chemical mechanical polishing or an etch-back process. During this storage node separation process, the bottom area of the TiN electrode can be damaged and electrical contact between the TiN bottom electrode and drain contact stud becomes problematic. To protect the contact region (indicated by arrow in Figure 8.4c) from damage, the inside space of the TiN bottom electrode is filled with SiO_2. ALD SiO_2 at low temperatures is indispensable due to the extremely high aspect ratio and potential oxidation of the TiN bottom electrode. After storage node separation, the remaining SiO_2 inside the

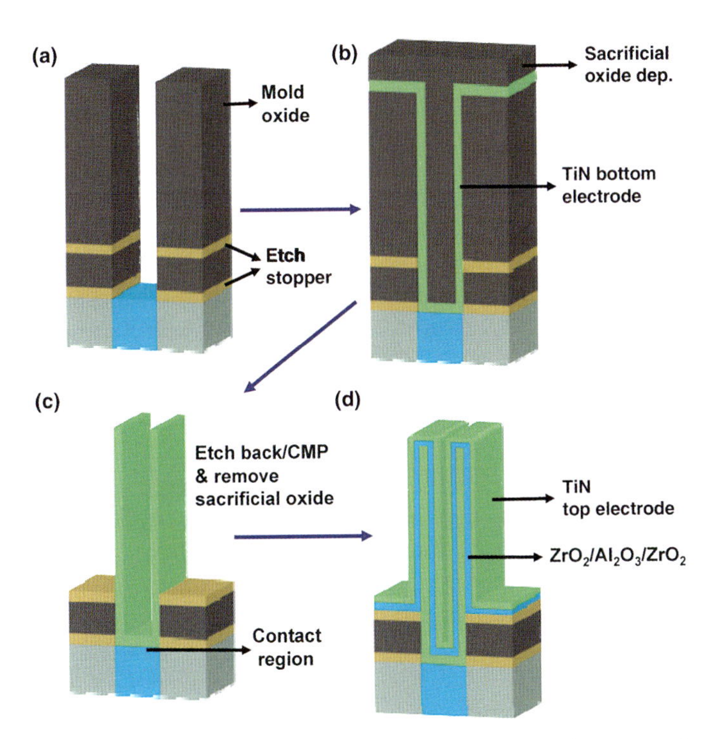

Figure 8.4 (a–d) Process flow schematics for the fabrication of $TiN/ZrO_2/Al_2O_3/ZrO_2/TiN$-type capacitors with a cylindrical cell structure. ALD SiO_2 layer can be used as the sacrificial layer to pattern the contact region.

storage node hole is removed concurrently with the mold oxide by wet etching. Therefore, this ALD SiO_2 layer is a sacrificial layer and does not require stringent electrical performance.

8.2.1.2 Flash Memory

The conventional flash memory cell is composed of double gate transistors. Charge transfer from channel to the floating gate (electron for "program" and hole for "erase" operation) corresponds to the write operation, which is accomplished by the high voltage (~18 V) between the control gate and the Si channel (or Si body). The stored charge in the floating gate increased (for programmed state) or decreased (for erased state) the cell V_{th}, which is read out when a reference read voltage is applied to the gate. Due to this unique V_{th} variation, each cell of the flash memory is composed of only one transistor, which makes the unit cell size of the NAND-type memory (~$5F^2$, where F is the minimum feature size) smaller than that of DRAM (~6–$10F^2$ depending on the bit line architecture). Therefore, NAND-type flash memory has dominantly led to the shrink technology and accomplished the highest packing density. However, the cross-coupling problem in highly scaled flash memory via the sidewall area of the floating poly-Si gate has hindered the easy scaling of flash memory. This problem becomes even more severe for multilevel cells. Therefore, charge trap flash (CTF), where a thin nonstoichiometric SiN_x layer works as the charge trapping layer instead of a floating gate, has been suggested [22]. This type of new flash memory relies more on new functional materials than the conventional flash memory does. Conventional flash memory is made almost exclusively of conventional materials: poly-Si floating and control gates, N-doped SiO_2 tunneling oxide, and SiO_2–Si_3N_4–SiO_2 (ONO) blocking oxide. The new functional materials in CTF include the Al_2O_3 blocking oxide, TaN control gate, and the nonstoichiometric SiN_x CT layer [23]. The blocking oxide should have a negligible leakage current (~10^{-9} A/cm^2) to ensure the safe operation of flash memory. Therefore, ALD at high temperature may be a viable option for growth of the blocking oxide and control metal gate. To ensure a low enough leakage current, the blocking oxide layer normally undergoes high-temperature annealing (~950–1000 °C), so that exceptional thermal stability is needed for this ALD layer. The higher capacitance density of the blocking oxide layer is another requirement for achieving better operation of the flash cells. Due to the very strict requirement for a low leakage current and thermal stability, other high-k dielectrics (HfO_2 or silicates) may be less likely to be used as the blocking oxide layer in the near term. However, there have been several attempts to adopt high-k dielectrics by further improving the ALD process conditions or combining different high-k layers [24, 25]. This application does not have the severe restriction on the temperature limit of the ALD process. Overall, these applications correspond to the active function of the ALD layers.

A very interesting application of the CTF cell is to construct a 3D stacked flash memory cell [26, 27]. The vertical integration of memory cells is considered the ultimate structure of highest density memory cells. However, this has been hindered by the relatively low performance and large scatter of V_{th} poly-Si channel materials. The stacking of laterally integrated poly-Si-based memory cell layers does not offer an optimal solution for the most economical integration schemes because each memory cell layer requires expensive photolithography steps. Figure 8.5a and b shows schematic diagrams of the two types of vertically integrated 3D stacked NAND-type flash cells [26, 27]. In Figure 8.5a, a hole was etched after alternative deposition of the control gate and insulating layers. Inside this hole, a blocking oxide, a charge trap layer, and tunneling oxide layers were deposited. Finally, the semiconductor layer was deposited. In some cases, the semiconductor layer was made very thin to make the V_{th} spread smaller, and the remaining core volume was filled with an insulator. In this case, the electronic structure of the semiconductor resembles the silicon-on-insulator (SOI) structure. Hence, the electrical characteristics also resemble it. When a standard 32-cell transistor string is formed along one vertical structure and the hole dimension is 50 nm, the aspect ratio of the hole easily exceeds 30 suggesting that ALD should play a key role in depositing various layers inside the hole. It is possible that the initial stage of these vertical cells would be fabricated with standard materials: ONO blocking oxide, SiN_x CT layer, and SiO_2 tunneling oxide. Among these materials, the tunneling SiO_2 layer can be formed either by the oxidation of a certain portion of the SiN_x CT layer or by ALD. With this regard, ALD SiO_2 must overcome the range of challenges from the very serious electrical requirements for the tunneling oxide layer as well as the blocking oxide layer. If some other materials, such as Al_2O_3, are considered, they must also have supreme quality. The poly-Si deposition process may not require an ALD process because the present low-pressure CVD still offers sufficient conformality. For the other structure in Figure 8.5b, the semiconductor and insulating layers are stacked first. They are then etched to form a vertical stud structure composed of alternating layers of semiconducting and insulating layers. The tunneling oxide, CT layer, and blocking oxide layers are subsequently formed, which is in the reverse order of deposition compared to the previous case. Again, the narrow width and high aspect ratio of the trench between the active studs highlight the need for ALD processes to deposit the layers. Here, the ALD layers are in direct contact with the etched side area of the semiconductor layers,

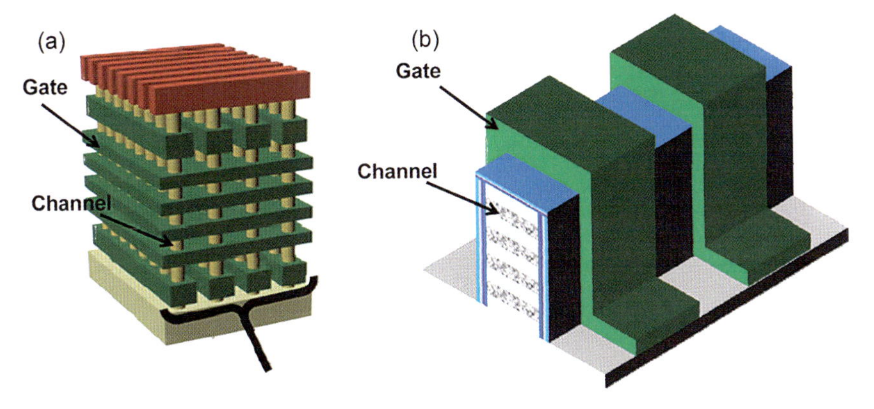

Figure 8.5 Schematic diagrams of the two types of vertically integrated 3D stacked NAND-type flash cells.

which will eventually form the channel. Therefore, extreme care of the ALD process is important to prevent damage to the sensitive semiconductor channel region.

For these applications, the ALD layers, that is, tunneling oxide, CT layer, blocking oxide, and hopefully the control gate, work as active materials. In general, CTF memory suffers from insufficient retention of trapped charges in the CT layer [28]. This has been attributed mainly to the not sufficiently deep trap level and its distribution over a certain energy range in the energy gap of the CT material. Therefore, there is still a need for the development of new CT materials using a better controlled ALD process and precursor chemistry. These 3D stacked memory cell technologies are discussed in Section 8.2.3.

Flash memory is another device where the sacrificial application of an ALD SiO_2 layer is taken seriously because of its smallest feature size for the given technology node. Figure 8.6 compares the design rule trends of NAND flash and DRAM, which reached the 30 nm design rule in 2010. To achieve the smallest feature size, particularly the gate and bit line pitch, the lithographic process requires revolutionary changes. The extensive use of DPT is a solution to pattern this sized features with the existing 193 nm immersion lithography. There are basically two types of DPT for making fine line/space patterns, which is the typical structure of control gates and bit lines in NAND flash memory: double exposure/double etch method and oxide spacer method [29]. The first method does not depend on any ALD process but requires the two times use of a very expensive lithographic exposure process. Therefore, the latter, which is critically dependent on the ALD of the passive layer on a prepatterned PR, is a more interesting solution. Figures 8.7 and 8.8 show the two process sequences of DPT using the ALD SiO_2 layer, which are used mainly as the shallow trench isolation/gate and bit line patterning, respectively. In either case, the final pitch is half of the first patterning pitch but the minimum feature size was determined to be the

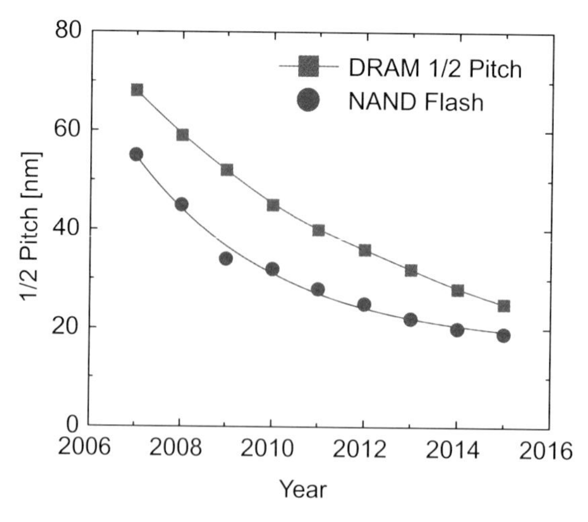

Figure 8.6 Design rule trends of NAND flash and DRAM, which reached the 20 nm design rule in 2010.

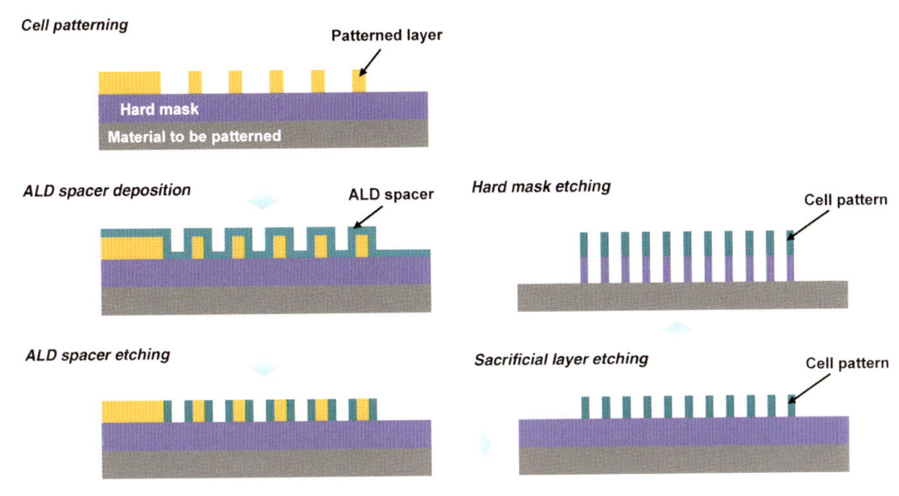

Figure 8.7 Process sequence of double patterning technology using the ALD SiO_2 layer, which is used mainly as the shallow trench isolation/gate patterning.

thickness of the deposited ALD SiO_2 layer (sidewall thickness). In Figure 8.7, the final mask line dimension (called bar dimension) is the ALD thickness, whereas in Figure 8.8 the distance between the final mask lines (called space) is the ALD thickness. Therefore, the constant critical dimension of the bar in Figure 8.7 and space in Figure 8.8 is achieved. In either case, the final mask layer is made of a bottom hard mask layer, which could be SiO_2, Si_3N_4, or TiN depending on the layer to be patterned. Therefore, the ALD layer works as a sacrificial layer for this application. Very tight thickness control (~1% variation within wafer and wafer-to-wafer), a step

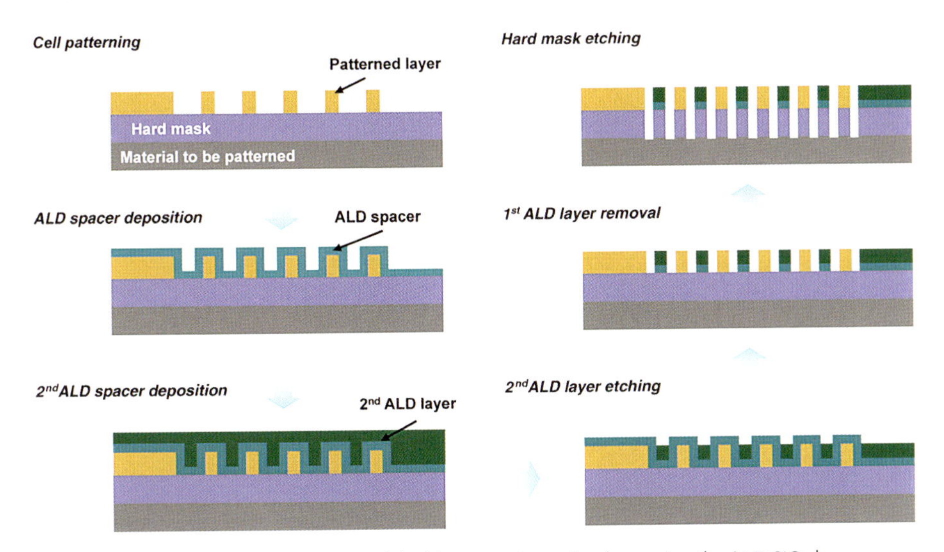

Figure 8.8 Another process sequence of double patterning technology using the ALD SiO_2 layer, which is used mainly as the bit line patterning.

coverage of ~100%, and low deposition temperature (<~200 °C) so as not to deform the finely patterned PR are required for this type of application. Low stress is another key factor. This is a very stringent requirement, even for ALD. ALD of SiO_2 has been quite challenging previously but it is now widely developed due mainly to the improvement in Si precursors [30, 31]. Various applications of ALD SiO_2 with different film characteristics have been reported. Some other materials with higher dry etching selectivity than SiO_2 might be necessary in the future depending on the type of patterning process.

8.2.1.3 Phase Change Memory

PCRAM employs a high resistance contrast (as high as 10^4) of crystalline and amorphous chalcogenide materials, most typically $Ge_2Sb_2Te_5$ (GST). The PC material is interposed between the very small bottom contact (diameter ≪50 nm) and large top electrode. The bottom contact connects the PC cell to the select transistor. (See Figure 8.9 for (a) conventional and (b) confined PCRAM structures [32].) The PC material changes its phase between crystalline and amorphous via the short (~50 ns)/intense (~0.2 mA) reset (from crystalline to amorphous) and longer (~150 ns)/mild (~0.1 mA) set (from amorphous to crystalline) current pulses. This corresponds to the write operation. For read, the bit line is precharged to a certain voltage, and word line of a selected cell is boosted. The drop level of the precharged bit line voltage is dependent on the resistance state of the PC cell.

There are two serious problems in PCRAM. One is related to the high reset current level, which is barely provided by a small-sized cell transistor (narrow width), and the other is related to the reliability of the write/read cycles. The former has been ascribed to the very inefficient use of Joule heat produced (only ~2% is used to melt the crystalline GST) and the latter appears to be related to the various process variables as well as the intrinsic ionic–covalent nature of the GST material. Considerable engineering efforts have been made to reduce the reset current by making a smaller bottom contact that concentrates the heat produced by the given reset current to a smaller volume. For this aim, ALD can contribute in two aspects. After contact hole etching, a thin dielectric layer is deposited conformally by ALD and etched back. The wafer surface is slightly lowered by a CMP to reduce the hole opening diameter.

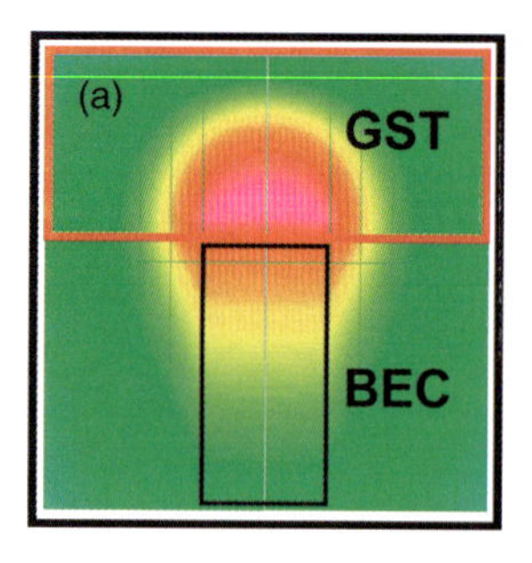

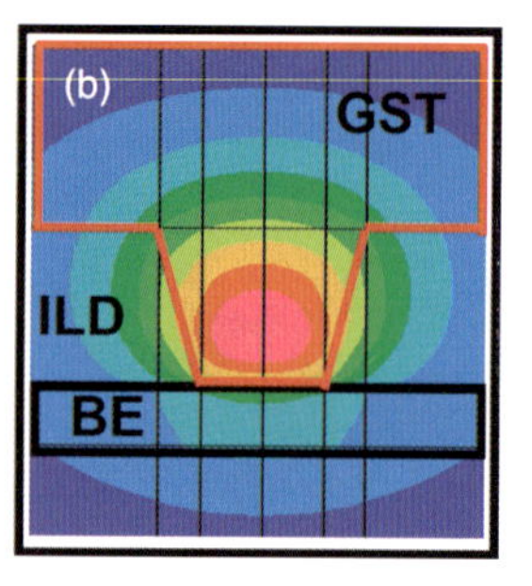

Figure 8.9 Schematic diagrams of (a) conventional and (b) confined PCRAM structure (BEC: bottom electrode contact; BE: bottom electrode; ILD: interlayer dielectric) [32].

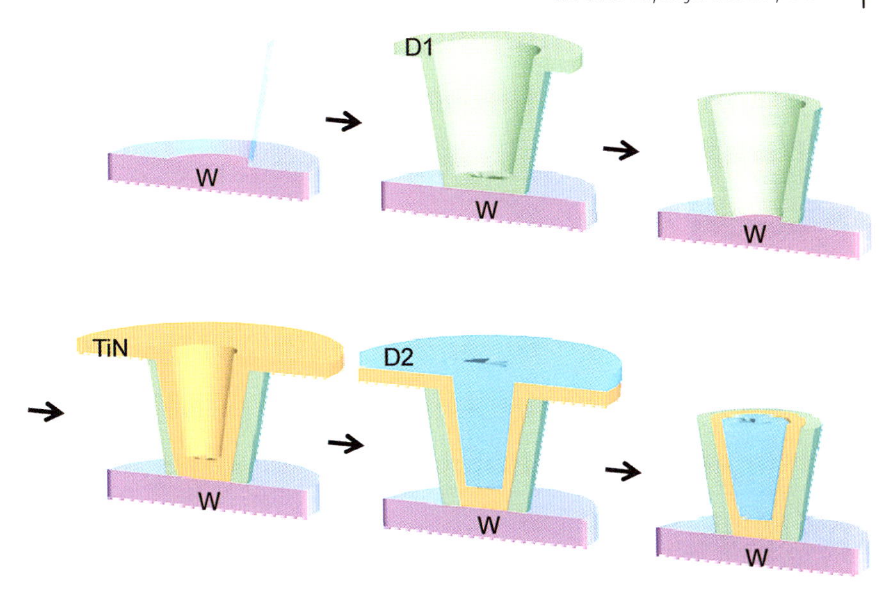

Figure 8.10 Schematic process sequence flow of ring-type bottom contact area (D1 and D2: dielectric layers 1 and 2 grown by ALD).

A thin ALD metal layer (typically TiN, and more recently ALD TiAlN) is then deposited. The central empty space is filled with another ALD dielectric and the entire structure is polished again to form a very fine ring-type bottom contact area. These process sequences are shown schematically in Figure 8.10.

Another more challenging area of ALD application in PCRAM is to fabricate a confined cell structure, as shown in Figure 8.9b. Here, the GST material is plugged into the small contact hole. It was reported that the heating efficiency of this confined cell was much higher than that of the conventional cell structure due to thermal simulation [33]. Fabricating this type of structure cannot be accomplished by the sputtering of GST, which is now widely used. CVD or ALD of GST is an indispensable technique. Several attempts have been made to fill such a small contact hole with stoichiometric GST conformally and uniformly [34, 35]. However, up to this moment, this has not been so successful. Thermal CVD films have a very rough surface and insufficient step coverage [34]. Plasma-enhanced ALD films showed several promising properties but the step coverage is also insufficient [35, 36]. Quite recently, Pore *et al.* reported ALD-type growth of the $Ge_2Sb_2Te_5$ alloy using silyl-Te and chloride-based Ge and Sb precursors [37]. Although this was a truly meaningful report from a materials chemistry point of view, several crucial issues need to be demonstrated from a material and processing point of view. Recently, the author's group acquired promising results using similar silylated Te precursors and Ge and Sb alkoxide precursors. The films grown in purely thermal ALD mode had a composition of $(GeTe_2)_x(Sb_2Te_3)_{1-x}$ $(0 < x < 1)$. As the film was grown by thermal ALD, conformal step coverage over a contact hole structure with an aspect ratio of 20 was achieved (see Figure 8.11). The films also exhibited promising PC properties. It is also hoped that

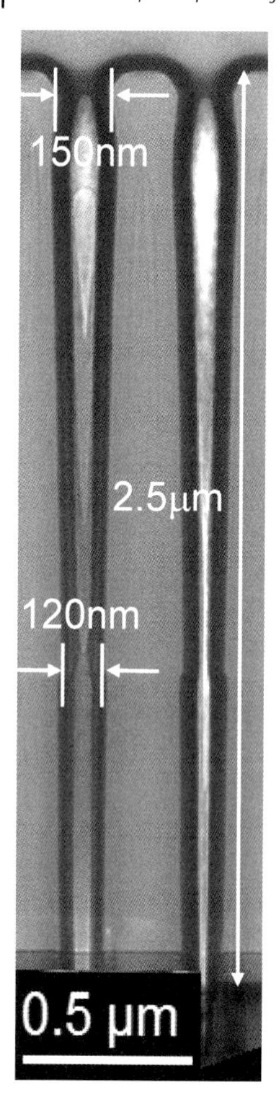

Figure 8.11 ALD GST film grown on hole structure with high aspect ratio.

the atomically uniform mixing of the constituent elements in the ALD GST material can contribute to the better reliability of the repeated PC operation. There are several other compositions in the Ge–Sb–Te alloy system with different features from the conventional $Ge_2Sb_2Te_5$ alloy. These are also good regions for material and process research into ALD processes. Again, these are the active application of ALD for PCRAM.

8.2.2 Emerging Memories

In this section, emerging memories, such as ferroelectric RAM (FeRAM), magnetic RAM (MRAM), and more recent resistive RAM (ReRAM), are reviewed briefly and the applications of ALD processes to these new memory devices are discussed.

FeRAM and MRAM were introduced more than 15 years ago. However, they are being produced only for some niche markets due to the many difficulties in device fabrication and limited functionality that have hindered the high-volume mass production. Therefore, they are still classified as emerging memories. Several new structures and operation modes, such as ferroelectric field effect transistor (FeFET) and spin-transfer torque RAM (STT-RAM), are emerging and have found some new applications. ReRAM conventionally indicates resistance-switching (RS) memory without an obvious phase change. This terminology is intended to discriminate a material system that shows reversible RS from PC materials but the distinction is rather vague as the RS mechanisms are unveiled [38]. Here, ReRAM means the RS memory that does not adopt the Ge–Sb–Te alloy as the active memory element. There are numerous material systems that show reversible RS with quite diverse RS mechanisms. Recent review by Waser *et al.* summarizes this aspect [39].

8.2.2.1 **Ferroelectric Memory**

The structure of FeRAM is similar to that of DRAM. The dielectric capacitor of DRAM is replaced with the ferroelectric capacitor in FeRAM, which makes nonvolatile data retention possible. The ferroelectric material is characterized by the presence of remnant polarization in the absence of an external electric field and the reversal of it with the reversal of an external field [40]. Ferroelectric switching is fast, reversible, low power consuming, nonvolatile, and with a large enough charge density for the stable memory operation [40]. With these characteristics, FeRAM appears to be the ideal charge-based memory device. However, the major drawbacks of FeRAM are related mainly to the difficult fabrication of small-sized ferroelectric capacitors. Typical ferroelectric thin film materials are $Pb(Zr, Ti)O_3$ (PZT) and $SrBi_2Ta_2O_9$ (SBT) of which the crystal structures are simple perovskite (PZT) or layered perovskite (SBT) [40]. As can be easily understood from the complicated chemical composition and crystal structures of the ferroelectric materials, their properties are easily affected by slight deviations from the stoichiometry (cation and oxygen stoichiometry) as well as the damaging effect on the crystal structure. The various process steps in fabricating FeRAM imposed a range of damaging effects on the ferroelectric layers. Dry etching has a heavy damaging effect. The electrode materials are limited to noble metals, such as Pt, Ir, or IrO_2, to prevent the formation of deleterious interfacial layers [41]. These materials are also difficult to etch. Therefore, making small-sized capacitors with a planar geometry suffers from the loss of stored charge as well as the various damaging effects from the fabrication process. Eventually, the ferroelectric capacitor structure should have a 3D structure as the DRAM capacitor does. This can largely relieve the process-induced damage. Therefore, CVD or more preferably ALD processes are needed for the ferroelectric and electrode layers. There have been many reports on the CVD of ferroelectric layers over the past few decades [42–45], but it has not been proven that the achieved technology level can be used to produce highly scaled ferroelectric capacitors with a 3D geometry and small design rule ($\ll$100 nm). Nonuniform deposition along the depth direction on a capacitor hole structure has hindered its adoption in FeRAM fabrication. The complete thermal decomposition of metal-organic precursors for PZT or SBT requires high temperatures

(>500–600 °C) during CVD, which is generally accompanied by poor thickness and composition step coverage [42, 46]. A high growth temperature is also necessary to ensure good crystalline quality of the ferroelectric layers.

Therefore, ALD of PZT or $PbTiO_3$ has been attempted at much lower temperatures [47–51]. The thermally fragile nature of the MO precursors limits the ALD temperature <~300 °C, but the impurity concentration is quite low and good thickness and composition step coverage has been achieved [50]. The atomically uniform incorporation of Zr into $PbTiO_3$ films to make a PZT phase is quite difficult due to the limited chemical reactivity between Zr and $PbTiO_3$ at such low temperatures [51]. An even more critical problem with the ALD PZT layer is that it has an amorphous structure at the as-deposited state with a relatively low density. Therefore, crystallization annealing at temperatures >600 °C is necessary to achieve the required ferroelectric properties. However, as in the case of the ALD STO layer, the initial low density results in the shrinkage of the film during annealing and the leakage current is very high [51]. This must be somewhat similar to sol–gel processing of a very thin oxide layer (<20 nm). Therefore, it is important to develop precursors that can be used in ALD at higher temperatures (>~350 °C) to achieve higher density with *in situ* crystallization.

The small-sized 3D ferroelectric capacitor also requires an ALD process for noble metal electrodes. Ir was reported to be a suitable bottom electrode material ensuring high remnant polarization, and fatigue-free and reliable long-term operation of the FeRAM for MOCVD PZT films [52]. This is due to the chemical inertness and high interfacial Schottky barrier with the ferroelectric layers. This must be the case for ALD ferroelectric thin films. Therefore, developing a precursor and process for the Ir electrode by ALD is another key application of ALD for FeRAM. However, this is quite premature compared to the Ru electrode for the DRAM capacitor. This can be attributed to the less active research in this field, primarily due to the smaller market size of FeRAM compared to DRAM.

Another important application of ALD for FeRAM is the hydrogen blocking layer. The ferroelectric performance of the capacitor is prone to degradation (decreasing remnant polarization, shift in the hysteresis loop along the voltage axis, and higher fatigue rate) during the back end of the line process. This was attributed mainly to the diffusion and incorporation of hydrogen into the ferroelectric layer during processing. Therefore, thin (~10 nm) ALD Al_2O_3 was deposited as a hydrogen blocking layer over the ferroelectric capacitor module [53]. This successfully protects the ferroelectric properties from degradation [53]. Figure 8.12 shows the cross-sectional structure of the ferroelectric capacitor module (Private communication with H.B. Kang).

The ferroelectric capacitor module is connected to the drain region of the select transistor via a contact plug. This contact plug material is made from TiN or TiAlN. TiAlN offers better oxidation resistance so that contact resistance degradation, which is caused by oxidation of the contacting portion of the plug material at the interface with the bottom electrode of the capacitor, can be decreased. As the contact size decreases, ALD of TiN and TiAlN is a suitable process for plugging the contact. This is commonly applied to the DRAM capacitor fabrication process with a MIM-type capacitor module.

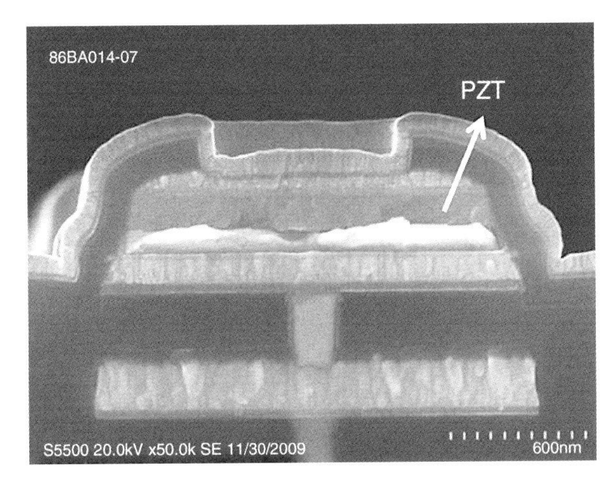

Figure 8.12 SEM image of FeRAM cell structure (Private communication with H.B. Kang).

Although there has been relatively little research into the very high density (>Gb) FeRAM, FeRAM still has very promising (perhaps the most promising) memory characteristics. Therefore, more active research in this field is necessary.

FeFET is another extremely interesting memory device of which the memory cell structure is similar to that of the flash memory cell, where the floating gate or charge trap layer in the flash cell is replaced with the ferroelectric layer in FeFET [54]. The remnant polarization of the ferroelectric layer affects the surface potential of the semiconductor channel, thereby controlling the resistance between the source and drain in a nonvolatile manner. In other words, the V_{th} of the transistor is modulated in a nonvolatile manner according to the polarization direction of the gate ferroelectric layer. The major problem of this type of ferroelectric device is the loss of data retention due to unwanted partial polarization reversal. This is basically caused by the incomplete screening of the ferroelectric surface charge by the low carrier density of the semiconductor surface, which is degraded further by the presence of a gate dielectric layer. The gate dielectric layer separates physically the ferroelectric surface charge and compensating channel charge, which produces the strong depolarization field. Charge injection through the thin gate dielectric layer by this strong depolarization field reorients the polarization direction of some part of the material into the opposite direction. Therefore, the high-k gate insulator, which can largely suppress carrier injection, is essential for this type of ferroelectric memory device. An ALD high-k layer was used for this purpose, which improved the device characteristics significantly [55]. Adopting ZnO oxide semiconductor channel layer was recently suggested by Kato *et al.* [56], which can form a 2D electron gas at the interface with the ferroelectric layer in FeFET. This structure showed highly improved retention properties. All these advanced applications of complex ferroelectric thin films require low temperature and less damaging (to the extremely delicate semiconductor layer) ALD processes.

8.2.2.2 Magnetic Random Access Memory

MRAM or STT-RAM employs the spin-dependent current transport across the thin dielectric film interposed between two magnetic electrode layers of a magnetic tunnel junction (MTJ). The MTJ is composed of two metallic magnetic electrode layers separated by a thin (1–2 nm) tunneling oxide, which is amorphous Al_2O_3 or highly textured crystalline MgO films [57–59]. The magnetization direction of one of the two magnetic layers can be switched (free layer) either by a stray magnetic field from the nearby electrical signal line (digit line), which carries a high switching current (MRAM), or by an intense injection of spin polarized current (STT-RAM), whereas the magnetization direction of the other layer is unaffected by these external stimuli (pinned layer). The tunneling current is high when the two magnetization directions are parallel, while it is low (by ~30% for Al_2O_3 and up to ~500% for MgO compared to the high current) [57] when the directions are antiparallel. Therefore, this is a type of resistance-based memory.

This type of switching requires minuscule amount of time, and the memory state is nonvolatile, which is quite robust to external disturbing noise. In addition, each memory cell is accessed randomly. Therefore, MRAM and STT-RAM have highly desirable memory properties. However, the presence of a digit line for MRAM and the very high current necessary for STT-RAM has impeded the rapid scaling of this type of magnetic memory device.

The key component of MRAM and STT-RAM is the MTJ. Although the thin dielectric layers are grown by metal deposition and (plasma) oxidation, it is highly desirable to deposit them by ALD in future memory. Because of the tunneling function of the dielectric layer, they must be pinhole free and extremely uniform (thickness variation <0.1 nm) over a large-area wafer (300 mm diameter). These can be readily achieved by ALD. In addition, the dielectric layers can be grown easily by ALD due to the simple structure and chemistry of the ALD material. The complicated magnetic metal layers are not expected to be grown by ALD. ALD may find even more importance in future MRAM or STT-RAM due to the following. One of the key difficulties in making such ultrasmall-scale MTJ is the etching damage or etching residue on the etched sidewall of the MTJ. No etching damage or conducting residue is allowed due to the extremely thin tunneling dielectric layer, which is quite difficult to achieve technically. One promising structure is the etching-free MTJ structure, where the stud-type bottom magnetic electrode layer is formed (as in the DRAM capacitor but with much lower height) and the ALD tunneling oxide is formed conformally on top of this stud structure. The top magnetic layer can be deposited with lower conformality, thereby minimizing the etching process. With this regard, ALD of the tunneling dielectric is essential. Even for conventional MRAM or STT-RAM, planar MTJ actually has a quite severe topography due to the presence of electrical contact. This again highlights the importance of ALD.

8.2.2.3 Resistive Random Access Memory

The memory cell structure of ReRAM is similar to that of PCRAM (Figure 8.9) while the PC cell is replaced with RS cells. There are many types of RS materials and underlying mechanisms, which comprise one of the most serious difficulties in

ReRAM presently. A recent review by Waser *et al.* summarized the diverse RS material classes and mechanisms [39, 60]. In any RS system, the RS cells are composed of an insulating (or resistance-switching) layer and two metal layers. Therefore, it is basically a MIM-type structure. In an anionic RS system, where anion migration causes RS, local structural or electronic changes in the insulating layer are responsible for RS. Typical materials are transition metal oxides, such as TiO_2, NiO, ZrO_2, HfO_2, ZnO, or doped/undoped perovskite oxides. These are the material classes that can be grown readily by ALD [61–64]. On the other hand, in cationic RS systems, where the cation channel is formed and ruptured repeatedly by the migration of one electrode element (active, typically Cu or Ag) to the counter (inert, typically Pt) electrode [65, 66]. The insulator layer works as a passive matrix and the high mobility of the diffusing cations is the necessary factor for the matrix layer in this case. The typical matrix layer comprises of sulfides [65], selenides [66], or even oxides [67]. These are also areas where ALD plays important role. The recent development of ALD for PCRAM chalcogenide materials also benefits this field [68].

The stacking of several RS layers [69] or doping [70] to improve the functionality becomes more common in this field. The structural modification of the cell to increase the efficiency of the driving power is also another key area for this type of memory device in the future. An increase in the active area of the RS cell would be necessary depending on the design and selection devices. In addition, a minimum feature size of $4F^2$ can be reached using a crossbar array structure in ReRAM [71]. The crossbar structure makes it easy to stack cells, which in turn makes it possible to achieve smaller feature sizes of $(4/n)F^2$ when n layers are stacked upon each other [72]. These structural modifications and 3D structural formation require conformal deposition of the active RS thin film material. For the 3D RS cells, the electrode must also be grown by a process with good conformality. Therefore, ALD is expected to play an increasingly important role in this field.

8.2.3
Three-Dimensional Stacked Memories

The eventual way of integrating the highest memory density must be 3D stacking. Chip level stacking is already in mass production (multichip packaging, MCP). However, MCP is not the optimum structure for the highest density memory nor is it most cost-effective fabrication method. Wafer level stacking by TSV technology has been highlighted. In TSV technology, the backside grounded wafers with the memory patterns were diced, and the fabricated chips were stacked to form multilevel memory. Each stacked layer was connected electrically through the TSV where large and deep via holes were dry etched and filled with interconnection metals. In TSV technology, the sidewall area of the via holes is insulated from the interconnection metal by adopting a SiO_2 insulating layer. ALD SiO_2 with a higher growth rate is necessary due to the high aspect ratio of the TSV hole and the necessary thickness of the SiO_2 layer. Processing of these steps on such small pieces of stacked chips is technically quite challenging. Meanwhile, the first stacking process may suffer from a low yield. In addition, expensive photolithography steps for the fabrication of each wafer layer are still necessary. Therefore, the wafer level stacking technology via TSV will not be the eventual method to reduce the cost of ultrahigh-density memory chips.

On the other hand, film or cell level stacking must be the ultimate structure for the highest memory devices without increasing the fabrication cost, which is related to the extremely fine patterning, to an uneconomically high level. This structure and related necessity for ALD were discussed in Section 8.2.1.2 for vertical NAND flash fabrication. In general, the usefulness of ALD increases as the memory structure becomes increasingly complex in the 3D structures to achieve a higher integration density. Oxide semiconductors, such as crystalline ZnO [73] or amorphous $InGaZnO_x$ [74], are attractive candidates as the channel material in vertically or laterally stacked 3D flash memory.

8.3 ALD for Logic Devices

8.3.1 Front End of the Line Process

Front end of the line fabrication process means the fabrication of MISFET. As the complementary MISFET is mainly used, this includes the fabrication of both the n- and p-type MISFETs, which actually requires several different process steps depending on the design and fabrication process. This will turn out to be even more evident as the high-k gate dielectric becomes the mainstream gate-related technology. Although the SiO_2 gate dielectric is still in major use for the logic chips with relatively long channel length ($L > \sim 50$ nm), it is evident that future devices will be heavily dependent on high-k gate dielectric technology. This must be more evident if the standard Si-based channel is replaced with Ge or III–V compound semiconductor materials, as these channel materials do not have stable thermal oxides. Figure 8.13 shows schematic cross sections of n- and p-MISFETs with high-k gate dielectrics. Here, the fabrication process is assumed to be "gate-first." When the fabrication process is assumed to be "gate-last," they are slightly changed to that shown in Figure 8.14 due to the different fabrication sequences. In Figures 8.13 and 8.14, the layers deposited by ALD are underlined.

The primary role of the logic MISFETs is fast data processing, which is achieved by a fast on/off speed. Therefore, the focus of the logic MISFET fabrication process lies on increasing the on current where an increase in the off current and gate leakage current to a certain level is allowed. This is quite contradictory to memory transistors where the low off current and low gate leakage current are of critical importance. The ratio between on and off currents is $\sim 10^8$ for memory transistors while it is only 10^4–10^6 for logic devices. To achieve these goals, the V_{th} of logic MISFETs is generally much lower (0.2–0.4 V) than that of the memory MISFETs (0.7–0.8 V). Other two key technologies for achieving high performance are high-k gate dielectrics (and related

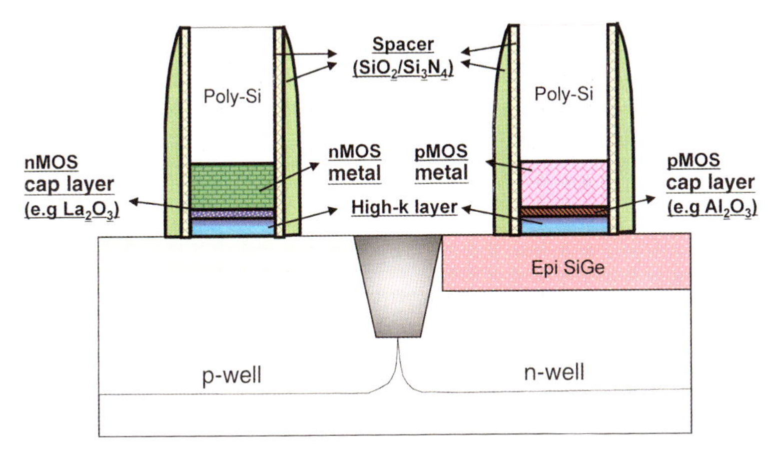

Figure 8.13 Schematic cross sections for the gate-first process of n- and p-type MISFETs with high-*k* gate dielectrics.

technologies) and strained channel technology. While the latter has a relatively poor relationship with ALD, the former is strongly dependent on it. Some high-stress ALD layers, such as SiN_x, may help achieve global stress control in n-type MISFET [75]. Achieving a high gate capacitance corresponds to increasing the gate controllability of the channel compared to that of the drain. However, increasing the gate capacitance also increases the gate delay time due to the RC effect. Therefore, a further decrease in *L* should be accompanied to compensate for the performance degradation by the RC effect. Nevertheless, the decreased gate leakage is the large impetus that drives high-*k* technology.

High-*k* dielectric layers are deposited by ALD. There are many technical and review reports on the materials, integration, and device characteristics in this field, all of

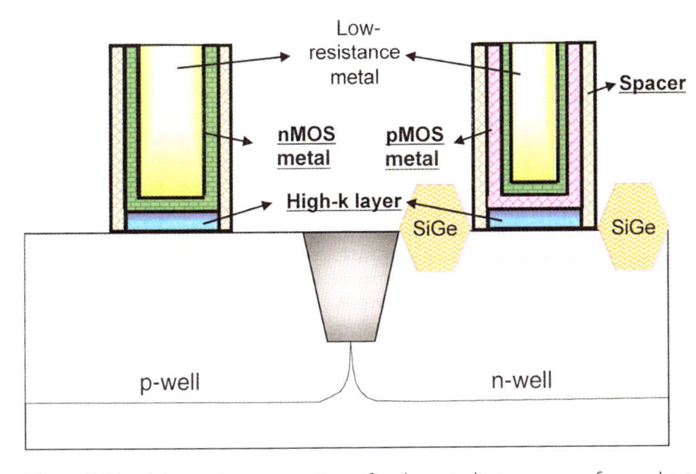

Figure 8.14 Schematic cross sections for the gate-last process of n- and p-type MISFETs with high-*k* gate dielectrics.

which cannot be referenced here. Interested readers can refer to several books and references therein [76–78]. Perhaps the most viable high-k dielectric layer is a HfO_2-based material. Intel already used it in their Penryn chips in 2007 [79]. It becomes clearer that the key component for a high-k dielectric is to achieve the appropriate balance between sacrifice of the gate capacitance by adopting a thicker interfacial layer (IL, mostly SiO_2) and sacrifice of the carrier mobility by adopting a thinner IL. A channel with a thinner IL suffers from a higher carrier scattering effect for many reasons, including the interface roughness and remote phonon scattering effects [80], which decrease the carrier mobility. In most ALD processes, the IL grows simultaneously with high-k dielectrics even on a previously grown IL using the optimized pretreatment techniques. This normally increases the EOT without improving the carrier mobility. Therefore, an optimized ALD process that minimizes the further growth of IL is essential. This can be achieved by the lower oxidation potential of the oxygen source during ALD but the lower oxidation potential can increase the defect concentration in high-k films. On the other hand, metal alkoxides and carboxylic acid-based ALD reaction routes, which potentially minimize interfacial oxidation, have been suggested [81]. However, the device characteristics using these new ALD mechanisms have not been reported.

Thicker high-k films usually show more severe reliability concerns, such as a V_{th} shift during the bias temperature instability tests, even with a lower gate leakage current suggesting that the defect concentration in the high-k film is still much higher than SiO_2 [78]. These reports suggest that several critical improvements in the ALD process and materials (high-k dielectric itself and precursor/oxygen source) are needed to achieve the optimum high-k gate dielectric layers.

In addition, further reduction of the EOT (currently it is $\sim$1 nm for $L \sim 30$ nm) is also necessary for the further scaled devices [5]. As is the case for SiO_2, decreasing the Hf-based high-k gate dielectric thickness would end up with a too high gate leakage current. Therefore, an even higher-k dielectric will be needed in the near future. Modifying the crystalline structure of HfO_2 from monoclinic ($k < \sim 20$) to tetragonal/cubic ($k \sim 30$–40) by doping [82, 83] will be a near-term approach. Eventually, different higher-k materials (perhaps La-based materials [83, 84] or even STO [85]) will be pursued. ALD should be an appropriate deposition method for doped high-k or higher-k dielectrics. However, these materials require ever-increasing complexity of film processing conditions involving multi-cation film growth.

The conformal film growth properties of ALD have received less attention in the MISFET area due to the much lower topological surface structure than memory devices. However, this may not be the case in future 3D MISFETs. In 3D MISFETs, such as FinFET, the gate dielectric and gate must surround the 3D channel, which increases the controllability of the channel by the gate compared to planar type devices. In such cases, the gate dielectric and gate metal must be deposited conformally over the 3D channel so that ALD must be the process of choice.

The gate in MISFET works as an electrode for applying V_g to the device. It should have metallic conductivity to minimize carrier depletion for the highly scaled EOT device. For CMISFET fabrication, the work function of the gate metal should be controllable up to $\sim$0.5 eV to achieve a low enough V_{th} for both n- and p-type

MISFETs. Achieving different work function metal gates for the n- and p-type MISFETs and processing them on the same wafer is quite complicated. Therefore, an alternative approach that employs the dipole inducing oxide layer in combination with the midgap metal gate (typically TiN) is being actively pursued [86] as discussed below.

Silicidation of the poly-Si gate and source/drain contact region was adopted to decrease the gate resistance and source/drain contact resistance. Conventional silicidation was achieved by depositing Ti, Co, or Ni metal by sputtering and rapid thermal annealing (RTA) under an inert atmosphere. After RTA, the residual metal layers over the sidewall area of the gate, where silicidation was prohibited by the underlying SiO_2/Si_3N_4 spacer, were removed by a wet etching process. This is a crucial step for achieving the good electrical isolation between the gate and source/drain. However, for very small scale MISFET devices, particularly for memory applications, the sputtering process is unsuitable for depositing metal layers with sufficient thickness on the source and drain regions. This is because the gap between the gates becomes too narrow to achieve sufficient deposition of the sputtered material on the source and drain regions. For 3D MISFET mentioned above, the 3D coverage of the silicides is certainly necessary. Therefore, ALD of metals, such as Co or Ni, is desirable for achieving a uniform and thin metal layer. However, ALD of such metals is a difficult task. There are several reports on this issue [87, 88].

Contamination of the gate dielectric and channel by the diffusion of any elements from the gate metal is strictly prohibited. For the gate-first integration scheme, the gate metal should withstand a very high temperature for source/drain activation. These requirements severely limit the choice of gate metal, and have ruled out fully silicided metal gate technology already [89, 90]. Transition metal nitrides, such as TiN and TaN, and their modification by doping with more cation-like Al and Si or anion-like C are the most probable candidates. TaC_xN_y is particularly interesting owing to its refractory metal nature, which provides the material with exceptional thermal stability, and controllability of the work function depending on the C content [91]. These materials are deposited readily by ALD processes using metal-organic precursors and H_2 plasma or NH_3 reaction gases [91, 92]. Carbide materials are especially interesting because they do not have a homologous insulating phase as opposed to the nitride material. However, suitable process optimization is still important for achieving a stable carbide phase because there is a risk of residual C, which acts as an impurity. Serious aging of the sheet resistance even at room temperature is observed when these are not controlled appropriately [93].

Even with the extensive engineering efforts to achieve an appropriate work function for the n- and p-type MISFETs with metal gate and high-k gate dielectrics, the work functions tend to decay to the midgap value as the EOT scaling proceeds. This might be related to the Fermi-level pinning effects between the metal gate and high-k gate dielectrics [94, 95]. The capping layer approach (typically Al_2O_3 and La_2O_3 for p- and n-type MISFETs, respectively) utilizing the interface dipole effect [96, 97] is useful for achieving the appropriate V_{th} values. The capping layer (or V_{th} control layer) needs to be thin enough not to increase the EOT but should be uniform enough to achieve the desired V_{th} control effect over a wide wafer surface area. ALD is particularly suitable for such purposes. It needs to be reminded that work function drift to the midgap point becomes more evident as the EOT scales down to $\ll$1 nm. This becomes even more severe after a high-temperature annealing process in the gate-first integration scheme. Therefore, a gate-last process (or called replacement metal gate process) will become increasingly dominant as the EOT scales down to $\ll$1 nm. For less performance-driven devices, for example, low standby power applications, the gate-first process with a slightly higher V_{th} will still be used.

Eventually, Si semiconductor channel will be replaced with Ge or III–V channel materials. The In_xGa_yAs channel has already been examined seriously by Intel [98, 99]. As these non-Si channel semiconductors do not have stable thermal oxides, the importance of the ALD process for the gate process will be even more significant. It is clear that the work on ALD high-k on non-Si channel materials is much less developed compared to Si. These non-Si channel materials will be grown epitaxially on Si, which will enable the use of preexisting Si fabrication infrastructure. Therefore, there is a need for ALD study in this field, particularly from the academic sector.

Other important application of ALD to the front-end process is the spacer formation. The spacer means (see Figure 8.13) an insulating layer composed of SiO_2/Si_3N_4 formed at the sidewall area of the gate. This layer has several important structural and electrical roles for the fabrication and electrical operation of MISFETs. It prohibits source-to-gate and drain-to-gate leakage current during MISFET operation. It works as a blocking layer for short path formation (between the gate-to-source and drain-to-source) during the silicidation (Ti, Co, and Ni silicides) process. Heavy doping of the channel with the counter dopant region during the source/drain implantation is also prevented by the spacer. The increased lateral electric field in highly scaled devices induces several unwanted short channel effects, which means a decrease in V_{th} with decreasing L. This effect has been reduced by adopting a lightly doped drain (LDD) between the channel and source/drain. The short channel effect can be reduced efficiently by the extremely shallow junction (between the source and channel, and drain and channel) due to a decrease in the 3D effect of the drain-induced barrier lowering effect. This requires an overall decrease in the total thermal budget after implantation and activation (spike) annealing. The thinner and more crystalline high-k dielectrics have also created concerns regarding dopant diffusion from the gate to the channel. This is especially the case for the gate-first process due to the adoption of doped poly-Si (most concern is on the B diffusion) on top of the thin metal gate layer. Therefore, an overall decrease in the thermal budget is essential for such a highly scaled device. The high thermal budget of thermal CVD for SiO_2/Si_3N_4 deposition is undesirable for these aspects. Therefore, ALD of SiO_2/Si_3N_4 layers is desirable for reducing the unwanted dopant diffusion in the LDD region. The spacer process has no effect on shallow source/drain junction formation because the spacers are formed before junction formation. The spacer is formed by the conformal deposition of the dielectric layers and etching back. As these are critical layers for the stable MISFET operation, they must have a very low defect density and high structural stability.

8.3.2
Back End of the Line Process

The back end of the line process contains an interconnection line and metal contact formations. Figures 8.15 and 8.16 [100, 101] show the cross section of logic device with eight-level Cu metallization and cell-to-periphery region of DRAMs. ALD is used in two areas for logic devices with Cu interconnects and low-k dielectrics: seed layer formation and pore sealing. For DRAMs, the CVD processes for deep W contacts and metal bit or word line require the aid of an ALD process.

As discussed above, the speed of the logic device is determined by metal line delay and MISFET channel delay. Here, the metallization process is focused on reducing the metal line delay, which is determined by the interconnect line resistance (R_l) and stray capacitance (C_s) near the metal lines. Cu has already replaced Al wires to reduce R_l. However, Cu deposition and making interconnection lines require subtle modification of the processing steps. While the Al interconnection lines are fabricated by sputtering and dry etching process, Cu interconnection lines require electrodeposition and damascene processes due to the difficulty in dry etching of Cu. In the damascene (also dual damascene process) process, the trench is first etched in the low-k dielectric layer, in which Cu was electrodeposited. Subsequently, the overgrown Cu layer is removed by CMP. The seed layer is necessary for the electrodeposition of Cu. This thin seed layer is normally deposited by sputtering. As Cu has very active diffusion properties in SiO_2, a diffusion barrier layer (typically Ta) is deposited onto the trench before Cu deposition. These integration processes require rather complicated etching/deposition/CMP steps. In addition, the step

Figure 8.15 SEM image of interconnect stack up to MT8 [100].

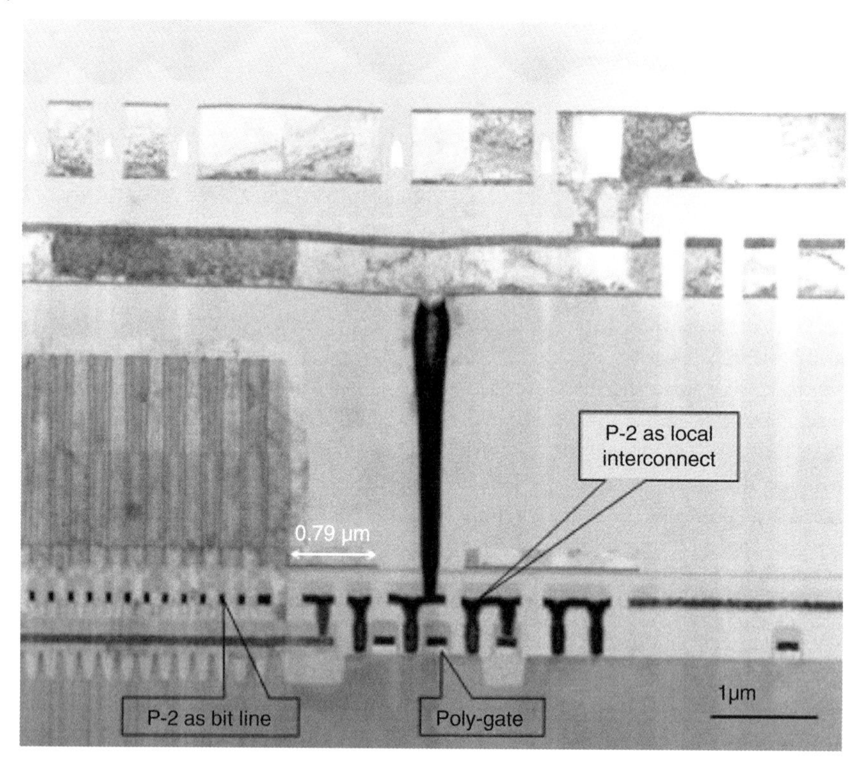

Figure 8.16 Memory cell-to-periphery region of DRAMs: Samsung 512Mb DDR SDRAM TEM image [101].

coverage of the sputtering process for the diffusion barrier and seed layers becomes insufficient with the decreasing line width and increasing trench depth (for compensating the increase of R_l by the decreased line width). Therefore, an ALD process of a certain metal layer that acts as both a diffusion barrier and a Cu seeding layer is necessary. Among the many candidate materials, ALD Ru appears to be the most viable barrier/seed layer owing to its promising material properties and process maturity. The Ru ALD process was discussed in Section 8.2.1.1.

Decreasing C_s has been attempted in several ways [102–104]. Recently, low-k dielectrics with sub-nm-scale pores were highlighted [103, 104]. However, these nanoporous low-k dielectric materials induced insulation failure between the interconnection lines due to severe Cu diffusion into the low-k dielectric layer along the possibly percolated pore channels. Therefore, sealing of the pore channel at the interface with the metal line is essential. Adopting a thin ALD SiO_2 layer may be a viable solution to this problem [105].

CVD W plug and lines are used in DRAMs for the metal contacts in the cell and periphery regions and word lines or bit lines. CVD W can be used to fill the deep contact holes and vias on account of its good step coverage and relatively easy processing based on the reduction of WF_6 source gas by H_2 or SiH_4 [106, 107]. With the smaller and deeper contact holes, the step coverage of CVD W process becomes insufficient. In particular, the seed layer (or nucleation layer) growth step with a lower deposition rate but better conformality suffers from the inadequate step coverage in highly scaled memory and logic devices. Therefore, a W ALD process for replacing the nucleation CVD W step is under development [108, 109].

The use of highly conducting W as the word line in flash memory and the bit line in DRAM is important for decreasing the stray capacitance. In particular, in DRAM, the bit line capacitance (C_b) really competes with the cell capacitance (C_c), so that reducing C_b is as important as increasing C_c. Decreasing C_b can be achieved by decreasing the line thickness, which may bring about a high bit line resistance issue. Therefore, a higher conductivity W is used as the metal bit line replacing the conventional heavily doped poly-Si bit line. The high thermal stability of W that can withstand the subsequent thermal budgets is another critical merit of W. For devices with W metal lines, the line width and trench depth are already too harsh for the standard CVD W processes. Therefore, the W ALD process requires another application in this field with the appropriate combination with the ALD steps of the barrier layers.

8.4 Concluding Remarks

This chapter reviewed the many applications of ALD processes and materials for the improvement and development of the semiconductor memory and logic devices. Due to the extreme diversity of the ALD processes in various semiconductor devices, the review proceeded on a rather arbitrary classification of semiconductor devices into mass production memory/emerging memory and logic devices. In this application field, the ALD layers work not only as functional materials in the final products but also as sacrificial layers that enhance several process capabilities. The former includes capacitor layers in DRAM, blocking oxide and charge trap layers in charge trap flash memory, phase change chalcogenide layers in PCRAM, ferroelectric layers in FeRAM, tunneling oxide layers in MRAM and STT-RAM, and various resistance-switching layers in ReRAM in memory devices. Fast emerging laterally stacked or vertically integrated 3D memory also requires several active layers that must be grown by ALD. In logic devices, the high-k gate stack, including the gate dielectric and metal gate, is the most typical area where the ALD layers play a key role. In the back end of the line process, ALD is essential for forming the metallization seed layer, W metal layer and contact, and porous low-k sealing layers. As a process capability enhancing factor, the double patterning process is a typical example of ALD, which critically improves the photolithographic capability without astronomical cost. These two applications require critically different properties of the film and process conditions. The active applications require supreme quality films that are generally accompanied by a higher process temperature and low growth rate, whereas the sacrificial applications require low temperatures but rather thick ALD films. These conditions are generally incompatible with each other. The generally low growth rate of the ALD films barely complies with the thick thickness requirements. Moreover, a high process temperature is normally difficult to achieve due to the limited thermal stability of the MO precursors. The high cost of several MO precursors is another serious concern for mass production. A typical example is the Ru precursor for the Ru or RuO_2 electrode. A batch- or semi-batch-type ALD reactor is used to comply with the low growth rate problem. Developing less bulky ligands and even adopting inorganic ligands can increase the growth rate further. New chemistry, such as cyclopentadienyl- or amidinate-based ligands, is still emerging and is expected to open the door for better ALD process windows. With the extreme requirements for the exceptionally precise thickness control on the nanometer scale over the 3D geometry, ALD appears to be the most suitable method for depositing the critical layers in the semiconductor field.

Acknowledgments

The author acknowledges the support of the National Research Program for the Nano Semiconductor Apparatus Development, IT R&D program of MKE/IITA, "Capacitor technology for next generation DRAMs having mass-production compatibility" 2009-F-013-01, the National Research Foundation of Korea (NRF) funded by the Ministry of Education, Science, and Technology (2009-0081961), and World Class University program through the Korea Science and Engineering Foundation funded by the Ministry of Education, Science, and Technology (R31-2008-000-10075-0). The author greatly appreciates the crucial help from postdoctoral fellows (Dr. Seong Keun Kim, Dr. Sang Woon Lee, and Dr. Kyung Min Kim) and graduate students (Mrs. Hyung-Suk Jung, Seok-Jun Won, Jeong Hwan Han, Sang Ho Rha, Seol Choi, Taeyong Eom, Hyo Kyeom Kim, Jeong Hwan Kim, Min Hyuk Park, Jun Yeong Seok, Misses Hyun Ju Lee, Sang Young Lee, and Seul Ji Song) of the author's group in Seoul National University.

References

1 Ahonen, M., Pessa, M., and Suntola, T. (1980) *Thin Solid Films*, **65**, 301.

2 Kim, K. (2005) International Electron Devices Meeting, Technical Digest, p. 323.

3 Kuesters, K.H., Beug, M.F., Schroeder, U., Nagel, N., Bewersdorff, U., Dallmann, G., Jakschik, S., Knoefler, R., Kudelka, S., Ludwig, C., Manger, D., Mueller, W., and Tilke, A. (2009) *Adv. Eng. Mater.*, **11**, 241.

4 Niinistö, J., Kukli, K., Heikkilä, M., Ritala, M., and Leskelä, M. (2009) *Adv. Eng. Mater.*, **11**, 223.

5 International Technology Roadmap for Semiconductors, 2009.

6 Kil, D.S., Song, H.S., Lee, K.J., Hong, K., Kim, J.H., Park, K.S., Yeom, S.J., Roh, J.S., Kwak, N.J., Sohn, H.C., Kim, J.W., and Park, S.W. (2006) Symposium on VLSI Technology, Digest of Technical Papers, p. 38.

7 Cho, H.J., Kim, Y.D., Park, D.S., Lee, E., Park, C.H., Jang, J.S., Lee, K.B., Kim, H.W., Ki, Y.J., Han, I.K., and Song, Y.W. (2007) *Solid-State Electron.*, **51**, 1529.

8 Kim, S.K., Kim, W.D., Kim, K., Hwang, C.S., and Jeong, J. (2004) *Appl. Phys. Lett.*, **85**, 4112.

9 Choi, G.-J., Kim, S.K., Lee, S.Y., Park, W.Y., Seo, M., Choi, B.J., and Hwang, C.S. (2009) *J. Electrochem. Soc.*, **156**, G71.

10 Kim, S.K., Choi, G.-J., Lee, S.Y., Seo, M., Lee, S.W., Han, J.H., Ahn, H.-S., Han, S., and Hwang, C.S. (2008) *Adv. Mater.*, **20**, 1429.

11 Hwang, C.S., Park, S.O., Cho, H.-J., Kang, C.S., Kang, H.-K., Lee, S.I., and Lee, M.Y. (1995) *Appl. Phys. Lett.*, **67**, 2821.

12 Hwang, C.S. (2002) *J. Appl. Phys.*, **92**, 432.

13 Black, C.T. and Welser, J.J. (1999) *IEEE Trans. Electron Devices*, **46**, 776.

14 Natori, K., Otani, D., and Sano, N. (1998) *Appl. Phys. Lett.*, **73**, 632.

15 No, S.Y., Oh, J.H., Jeon, C.B., Schindler, M., Hwang, C.S., and Kim, H.J. (2005) *J. Electrochem. Soc.*, **152**, C435.

16 Menou, N., Popovici, M., Clima, S., Opsomer, K., Polspoel, W., Kaczer, B., Rampelberg, G., Tomida, K., Pawlak, M.A., Detavernier, C., Pierreux, D., Swerts, J., Maes, J.W., Manger, D., Badylevich, M., Afanasiev, V., Conard, T., Favia, P., Bender, H., Brijs, B., Vandervorst, W., Van Elshocht, S., Pourtois, G., Wouters, D.J., Biesemans, S., and Kittl, J.A. (2009) *J. Appl. Phys.*, **106**, 094101.

17 Kawahara, T., Yamamuka, M., Yuuki, A., and Ono, K. (1996) *Jpn. J. Appl. Phys.*, **35**, 4880.

18 Tsai, M.S., Sun, S.C., and Tseng, T.-Y. (1999) *IEEE Trans. Electron Devices*, **46**, 1829.

19 Popovici, M., Van Elshocht, S., Menou, N., Swerts, J., Pierreux, D., Delabie, A., Brijs, B., Conard, T., Opsomer, K., Maes, J.W., Wouters, D.J., and Kittl, J.A. (2010) *J. Electrochem. Soc.*, **157**, G1.

20 Kim, S.K., Lee, S.W., Han, J.H., Lee, B., Han, S., and Hwang, C.S. (2010) *Adv. Funct. Mater.*, **20**, 2989.

21 Han, J.H., Lee, S.W., Choi, G.J., Lee, S.Y., Hwang, C.S., Dussarrat, C., and Gatineau, J. (2009) *Chem. Mater.*, **21**, 207.

22 Reisinger, H., Franosch, M., Hasle, B., and Bohm, T. (1997) Symposium on VLSI Technology, Technical Digest, p. 113.

23 Lee, C.H., Choi, K.I., Cho, M.K., Song, Y.H., Park, K.C., and Kim, K. (2003) International Electron Devices Meeting, Technical Digest, p. 26.

24 Sung, S.-K., Lee, S.-H., Choi, B.Y., Lee, J.J., Choe, J.-D., Cho, E.S., Ahn, Y.J., Choi, D., Lee, C.-H., Kim, D.H., Lee, Y.-S., Kim, S.B., Park, D., and Ryu, B.-I. (2006) Symposium on VLSI Technology, Digest of Technical Papers, p. 86.

25 Lue, H.-T., Lai, S.-C., Hsu, T.-H., Du, P.-Y., Wang, S.-Y., Hsieh, K.-Y., Liu, R., and Lu, C.-Y. (2009) Proceedings of IEEE International Reliability Physics Symposium, p. 847.

26 Tanaka, H., Kido, M., Yahashi, K., Oomura, M., Katsumata, R., Kito, M., Fukuzumi, Y., Sato, M., Nagata, Y., Matsuoka, Y., Iwata, Y., Aochi, H., and Nitayama, A. (2007) Symposium on VLSI Technology, Digest of Technical Papers, p. 14.

27 Kim, W., Choi, S., Sung, J., Lee, T., Park, C., Ko, H., Jung, J., Yoo, I., and Park, Y. (2009) Symposium on VLSI Technology, Digest of Technical Papers, p. 188.

28 Hong, S.H., Jang, J.H., Park, T.J., Jeong, D.S., Kim, M., Won, J.Y., and Hwang, C.S. (2005) *Appl. Phys. Lett.*, **87**, 152106.

29 Hwang, B., Han, J., Kim, M.C., Jung, S., Lim, N., Jin, S., Yim, Y., Kwak, D., Park, J., Choi, J., and Kim, K. (2009) Proceedings of European Solid-State Device Research Conference, p. 269.

30 Kamiyama, S., Miura, T., and Nara, Y. (2006) *Thin Solid Films*, **515**, 1517.

31 Katamreddy, R., Feist, B., and Takoudisb, C. (2008) *J. Electrochem. Soc.*, **155**, G163.

32 Kim, Y.-T., Hwang, Y.-N., Lee, K.-H., Lee, S.-H., Jeong, C.-W., Ahn, S.-J., Yeung, F., Koh, G.-H., Jeong, H.-S., Chung, W.-Y., Kim, T.-K., Park, Y.-K., Kim, K.-N., and Kong, J.-T. (2007) International Conference on Solid State Devices and Materials, Extended Abstracts, p. 244.

33 Kim, Y.T., Hwang, Y.N., Lee, K.H., Lee, S.H., Jeong, C.W., Ahn, S.J., Yeung, F., Koh, G.H., Jeong, H.S., Chung, W.Y., Kim, T.K., Park, Y.K., Kim, K.N., and Kong, J.T. (2005) *Jpn. J. Appl. Phys.*, **44**, 2701.

34 Kim, R.Y., Kim, H.G., and Yoon, S.G. (2006) *Appl. Phys. Lett.*, **89**, 102107.

35 Choi, B.J., Choi, S., Shin, Y.C., Hwang, C.S., Lee, J.W., Jeong, J., Kim, Y.J., Hwang, S.Y., and Hong, S.K. (2007) *J. Electrochem. Soc.*, **154**, H318.

36 Choi, B.J., Choi, S., Shin, Y.C., Kim, K.M., Hwang, C.S., Kim, Y.J., Son, Y.J., and Hong, S.K. (2007) *Chem. Mater.*, **19**, 4387.

37 Pore, V., Hatanpää, T., Ritala, M., and Leskelä, M. (2009) *J. Am. Chem. Soc.*, **131**, 3478.

38 Kwon, D.H., Kim, K.M., Jang, J.H., Jeon, J.M., Lee, M.H., Kim, G.H., Li, X.S., Park, G.S., Lee, B., Han, S., Kim, M., and Hwang, C.S. (2010) *Nat. Nanotechnol.*, **5**, 148.

39 Waser, R., Dittmann, R., Staikov, G., and Szot, K. (2009) *Adv. Mater.*, **21**, 2632.

40 Scott, J.F. (2000) *Ferroelectric Memories*, Springer, Heidelberg.

41 Nakamura, T., Nakao, Y., Kamisawa, A., and Takasu, H. (1994) *Appl. Phys. Lett.*, **65**, 1522.

42 Zhao, J.S., Lee, H.J., Sim, J.S., Lee, K., and Hwang, C.S. (2006) *Appl. Phys. Lett.*, **88**, 172904.

43 Zhao, J.S., Sim, J.S., Lee, H.J., Park, D.Y., Hwang, G.W., Lee, K., and Hwang, C.S. (2006) *J. Electrochem. Soc.*, **153**, F81.

44 Nagashima, K., and Funakubo, H. (2000) *J. Appl. Phys.*, **39**, 212.

45 Scott, J.F., Paz de Araujo, C.A., Mcmillan, L.D., Yoshimori, H., Watanabe, H., Mihara, T., Azuma, M., Ueda, T., Ueda, T., Ueda, D., and Kano, G. (1992) *Ferroelectrics*, **133**, 47.

46 Wouters, D.J., Maes, D., Goux, L., Lisoni, J.G., Paraschiv, V., Johnson, J.A., Schwitters, M., Everaert, J.-L., Boullart, W., Schaekers, M., Willegems, M., Vander Meeren, H., Haspeslagh, L., Artoni, C., Caputa, C., Casella, P., Corallo, G., Russo, G., Zambrano, R., Monchoix, H., Vecchio, G., and Van Autryve, L. (2006) *J. Appl. Phys.*, **100**, 051603.

47 Harjuoja, J., Kosola, A., Putkonen, M., and Niinistö, L. (2006) *Thin Solid Films*, **496**, 346.

48 Hwang, G.W., Lee, H.J., Lee, K., and Hwang, C.S. (2007) *J. Electrochem. Soc.*, **154**, G69.

49 Watanabe, T., Hoffmann-Eifert, S., Mi, S., Jia, C., Waser, R., and Hwang, C.S. (2007) *J. Appl. Phys.*, **101**, 014114.

50 Watanabe, T., Hoffmann-Eifert, S., Peter, F., Mi, S., Jia, C., Hwang, C.S., and Waser, R. (2007) *J. Electrochem. Soc.*, **152**, G262.

51 Watanabe, T., Hoffmann-Eifert, S., Hwang, C.S., and Waser, R. (2007) *J. Electrochem. Soc.*, **155**, D715.

52 Lee, M.S., Park, K.S., Nam, S.D., Lee, K.M., Seo, J.S., Joo, S.H., Lee, S.W., Lee, Y.T., An, H.G., Kim, H.J., Cho, S.L., Son, Y.H., Kim, Y.D., Jung, Y.J., Heo, J.E., Park, S.O., Chung, U.I., and Moon, J.T. (2002) *Jpn. J. Appl. Phys.*, **41**, 6709.

53 Kang, Y.M., Kim, J.-H., Joo, H.J., Kang, S.K., Rhie, H.S., Park, J.H., Choi, D.Y., Oh, S.G., Koo, B.J., Lee, S.Y., Jeong, H.S., and Kim, K. (2005) Symposium on VLSI Technology, Digest of Technical Papers 6B, p. 102.

54 Waser, R. (2003) *Nanoelectronics and Information Technology*, Wiley-VCH Verlag GmbH, Weinheim.

55 Sakai, S., Ilangovan, R., and Takahashi, M. (2004) *Jpn. J. Appl. Phys.*, **43**, 7876.

56 Kato, Y., Kaneko, Y., Tanaka, H., and Shimada, Y. (2008) *Jpn. J. Appl. Phys.*, **47**, 2719.

57 Jullière, M. (1975) *Phys. Lett. A*, **54**, 225.

58 Moodera, J.S., Kinder, L.R., Wong, T.M., and Meservey, R. (1995) *Phys. Rev. Lett.*, **74**, 3273.

59 Parkin, S.S.P., Kaiser, C., Panchula, A., Rice, P.M., Huges, B., Samanti, M., and Yang, S.H. (2004) *Nat. Mater.*, **3**, 862.

60 Waser, R. and Aono, M. (2007) *Nat. Mater.*, **6**, 833.

61 Choi, B.J., Jeong, D.S., Kim, S.K., Choi, S., Oh, J.H., Rohde, C., Kim, H.J., Hwang, C.S., Szot, K., Waser, R., Reichenberg, B., and Tiedke, S. (2005) *J. Appl. Phys.*, **98**, 033715.

62 Lee, M.-J., Kim, S.I., Lee, C.B., Yin, H., Ahn, S.-E., Kang, B.S., Kim, K.H., Park, J.C., Kim, C.J., Song, I., Kim, S.W., Stefanovich, G., Lee, J.H., Chung, S.J., Kim, Y.H., and Park, Y.S. (2009) *Adv. Funct. Mater.*, **19**, 1587.

63 Yousfi, E.B., Fouache, J., and Lincot, D. (2000) *Appl. Surf. Sci.*, **153**, 223.

64 Lee, H.Y., Chen, P.S., Wu, T.Y., Wang, C.C., Tzeng, P.J., Lin, C.H., Chen, F., Tsai, M.J., and Lien, C. (2008) *Appl. Phys. Lett.*, **92**, 143911.

65 Sakamoto, T., Sunamura, H., Kawaura, H., Hasegawa, T., Nakayama, T., and Aono, M. (2003) *Appl. Phys. Lett.*, **82**, 3032.

66 Kozicki, M.N., Mitkova, M., Park, M., Balakrishnan, M., and Gopalan, C. (2003) *Superlatt. Microstruct.*, **34**, 459.

67 Guo, X., Schindler, C., Menzel, S., and Waser, R. (2007) *Appl. Phys. Lett.*, **91**, 133513.

68 Choi, B.J., Choi, S., Eom, T., Ryu, S.W., Cho, D.Y., Heo, J., Kim, H.J., Hwang, C.S., Kim, Y.J., and Hong, S.K. (2009) *Chem. Mater.*, **21**, 2386.

69 Terai, M., Sakotsubo, Y., Kotsuji, S., and Hada, H. (2010) *IEEE Electron Device Lett.*, **31**, 3.

70 Janousch, M., Meijer, G.I., Staub, U., Delley, B., Karg, S.F., and Andreasson, B.P. (2007) *Adv. Mater.*, **19**, 2232.

71 Lee, M.-J., Seo, S., Kim, D.-C., Ahn, S.-E., Seo, D.H., Yoo, I.-K., Baek, I.G., Kim, D.-S., Byun, I.-S., Kim, S.-H., Hwang, I.-R., Kim, J.-S., Jeon, S.-H., and Park, B.H. (2007) *Adv. Mater.*, **19**, 73.

72 Baek, I.G., Kim, D.C., Lee, M.J., Kim, H.J., Yim, E.K., Lee, M.S., Lee, J.E., Ahn, S.E., Seo, S., Lee, J.H., Park, J.C., Cha, Y.K., Park, S.O., Kim, H.S., Yoo, I.K., Chung, U.-I., Moon, J.T., and Ryu, B.I. (2005) International Electron Devices Meeting, Technical Digest, p. 750.

73 Hoffman, R.L., Norris, B.J., and Wager, J.F. (2003) *Appl. Phys. Lett.*, **82**, 733.

74 Nomura, K., Ohta, H., Takagi, A., Kamiya, T., Hirano, M., and Hosono, H. (2004) *Nature*, **432**, 488.

75 Thompson, S.E., Armstrong, M., Auth, C., Alavi, M., Buehler, M., Chau, R., Cea, S., Ghani, T., Glass, G., Hoffman, T., Jan, C.-H., Kenyon, C., Klaus, J., Kuhn, K., Ma, Z., McIntyre, B., Mistry, K., Murthy, A., Obradovic, B., Nagisetty, R., Nguyen, P., Sivakumar, S., Shaheed, R., Shifren, L., Tufts, B., Tyagi, S., Bohr, M., and El-Mansy, Y. (2004) *IEEE Trans. Electron Devices*, **51**, 1790.

76 Houssa, M. (2004) *High k Gate Dielectrics*, IOP, London.

77 Huff, H. and Gilmer, D. (2004) *High Dielectric Constant Materials*, Springer, New York.

78 Gusev, E. (2005) *Defects in High-k Gate Dielectric Stacks*, Springer, New York.

79 Intel, Hafnium-based Intel® 45nm Process Technology, http://www.intel.com/technology/45nm/index.htm

80 Fischetti, M.V., Neumayer, D.A., and Cartier, E.A. (2001) *J. Appl. Phys.*, **90**, 4587.

81 Rauwel, E., Clavel, G., Willinger, M.-G., Rauwel, P., and Pinna, N. (2008) *Angew. Chem., Int. Ed.*, **120**, 3648.

82 Zhao, X. and Vanderbilt, D. (2002) *Phys. Rev. B*, **65**, 233106.

83 Rignanese, G.-M., Gonze, X., Jun, G., Cho, K., and Pasquarello, A. (2004) *Phys. Rev. B*, **69**, 184301.

84 Lopes, J.M.J., Özben, E.D., Roeckerath, M., Littmark, U., Lupták, R., Lenk, St. Luysberg, M., Besmehn, A., Breuer, U., Schubert, J., and Mantl, S. (2009) *Microelectron. Eng.*, **86**, 1646.

85 Park, T.J., Kim, J.H., Jang, J.H., Lee, J., Lee, S.W., Lee, S.Y., Jung, H.-S., and Hwang, C.S. (2009) *J. Electrochem. Soc.*, **156**, G129.

86 Park, C.S., Hussain, M.M., Huang, J., Park, C., Tateiwa, K., Young, C., Park, H.K., Cruz, M., Gilmer, D., Rader, K., Price, J., Lysaght, P., Heh, D., Bersuker, G., Kirsch, P.D., Tseng, H.-H., and Jammy, R. (2009) Symposium on VLSI Technology, Digest of Technical Papers, p. 208.

87 Utriainen, M., Kröger-Laukkanen, M., Johansson, L.-S., and Niinistö, L. (2000) *Appl. Surf. Sci.*, **157**, 151.

88 Do, K.-W., Yang, C.-M., Kang, I.-S., Kim, K.-M., Back, K.-H., Cho, H.-I., Lee, H.-B., Kong, S.-H., Hahm, S.-H., Kwon, D.-H., Lee, J.-H., and Lee, J.-H. (2006) *Jpn. J. Appl. Phys.*, **45**, 2975.

89 Kittl, J.A., Veloso, A., Lauwers, A., Anil, K.G., Demeurisse, C., Kubicek, S., Niwa, M., van Dal, M.J.H., Richard, O., Pawlak, M.A., Jurczak, M., Vrancken, C., Chiarella, T., Brus, S., Maex, K., and Biesemans, S. (2005) Symposium on VLSI Technology, Digest of Technical Papers, p. 72.

90 Shickova, A., Kauerauf, T., Rothschild, A., Aoulaiche, M., Sahhaf, S.A., Kaczer, B., Veloso, A., Torregiani, C., Pantisano, L., Lauwers, A., Zahid, M., Rost, T., Tigelaar, H., Pas, M., Fretwell, J., McCormack, J., Hoffmann, T., Kemer, C., Chiarella, T., Bras, S., Harada, Y., Niwa, M., Kaushik, V., Maes, H., Absil, P.P., Groeseneken, G., Biesemans, S., and Kittl, J.A. (2007) Symposium on VLSI Technology, Digest of Technical Papers, p. 158.

91 Park, T.J., Kim, J.H., Jang, J.H., Na, K.D., Hwang, C.S., Kim, G.-M., Choi, K.J., and Jeong, J.H. (2008) *Appl. Phys. Lett.*, **92**, 202902.

92 Lemberger, M., Baunemann, A., and Bauer, A.J. (2007) *Microelectron. Reliab.*, **47**, 635.

93 Hou, Y.T., Yen, F.Y., Hsu, P.F., Chang, V.S., Lim, P.S., Hung, C.L., Yao, L.G., Jiang, J.C., Lin, H.J., Jin, Y., Jang, S.M., Tao, H.J., Chen, S.C., and Liang, M.S. (2005) International Electron Devices Meeting, Technical Digest, p. 31.

94 Hobbs, C., Fonseca, L., Dhandapani, V., Samavedam, S., Taylor, B., Grant, J., Dip, L., Triyoso, D., Hegde, R., Gilmer, D., Garcia, R., Roan, D., Lovejoy, L., Rai, R., Hebert, L., Tseng, H., White, B., and Tobin, P. (2003) Symposium on VLSI Technology, Digest of Technical Papers, p. 9.

95 Samavedam, S.B., La, L.B., Tobin, P.J., White, B., Hobbs, C., Fonseca, L.R.C., Demkov, A.A., Schaeffer, J., Luckowski, E., Martinez, A., Raymond, M., Triyoso, D., Roan, D., Dhandapani, V., Garcia, R., Anderson, S.G.H., Moore, K., Tseng, H.H., Capasso, C., Adetutu, O., Gilmer, D.C., Taylor, W.J., Hegde, R., and Grant, J. (2005) International Electron Devices Meeting, Technical Digest, p. 13.1.

96 Abe, Y., Miyata, N., Shiraki, Y., and Yasuda, T. (2007) *Appl. Phys. Lett.*, **90**, 172906.

97 Kita, K. and Toriumi, A. (2008) International Electron Devices Meeting, Technical Digest, p. 1.

98 Goel, N., Majhi, P., Tsai, W., Warusawithana, M., Scholom, D.G., Santos, M.B., Harris, J.S., and Nishi, Y. (2007) *Appl. Phys. Lett.*, **91**, 093509.

99 Radosavljevic, M., Chu-Kung, B., Corcoran, S., Dewey, G., Hudait, M.K., Fastenau, J.M., Kavalieros, J., Liu, W.K., Lubyshev, D., Metz, M., Millard, K., Mukherjee, N., Rachmady, W., Shah, U., and Chau, R. (2009) International Electron Devices Meeting, Technical Digest, p. 1.

100 Intel, Process and Electrical Results for the On-die Interconnect Stack for Intel's 45nm Process Generation, http://www.intel.com/technology/itj/2008/v12i2/2-process/2-intro.htm

101 MA-tek, Samsung 512Mb C-die DDR SDRAM Constructional Analysis, http://www.ma-tek.com/industry_detail.php?cpath=23

102 Hayashi, Y. (2002) Proceedings of IEEE International Interconnect Technology Conference, **145**

103 Ohtake, H., Tagami, M., Tada, M., Ueki, M., Abe, M., Saito, S., Ito, F., and Hayashi, Y. (2006) *IEEE Tans. Semicond. Manuf.*, **19**, 455.

104 Tada, M., Tamura, T., Ito, F., Ohtake, H., Narihiro, M., Tagami, M., Ueki, M., Hijioka, K., Abe, M., Inoue, N., Takeuchi, T., Saito, S., Onodera, T., Furutake, N., Arai, K., Sekine, M., Suzuki, M., and Hayashi, Y. (2006) *IEEE Trans. Electron Devices*, **53**, 1169.

105 de Rouffignac, P., Li, Z.W., and Gordon, R.G. (2004) *Electrochem. Solid-State Lett.*, **7**, G306.

106 Rosenberg, R., Edelstein, D.C., Hu, C.K., and Rodbell, K.P. (2000) *Annu. Rev. Mater. Sci.*, **30**, 229.

107 Ivanov, I.P., Sen, I., and Keswick, P. (2006) *J. Vac. Sci. Technol. B*, **24**, 523.

108 Klaus, J.W., Ferro, S.J., and George, S.M. (2000) *Thin Solid Films*, **360**, 145.

109 Kim, S.-H., Yeom, S.-J., Kwak, N., and Sohn, H. (2008) *J. Electrochem. Soc.*, **155**, D148.

9
Nanopatterning by Area-Selective Atomic Layer Deposition

Han-Bo-Ram Lee and Stacey F. Bent

In conventional device fabrication, patterning is a top-down process based largely on photolithography and etching and is a main bottleneck for device downscaling. Atomic layer deposition (ALD) can provide an alternative bottom-up method for patterning when used in conjunction with self-assembled materials and selective chemistries. As presented in previous chapters, ALD is a powerful technique for depositing thin films for nanoscale device fabrication thanks to its excellent conformality, atomic scale thickness controllability, and large-area uniformity. Film growth controlled by surface reactions is one of the inherent properties of ALD. Because in an ideal ALD process, all of the precursors used for ALD react with each other only at the surface, highly conformal films can be deposited even inside complex 3D structures. By taking advantage of this inherent surface reaction property, ALD can be utilized for patterning based on a bottom-up process. In the following chapter, selective deposition methods will be presented. The contents include the surface modification techniques that are employed to exploit the surface reaction properties of ALD and related patterning processes with reported results.

9.1
Concept of Area-Selective Atomic Layer Deposition

Film growth by ALD begins with formation of nuclei through a reaction of precursor molecules and surface species. Once nuclei are formed on a surface, the growth of the film may proceed by growth and coalescence of the nuclei, whereas if no nucleation occurs no film will be deposited. Therefore, the deposition characteristics of ALD strongly depend on the surface properties of the substrate. For example, in many cases, the nucleation of ALD is easy on hydrophilic OH-terminated substrates (e.g., SiO_2), while it is difficult on hydrophobic H-terminated surfaces (e.g., Si) [1–4]. Similarly, a nucleation delay in ALD, typically called the incubation time, is due to difficulty of nucleation at the surface. The nucleation delay and the variability of the growth characteristics depending on the surface can be problematic in typical thin

Atomic Layer Deposition of Nanostructured Materials, First Edition. Edited by Nicola Pinna and Mato Knez.

film deposition by ALD. However, this apparent disadvantage can also provide an opportunity to exploit ALD as a nanopatterning tool, as described in this chapter.

If we can control the surface properties, we can control the film deposition. Figure 9.1 provides a simple drawing to illustrate this concept, for an exemplary case where nucleation occurs preferentially on an OH-terminated surface. There may be two regions of the surface, OH-terminated (hydrophilic) and H-terminated (hydrophobic), which are separately formed as shown in Figure 9.1a and b. The precursor molecules come into contact with the two different surface terminations during the ALD cycles. Although the precursors are exposed on the entire surface, the precursors chemisorb only on the OH-terminated surface and not on the H-terminated surface, as shown in Figure 9.1c. Subsequently, the self-saturating surface reaction occurs only on the OH-terminated surface region, leading to nucleation (see Figure 9.1d). As the ALD cycle is repeated, the film is selectively grown on the regions of the substrate that were initially OH-terminated, as shown in Figure 9.1e. Finally, patterned films are formed, and the film pattern is directly transferred from

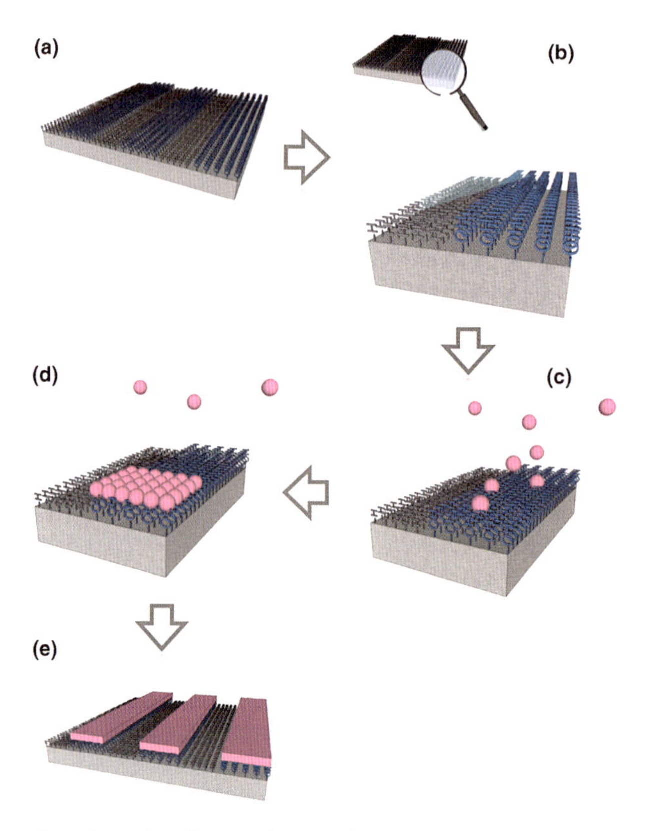

Figure 9.1 The schematic drawings for the concept of AS-ALD. (a and b) Preparation of a substrate with two kinds of surfaces, OH- and H-terminated surfaces, (c) adsorption of precursor only on the OH-terminated surface, (d) self-saturated reaction only on the OH-terminated surface, and finally (e) self-patterned thin film formation.

the pattern of surface termination. In this example, starting with nucleation-active (OH-terminated) and nucleation-inactive (H-terminated) regions of the surface, the deposited films are spontaneously patterned without the need for photolithography and etching because of the surface-specific growth nature of ALD. This patterning method is called "area-selective ALD" (AS-ALD). It is also referred to as selective area ALD. Selective chemical vapor deposition (CVD) using a similar approach was previously demonstrated for Pd and Pt [5]. Because gas-phase chemistry in addition to surface reactions can occur in CVD, in certain cases ALD may be more effective for a selective deposition process due to its surface-sensitive growth mechanism.

For AS-ALD, two requirements must be considered. First, how can the surface properties of a substrate be controlled? Although the example shown in Figure 9.1 was chosen to simplify the concept of AS-ALD, real substrates used for thin film deposition have a wide variety of surface properties. Moreover, the controlled surfaces must exhibit selectivity in their reaction with the precursors used. Second, the controlled surfaces should contain patterns like the cartoon of Figure 9.1a. Because the pattern of the substrate is directly transferred to the final film pattern in AS-ALD, the preparation of the surface property pattern is important. To exploit AS-ALD for various applications, control over pattern shape and minimum feature size is required. The following sections are dedicated to these considerations: change of surface properties and patterning of surfaces.

9.2 Change of Surface Properties

9.2.1 Self-Assembled Monolayers

Self-assembled monolayers (SAMs) are organic films of single molecular thickness. Typically, SAMs consist of three parts: the head group, backbone, and tail group [6, 7]. Figure 9.2a shows the molecular structure of octadecyltrichlorosilane (ODTS) [$CH_3(CH_2)_{17}SiCl_3$], which is one of the commonly used molecules for SAMs on SiO_2 substrates. The head group is directly bonded to the surface of the substrate through chemisorption. The alkyl chain as a backbone is composed of C–C and C–H bonds, and van der Waals interactions between chains further contribute to the ordering of the monolayer. The tail group is exposed at the outer interface after the formation of a well-packed SAM on the substrate. For this case, because ODTS has methyl ($-CH_3$) as a tail group, the surface is converted into CH_3 termination after deposition of ODTS. If the original substrate is SiO_2, the surface property is hydrophilic due to the presence of OH termination; after deposition of ODTS on the SiO_2 substrate, the final CH_3-terminated surface is hydrophobic as shown in Figure 9.2b. Contact angle pictures of water droplets in Figure 9.2c clearly show the difference of hydrophobicity before and after the formation of an ODTS SAM.

SAMs can be categorized into three main types based on the bonding between their head groups and the substrate surface [6]. The *n*-alkanoic acid ($C_nH_{2n+1}COOH$)

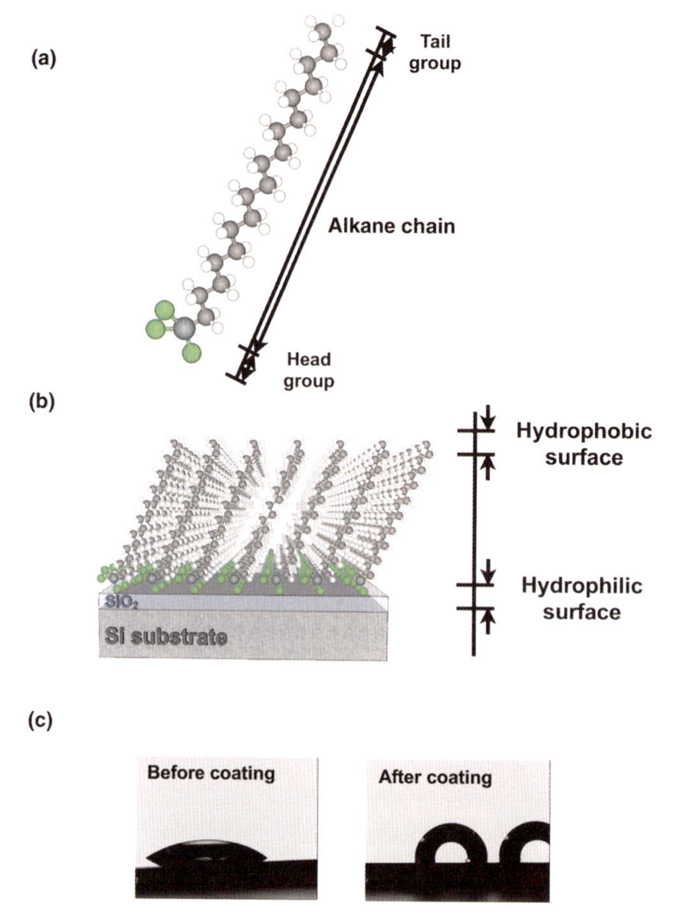

Figure 9.2 (a) Molecular structure of ODTS SAM. (b) Schematic drawing of ODTS SAM coated on SiO_2 surface. (c) Contact angle pictures before and after ODTS coating.

SAMs bond to the surface through ionic bonding. It was reported that *n*-alkanoic acid forms on native oxide surfaces of Al [8, 9], Ag [10, 11], and Cu [12] by reaction between the metal cation and carboxylate anion. Another type of SAM, composed of organosulfur head groups, is formed by charge transfer complex. The formation of organosulfur SAMs has been mainly studied for Au surfaces. It was found that the organosulfur SAM molecules are formed by lateral alignments up to a critical coverage value, followed by realignment of molecules into the vertical direction [13, 14]. Many researchers have investigated the formation of organosulfur SAMs on various surfaces, such as Ag [15, 16], Cu [16, 17], Pt [18], Fe_2O_3 [19], GaAs [20], InP [21], and Ge [22]. The ODTS SAMs mentioned above consist of organosilicon, which is chemisorbed on the surface through covalent bonding. The hydroxyl group, formed from the reaction of silicon chloride bonds and water, reacts with the silanol groups present at the surface, resulting in a covalent Si–O–Si bond [6]. Cross-linking can also occur between neighboring hydroxyl groups in the adsorbates. Because the resulting SAM contains molecules that are covalently linked both to the substrate and to their neighbors, the ODTS SAM shows robust bonding [23–25]. Organosilicon SAMs have been successfully formed on various substrates, including SiO_2 [25, 26], Al_2O_3 [27, 28], quartz [29, 30], ZnSe [27], GeO [27], TiN [5], and Au [31]. Instead of an organosilicon head group, C in molecules such as alkenes and alkynes can also be directly bonded to H-terminated Si surface with covalent bonding [32, 33].

Because of the spontaneous formation of SAMs with close-packed structure as well as strong chemical bonding to the substrate, SAMs have been shown to be useful for many examples of nanofabrication including soft lithography, molecular electronics, control of wetting and friction behavior, and protection of surfaces against corrosive environments [34–36]. SAMs also provide a suitable method to change the surface properties for AS-ALD. Because SAMs of many kinds of head groups and tail groups can be formed, the surface can be easily modified to incorporate various properties just by coating SAMs on the original substrate. For instance, a single surface can be modified to have regions of two or more different properties depending on the choice of SAM. Seo *et al.* demonstrated a change of the properties of a Au surface by using two different SAMs, octadecanethiol (ODT) and mercaptoundecanol (MOU) [37]. These two SAM molecules have same thiol head group but different tail groups: CH_3 termination and OH termination for ODT and MOU, respectively. Consequently, the original Au metal surface was separately modified into hydrophobic and hydrophilic regions. After an ALD process on this substrate, TiO_2 films were found to deposit only on the hydrophilic surfaces.

To date, many SAM molecules have been investigated for AS-ALD. The reported SAM molecules are summarized with the corresponding ALD film materials in Table 9.1. Although there are many different SAMs reported, it is noted that the SAMs used for AS-ALD have been limited to a few categories of molecules. As shown in Table 9.1, nearly all the SAMs have same head group, organosilane, and most of those are chlorosilanes. As mentioned earlier, the head group plays the role of linker to the substrate. Because Si is the most commonly used substrate for many applications, SAMs composed of organosilane head groups that easily form bonds to the silicon oxide surface at the Si substrate are the most intensively studied.

The change of the surface properties caused by the presence of the SAM directly affects the deposition characteristics of ALD. The SAMs may act as either an activator or a deactivator for film deposition. Several studies have proposed and reported a model to explain the deactivation mechanism related to surface reactions in ALD. In many reports, the CH_3-terminated surface formed by SAMs is a deposition deactivator while the OH-terminated surface (typically the uncoated original surface) is a relatively active surface for deposition. In a previous report, Lee *et al.* clearly showed the role of the CH_3-terminated surface as a deposition deactivator [41]. To fabricate various ratios of CH_3 to OH termination at the surface, two SAM molecules composed of different tail groups, CH_3 and OH, were mixed with different ratios. The resulting surfaces coated by the mixed SAMs showed various values of hydrophobicity according to the mixing ratio, with the calculated surface energies decreasing with increasing CH_3 termination ratio. The results on TiO_2 ALD, as summarized

Table 9.1 Reported SAMs for AS-ALD.

	SAMs	Film material	Substrates
Organosulfur	11-Mercapto-1-undecanol [$HOCH_2(CH_2)_{10}SH$]	TiO_2	Au [37]
	1-Octadecanethiol [$CH_3(CH_2)_{17}SH$]	TiO_2	Au [37], Ge [38]
Organosilicon	Benzyltrichlorosilane [$C_7H_7SiCl_3$]	HfO_2	SiO_2 [39]
	Bromotrimethylsilane [$(CH_3)_3SiBr$]	HfO_2	SiO_2 [39]
	Chlorotrimethylsilane [$(CH_3)_3SiCl$]	TiO_2	SiO_2 aerogels [40]
		HfO_2	SiO_2 [39]
	Decyltrichlorosilane [$CH_3(CH_2)_9SiCl_3$]	TiO_2	SiO_2 [41]
	Dimethyldichlorosilane [$(CH_3)_2SiCl_2$]	HfO_2	SiO_2 [39]
	Hexadecyltrichlorosilane [$CH_3(CH_2)_{16}SiCl_3$]	WN_xC_y	SiO_2 [42]
	Iodotrimethylsilane [$(CH_3)_3SiI$]	HfO_2	SiO_2 [39]
	iso-Butyltrichlorosilane [*iso*-$BuSiCl_3$]	HfO_2	SiO_2 [39]
	Methyltrichlorosilane [$(CH_3)SiCl_3$]	HfO_2	SiO_2 [39]
	Methyl-10-(trichlorosilyl) decanoate [$H_3CO_2C(CH_2)_9SiCl_3$]	TiO_2	SiO_2 [41]
	N-Butyltrichlorosilane [*n*-$BuSiCl_3$]	HfO_2	SiO_2 [39]
	Octadecyltrichlorosilane [$CH_3(CH_2)_{17}SiCl_3$]	TiO_2	SiO_2 [35, 43, 44], silica colloidal sphere [43]
		ZrO_2	SiO_2 [45–47]
		HfO_2	SiO_2 [39, 45, 48–51], YSZ [48]
		Al_2O_3	SiO_2 [52]
		ZnO	SiO_2 [53], glass [54]
		TiN	SiO_2 [55]
		Pt	SiO_2 [48, 50, 56, 57], YSZ [48, 56, 57]
		Ru	SiO_2 [58], HfO_2 [58]
		PbS	SiO_2 [59]
		Co	SiO_2 [60, 61]
	Octadecyltriethoxysilane [$CH_3(CH_2)_{17}Si(OCH_2CH_3)_3$]	HfO_2	SiO_2 [39]
	Octadecyltrimethoxysilane [$CH_3(CH_2)_{17}Si(OCH_3)_3$]	Ir	SiO_2 [62]
		HfO_2	SiO_2 [39]

Table 9.1 *(Continued)*

	SAMs	Film material	Substrates
	Octyltrichlorosilane	WN_xC_y	SiO_2 [42]
	$[CH_3(CH_2)_6SiCl_3]$	HfO_2	SiO_2 [39]
	tert-Butyltrichlorosilane [*tert*-$BuSiCl_3$]	HfO_2	SiO_2 [39]
	Tetrahydrooctyltrichlorosilane	HfO_2	SiO_2 [39, 50]
	$[CF_3(CF_2)_5(CH_2)_2SiCl_3]$	Pt	SiO_2 [50, 56], YSZ [56]
		TiN	SiO_2 [55]
	Triacontyltrichlorosilane $[CH_3(CH_2)_{29}SiCl_3]$	TiN	SiO_2 [55, 63]
Alkyl monolayer	1-Decene	HfO_2	Ge [64]
	$[(CH)_2(CH)(CH_2)_7(CH)_3]$	Pt	Ge [64]
	1-Decyne	HfO_2	Ge [64]
	$[(CH)C(CH_2)_7(CH)_3]$	Pt	Ge [64]
	1-Dodecene	HfO_2	Ge [64]
	$[(CH)_2(CH)(CH_2)_9(CH)_3]$	Pt	Ge [64]
	1-Dodecyne	HfO_2	Ge [64]
	$[(CH)C(CH_2)_9(CH)_3]$	Pt	Ge [64]
	1-Hexadecene	Pt	Ge [64], glass [65]
	$[(CH)_2(CH)(CH_2)_{13}(CH)_3]$	HfO_2	Ge [64]
	1-Octadecene	HfO_2	Si [50, 64], Ge [64]
	$[(CH)_2(CH)(CH_2)_{15}(CH)_3]$	Pt	Si [49, 64], Ge [64]
	1-Octene	HfO_2	Ge [64]
	$[(CH)_2(CH)(CH_2)_5(CH)_3]$	Pt	Ge [64]
	1-Octyne	HfO_2	Ge [64]
	$[(CH)_2(CH)(CH_2)_5(CH)_3]$	Pt	Ge [64]
	1-Tetradecene	HfO_2	Ge [64]
	$[(CH)_2(CH)(CH_2)_{11}(CH)_3]$	Pt	Ge [66]
Alkanoic acid	1-Undecylenic acid $[(CH_2)(CH)(CH_2)_8COOH]$	Al_2O_3	Si [64]

in Figure 9.3 [41], showed that TiO_2 nucleation occurred easily on the OH-terminated surface, while the deposition mode changed to island growth with an increasing ratio of CH_3 termination [41]. This research suggests that the CH_3 termination in a SAM lowers the surface energy, resulting in reduced nucleation of the ALD reaction. Similarly, Xu and Musgrave [52] reported thermodynamic and kinetic calculations using density functional theory for the reaction of trimethylaluminum (TMA) on various SAM-terminated surfaces. The TMA precursor for Al_2O_3 ALD had much higher reactivity on the OH-terminated surface than either the NH_2- or CH_3-terminated surface [52]. An *ab initio* study based on density functional theory (DFT)

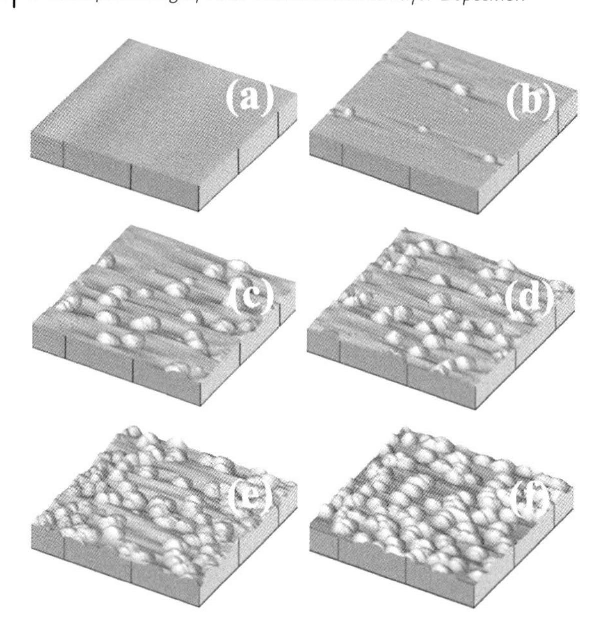

Figure 9.3 Three-dimensional AFM images (5 μm × 5 μm) for the TiO_2 films deposited on the SAMs. R_{SAMs} is the ratio of OH to CH_3 termination. (a) Pure OH termination (rms = 2.5 Å); (b) $R_{SAMs} = 4:1$ (rms = 20.4 Å); (c) $R_{SAMs} = 2:1$ (rms = 128.4 Å); (d) $R_{SAMs} = 1:1$ (rms = 175.5 Å); (e) $R_{SAMs} = 1:2$ (rms = 256.8 Å); (f) $R_{SAMs} = 1:3$ (rms = 326.3 Å). Reproduced with permission from Ref. [41]. Copyright 2003, Wiley-VCH Verlag GmbH.

has also shown that the reaction of several Ti and Zr precursors including tetrakis (dimethylamido)titanium (TDMAT), tetrakis(diethylamido)titanium (TDEAT), titanium tetramide ($Ti(NH_2)_4$), tetrakis(dimethylamido)zirconium (TDMAZ), and zirconium tetramide ($Zr(NH_2)_4$) has reactivity in the order $OH > SH > NH_2$ [67]. Dube *et al.* investigated the growth mode of TiN ALD experimentally using TDMAT and NH_3 on SAMs with several different tail groups such as OH, NH_2, and CH_3, and found that the CH_3-terminated surface much more effectively attenuated TiN growth than OH- and NH_2-terminated surface, which was strongly dependent on the structure and thickness of SAMs possessing a CH_3 tail group [55, 63].

If the SAM is used as a deposition deactivator, the ability to block the film deposition during the ALD cycle is important. In other words, because the deactivated region formed by the SAM becomes the film-free region of the final pattern, incomplete deactivation may cause failures in final device fabrications. Chen *et al.* focused on the effects of alkyl chain length, reactive head group, and tail group structure in their studies [39]. From a systematic study, it was found that the major factor affecting deactivation of film deposition is strongly related to the quality of packing. Because a long alkyl chain length produces higher van der Waals forces between SAM molecules than a short alkyl chain, it makes the packing of SAM molecules dense. In addition, a smaller profile of the tail group increases the packing quality because of less steric hindrance, so less bulky tails such as linear alkanes are more effective than branched or aryl tails. The nature of the ALD precursor is another factor that may affect the deactivation of film deposition by SAMs. In one study, even though the same ODTS SAM was used as a deactivation agent for both HfO_2 and Pt ALD, the ability to block the deposition of Pt was higher than that of HfO_2 [48]. This may be due to precursor size or inherent reactivity [48]. Bent and coworkers have in a series of studies reported the effects of the SAM coating quality on the blocking properties [39, 52, 56, 59, 64, 68, 69]. Although a surface coated by ODTS may show high hydrophobicity even with a short coating time, its property was not sufficient to block the deposition of HfO_2 and ZrO_2 by ALD [70]. To achieve effective blocking of ALD films, sufficient coating time was required to form densely packed SAMs without pinholes. Because selective deposition in many reported AS-ALD results relies on a change of hydrophobicity, namely, surface energy, the quality of the SAM is often estimated by static water contact angle measurement [71–73]. However, the contact angle analyzes the overall quality of the SAM coating, so it is hard to detect microscopic defects such as pinholes. Therefore, an optimization of the packing condition for the SAM is required prior to performing the ALD process on the substrate.

9.2.2
Polymers

Polymers have been intensively investigated for various applications, especially for Si device fabrication, since polymers as a photoresist (PR) are essential for the photolithography process [74, 75]. For photolithography, a polymeric PR serves as either a mask for subsequent etching processes or a passivation layer for liftoff processes [74, 75]. Whereas SAMs are formed on a surface through chemical reactions and molecular ordering with formation of chemical bonds to surface species, a polymer usually only physisorbs on a surface. So, formation and removal of the polymer is easier than that of SAMs. Therefore, polymers can be applied by spin coating, which is an inexpensive, convenient, and simple method. In many cases, surfaces coated by a polymer exhibit hydrophobic properties due to the presence of hydrocarbon chains. Because of the nature of the polymer surface, an adhesion promoter such as hexamethyldisilazane (HMDS) is required to coat PR on the hydrophilic SiO_2 surface for the photolithography process to enhance adhesion between the PR and SiO_2 [74, 75]. By exploiting this surface property, a polymer can be an effective deactivator for ALD film growth. In addition, the readily available knowledge of polymer properties and simple fabrication procedures are two big advantages to applying AS-ALD into mass production.

Reported AS-ALD processes using polymers are summarized in Table 9.2 [39, 54, 55, 76–83]. In early studies, researchers investigated the blocking ability of polymer layers against various ALD systems. Sinha *et al.* have published several papers on TiO_2 AS-ALD using poly(methyl methacrylate) (PMMA) as a deposition deactivator [76–79]. They employed titanium tetrachloride ($TiCl_4$) as precursor and water as a reactant [77]. TiO_2 deposition was observed even on polymer-covered regions of the

Table 9.2 Reported polymer deactivator for AS-ALD.

Deposition deactivator	Materials	References
Hexamethyldisilazane [HN[Si(CH$_3$)$_3$]$_2$]	HfO_2	[39]
	Rh	[82]
	TiN	[55]
Hexafluoroisopropyl alcohol	TiO_2	[76]
Photoresist	Rh	[82]
	Ru	[84]
Polyhydroxystyrene	TiO_2	[76]
Polymethacrylamide	Pt	[85]
Poly(methyl methacrylate)	TiO_2	[76–78, 81, 86]
	Ir	[81]
	Pt	[81]
	Ru	[81]
	Al_2O_3	[81]
	ZnO	[54]
Polystyrene	ZnO	[54]
Poly(*tert*-butyl methacrylate)	TiO_2	[79]
Polyvinylpyrrolidone	Pt	[80]
	ZrO_2	[80]
	Ru	[80]
	Ir	[80]
	Al_2O_3	[80]
Wax in cicida's wing	ZnO	[83]
	Al_2O_3	[83]

substrate, and the amount of Ti was dependent on the thickness of the polymer, indicating that precursors penetrated into the PMMA layer, and then reacted at the interface between PMMA and the substrate [77]. In a subsequent paper, they focused on effects of precursor chemistries on the blocking capability of PMMA by using two different Ti precursors, $TiCl_4$ and titanium isopropoxide ($Ti(O^iPr)_4$) [Ti[OCH(CH$_3$)$_2$]$_4$] [78]. A PMMA layer with 10 nm thickness effectively attenuates TiO_2 deposition from the $Ti(O^iPr)_4$ precursor while TiO_2 ALD using $TiCl_4$ required at least 200 nm of PMMA to block the deposition. Furthermore, evidence of Ti was found inside the PMMA layer after exposure of $TiCl_4$ precursor without reactant water, indicating that the high reactivity of $TiCl_4$ results in a degradation of the blocking property of PMMA [78]. By using a quartz crystal microbalance (QCM), adsorption and desorption behavior of the two Ti precursors on several polymers, PMMA, polyhydroxystyrene (PHOST), and hexafluoroisopropyl alcohol (HFA-PNB), was investigated [76]. The results showed that the $TiCl_4$ precursor had higher diffusivity than $Ti(O^iPr)_4$, which was consistent with the experimental results on TiO_2 ALD [76]. In another study on AS-ALD, the blocking ability of PMMA toward various ALD systems – Pt, Ir, Ru, Al_2O_3, and TiO_2 – was investigated [81]. PMMA successfully prevented the growth of Pt, Ir, Ru, and TiO_2 even at a relatively high deposition temperature of 300 °C [81]. For Al_2O_3 AS-ALD using tetramethylaluminum (TMA) and $AlCl_3$ as a precursor, growth of Al_2O_3 films was observed even on PMMA-coated regions. Compared to the TMA precursor, $AlCl_3$ was more reactive with PMMA. In subsequent research, the authors tried to use another polymer, polyvinylpyrrolidone (PVP), for deactivation and investigated the blocking capability according to precursors [80]. In contrast to PMMA, TMA is more reactive than $AlCl_3$ on a PVP deactivation layer. This opposite behavior can be attributed to the solubility difference of the precursors in the polymer layers [80].

9.2.3
Inherent Surface Reactivity

Although SAMs have the ability to produce various surface properties thanks to the large variety of head and tail groups, the inherent properties of surfaces can also be used to achieve AS-ALD. In fact, control of surface termination has been utilized for selective deposition with other vapor-phase deposition techniques, such as CVD [87]. H-terminated Si and OH-terminated SiO_2 are typical surfaces used for area-selective CVD (AS-CVD) because of large differences in their bonding and reactivity [87–89]. Al and W deposition for metallization is one of the well-established AS-CVD processes where W films are deposited only on the Si substrate but not on SiO_2 and Si_3N_4 insulators [89]. In this scheme, the precursors exhibit reaction selectivity with surface species, so reduction of the precursors occurs only on specific surfaces. Similarly, AS-ALD was investigated on H-terminated Si and CVD-grown Si_3N_4 surfaces by using SiH_2Cl_2 and NH_3 as a precursor and a reactant, respectively [90]. SiN_x films were formed only on H-terminated surfaces. The growth of SiN_x begins from the formation of N–H bonding from the reaction of surface H and NH_3. The SiH_2Cl_2 precursor reacts only at N–H terminated sites, resulting in the growth of SiN_x [90]. Recently, AS-ALD HfO_2 was reported on two different surface materials, Si and Cu [91]. No HfO_2 film was formed on the Cu surface up to 25 cycles; beyond 25 cycles, HfO_2 was deposited. This deposition selectivity was attributed to the incubation time of HfO_2 deposition on the Cu surface [91]. In fact, surface-dependent growth has been reported several times in ALD [92–95]. Moreover, the dependence of ALD incubation time on the substrate has been reported many times [96–99]. For small film thicknesses, AS-ALD using inherent surface properties, namely, exploiting differences in incubation time, can be an effective method.

Another approach is to combine SAMs with inherent surface property differences. The appropriate head group for the formation of a stable monolayer depends sensitively on the substrate. For example, SAMs of alkyltrichlorosilanes $[CH_3(CH_2)_{n-1}SiCl_3]$ are preferably adsorbed on Si_3N_4, SiO_2, and other oxide surfaces. This specificity to the particular substrate suggests that SAMs may be selectively deposited on different surfaces, or, more relevantly, on different areas on the same substrate. For example, ODTS preferentially reacts with an OH-terminated surface while an alkene preferentially reacts with a H-terminated surface [33, 100, 101]. If a substrate with patterns of OH- and H-terminated regions is exposed to ODTS or alkenes, under proper conditions a SAM can be selectively formed for each surface termination [49, 68]. This allows the patterns of surface termination to be propagated into a pattern of SAMS, which in turn provide a pathway to fabricate the final film pattern through AS-ALD. In such an approach, AS-ALD can be used without an additional patterning step for the SAM.

9.2.4 Vapor-Phase Deposition

Although liquid-phase deposition is an easy and simple route to form SAMs on a substrate, it is not easily scaled up nor readily applied to current unit operations for Si device fabrication that are based on vacuum processes. To address this limitation, several researchers have investigated the formation of SAMs by vapor-phase deposition [50, 52, 62, 102–105]. The vapor phase provides several advantages over a liquid-phase process. One advantage is less consumption of the SAM molecules. Furthermore, aggregation of the SAM molecules prior to deposition at the substrate, a process that leads to a deterioration of the quality of the SAM in liquid-phase processes, is significantly reduced in vapor-phase deposition processes [106, 107]. In addition, since SAMs can be formed *in situ* inside of the ALD chamber, the vapor-phase deposition of SAMs has benefits for an integrated tool for mass production. Similar to the delivery process used for the ALD precursors, the SAM solution may be contained in a separate bubbler and attached to the deposition chamber where the substrate is placed. Alternatively, the SAM solution may be placed in the chamber, so the vaporized SAM molecules are directly adsorbed on the substrate.

In the vapor-phase deposition process, because the formation of SAMs occurs under vacuum conditions, the water concentration, which is one of the key factors for the formation of high-quality SAMs, must be controlled differently from other deposition methods [25, 102, 108]. Typically, controlled amounts of water vapor may be deliberately dosed together with the SAM molecules. Leskelä's group at University of Helsinki investigated the vapor-based deposition of octadecyltrimethoxysilane ($CH_3(CH_2)_{17}Si(OCH_3)_3$, ODS) with water to form a SAM blocking layer for AS-ALD of Ir [64]. In this study, the vaporized ODS was deposited on Al patterns and the patterning of ODS was performed by a liftoff process though etching of the Al. The selectivity was found to be better for the ODS SAMs formed by alternating exposure of SAM molecules and water during vapor-phase delivery.

Hong *et al.* studied the use of vapor-deposited ODTS, tridecafluoro-1,1,2,2-tetrahydrooctyltrichlorosilane (FOTS), and 1-octadecene as deactivation layers for ALD [50]. The ODTS and FOTS were used to form SAMs on OH-terminated silicon oxide surfaces while 1-octadecene was used to form monolayers on H-terminated silicon surfaces. The SAMs were shown to effectively deactivate HfO_2 and Pt ALD. Furthermore, the ODTS SAMs could be selectively adsorbed on OH-terminated regions of a surface and not on H-terminated regions, even with exposure over the entire surface. This result shows the feasibility for vapor-phase deposition to propagate the transfer of a pattern on the surface materials into a SAM pattern. In addition, it was found that the formation time for densely packed SAM molecules was similar to that of liquid-phase deposition.

9.3 Patterning

9.3.1 Surface Modification without Patterning

The most common protocol for preparing SAMs on a substrate is liquid-phase deposition. The SAM molecules are diluted in solvents, such as hexane, toluene, and chloroform, and then freshly prepared and clean substrates are fully immersed in the solution for a controlled period of time. The mobile SAM molecules adsorb on the surface and spontaneously assemble into a monolayer. Many early researchers employed this solution-based deposition as summarized in Table 9.3 because it is very simple as well as easy to form high-quality SAMs; in those studies, the SAMs fully coated the entire surface without patterns, allowing the efficacy of deactivation to be investigated [39–42, 45, 51, 52, 54, 60, 61]. For polymer film fabrication, spin coating has typically been used [39, 54, 55, 76–83]. In spin coating, a polymer is dissolved in an appropriate solvent, and the solution is dropped onto a surface undergoing rotation. As the solvent is evaporated during spin coating and subsequent baking, thin films of the polymer are formed. The polymer layer thickness is determined by rotation speed and viscosity of the solution. However, the ability to impart a pattern in AS-ALD is required for various applications, so much research has investigated and developed various patterning methods, as described below.

9.3.2 Microcontact Printing

Microcontact printing (μCP) is one of the most commonly used methods for the patterning of SAM molecules [112, 113]. Figure 9.4a shows a schematic drawing of the μCP process. The concept of μCP is simple, and it can be compared to stamping. A patterned stamp is immersed in a SAM solution as an ink and then the inked stamp is contacted with a substrate. Consequently, the ink is directly transferred onto the substrate with the pattern of the stamp. The SAMs are thus formed within the regions defined by the stamp patterns. Generally, the stamp is made of elastomeric materials, such as polydimethylsiloxane (PDMS), through photolithography-assisted processes as described in Figure 9.4b [114–117]. A Si master with the opposite pattern to that desired for the stamp is made by photolithography and etching processes, leading to the master as shown at the top of Figure 9.4b. The PDMS solution is cured onto the

Table 9.3 Reported patterning method for AS-ALD.

Patterning method	Materials	References
Liquid-phase deposition	TiO_2	[40, 41]
	HfO_2	[39, 45, 51]
	ZrO_2	[45]
	Al_2O_3	[52]
	Co	[60, 61]
	WN_xC_y	[42]
	ZnO	[54]
Microcontact printing	TiO_2	[35, 37, 109]
	ZnO	[53, 109]
	Ru	[58]
	Pt	[48, 56, 57]
	HfO_2	[48]
	ZrO_2	[109]
Vapor-phase deposition	HfO_2	[50]
	Ir	[62]
	Pt	[50]
Photolithography	ZrO_2	[46]
	Co	[60]
	TiO_2	[77–81]
	Pt	[80, 81]
	Ru	[80, 81]
	Ir	[80, 81]
	Al_2O_3	[80, 81]
Nanotemplating	TiO_2	[43, 44, 110, 111]
	ZrO_2	[110, 111]
	ZnO	[111]
Scanning probe lithography	ZrO_2	[47]
	TiO_2	[86]
	PbS	[59]

surface relief pattern of the Si master, and the pattern is thus transferred to the stamp. The μCP process provides a convenient, quick, and inexpensive way of patterning SAMs. In addition, the SAMs can be coated onto curved surfaces by μCP because of the flexibility of the elastomeric stamp [118, 119]. So, this method has become one of the most widely adopted approaches in AS-ALD.

Park *et al.* reported a process using μCP with a PDMS stamp and ODTS for AS-ALD of TiO_2 in 2004 [35]. The results showed that TiO_2 patterns could be obtained down to the submicron scale by μCP, indicating that AS-ALD using a PDMS stamp is feasible in the nanoscale regime. This work followed an earlier demonstration in 1997 of area-selective CVD of Pt using μCP SAMs by Jeon *et al.* [5]. Jiang *et al.* also examined

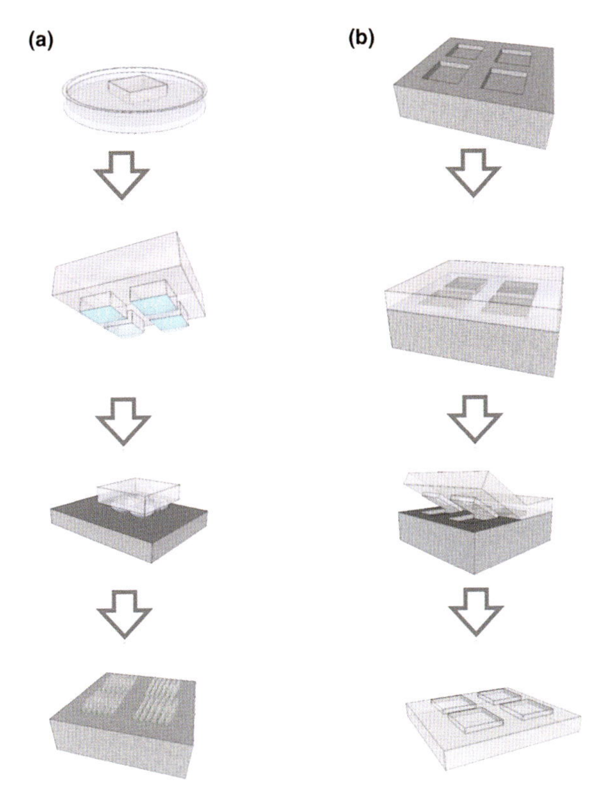

Figure 9.4 The schematic drawings for (a) μCP printing for SAM patterning and (b) fabrication steps for elastomeric stamp.

AS-ALD of Pt and HfO_2 through μCP in a series of papers [48, 56]. In one of their studies, the pattern transfer was performed onto yttrium-stabilized zirconia (YSZ) for fuel cell applications. Figure 9.5 shows the mapping data obtained in that work of AS-ALD Pt on a YSZ substrate analyzed by scanning Auger electron spectroscopy [56]. The deposition selectivity of Pt AS-ALD using μCP with a PDMS stamp was confirmed with up to 1 μm resolution. In addition, it was found that the contact time is essential [48, 56]. Because the organosilane molecules need a certain length of time to form a quality SAM on the substrate, the stamp should be kept on the substrate for above a minimum time during ink transfer. The optimized contact time was estimated by measurement of contact angle and film thickness of the SAMs [48, 56].

9.3.3 Photolithography

Photolithography is another approach for defining polymer patterns as well as SAM patterns. For this method, the surface modification layers, such as the SAM or the

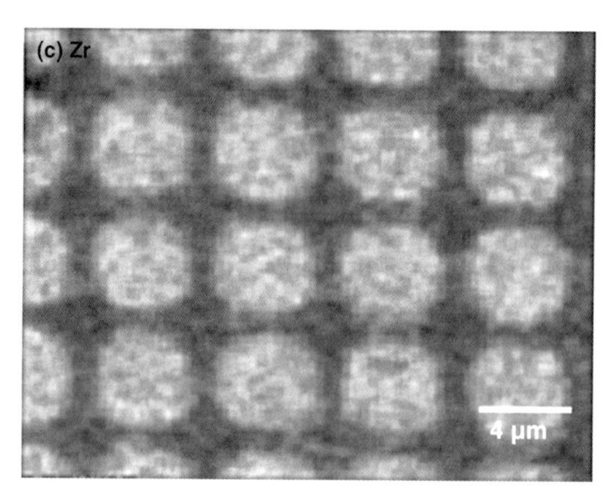

Figure 9.5 Auger elemental maps for (a) platinum, (b) carbon, and (c) zirconium on the grid structure patterned by μCP with ODTS SAM after AS-ALD of Pt on YSZ. Reproduced with permission from Ref. [56]. Copyright 2007, the Electrochemical Society.

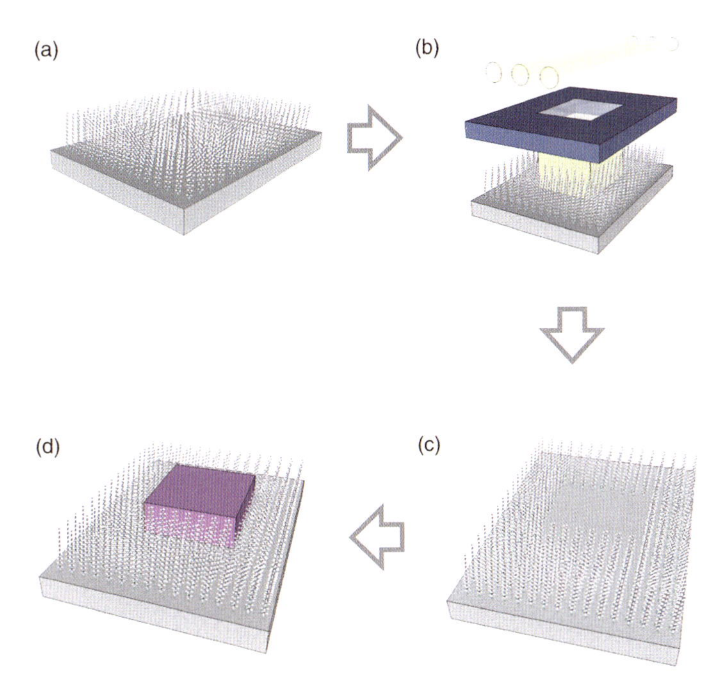

Figure 9.6 The schematic drawings for fabrication process of SAM patterns by using photolithography: (a) after SAM coating on a substrate, (b) exposure of SAM to light through the mask, (c) removal of SAM exposed, and (d) AS-ALD thin film formation on the SAM-free region.

polymer, play the role of the photoresist used in a conventional photolithography process. After coating the SAM on the substrate, a mask is placed over the substrate as shown in Figure 9.6b. Upon irradiation, light is passed through the openings of the mask and removes the SAM in the exposed regions. This method is quite similar to conventional photolithography. For polymer layers, exactly the same scheme is used except the change of modification layer from SAM to polymer. Lee and Sung investigated the use of a specialized mask containing photocatalytic materials [46]. Photocatalytic TiO_2 material was deposited onto the open regions of the mask, and upon exposure to UV light the TiO_2 produced activated O_2 that was effective in removing the SAM material (see Figure 9.7a). By using this photocatalytic mask, ODTS was partially removed in the pattern, and ZrO_2 was deposited by ALD on these modified regions as shown in the AFM data of Figure 9.7b [46]. In another report, photolithography using a PR combined with liquid-phase deposition of ODTS was carried out [60]. After PR (AZ4330) patterning by photolithography, the substrate was immersed in a SAM solution. Because ODTS did not adsorb on the PR region, the ODTS SAM was formed only on the PR-free regions of the substrate. After removal of the PR, a patterned ODTS surface was generated, and this pattern was transferred into the final Co films by Co AS-ALD [60].

Another approach is to use a photolithographically patterned surface followed by two selective steps: selective SAM formation and AS-ALD. The SAM

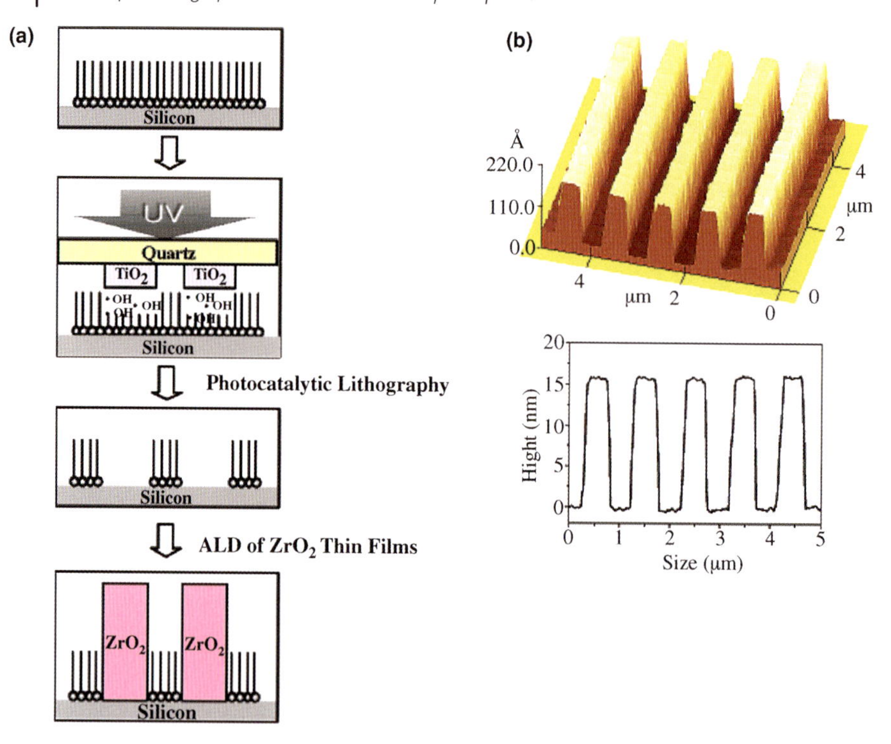

Figure 9.7 (a) Schematic outline of the procedure to fabricate patterned thin films by using photocatalytic lithography and ALD. (b) Three-dimensional AFM images (5 μm × 5 μm) and cross sections for the patterned ZrO_2 thin films fabricated by using photocatalytic lithography and ALD. Reprinted with permission from Ref. [46]. Copyright 2004, the American Chemical Society.

molecules have inherently selective adsorption behavior depending on their head groups and the surface termination, as mentioned previously in Section 9.2.3. By utilizing this property of the SAMs, the patterning step can be simplified through the use of predefined substrate patterns. Chen and Bent have used this approach with a prepatterned substrate combined with selective SAM formation to carry out AS-ALD [49]. They reported both positive and negative patterning using this concept as shown in Figure 9.8. The starting substrate was a silicon wafer with a pattern alternating between thermally deposited SiO_2 and H-terminated bare silicon. After dipping the substrate in ODTS solution, ODTS SAMs formed only on the SiO_2 regions, allowing a patterned ODTS substrate to be obtained without additional patterning. Similarly, exposure to 1-octadecene formed a monolayer resist selectively on the H-terminated regions of the surfaces through a hydrosilylation reaction. Subsequent introduction of the SAM patterned surfaces into an ALD reactor achieved selective deposition of Pt and HfO_2 [49].

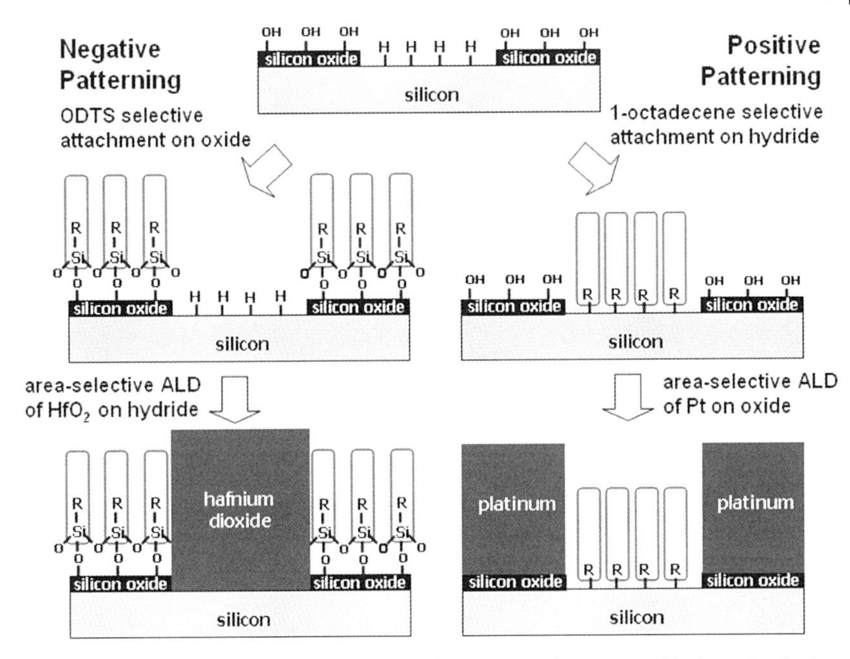

Figure 9.8 Schematic illustration of two area-selective ALD schemes possible through selective surface modifications. Reproduced with permission from Ref. [49]. Copyright 2006, Wiley-VCH Verlag GmbH.

Until now, most AS-ALD processes using a polymer have employed photolithography to make patterns. In fact, PMMA, which is used as a deposition deactivator for AS-ALD, has been utilized as a PR for many applications, such as Si nanowire transistor fabrications [120–122]. Sinha *et al.* investigated TiO_2 AS-ALD with PMMA patterns defined by photolithography [77, 78]. In their subsequent paper, the inherently inactive surface of poly(*tert*-butyl methacrylate) (P*t*BMA) was selectively converted into activated OH termination by exposure of UV [79]. TiO_2 patterns were formed on the activated regions, and then they were used for an etching mask to transfer the patterns to the original substrate [79]. Farm *et al.* showed the feasibility of polymer-based, photolithographic AS-ALD with various material systems, such as Pt, Ru, Ir, Al_2O_3, and TiO_2 [80, 81]. They observed no deposition on PMMA and PVP with Pt, Ru, Ir, and TiO_2, but deposition did occur with Al_2O_3 [80, 81]. For Al_2O_3, however, a liftoff process could be used for making Al_2O_3 patterns because of the easier removal process of the polymer compared to SAMs. Recently, photolithography-assisted patterning with polymer was successfully applied to large-area processing [54], in which PMMA patterns were defined by photolithography, and then transferred into ZnO patterns for display device fabrication by spatial ALD.

In another study, Park and Parsons directly applied toward AS-ALD of Rh a standard PR used for the photolithography process in Si device fabrication [82]. After typical photolithography to pattern the PR, Rh ALD was carried out on the substrate. The Rh films were found to selectively deposit only on the PR-free region. In addition, the researchers observed the ability of HMDS, which is generally used as an adhesion promoter for PR, to block Rh film deposition by ALD. The prevention of Rh nucleation on the polymers is attributed to their hydrophobicity, because typical PRs and HMDS are hydrophobic in nature. Similarly, another group also fabricated Ru patterns by AS-ALD using PR and investigated MOS characteristics with the resulting Ru metal gate [84].

9.3.4 Nanotemplating

Nanotemplates are a type of scaffold with nanoscale features used to fabricate nanostructures. Generally, the nanotemplates are fabricated by self-assembly processes and they include ordered patterns of various size and shape. Because of the self-assembly process, the size of nanotemplate patterns is easily controlled down to tens of nanometers, a length scale that is difficult to fabricate by conventional patterning tools. Therefore, nanotemplates such as anodic aluminum oxide (AAO) and diblock copolymers have been intensively investigated for various applications [2, 3, 123, 124]. Combined with nanotemplates, AS-ALD provides another way to fabricate nanostructures.

Shin and coworkers reported a method for fabricating oxide nanotubes using a polycarbonate (PC) nanotemplate and AS-ALD [110, 125]. The process scheme is presented in Figure 9.9. ODTS was transferred onto the top of the PC membrane via μCP (see Section 9.3.2). Because the μCP stamp contacts only the top of the PC, only the top of the structure, and not the inside of the holes, is deactivated by ODTS molecules. Subsequently, oxide films are deposited inside the unmodified holes during an ALD process. After removal of the PC template, oxide nanotubes are fabricated as a replica of the hole structure. Because high conformality and uniformity are inherent properties of ALD, oxide films are deposited on the whole template if the top of the structure is not deactivated by the SAM. From the same group, TiO_2, ZrO_2, and ZnO nanotube fabrication was also reported using another nanotemplate, AAO [111].

Another widely used self-assembly template consists of nanospheres. A monolayer of nanospheres, made of polystyrene (PS) or silica, can be easily formed on a flat surface with relatively large-scale uniformity, which provides an economical way of patterning [126]. Recently, a new patterning method for SAM layers using nanospheres was reported [43, 44]. The silica nanospheres, which have a diameter of 200–500 nm, are spontaneously arranged into a close-packed structure during a liquid-phase coating process [127, 128]. In this scheme, the substrate containing the well-arranged silica nanospheres is then dipped into a solution of ODTS. The

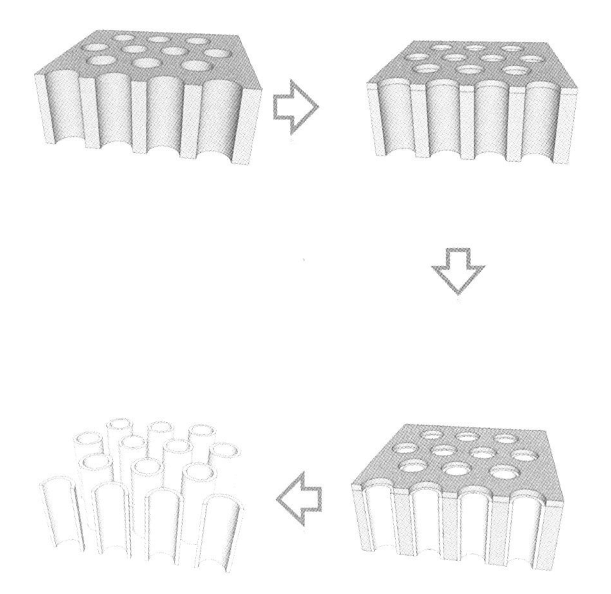

Figure 9.9 Schematic drawings of the nanotube fabrication process using AS-ALD and nanotemplates.

surface is coated by ODTS everywhere except the regions contacting the nanospheres, resulting in the formation of a periodically uncoated pattern several tens of nanometer in size. After removal of nanospheres, TiO_2 nanohemispheres with various sizes were formed on this substrate by using AS-ALD of TiO_2 [43, 44]. In subsequent research, this patterning process was extended to 3D nanostructure formation on the nanospheres. After the formation of a multilayer of nanospheres, the SAM was coated on the uncontacted region between nanospheres. Subsequent AS-ALD of TiO_2 formed the periodic points on the nanospheres as shown in Figure 9.10.

9.3.5 Scanning Probe Microscopy

Nanolithography by scanning probe microscopy (SPM) has also been explored with AS-ALD. Figure 9.11 shows a schematic drawing for this concept. After forming a SAM coating on a substrate, the SPM tip sequentially changes the surface property by applying an external field, similar to drawing. When the ALD process proceeds on this substrate, the film is selectively deposited due to differences between intact and changed regions. Recently, Lee *et al.* demonstrated AS-ALD of ZrO_2 through SPM lithography [47, 59]. An ODTS SAM was removed followed by the formation of SiO_2

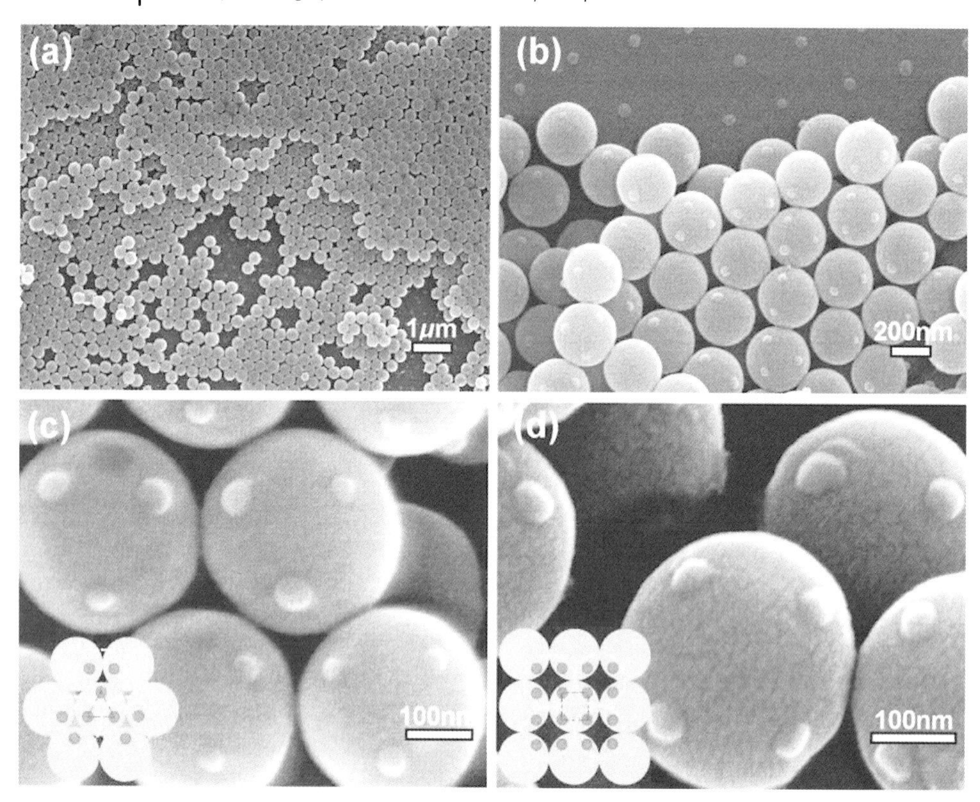

Figure 9.10 SEM micrographs of asymmetric colloidal clusters (ACCs) consisting of titania/silica. (a) An experimental result is shown as proof of concept; a monolayer of ACCs was prepared on a planar surface from the silica colloidal crystals with double layers in which the upper layers of silica were removed before the site-selective growth of titania. (b) ACCs in a certain region are more than two layers; the pieces having one or two titania dots result from the insecure contacts with the upper silica layers. Panel (c) is from the (111) surface of the underlying colloidal films; panel (d) is from the (100) facet. The solid bright grey circles in the insets indicate the underlying silica colloids, and the dotted white lines indicate the upper layer detached to open the nucleation sites; the solid dark grey circles represent titania dots selectively grown on the oxide openings of the silica spheres. Reprinted with permission from Ref. [43]. Copyright 2007, the American Chemical Society.

on the regions where an electric field was applied via atomic force microscopy (AFM) tip. After the SiO_2 patterns were removed by HF etching, subsequent ALD led to selective deposition on the SAM-free regions.

In other work, the PMMA patterning was carried out by using thermal writing by AFM. After coating PMMA on a substrate, the AFM tip locally removed PMMA in a pattern by applying a bias. PMMA was partially removed by localized heating from the AFM tip along the pattern lines and TiO_2 was selectively deposited on the PMMA-depleted region [86].

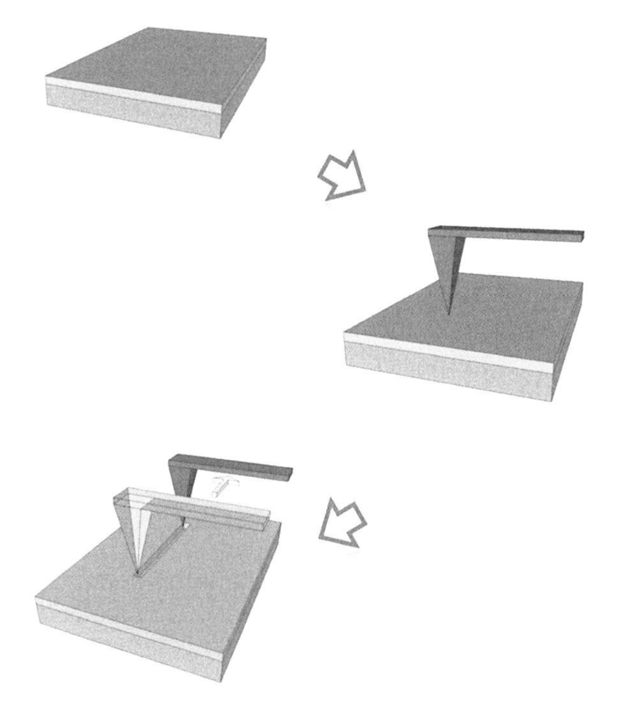

Figure 9.11 Schematic drawings for SAM patterning by using SPM.

9.4 Applications of AS-ALD

Microelectronics device fabrication has been one of the major research fields for ALD. For example, Ru, HfO_2, and TaN have been intensively investigated by ALD for metal gate, high-k, and diffusion barrier materials, respectively [1, 4]. As an application for device fabrication, Parson's group demonstrated the fabrication and characterization of metal-insulator-metal capacitors using AS-ALD of Ru by $RuCp_2$ and O_2, utilizing ODTS SAMs patterned on SiO_2 and HfO_2 surfaces [58]. The Ru gate electrode pattern was defined by AS-ALD with PDMS stamping without the need for an etching process.

Bent's group has actively investigated AS-ALD for energy applications, especially for fuel cell electrodes. They demonstrated selective deposition of Pt films on a YSZ substrate, which is an electrolyte for solid oxide fuel cells (SOFCs), and successfully fabricated Pt patterns without etching by using AS-ALD and μCP [48, 56]. In a subsequent report, they showed the feasibility of AS-ALD Pt patterns as a current collector [57]. By using AS-ALD and μCP, Pt grid patterns with high lateral resolution were fabricated without an etching process, and the Pt grid current collector showed better fuel cell performance than did blanket Pt films of the same loading.

ZnO is one the most investigated materials for transparent conducting electrodes, and many studies on ZnO ALD have been reported. Levy *et al.* developed roll-to-roll-type spatial ALD and applied it to AS-ALD of ZnO with three deactivation agents including PMMA, ODTS, and polystyrene [54]. PMMA and ODTS showed similar deactivation quality for ZnO ALD while PS did not block the deposition even with larger thickness than the others. The importance of this research is that it showed the feasibility for AS-ALD to be applied to large-area and high-speed deposition processes, especially for display device fabrication.

9.5 Current Challenges

Although AS-ALD has been intensively studied and applied to various ALD systems, there are still several challenges. One of the most frequently reported challenges is degradation of the SAM or polymer during the ALD process [50, 56, 60]. Generally, the film deposition by ALD is performed at a temperature range from 100 to 400 °C [1, 4]. Therefore, the SAM or polymer should be stable and maintain its deactivating properties toward ALD at those temperatures. However, SAMs, since they are based on organic molecules, have relatively low thermal stability [110, 111, 125]. Although no degradation of SAMs was reported up to 400 °C under vacuum conditions in a previous report [129], the loss of the ability to block film deposition above a certain number of ALD cycles was noted in several reports, and the degradation was found to be a time-dependent process [56, 60, 130]. For example, the deposition of Pt films was found to be significantly deactivated up to 400 cycles at a substrate temperature of 300 °C; above 400 cycles, the Pt films began to deposit even on ODTS-coated surface due to thermal degradation [56]. Furthermore, the ALD process employs various reactants, such as gases and plasma. Meanwhile, the reaction between SAMs and ALD reactants affects the degradation of SAMs, as the SAM molecules at the substrate encounter various gases and chemical species during each ALD cycle. Recently, Lee *et al.* reported the degradation of ODTS SAMs by NH_3 plasma reactants during plasma-enhanced ALD (PE-ALD) of Co [61]. In addition, there are many reports on the degradation of SAMs by plasma, indicating that the PE-ALD process cannot be easily applied to AS-ALD [131, 132].

Another important challenge is the size effect of the patterns. The deactivation of thin film deposition by SAMs takes place by inhibition of the nucleation on the SAM-coated regions of the substrate. As the thickness of the deposited film on the SAM-free region grows to be above the height of the SAM, the individual patterns become a 3D structure and not a planar 2D film anymore. Consequently, new surfaces of the film materials will begin to form around the pattern. Because the ALD growth is based on the surface reaction and isotropically occurs on the entire surface, the lateral size of the pattern will increase by this process (see Figure 9.12). This situation becomes significant when the size of the pattern is smaller than the film thickness. In a previous report, although a small size of TiO_2 patterns under 100 nm was achieved from AS-ALD and colloidal particles, a widening of the TiO_2 patterns was observed

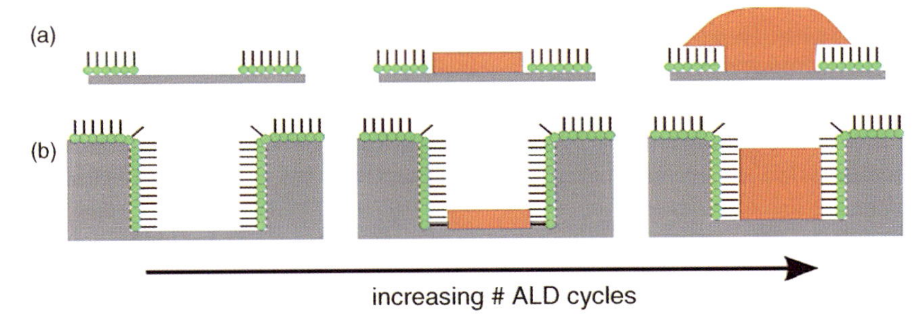

Figure 9.12 (a) Conventional area-selective ALD in which the substrate is planar and contains patterns of self-assembled monolayers. With increasing number of deposition cycles, sideways film growth also occurs originating from adsorption of ALD reactants on the previously deposited ALD film. (b) Blocking the lateral ALD growth independent of deposited film thickness by combining surface modification and topographical features. Reprinted with permission from Ref. [83]. Copyright 2008, the American Chemical Society.

due to the reason mentioned above [43]. To overcome this challenge, topographical confinement as well as surface modification by SAMs was proposed for AS-ALD in the nanoscale regime as shown in Figure 9.12b [83]. In the report, an insect's wing, which had ordered hole patterns with 60 nm diameter and 250 nm height and an inherent coating of waxy material, was employed for nanopatterns. The authors showed that they could use biomimetic nanopatterns with lateral confinement as well as deactivated surfaces without additional processes. The ZnO was grown by ALD inside the holes confined by topographical feature and surface modification without lateral widening.

Although AS-ALD can minimize and change conventional patterning process steps, it still requires a patterning process. Therefore, future applications of AS-ALD depend on how patterns are fabricated effectively. The site-specific adsorption property of SAMs gives us the opportunity for AS-ALD to reuse patterns defined during previous steps. For example, silicide fabrication step for metal-oxide-semiconductor field effect transistor device fabrication can be simplified by using AS-ALD with a SAM adsorbed only on the SiO_2 surface. A typical silicide fabrication process consists of three steps including metal deposition, annealing, and removal of unreacted metal as shown in Figure 9.13a. If a SAM was used to deactivate the SiO_2 spacer region, the metal etching process would not be needed anymore and the original patterns could be reused. This concept can be successfully adapted to nanoscale device fabrication, such as Si nanowire transistors. By using a pattern of insulating material that surrounds part of the nanowires, selective adsorption of a deactivating agent on the insulator followed by AS-ALD could be used to define the metal contact, which would reduce patterning steps including pattern alignment, as shown in Figure 9.13b. In addition, vapor-phase deposition of the SAM could be an advantage for this. Moreover, by combining the large-area uniformity of ALD, AS-ALD can be effective in a large-area process, such as in the fabrication of display devices (see Figure 9.13c). Because the AS-ALD process is

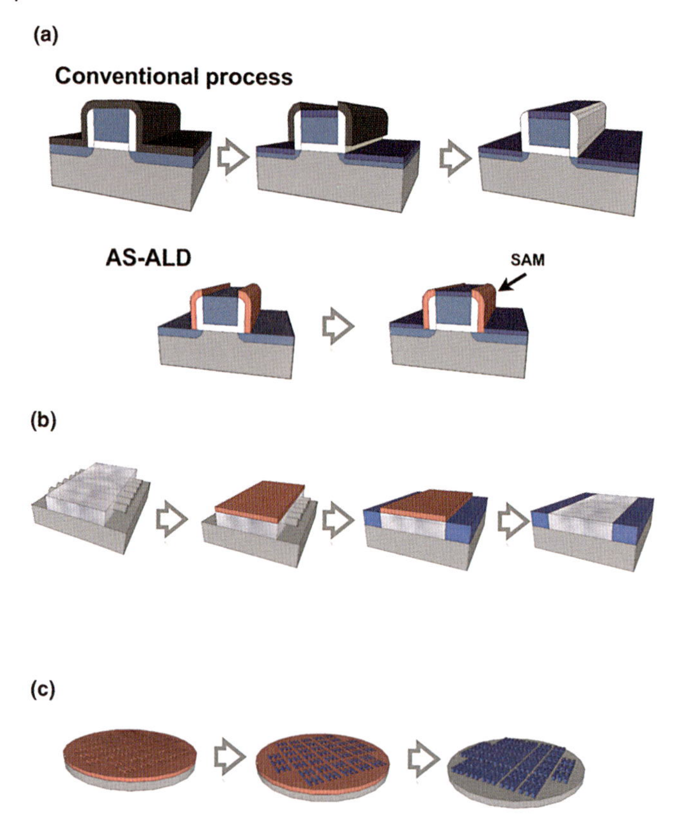

Figure 9.13 Schematic drawings for AS-ALD applications on (a) silicide fabrication, (b) nanowire device fabrication, and (c) large-area process.

an etch-free patterning process, it has the potential to impact device fabrication for a number of emerging applications.

Acknowledgment

This work was supported by the US Department of Energy Hydrogen, Fuel Cells, and Infrastructure Program through the National Renewable Energy Laboratory under Contract No. DE-AC36-08-GO28308.

References

1 Kim, H. (2003) Atomic layer deposition of metal and nitride thin films: current research efforts and applications for semiconductor device processing. *J. Vac. Sci. Technol. B*, **21** (6), 2231.

2 George, S.M. (2009) Atomic layer deposition: an overview. *Chem. Rev.*, **110** (1), 111.

3 Kim, H., Lee, H.-B.-R., and Maeng, W.J. (2009) Applications of atomic layer

deposition to nanofabrication and emerging nanodevices. *Thin Solid Films*, **517** (8), 2563.

4 Leskelä, M. and Ritala, M. (2003) Atomic layer deposition chemistry: recent developments and future challenges. *Angew. Chem., Int. Ed.*, **42** (45), 5548.

5 Jeon, N.L., Lin, W., Erhardt, M.K., Girolami, G.S., and Nuzzo, R.G. (1997) Selective chemical vapor deposition of platinum and palladium directed by monolayers patterned using microcontact printing. *Langmuir*, **13** (14), 3833.

6 Ulman, A. (1996) Formation and structure of self-assembled monolayers. *Chem. Rev.*, **96** (4), 1533.

7 Schreiber, F. (2000) Structure and growth of self-assembling monolayers. *Prog. Surf. Sci.*, **65** (5–8), 151.

8 Allara, D.L. and Nuzzo, R.G. (1985) Spontaneously organized molecular assemblies. 2. Quantitative infrared spectroscopic determination of equilibrium structures of solution-adsorbed *n*-alkanoic acids on an oxidized aluminum surface. *Langmuir*, **1** (1), 52.

9 Allara, D.L. and Nuzzo, R.G. (1985) Spontaneously organized molecular assemblies. 1. Formation, dynamics, and physical properties of *n*-alkanoic acids adsorbed from solution on an oxidized aluminum surface. *Langmuir*, **1** (1), 45.

10 Schlotter, N.E., Porter, M.D., Bright, T.B., and Allara, D.L. (1986) Formation and structure of a spontaneously adsorbed monolayer of arachidic on silver. *Chem. Phys. Lett.*, **132** (1), 93.

11 Samant, M.G., Brown, C.A., and Gordon, J.G. (1993) An epitaxial organic film. The self-assembled monolayer of docosanoic acid on silver(111). *Langmuir*, **9** (4), 1082.

12 Tao, Y.T. (1993) Structural comparison of self-assembled monolayers of *n*-alkanoic acids on the surfaces of silver, copper, and aluminum. *J. Am. Chem. Soc.*, **115** (10), 4350.

13 Poirier, G.E. and Pylant, E.D. (1996) The self-assembly mechanism of alkanethiols on Au(111). *Science*, **272** (5265), 1145.

14 Love, J.C., Estroff, L.A., Kriebel, J.K., Nuzzo, R.G., and Whitesides, G.M. (2005) Self-assembled monolayers of thiolates on metals as a form of nanotechnology. *Chem. Rev.*, **105** (4), 1103.

15 Laibinis, P.E., Whitesides, G.M., Allara, D.L., Tao, Y.T., Parikh, A.N., and Nuzzo, R.G. (1991) Comparison of the structures and wetting properties of self-assembled monolayers of *n*-alkanethiols on the coinage metal surfaces, copper, silver, and gold. *J. Am. Chem. Soc.*, **113** (19), 7152.

16 Walczak, M.M., Chung, C., Stole, S.M., Widrig, C.A., and Porter, M.D. (1991) Structure and interfacial properties of spontaneously adsorbed *n*-alkanethiolate monolayers on evaporated silver surfaces. *J. Am. Chem. Soc.*, **113** (7), 2370.

17 Laibinis, P.E. and Whitesides, G.M. (1992) ω-Terminated alkanethiolate monolayers on surfaces of copper, silver, and gold have similar wettabilities. *J. Am. Chem. Soc.*, **114** (6), 1990.

18 Laibinis, P.E. and Whitesides, G.M. (1992) Self-assembled monolayers of *n*-alkanethiolates on copper are barrier films that protect the metal against oxidation by air. *J. Am. Chem. Soc.*, **114** (23), 9022.

19 Liu, Q. and Xu, Z. (1995) Self-assembled monolayer coatings on nanosized magnetic particles using 16-mercaptohexadecanoic acid. *Langmuir*, **11** (12), 4617.

20 Sheen, C.W., Shi, J.X., Maartensson, J., Parikh, A.N., and Allara, D.L. (1992) A new class of organized self-assembled monolayers: alkane thiols on gallium arsenide(100). *J. Am. Chem. Soc.*, **114** (4), 1514.

21 Gu, Y., Lin, Z., Butera, R.A., Smentkowski, V.S., and Waldeck, D.H. (1995) Preparation of self-assembled monolayers on InP. *Langmuir*, **11** (6), 1849.

22 Ardalan, P., Sun, Y., Pianetta, P., Musgrave, C.B., and Bent, S.F. (2010) Reaction mechanism, bonding, and thermal stability of 1-alkanethiols self-assembled on halogenated Ge surfaces. *Langmuir*, **26** (11), 8419.

23 van Zanten, T.S., Lammertink, R.G.H., Borneman, Z., Nijdam, W., van Rijn, C.J.M., Wessling, M., and Geerken, M.J. (2004) Chemical and thermal stability of alkylsilane based coatings for membrane emulsification. *Adv. Eng. Mater.*, **6** (9), 749.

24 Allara, D.L., Parikh, A.N., and Rondelez, F. (1995) Evidence for a unique chain organization in long-chain silane monolayers deposited on two widely different solid substrates. *Langmuir*, **11** (7), 2357.

25 Silberzan, P., Leger, L., Ausserre, D., and Benattar, J.J. (1991) Silanation of silica surfaces: a new method of constructing pure or mixed monolayers. *Langmuir*, **7** (8), 1647.

26 Sagiv, J. (1980) Organized monolayers by adsorption. 1. Formation and structure of oleophobic mixed monolayers on solid surfaces. *J. Am. Chem. Soc.*, **102** (1), 92.

27 Gun, J., Iscovici, R., and Sagiv, J. (1984) On the formation and structure of self-assembling monolayers. II. A comparative study of Langmuir–Blodgett and adsorbed films using ellipsometry and IR reflection–absorption spectroscopy. *J. Colloid Interface Sci.*, **101** (1), 201.

28 Tillman, N., Ulman, A., Schildkraut, J.S., and Penner, T.L. (1988) Incorporation of phenoxy groups in self-assembled monolayers of trichlorosilane derivatives. Effects on film thickness, wettability, and molecular orientation. *J. Am. Chem. Soc.*, **110** (18), 6136.

29 Mathauer, K. and Frank, C.W. (1993) Naphthalene chromophore tethered in the constrained environment of a self-assembled monolayer. *Langmuir*, **9** (11), 3002.

30 Brandriss, S. and Margel, S. (1993) Synthesis and characterization of self-assembled hydrophobic monolayer coatings on silica colloids. *Langmuir*, **9** (5), 1232.

31 Finklea, H.O., Robinson, L.R., Blackburn, A., Richter, B., Allara, D., and Bright, T. (1986) Formation of an organized monolayer by solution adsorption of octadecyltrichlorosilane on gold: electrochemical properties and structural characterization. *Langmuir*, **2** (2), 239.

32 Linford, M.R. and Chidsey, C.E.D. (1993) Alkyl monolayers covalently bonded to silicon surfaces. *J. Am. Chem. Soc.*, **115** (26), 12631.

33 Linford, M.R., Fenter, P., Eisenberger, P.M., and Chidsey, C.E.D. (1995) Alkyl monolayers on silicon prepared from 1-alkenes and hydrogen-terminated silicon. *J. Am. Chem. Soc.*, **117** (11), 3145.

34 Park, M.H., Jang, Y.J., Sung-Suh, H.M., and Sung, M.M. (2004) Selective atomic layer deposition of titanium oxide on patterned self-assembled monolayers formed by microcontact printing. *Langmuir*, **20** (6), 2257.

35 Akkerman, H.B., Blom, P.W.M., de Leeuw, D.M., and de Boer, B. (2006) Towards molecular electronics with large-area molecular junctions. *Nature*, **441** (7089), 69.

36 Zamborini, F.P. and Crooks, R.M. (1998) Corrosion passivation of gold by *n*-alkanethiol self-assembled monolayers: effect of chain length and end group. *Langmuir*, **14** (12), 3279.

37 Seo, E.K., Lee, J.W., Sung-Suh, H.M., and Sung, M.M. (2004) Atomic layer deposition of titanium oxide on self-assembled-monolayer-coated gold. *Chem. Mater.*, **16** (10), 1878.

38 Ardalan, P., Musgrave, C., and Bent, S. (2009) Effects of surface functionalization on titanium dioxide atomic layer deposition on Ge surfaces. *ECS Trans.*, **25** (4), 131.

39 Chen, R., Kim, H., McIntyre, P.C., and Bent, S.F. (2005) Investigation of self-assembled monolayer resists for hafnium dioxide atomic layer deposition. *Chem. Mater.*, **17** (3), 536.

40 Ghosal, S., Baumann, T.F., King, J.S., Kucheyev, S.O., Wang, Y.M., Worsley, M.A., Biener, J., Bent, S.F., and Hamza, A.V. (2009) Controlling atomic layer deposition of TiO_2 in aerogels through surface functionalization. *Chem. Mater.*, **21** (9), 1989.

41 Lee, J.P., Jang, Y.J., and Sung, M.M. (2003) Atomic layer deposition of TiO_2 thin films on mixed self-assembled monolayers studied as a function of surface free energy. *Adv. Funct. Mater.*, **13** (11), 873.

42 Hoyas, A.M., Schuhmacher, J., Whelan, C.M., Landaluce, T.F., Vanhaeren, D., and Maex, K. (2006) Using scaling laws to understand the growth mechanism of atomic layer deposited WN_xC_y films on methyl-terminated surfaces. *J. Appl. Phys.*, **100** (11), 114903.

43 Bae, C., Moon, J., Shin, H., Kim, J., and Sung, M.M. (2007) Fabrication of monodisperse asymmetric colloidal clusters by using contact area lithography (CAL). *J. Am. Chem. Soc.*, **129** (46), 14232.

44 Bae, C., Shin, H., Moon, J., and Sung, M.M. (2006) Contact area lithography (CAL): a new approach to direct formation of nanometric chemical patterns. *Chem. Mater.*, **18** (5), 1085.

45 Chen, R., Kim, H., McIntyre, P.C., and Bent, S.F. (2004) Self-assembled monolayer resist for atomic layer deposition of HfO and ZrO high-*k* gate dielectrics. *Appl. Phys. Lett.*, **84**, 4017.

46 Lee, J.P. and Sung, M.M. (2004) A new patterning method using photocatalytic lithography and selective atomic layer deposition. *J. Am. Chem. Soc.*, **126** (1), 28.

47 Lee, W. and Prinz, F.B. (2009) Area-selective atomic layer deposition using self-assembled monolayer and scanning probe lithography. *J. Electrochem. Soc.*, **156** (9), G125.

48 Jiang, X.R., Chen, R., and Bent, S.F. (2007) Spatial control over atomic layer deposition using microcontact-printed resists. *Surf. Coat. Technol.*, **201** (22–23), 8799.

49 Chen, R. and Bent, S.F. (2006) Chemistry for positive pattern transfer using area-selective atomic layer deposition. *Adv. Mater.*, **18** (8), 1086.

50 Hong, J., Porter, D.W., Sreenivasan, R., McIntyre, P.C., and Bent, S.F. (2007) ALD resist formed by vapor-deposited self-assembled monolayers. *Langmuir*, **23** (3), 1160.

51 Chen, R., Kim, H., McIntyre, P.C., Porter, D.W., and Bent, S.F. (2005) Achieving area-selective atomic layer deposition on patterned substrates by selective surface modification. *Appl. Phys. Lett.*, **86** (19), 191910.

52 Xu, Y. and Musgrave, C.B. (2004) A DFT study of the Al_2O_3 atomic layer deposition on SAMs: effect of SAM termination. *Chem. Mater.*, **16** (4), 646.

53 Yan, M., Koide, Y., Babcock, J.R., Markworth, P.R., Belot, J.A., Marks, T.J., and Chang, R.P.H. (2001) Selective-area atomic layer epitaxy growth of ZnO features on soft lithography-patterned substrates. *Appl. Phys. Lett.*, **79** (11), 1709.

54 Levy, D.H., Nelson, S.F., and Freeman, D. (2009) Oxide electronics by spatial atomic layer deposition. *J. Display Technol.*, **5** (12), 484.

55 Dube, A., Sharma, M., Ma, P.F., Ercius, P.A., Muller, D.A., and Engstrom, J.R. (2007) Effects of interfacial organic layers on nucleation, growth, and morphological evolution in atomic layer thin film deposition. *J. Phys. Chem. C*, **111** (29), 11045.

56 Jiang, X. and Bent, S.F. (2007) Area-selective atomic layer deposition of platinum on YSZ substrates using microcontact printed SAMs. *J. Electrochem. Soc.*, **154** (12), D648.

57 Jiang, X., Huang, H., Prinz, F.B., and Bent, S.F. (2008) Application of atomic layer deposition of platinum to solid oxide fuel cells. *Chem. Mater.*, **20** (12), 3897.

58 Park, K.J., Doub, J.M., Gougousi, T., and Parsons, G.N. (2005) Microcontact patterning of ruthenium gate electrodes by selective area atomic layer deposition. *Appl. Phys. Lett.*, **86**, 051903.

59 Lee, W., Dasgupta, N.P., Trejo, O., Lee, J.R., Hwang, J., Usui, T., and Prinz, F.B. (2010) Area-selective atomic layer deposition of lead sulfide: nanoscale patterning and DFT simulations. *Langmuir*, **26** (9), 6845.

60 Lee, H.B.R., Kim, W.H., Lee, J.W., Kim, J.M., Heo, K., Hwang, I.C., Park, Y., Hong, S., and Kim, H. (2010) High quality area-selective atomic layer deposition Co using ammonia gas as a reactant. *J. Electrochem. Soc.*, **157** (1), D10.

61 Lee, H.B.R., Kim, J., Kim, H., Kim, W.H., Lee, J.W., and Hwang, I. (2010) Degradation of the deposition blocking layer during area-selective plasma-enhanced atomic layer deposition of cobalt. *J. Korean Phys. Soc.*, **56** (1), 104.

62 Farm, E., Kernell, M., Ritala, M., and Leskelä, M. (2006) Self-assembled octadecyltrimethoxysilane monolayers enabling selective-area atomic layer deposition of iridium. *Chem. Vapor Depos.*, **12** (7), 415.

63 Dube, A., Sharma, M., Ma, P.F., and Engstrom, J.R. (2006) Effects of interfacial organic layers on thin film nucleation in atomic layer deposition. *Appl. Phys. Lett.*, **89** (16), 164108.

64 Chen, R. and Bent, S.F. (2006) Highly stable monolayer resists for atomic layer deposition on germanium and silicon. *Chem. Mater.*, **18** (16), 3733.

65 ter Maat, J., Regeling, R., Yang, M., Mullings, M.N., Bent, S.F., and Zuilhof, H. (2009) Photochemical covalent attachment of alkene-derived monolayers onto hydroxyl-terminated silica. *Langmuir*, **25** (19), 11592.

66 Li, M., Dai, M., and Chabal, Y.J. (2009) Atomic layer deposition of aluminum oxide on carboxylic acid-terminated self-assembled monolayers. *Langmuir*, **25** (4), 1911.

67 Haran, M., Engstrom, J.R., and Clancy, P. (2006) *Ab initio* calculations of the reaction mechanisms for metal-nitride deposition from organo-metallic precursors onto functionalized self-assembled monolayers. *J. Am. Chem. Soc.*, **128** (3), 836.

68 Chen, R., Kim, H., McIntyre, P.C., Porter, D.W., and Bent, S.F. (2005) Achieving area-selective atomic layer deposition on patterned substrates by selective surface modification. *Appl. Phys. Lett.*, **86** (19), 191910.

69 Jiang, X. and Bent, S.F. (2009) Area-selective ALD with soft lithographic methods: using self-assembled monolayers to direct film deposition. *J. Phys. Chem. C*, **113** (41), 17613.

70 Mirji, S.A. (2006) Octadecyltrichlorosilane adsorption kinetics on Si (100)/SiO_2 surface: contact angle, AFM, FTIR and XPS analysis. *Surf. Interface Anal.*, **38** (3), 158.

71 Dubois, L.H., Zegarski, B.R., and Nuzzo, R.G. (1990) Fundamental studies of microscopic wetting on organic surfaces. 2. Interaction of secondary adsorbates with chemically textured organic monolayers. *J. Am. Chem. Soc.*, **112** (2), 570.

72 Drelich, J., Wilbur, J.L., Miller, J.D., and Whitesides, G.M. (1996) Contact angles for liquid drops at a model heterogeneous surface consisting of alternating and parallel hydrophobic/hydrophilic strips. *Langmuir*, **12** (7), 1913.

73 Wang, Y. and Lieberman, M. (2003) Growth of ultrasmooth octadecyltrichlorosilane self-assembled monolayers on SiO_2. *Langmuir*, **19** (4), 1159.

74 Jaeger, R. (1988) *Introduction to Microelectronic Fabrication*, Addison-Wesley, Reading, MA.

75 Taur, Y. and Ning, T. (1998) *Fundamentals of Modern VLSI Devices*, Cambridge University Press, Cambridge.

76 Sinha, A., Hess, D.W., and Henderson, C.L. (2007) Transport behavior of atomic layer deposition precursors through polymer masking layers: influence on area selective atomic layer deposition. *J. Vac. Sci. Technol. B*, **25** (5), 1721.

77 Sinha, A., Hess, D.W., and Henderson, C.L. (2006) Area-selective ALD of titanium dioxide using lithographically defined poly(methyl methacrylate) films. *J. Electrochem. Soc.*, **153** (5), G465.

78 Sinha, A., Hess, D.W., and Henderson, C.L. (2006) Area selective atomic layer deposition of titanium dioxide: effect of precursor chemistry. *J. Vac. Sci. Technol. B*, **24** (6), 2523.

79 Sinha, A., Hess, D.W., and Henderson, C.L. (2006) A top surface imaging method using area selective ALD on chemically amplified polymer photoresist films. *Electrochem. Solid-State Lett.*, **9**, G330.

80 Farm, E., Kemell, M., Santala, E., Ritala, M., and Leskelä, M. (2010) Selective-area atomic layer deposition using poly(vinyl pyrrolidone) as a passivation layer. *J. Electrochem. Soc.*, **157** (1), K10.

81 Färm, E., Kemell, M., Ritala, M., and Leskelä, M. (2008) Selective-area atomic layer deposition using poly(methyl methacrylate) films as mask layers. *J. Phys. Chem. C*, **112** (40), 15791.

82 Park, K.J. and Parsons, G.N. (2006) Selective area atomic layer deposition of rhodium and effective work function

characterization in capacitor structures. *Appl. Phys. Lett.*, **89** (4), 043111.

83 Ras, R.H.A., Sahramo, E., Malm, J., Raula, J., and Karppinen, M. (2008) Blocking the lateral film growth at the nanoscale in area-selective atomic layer deposition. *J. Am. Chem. Soc.*, **130** (34), 11252.

84 Dey, S.K., Goswami, J., Gu, D., de Waard, H., Marcus, S., and Werkhoven, C. (2004) Ruthenium films by digital chemical vapor deposition: selectivity, nanostructure, and work function. *Appl. Phys. Lett.*, **84** (9), 1606.

85 Mullings, M.N., Lee, H.-B.-R., Marchack, N., Jiang, X., Chen, Z., Gorlin, Y., Lin, K.-P., and Bent, S.F. (2010) Area selective atomic layer deposition by microcontact printing with a water-soluble polymer. *J. Electrochem. Soc.*, **157** (12), D600.

86 Hua, Y.M., King, W.P., and Henderson, C.L. (2008) Nanopatterning materials using area selective atomic layer deposition in conjunction with thermochemical surface modification via heated AFM cantilever probe lithography. *Microelectron. Eng.*, **85** (5–6), 934.

87 Carlsson, J.-O. (1990) Selective vapor-phase deposition on patterned substrates. *Crit. Rev. Solid State Mater. Sci.*, **16** (3), 161.

88 Gates, S.M. (1996) Surface chemistry in the chemical vapor deposition of electronic materials. *Chem. Rev.*, **96** (4), 1519.

89 Tsubouchi, K. and Masu, K. (1993) Area-selective CVD of metals. *Thin Solid Films*, **228** (1–2), 312.

90 Yokoyama, S., Ikeda, N., Kajikawa, K., and Nakashima, Y. (1998) Atomic-layer selective deposition of silicon nitride on hydrogen-terminated Si surfaces. *Appl. Surf. Sci.*, **130**, 352.

91 Tao, Q., Jursich, G., and Takoudis, C. (2010) Selective atomic layer deposition of HfO_2 on copper patterned silicon substrates. *Appl. Phys. Lett.*, **96** (19), 192105.

92 Green, M.L., Ho, M.Y., Busch, B., Wilk, G.D., Sorsch, T., Conard, T., Brijs, B., Vandervorst, W., Raisanen, P.I., Muller, D., Bude, M., and Grazul, J. (2002) Nucleation and growth of atomic layer deposited HfO_2 gate dielectric layers on chemical oxide (Si–O–H) and thermal oxide (SiO_2 or Si–O–N) underlayers. *J. Appl. Phys.*, **92** (12), 7168.

93 Kim, J.-H., Kim, J.-Y., and Kang, S.-W. (2005) Film growth model of atomic layer deposition for multicomponent thin films. *J. Appl. Phys.*, **97** (9), 093505.

94 Green, M.L., Allen, A.J., Li, X., Wang, J., Ilavsky, J., Delabie, A., Puurunen, R.L., and Brijs, B. (2006) Nucleation of atomic-layer-deposited HfO_2 films, and evolution of their microstructure, studied by grazing incidence small angle X-ray scattering using synchrotron radiation. *Appl. Phys. Lett.*, **88** (3), 032907.

95 Yim, S.-S., Lee, D.-J., Kim, K.-S., Kim, S.-H., Yoon, T.-S., and Kim, K.-B. (2008) Nucleation kinetics of Ru on silicon oxide and silicon nitride surfaces deposited by atomic layer deposition. *J. Appl. Phys.*, **103** (11), 113509.

96 Lim, J.-W., Park, H.-S., and Kang, S.-W. (2000) Analysis of a transient region during the initial stage of atomic layer deposition. *J. Appl. Phys.*, **88** (11), 6327.

97 Kim, H. and Rossnagel, S.M. (2002) Growth kinetics and initial stage growth during plasma-enhanced Ti atomic layer deposition. *J. Vac. Sci. Technol. A*, **20** (3), 802.

98 Frank, M.M., Chabal, Y.J., Green, M.L., Delabie, A., Brijs, B., Wilk, G.D., Ho, M.-Y., da Rosa, E.B.O., Baumvol, I.J.R., and Stedile, F.C. (2003) Enhanced initial growth of atomic-layer-deposited metal oxides on hydrogen-terminated silicon. *Appl. Phys. Lett.*, **83** (4), 740.

99 Puurunen, R.L. (2003) Growth per cycle in atomic layer deposition: a theoretical model. *Chem. Vapor Depos.*, **9** (5), 249.

100 Buriak, J.M. (2002) Organometallic chemistry on silicon and germanium surfaces. *Chem. Rev.*, **102** (5), 1271.

101 Lopinski, G.P., Wayner, D.D.M., and Wolkow, R.A. (2000) Self-directed growth of molecular nanostructures on silicon. *Nature*, **406** (6791), 48.

102 Mayer, T.M., de Boer, M.P., Shinn, N.D., Clews, P.J., and Michalske, T.A. (2000) Chemical vapor deposition of fluoroalkylsilane monolayer films for adhesion control in

microelectromechanical systems. *J. Vac. Sci. Technol. B*, **18** (5), 2433.

103 Mayer, T.M., Elam, J.W., George, S.M., Kotula, P.G., and Goeke, R.S. (2003) Atomic-layer deposition of wear-resistant coatings for microelectromechanical devices. *Appl. Phys. Lett.*, **82** (17), 2883.

104 Ashurst, W.R., Carraro, C., and Maboudian, R. (2003) Vapor phase anti-stiction coatings for MEMS. *IEEE Trans. Device Mater. Reliab.*, **3** (4), 173.

105 Jung, G.-Y., Li, Z., Wu, W., Chen, Y., Olynick, D.L., Wang, S.-Y., Tong, W.M., and Williams, R.S. (2005) Vapor-phase self-assembled monolayer for improved mold release in nanoimprint lithography. *Langmuir*, **21** (4), 1158.

106 Ashurst, W.R., Yau, C., Carraro, C., Lee, C., Kluth, G.J., Howe, R.T., and Maboudian, R. (2001) Alkene based monolayer films as anti-stiction coatings for polysilicon MEMS. *Sens. Actuators A*, **91** (3), 239.

107 Ashurst, W.R., Yau, C., Carraro, C., Maboudian, R., and Dugger, M.T. (2001) Dichlorodimethylsilane as an anti-stiction monolayer for MEMS: a comparison to the octadecyltrichlorosilane self-assembled monolayer. *J. Microelectromech. Syst.*, **10** (1), 41.

108 Kumar, A., Biebuyck, H.A., and Whitesides, G.M. (1994) Patterning self-assembled monolayers: applications in materials science. *Langmuir*, **10** (5), 1498.

109 Lee, B.H. and Sung, M.M. (2007) Selective atomic layer deposition of metal oxide thin films on patterned self-assembled monolayers formed by microcontact printing. *J. Nanosci. Nanotechnol.*, **7** (11), 3758.

110 Shin, H.J., Jeong, D.K., Lee, J.G., Sung, M.M., and Kim, J.Y. (2004) Formation of TiO_2 and ZrO_2 nanotubes using atomic layer deposition with ultraprecise control of the wall thickness. *Adv. Mater.*, **16** (14), 1197.

111 Bae, C.D., Kim, S.Y., Ahn, B.Y., Kim, J.Y., Sung, M.M., and Shin, H.J. (2008) Template-directed gas-phase fabrication of oxide nanotubes. *J. Mater. Chem.*, **18** (12), 1362.

112 Gates, B.D. (2005) Nanofabrication with molds & stamps. *Mater. Today*, **8** (2), 44.

113 Rogers, J.A. and Nuzzo, R.G. (2005) Recent progress in soft lithography. *Mater. Today*, **8** (2), 50.

114 Zhao, X.M. (1997) Soft lithographic methods for nano-fabrication. *J. Mater. Chem.*, **7** (7), 1069.

115 Xia, Y. and Whitesides, G.M. (1998) Soft lithography. *Annu. Rev. Mater. Sci.*, **28** (1), 153.

116 Bietsch, A. and Michel, B. (2000) Conformal contact and pattern stability of stamps used for soft lithography. *J. Appl. Phys.*, **88** (7), 4310.

117 Michel, B., Bernard, A., Bietsch, A., Delamarche, E., Geissler, M., Juncker, D., Kind, H., Renault, J.P., Rothuizen, H., Schmid, H., Schmidt-Winkel, P., Stutz, R., and Wolf, H. (2001) Printing meets lithography: soft approaches to high-resolution patterning. *IBM J. Res. Dev.*, **45** (5), 697.

118 Childs, W.R. and Nuzzo, R.G. (2004) Patterning of thin-film microstructures on non-planar substrate surfaces using decal transfer lithography. *Adv. Mater.*, **16** (15), 1323.

119 Lima, O., Tan, L., Goel, A., Negahban, M., and Li, Z. (2007) Creating micro- and nanostructures on tubular and spherical surfaces. *J. Vac. Sci. Technol. B*, **25**, 2412.

120 Xiang, J., Lu, W., Hu, Y., Wu, Y., Yan, H., and Lieber, C.M. (2006) Ge/Si nanowire heterostructures as high-performance field-effect transistors. *Nature*, **441** (7092), 489.

121 Goldberger, J., Hochbaum, A.I., Fan, R., and Yang, P. (2006) Silicon vertically integrated nanowire field effect transistors. *Nano Lett.*, **6** (5), 973.

122 Thelander, C., Agarwal, P., Brongersma, S., Eymery, J., Feiner, L.F., Forchel, A., Scheffler, M., Riess, W., Ohlsson, B.J., Gŝele, U., and Samuelson, L. (2006) Nanowire-based one-dimensional electronics. *Mater. Today*, **9** (10), 28.

123 Steinhart, M. (2008) Supramolecular organization of polymeric materials in nanoporous hard templates. *Adv. Polym. Sci.*, **220**, 123.

124 Van Gough, D., Juhl, A.T., and Braun, P.V. (2009) Programming structure into 3D nanomaterials. *Mater. Today*, **12** (6), 28.

125 Jeong, D., Lee, J., Shin, H., Lee, J., Kim, J., and Sung, M. (2004) Synthesis of metal-oxide nanotubular structures by using atomic layer deposition on nanotemplates. *J. Korean Phys. Soc.*, **45** (5), 1249.

126 Hulteen, J.C. and Van Duyne, R.P. (1995) Nanosphere lithography: a materials general fabrication process for periodic particle array surfaces. *J. Vac. Sci. Technol. A*, **13** (3), 1553.

127 Denkov, N.D., Ivanov, I.B., Kralchevsky, P.A., and Wasan, D.T. (1992) A possible mechanism of stabilization of emulsions by solid particles. *J. Colloid Interface Sci.*, **150** (2), 589.

128 Stober, W., Berner, A., and Blaschke, R. (1969) Aerodynamic diameter of aggregates of uniform spheres. *J. Colloid Interface Sci.*, **29** (4), 710.

129 Zhuang, Y.X., Hansen, O., Knieling, T., Wang, C., Rombach, P., Lang, W., Benecke, W., Kehlenbeck, M., and Koblitz, J. (2006) Thermal stability of vapor phase deposited self-assembled monolayers for MEMS anti-stiction. *J. Micromech. Microeng.*, **16** (11), 2259.

130 Kulkarni, S.A., Mirji, S.A., Mandale, A.B., and Vijayamohanan, K.P. (2006) Thermal stability of self-assembled octadecyltrichlorosilane monolayers on planar and curved silica surfaces. *Thin Solid Films*, **496** (2), 420.

131 Raiber, K., Terfort, A., Benndorf, C., Krings, N., and Strehblow, H.-H. (2005) Removal of self-assembled monolayers of alkanethiolates on gold by plasma cleaning. *Surf. Sci.*, **595** (1–3), 56.

132 Tatoulian, M., Bouloussa, O., Moriere, F., Arefi-Khonsari, F., Amouroux, J., and Rondelez, F. (2004) Plasma surface modification of organic materials: comparison between polyethylene films and octadecyltrichlorosilane self-assembled monolayers. *Langmuir*, **20** (24), 10481.

10
Coatings on High Aspect Ratio Structures

Jeffrey W. Elam

10.1
Introduction

Atomic layer deposition (ALD) utilizes self-limiting chemical reactions to achieve atomic-level control over film thickness and composition without the need for line-of-site access to the precursor source required for most other deposition techniques [1, 2]. Consequently, ALD allows complex surface features such as deep trenches and narrow pores to be coated with great precision. This capability benefits a remarkably broad range of applications, some of which have been realized and many others that are currently topics of active research.

Since this chapter is devoted to ALD coatings on high aspect ratio structures, it is helpful to define what is meant by aspect ratio. Any topographical surface feature such as a trench or pore can be characterized by a dimensionless aspect ratio. In the case of a rectangular trench, the aspect ratio is given by the trench depth divided by the trench width. Likewise, the aspect ratio of a cylindrical pore that is open on one end is equal to the length of the pore divided by the pore diameter. A thin film that has a uniform thickness on all surfaces of a coated object is called "conformal." ALD excels at producing conformal coatings on high aspect ratio features. This capability provides a straightforward means for imparting the desired chemical properties (e.g., surface acidity, corrosion resistance, and catalytic activity) or physical properties (e.g., hardness, optical reflectivity, coefficient of friction, and electrical resistance), to nearly any surface. In this way, the three-dimensional shape of an object can be decoupled from the surface properties of the object. Consequently, one has the freedom to select a convenient template with the desired shape, porosity, or structure without worrying about the physical or chemical properties of the template because these traits can be tuned later by applying ALD layers. This versatility is made even more powerful when applied to nanoscale materials, and currently ALD is being utilized to synthesize nanostructured materials for a diverse range of technologies [3].

Some of the most vexing problems facing society today relate to energy. The future of our economy and climate depend upon developing new means for producing,

Atomic Layer Deposition of Nanostructured Materials, First Edition. Edited by Nicola Pinna and Mato Knez.

storing, and conserving energy in renewable and economically viable ways. Solutions to these problems are likely to result from revolutionary rather than incremental changes in technology, and nanomaterials with tunable structure, porosity, and composition hold tremendous promise in this respect [4]. For this reason, the topic of ALD coatings on high aspect ratio structures is more than purely academic. An appreciation and understanding of this subject, particularly by practitioners in energy disciplines that may be new to ALD, could lead to the breakthroughs in materials synthesis necessary to solve our energy problems.

This chapter will survey the field of ALD as it applies to the coating of high aspect ratio structures. In the first section, we will introduce several mathematical models useful for understanding ALD in porous structures and for predicting the necessary exposure conditions required to achieve conformal coverage. Next, we will review some of the analytical characterization techniques used to evaluate the degree of conformality or infiltration of ALD coatings in high aspect ratio structures. After this, we will give examples to illustrate ALD at high aspect ratios and the technological applications that they address. These examples will be presented in the order of increasing aspect ratio: 10, 100, 1000, and beyond. Next, we will describe some nonideal situations in which limitations in the coating process or violations of the simplifying assumptions made in the models lead to nonconformal films. Finally, we will summarize and briefly touch on the future of ALD in high aspect ratio structures.

10.2
Models and Analysis

A fundamental question when coating high aspect ratio structures by ALD is: "How long must the exposure times be?" The answer to this question depends on the particular conditions used that will dictate the coating regime. In the case of a very high surface area substrate coated in a relatively small ALD reactor, nearly all of the precursor will be consumed immediately upon entering the reactor and the coating process will be limited by the rate of precursor delivery. In this transport-limited regime, the exposure time is independent of the particular nanostructure of the support and can be calculated by dividing the number of sites to be coated by the rate of introduction of precursor molecules into the ALD reactor. In the reaction-limited regime, the ALD surface reactions are exceedingly slow and the precursor supply is constant and abundant. In this regime, the nanostructure of the support is inconsequential and the exposure time can be calculated from the Langmuir adsorption model. However, most cases of interest occur in the diffusion-limited regime where both the ALD reaction rate and the precursor supply are sufficiently high. In the diffusion-limited regime, the exposure time required to achieve saturation increases as the square of the aspect ratio of the object being coated. The following models apply to the diffusion-limited regime, but a criterion will be given for differentiating between the reaction- and diffusion-limited cases.

10.2.1
Analytical Method

Gordon analytically derived an expression to calculate the saturation exposure time based on gas conductance equations [5]. This derivation assumes Knudsen diffusion of the ALD precursors, which means that gas-phase collisions between molecules can be ignored. Knudsen diffusion requires that the mean free path of the precursor molecule, λ, is much larger than the characteristic feature size such as the pore diameter, d. For instance, consider trimethylaluminum (TMA) in a 50 nm pore under a nitrogen pressure of 1 Torr at 177 °C. In this case, $\lambda \sim 50\,\mu m$ and $\lambda >> d$, so Knudsen diffusion applies. For Knudsen diffusion conditions, Gordon derives the following quantity:

$$t = S(2\pi mkT)^{1/2}/P[1 + (19/4)a + 3/2a^2], \quad (10.1)$$

where P is the pressure (Pa), t is the time (s), S is the saturation coverage of the molecule in question (molecules/m^2), m is the mass of the precursor (kg), k is the Boltzmann constant (J/K), T is the temperature (K), and a is the aspect ratio of the pore. The quantity $S(2\pi mkT)^{1/2}/P$ is the exposure time required to reach saturation on a planar substrate assuming a reactive sticking coefficient of 1. For very large aspect ratios, the a^2 term in Eq. (10.1) dominates, so the exposure time increases as the square of the aspect ratio. This model provides a simple analytical expression for the exposure time but makes a number of simplifying assumptions that are sometimes violated in practical situations. For instance, this analysis assumes a reactive sticking coefficient of 1 and neglects the effect of coverage on sticking coefficient (i.e., Langmuir adsorption). Furthermore, this model assumes that the aspect ratio of the pore does not change during the coating process.

Improvements on this model have been made in which the expressions for coverage as a function of exposure time and depth into a surface feature were allowed to depend on the sticking coefficient and the surface coverage [6, 7]. The primary difference between these methods and the Gordon method is that the coverage profiles obtained exhibited a more gradual transition between saturation coverage and zero coverage as the sticking coefficient decreased. In other words, for slower ALD surface reactions (lower initial sticking probability), the thickness profiles become broadened.

10.2.2
Monte Carlo Simulations

Elam used Monte Carlo simulations to analyze ALD in nanopores and derived an expression for the saturating exposure time [8]. This method also assumed Knudsen diffusion, and furthermore that the ALD metal precursor (e.g., TMA) was the limiting reagent. However, this simulation explicitly included the sticking coefficient and also accounted for pore shrinkage during coating so that the effective aspect ratio changed during the deposition. To simulate Knudsen diffusion, a particle was allowed to “hop” a distance d, equal to the local pore diameter, randomly either into or out of the pore. The Monte Carlo simulations yielded the following expression for the minimum exposure time:

$$t = 9.2 \times 10^{-7}\left(\sqrt{m}Na^2\right)/P, \qquad (10.2)$$

where N is the density of ALD reactive sites (sites/cm^2), m is the molecular mass of the metal precursor (amu), and P is the partial pressure of the metal precursor (Torr). The original study used a different definition for the aspect ratio, so the constant term in Eq. (10.2) has been adjusted for consistency with Eq. (10.1). In the limit of a thin ALD coating and a high sticking coefficient, Eq. (10.2) can be shown to be equivalent to Eq. (10.1).

To differentiate between the reaction- and diffusion-limited regimes, Elam defined the hopping coefficient, $H = 16/a^2$, as the fraction of hops that reach an empty site in the pore. The hopping confident can then be compared with the reactive sticking coefficient, S. In the diffusion-limited regime, $S \gg H$ and the surface sites along each pore fill up sequentially in order from the pore entrance. In the reaction-limited regime, $H \gg S$ and the precursor molecules essentially equilibrate along the pore axis before reaction occurs. Consequently, the surface sites fill up uniformly along each pore. Both Eqs. (10.1) and (10.2) apply to the diffusion-limited regime, which is the most typical situation for ALD in high aspect ratio structures.

10.3
Characterization Methods for ALD Coatings in High Aspect Ratio Structures

The analytical methods described above are very useful for predicting the necessary partial pressures and exposure times required to achieve self-limiting, saturating growth in high aspect ratio structures. But to validate these models, it is necessary to have reliable methods for quantifying the degree of infiltration of the ALD materials. A wide variety of characterization methods have been developed to accomplish this objective. Although some of these methods apply equally well to thin films on both porous and planar supports, other techniques are useful only for porous materials because they rely on the tremendous surface area afforded by high aspect ratio nanoporous materials. For instance, the infiltration of a nanoporous template by an optically absorbing material such as a metal can often be confirmed by eye after a single ALD cycle. Similarly, visual inspection under an optical microscope of cross-sectional specimens of a high surface area support such as silica gel or nanoporous extrudate can reveal the extent of infiltration of even submonolayer ALD coatings. Simple weight gain measurements are very effective for quantifying the saturation ALD conditions or for determining the ALD growth rate in nanoporous media [9, 10].

Electron microscopy is another very useful technique for characterizing ALD coatings in high aspect ratio structures. Transmission electron microscopy (TEM) is used very often to evaluate the conformality of ALD coatings in high aspect ratio trench structures [11]. Typically, this method requires the careful preparation of cross-sectional specimens using ion milling or a focused ion beam (FIB). However, a simpler method has been developed in which nanoporous anodic aluminum oxide (AAO) is infiltrated by ALD and the alumina is dissolved to facilitate TEM analysis of the resulting nanotubes [12]. Cross-sectional scanning electron microscopy (SEM) can also be quite revealing. However, this technique requires destroying the sample to expose a cross-sectional view. While the resolution of SEM (~1–10 nm) is typically lower than that of TEM (~0.1–1 nm), it has the advantage of allowing a wider range of view to facilitate the analysis of larger specimens such as entire 100 μm silica gel beads coated by ALD [9].

Because aspect ratio is a dimensionless quantity, the analysis of ALD films in macroscopic pores or features can be used for understanding ALD in microscopic pores. For instance, one useful and convenient technique for evaluating ALD in trenches is to fabricate an "artificial trench" by clamping two pieces of silicon together separated by a thin metal foil to define a precise gap. The thickness of an ALD coating can then be evaluated as a function of distance into the "trench" using macroscopic measurement techniques such as ellipsometry allowing a direct comparison with models [7]. Another strategy used glass capillaries with an inside diameter of 20 μm and a length of 2 mm to evaluate ALD conformality at high aspect ratios. Following WN ALD, the capillaries were examined under an optical microscope to determine the extent of infiltration [13].

A variety of methods exist that provide element-specific information about the spatial distribution of ALD films in high aspect ratio materials. For instance, TEM and SEM can be equipped with energy- or wavelength-dispersive X-ray detectors (e.g., EDAX) to image the spatial distribution of different elements in ALD-coated nanomaterials [10]. Rutherford backscattering spectroscopy (RBS) is commonly used to evaluate the elemental composition of thin films on planar substrates as a function of depth into the film. However, this technique has also been used with great success to evaluate the degree of infusion of ALD thin films into porous templates such as porous silicon [14] and TiO_2 nanoparticle films [15]. Secondary ion mass spectrometry (SIMS) is another element-specific analytical technique that has been utilized to analyze cross-sectional specimens of high aspect ratio materials coated by ALD including aerogels [14, 16].

The methods described above are all performed *ex situ* so that the information is obtained only after the film is deposited. However, a variety of *in situ* analytical methods have been utilized to analyze ALD films in porous materials during the coating process. For example, Fourier transform infrared (FTIR) spectroscopy has been used to examine the infiltration of ALD Al_2O_3 and SiO_2 into high aspect ratio AAO membranes [8]. FTIR has also been used extensively to evaluate ALD processes in samples comprised of pressed nanopowders [17]. The quartz crystal microbalance (QCM) is another *in situ* characterization tool that has been utilized to explore ALD in high aspect ratio substrates. In these experiments, a film of colloidal semiconducting nanocrystals (quantum dots) was prepared on the surface of the QCM sensor. The mass gain per ZnO ALD cycle observed by QCM for the quantum dot film was 13 × larger than the corresponding mass gain on an uncoated sensor, and provided a quantitative measure of the internal surface area of the film [18]. Finally, quadrupole mass spectrometry (QMS) has been used to elucidate the infiltration of high aspect ratio nanoporous silica gel [9]. In these experiments, the QMS could resolve subtle changes in the surface chemistry as a function of the ALD Al_2O_3 film thickness in addition to establishing the necessary exposure conditions to achieve saturation.

10.4 Examples of ALD in High Aspect Ratio Structures

The following examples of ALD coatings in high aspect ratio supports are intended to illustrate the unique capabilities of ALD for coating porous materials. We will present these examples in the order of increasing aspect ratio (10, 100, 1000, and beyond).

10.4.1 Aspect Ratio of 10

10.4.1.1 Trenches

The classic example of a high aspect ratio structure is a micromachined trench in single-crystal silicon. Silicon trenches are ubiquitous in the microelectronics industry. For instance, ALD coatings of high dielectric constant materials are used in dynamic random access memory [19]. Trenches are also commonly used during ALD process development as a convenient substrate for assaying the conformality of the coating. In such cases, the wafer is typically cleaved to reveal a cross section for examination by SEM. For example, Figure 10.1 shows an SEM image of an ALD copper sulfide (Cu_2S) film prepared in a trench ~1 μm wide with an aspect ratio of 9 [20]. As shown by the magnified insets, the 45 nm thick Cu_2S film has the same

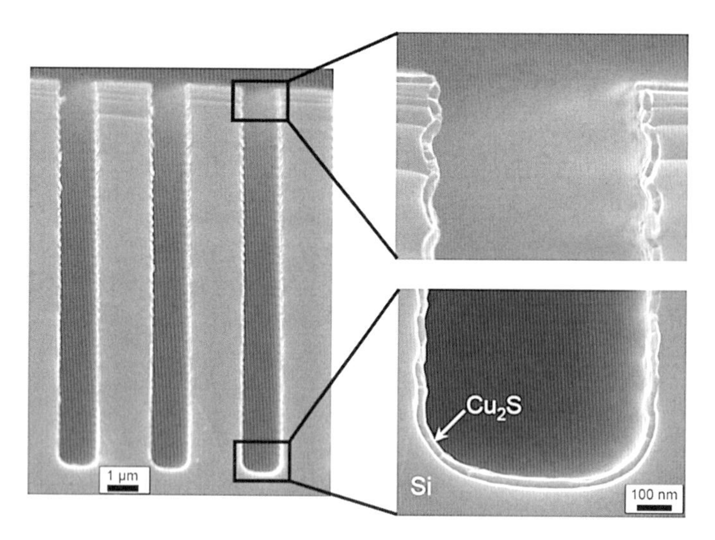

Figure 10.1 Cross-sectional SEM images of 45 nm ALD Cu_2S film on silicon trench structure. Reprinted with permission from Ref. [20]. Copyright 2009, the American Institute of Physics.

thickness on the top and bottom of the trench and therefore exhibits high conformality.

10.4.1.2 MEMS

Another example of ALD films in structures with aspect ratio of order 10 is microelectromechanical systems (MEMS). MEMS devices consist of micromachined structures, typically comprised of silicon, with dimensions on the micron scale designed to act as resonators, mirrors, actuators, motors, and so on. A common problem encountered in MEMS devices is stiction, a phenomenon whereby adjacent moving parts become permanently bonded due to hydrophobic, electrostatic, or other forces. Another problem is friction and wear between moving parts. These problems can be alleviated by coating all surfaces with antistiction [21], lubricating [22], or wear-resistant coatings [23], and ALD is an attractive means for accomplishing this. Figure 10.2 shows SEM images of a MEMS cantilever that has been coated using ALD Al_2O_3 and subsequently cut using a focused ion beam to reveal a cross section [24]. The gap between the cantilever beam and the underlying contact pad is 1 μm, and the half-width of the beam is 25 μm yielding an aspect ratio of 25. The lower SEM image

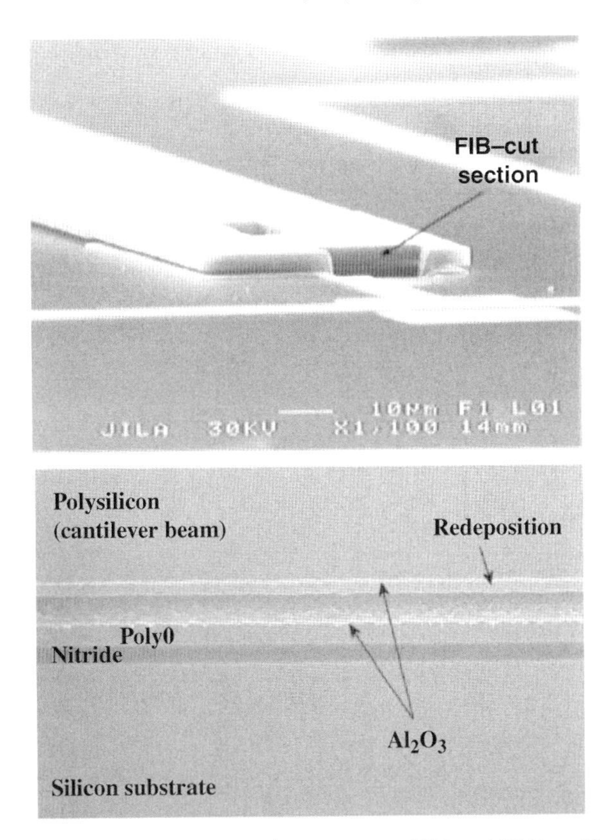

Figure 10.2 Cross-sectional SEM images of FIB-cut MEMS cantilever coated with 60 nm ALD Al_2O_3. Reprinted with permission from Ref. [24]. Copyright 2003, Elsevier.

demonstrates that the Al_2O_3 coating extends uniformly over all exposed surfaces inside of this narrow gap.

10.4.2 Aspect Ratio of 100

10.4.2.1 Anodic Aluminum Oxide

AAO is a common platform for synthesizing nanostructured materials by ALD. AAO is made by the simultaneous electrochemical oxidation and etching of aluminum metal in an acid solution and results in the formation of nanometer-sized, circular pores arranged in a hexagonal close-packed pattern that extend into the aluminum oxide perpendicular to the surface (Figure 10.3). The AAO can be left bonded to the aluminum substrate, or it can be removed and further processed to form free-standing membranes in which the pores are accessible from both sides. A typical pore diameter of 50 nm and a membrane thickness of 25 μm yield an aspect ratio of 250. AAO membranes have been coated by ALD with a wide variety of materials including metal oxides [8, 25], metal nitrides [26, 27], and metals [28–30]. In some cases, the AAO template can be subsequently dissolved yielding nanotubes of the ALD coating [31]. Track etched polycarbonate membranes have a similar nanostructure and have also been used as a template for fabricating nanotubes by ALD [32].

AAO membranes provide a convenient scaffold for synthesizing functional nanostructured materials by ALD for use in applications such as photovoltaics [33], catalysis [34], and magnetic devices [28]. In addition, the regular geometry and high aspect ratio make AAO an excellent substrate for evaluating ALD conformality and

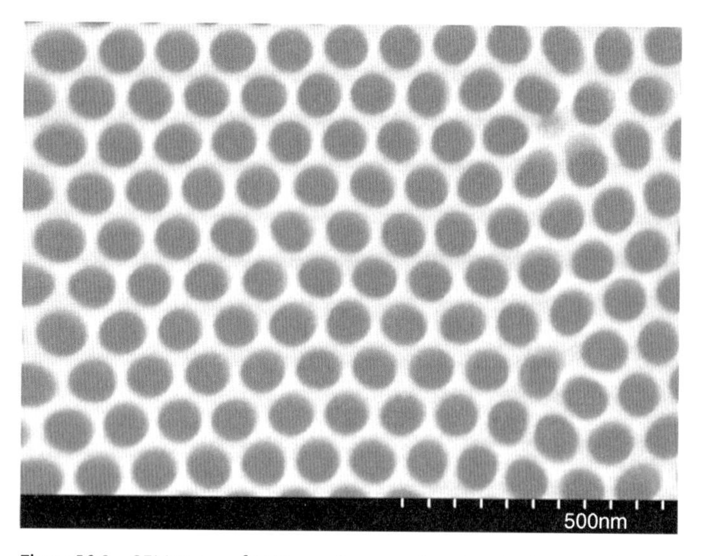

Figure 10.3 SEM image of AAO membrane surface showing hexagonal arrangement of $d = 75$ nm nanopores. Reprinted with permission from J.W. Elam, unpublished.

for testing infiltration models. In one such investigation, freestanding AAO samples were coated with ALD ZnO and the exposure times for the ALD ZnO precursors were varied [8]. Cross-sectional specimens of the coated AAO were examined using EDAX to determine the spatial location of the Zn. Figure 10.4a demonstrates that as the exposure times increased, the ZnO migrated progressively further into the AAO membrane. This behavior is quantified in Figure 10.4b in which the two-dimensional Zn maps of Figure 10.4a were integrated to show line scans. These results were in excellent quantitative agreement with Monte Carlo simulations of the ALD infiltration process. Figure 10.4c shows the integration of the areas under the curves in Figure 10.4b and demonstrates that the ALD ZnO growth in the AAO nanopores follows the expected $t^{1/2}$ time dependence. Furthermore, the exposure times in Figure 10.4c are in good quantitative agreement with Eqs. (10.1) and (10.2).

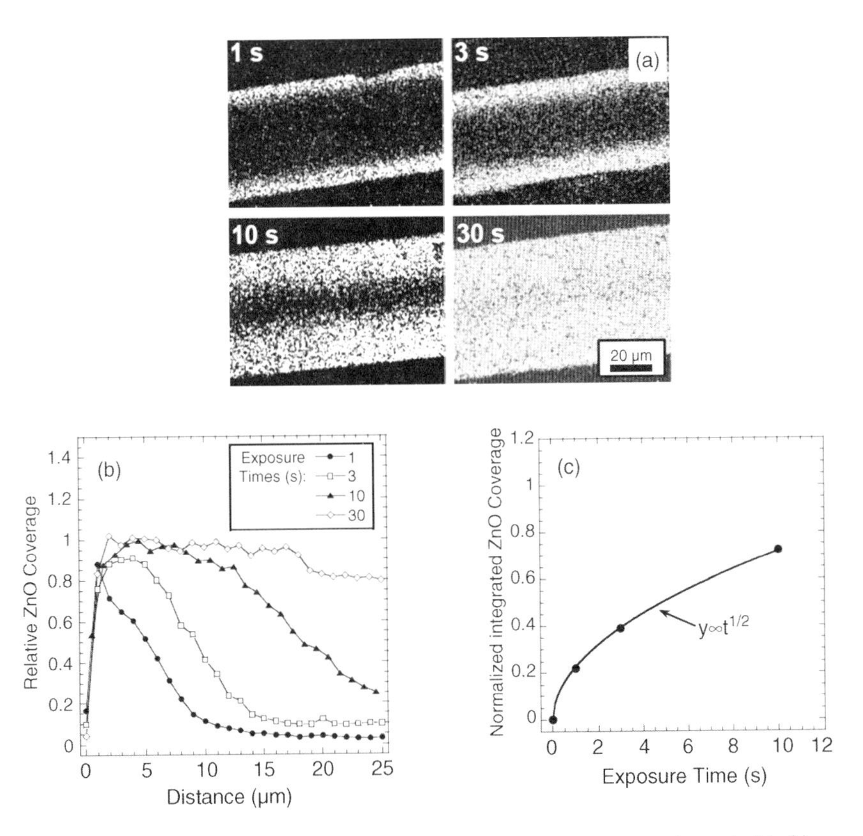

Figure 10.4 (a) Cross-sectional EDAX images of Zn concentration in AAO membranes coated with ZnO ALD films using precursor exposures times of 1, 3, 10, and 30 s. (b) Line scan profiles of Zn coverage versus distance into AAO membrane generated by integrating the EDAX images in (a). (c) Normalized ZnO coverage versus precursor exposure time calculated by integrating the line scan profiles in (b) demonstrating the $t^{1/2}$ functional dependence of the ZnO infiltration. Reprinted with permission from Ref. [8]. Copyright 2003, the American Chemical Society.

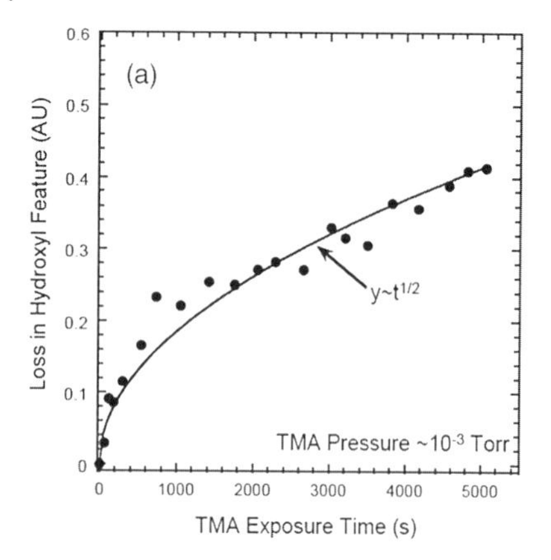

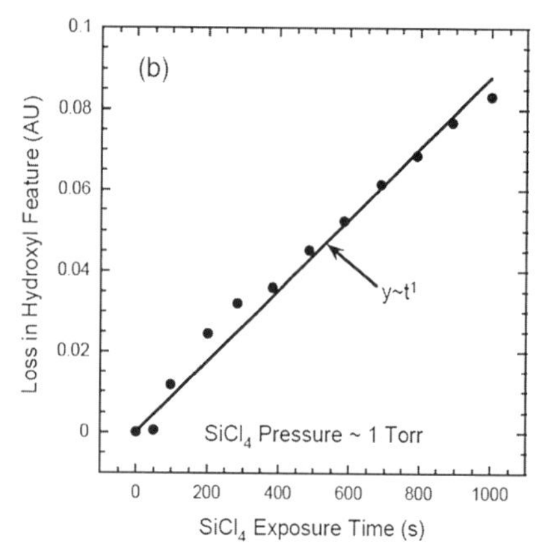

Figure 10.5 (a) Loss in surface hydroxyl concentration versus TMA exposure time during Al_2O_3 ALD measured using *in situ* FTIR. The $t^{1/2}$ functional dependence is characteristic of a diffusion-limited process. (b) *In situ* FTIR measurements of surface hydroxyl loss during SiO_2 ALD versus $SiCl_4$ exposure, where the $t^{1/2}$ functional form signals a reaction-limited process. Reprinted with permission from Ref. [8]. Copyright 2003, the American Chemical Society.

The study described above also explored the different infiltration behaviors for Al_2O_3 and SiO_2 ALD in AAO corresponding to the diffusion- and reaction-limited regimes explained in Section 10.2.2 [8]. During the TMA exposures for Al_2O_3 ALD, hydroxyl groups are consumed as the TMA chemisorbs to the surface. By monitoring the decrease in the OH intensity using FTIR, the extent of the TMA reaction was determined. Figure 10.5a plots the hydroxyl loss as a function of TMA exposure time in a high aspect ratio AAO membrane. In this case, the sticking coefficient, S, for the TMA chemisorption is $\sim 10^{-2}$, while the hopping coefficient $H = 16/a^2 \sim 10^{-5}$. Since $S >> H$, the process is diffusion limited so that the reactive sites in each AAO pore fill up in order and the $t^{1/2}$ time dependence is observed. In contrast, Figure 10.5b shows a similar measurement of the $SiCl_4$ adsorption during SiO_2 ALD and exhibits a linear dependence of coverage versus exposure time. In this case, $S \sim 10^{-9}$, so $S << H$ meaning the process is reaction limited so that the reactive sites in the pores fill up uniformly versus time.

The fact that the reactive sites in the AAO pores fill up in order in the diffusion-limited case facilitates a technique known as "stripe coating" in which ALD can be used to deposit materials selectively at specific depth locations in the AAO pores [35]. In this process, a masking exposure is performed prior to each ALD cycle in order to render the sites near the ends of the nanopores nonreactive to the ALD metal precursor such as diethylzinc for ZnO ALD. For instance, if the AAO is first exposed to a subsaturating TMA exposure, then each pore will be methyl-terminated to a pore depth determined by the duration of the TMA exposure. During the subsequent DEZ exposure, these sites will be nonreactive so that the DEZ will only chemisorb deeper down the pores. The duration of the DEZ exposure then determines the width of the resulting ZnO "stripe." After the DEZ exposure, the sample is exposed to H_2O to repopulate the surface with hydroxyl groups so that the TMA–DEZ–H_2O exposures

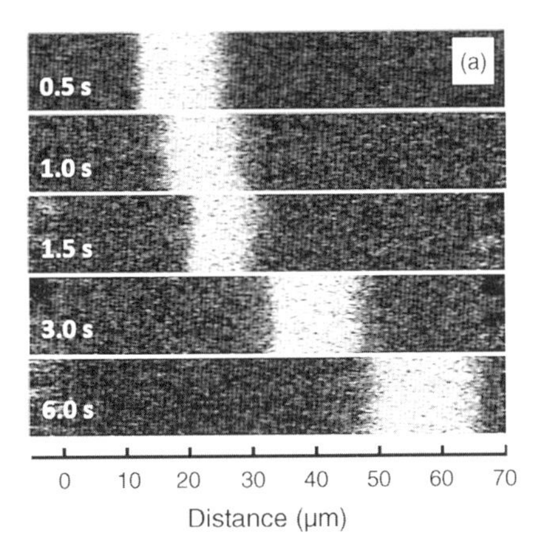

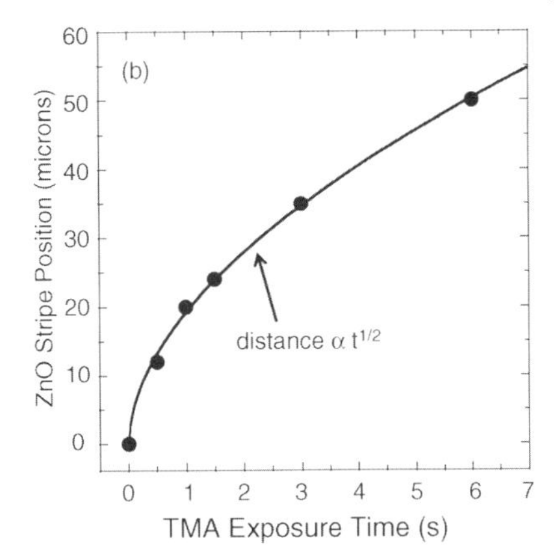

Figure 10.6 (a) Cross-sectional EDAX maps of Zn concentration following ZnO stripe coating in AAO using a TMA/DEZ/H_2O dosing sequence and varying only the TMA exposure time. (b) Position of trailing edge of ZnO stripes in (a) versus TMA exposure time demonstrating the $t^{1/2}$ dependence expected for this diffusion-limited infiltration process. Reprinted with permission from Ref. [35]. Copyright 2007, the American Institute of Physics.

can be repeated. Figure 10.6a shows two-dimensional EDAX maps of the spatial distribution of Zn for polished cross sections of AAO membranes prepared using TMA masking exposures of increasing exposure time. As seen in this figure, the stripes move progressively farther into the AAO pores for longer TMA exposure times. The position of the trailing edge of the ZnO stripe is plotted in Figure 10.6b versus the TMA exposure time, and exhibits the $t^{1/2}$ time dependence characteristic of this diffusion-limited coating process.

10.4.2.2 **Inverse Opals**

Inverse opals represent another class of high aspect ratio materials that have been fabricated successfully by ALD infiltration (see Chapter 15). In one example, a suspension of monodisperse silica nanospheres with a diameter of 400 nm was cast on a planar substrate to generate a film with a thickness of 10 μm in which the nanospheres formed a face-centered cubic lattice [36]. The resulting gap between the spheres was ~60 μm yielding an initial aspect ratio of 175. The opal was then infiltrated by TiO_2 ALD, and the silica spheres were subsequently removed by etching in hydrofluoric acid leaving behind the inverse opal structure in Figure 10.7a. Reflectivity spectra recorded for a series of such materials with increasing TiO_2 thickness exhibited a smooth and continuous variation in the location of a peak in the reflectance spectrum known as the pseudophotonic band gap. It is interesting to note that as the ALD TiO_2 thickness increased, the gaps narrowed and gradually closed down so that the aspect ratio increased greatly. Nevertheless, the highly conformal TiO_2 ALD process allowed for a nearly complete filling of the voids leaving behind only small air pockets (Figure 10.7b). Inverse opals have been fabricated in this way for a variety of ALD materials including ZnO [37], WN [38], GaAs [39], and ZnS [40], as well as multilayers of different materials [41].

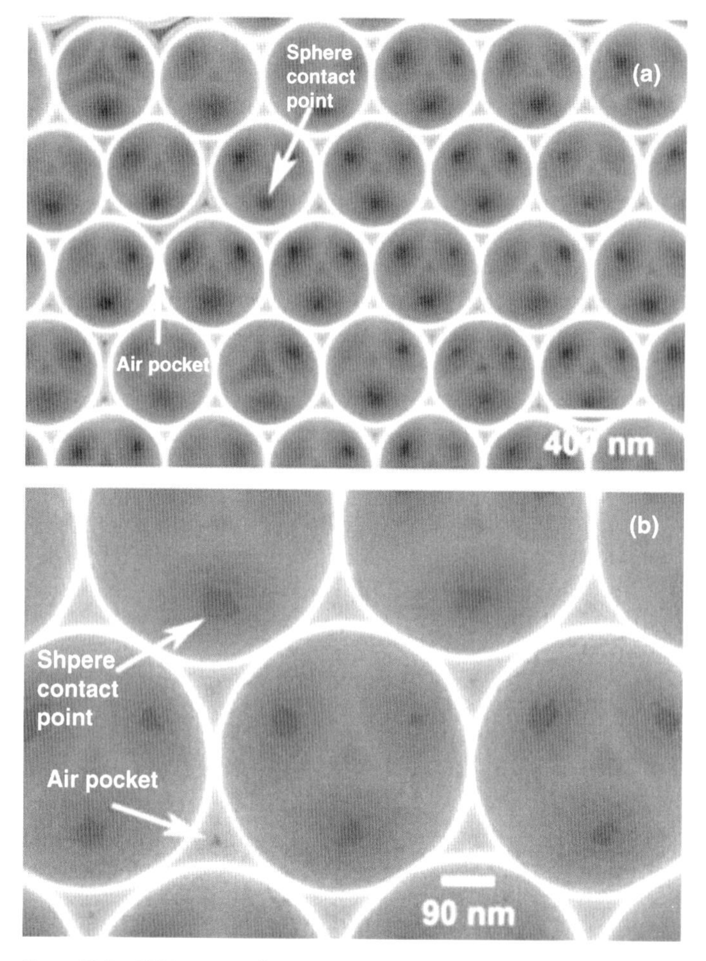

Figure 10.7 SEM images of cross sections of 433 nm ALD TiO_2 inverse opals prepared by FIB to expose the (111) surface. Reprinted with permission from Ref. [36]. Copyright 2005, Wiley.

10.4.3
Aspect Ratio of 1000 and Beyond

10.4.3.1 Silica gel

Silica gel is a granular, mesoporous material that is widely utilized as a support material for heterogeneous catalysts because of its very high specific surface area (50–500 m^2/g). A typical silica gel particle has a radius of 50 μm and a pore size of 30 nm yielding an aspect ratio of ~1600. In a previous study exploring Al_2O_3 ALD on this challenging substrate [9], a series of 1 g specimens were treated using three ALD cycles and the weight changes were recorded as the TMA exposure time was varied (Figure 10.8). Saturation was observed for TMA exposures of ~60 s demonstrating self-limiting behavior even on this high aspect ratio, high surface area support. *In situ*

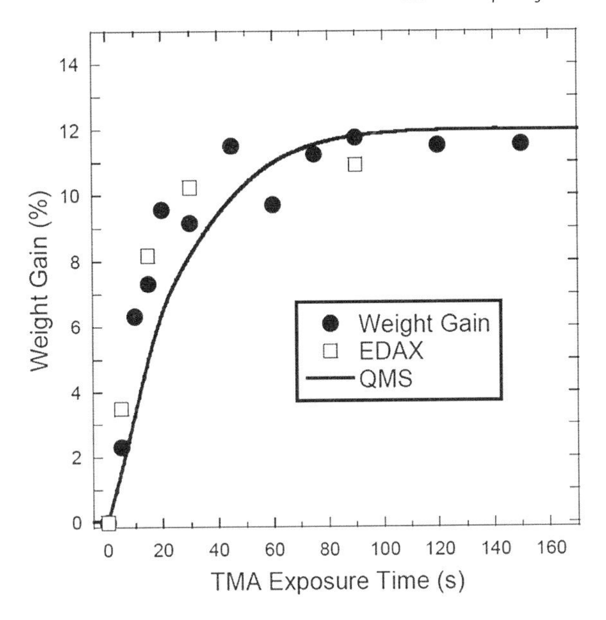

Figure 10.8 Measurements of ALD Al_2O_3 infiltration of silica gel versus TMA exposure time obtained from weight gain measurements (solid circles), EDAX measurements (open squares), and integrated QMS measurements (solid line). Reprinted with permission from Ref. [9]. Copyright 2010, Elsevier.

quadrupole mass spectrometry (QMS) was used to monitor the amount of CH_4 produced during a single TMA exposure versus time. Integrating this signal yielded the accumulated amount of CH_4 versus TMA exposure time, and this signal agreed well with the weight gain measurements (solid line in Fig. 10.2) demonstrating that QMS can accurately probe ALD processes in porous media. The weight gain and Al content as determined by EDAX measurements increased linearly with ALD cycles for Al_2O_3 film thicknesses of up to 2.4 nm. These findings were consistent with layer-by-layer growth where the silica gel pores remained open and the surface area stayed relatively constant as expected for conformal growth in the 30 nm pores.

To gain more insight into the Al_2O_3 ALD process on the silica gel, cross-sectional specimens of individual silica gel beads were prepared and examined using EDAX measurements of the Al concentration for various TMA exposure times (Figure 10.9).

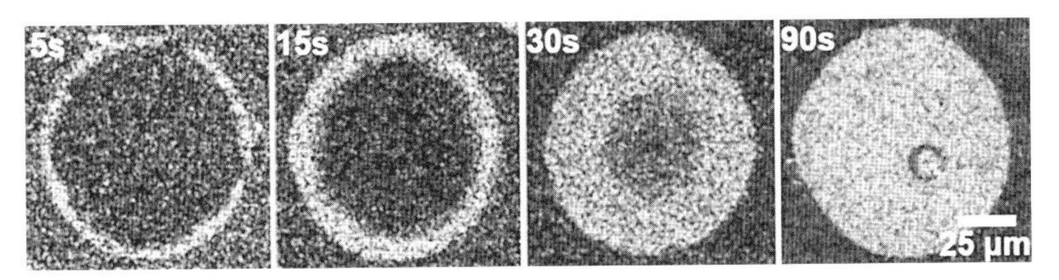

Figure 10.9 EDAX elemental maps for Al obtained from polished cross sections of silica gel following Al_2O_3 ALD using TMA exposure times of 5, 15, 30, and 90 s. Reprinted with permission from Ref. [9]. Copyright 2010, Elsevier.

The Al_2O_3 ALD infiltrated progressively from the outside toward the middle with increasing TMA exposure times. Recall that a similar behavior was observed for the diffusion-limited ZnO ALD in AAO membranes (Figure 10.4a). In this case, the diffusion-limited regime was confirmed by plotting the integrated Zn coverage against the DEZ exposure time and observing a $t^{1/2}$ dependence (Figure 10.4c). Al coverages were obtained for the silica gel by radially integrating the two-dimensional Al EDAX profiles and using an r^3 weighting factor, where r is the radial distance from the center of the silica gel particle. These data are plotted as the open squares in Figure 10.8. The EDAX measurements agree with the direct weight gain measurements, and both data sets exhibit a linear dependence of the Al coverage with TMA exposure time for coverages below ~10 wt%, where the Al_2O_3 ALD is below saturation. This linear dependence contradicts the anticipated $t^{1/2}$ dependence expected for a diffusion-limited process. However, in these experiments the very high surface area of the silica gel forced the process to be in a transport-limited regime and consequently the Al coverage varied linearly with exposure time [9].

10.4.3.2 **Aerogels**

Aerogels are another form of high surface area, high aspect ratio solid and can be manufactured from a variety of materials including silica, carbon, and titania [42, 43]. For instance, silica aerogels are fabricated by the supercritical drying of a silica sol–gel and consist of a randomly interconnected network of fine filaments and particles. The silica sol–gel can be spin or dip coated onto a planar substrate to produce a thin film aerogel or cast to form an aerogel monolith. The ability to coat ultralow-density silica aerogels with conformal layers of different materials using ALD can benefit technologies such as gas sensing, catalysis, and photovoltaics [16, 44–49].

Aerogels can have extremely high aspect ratios. For instance, one study examined ZnO ALD on monolithic silica aerogel specimens with a thickness $t = 1$ mm and a pore size $d = 30$ nm yielding an aspect ratio $t/(2d) \sim 17\,000$ [47]. The silica aerogels had an open, filamentous structure and SEM images revealed that the conformal ZnO coatings increased the filament diameter with increasing number of ALD ZnO cycles (Figure 10.10). Curiously, both the ZnO weight gain and the Zn content inferred from EDAX measurements increased quadratically versus the number of ZnO ALD cycles (Figure 10.11). This finding is surprising given the highly linear ALD ZnO growth observed on flat substrates. This discrepancy is explained by the linear increase in filament diameter with ZnO cycles (Figure 10.10) since the amount of ZnO deposited per cycle varies with the square of the filament diameter. This behavior is in contrast to the linear increase in mass with increasing ALD cycles observed on silica gel [9]. However, in the silica gel example, the 2.4 nm ALD Al_2O_3 coating was much smaller than the 30 nm silica gel primary particles so that the surface area remained nearly constant. But in the silica aerogel study, the 12 nm ZnO coating was much larger than the ~2 nm silica aerogel filaments causing the surface area to change dramatically.

A different study explored W ALD in carbon aerogels for use as spallation targets in rare isotope facilities [50]. In these experiments, carbon aerogel monoliths were cut into 1 mm thick slabs comprised of an interconnected network of 10 nm filaments

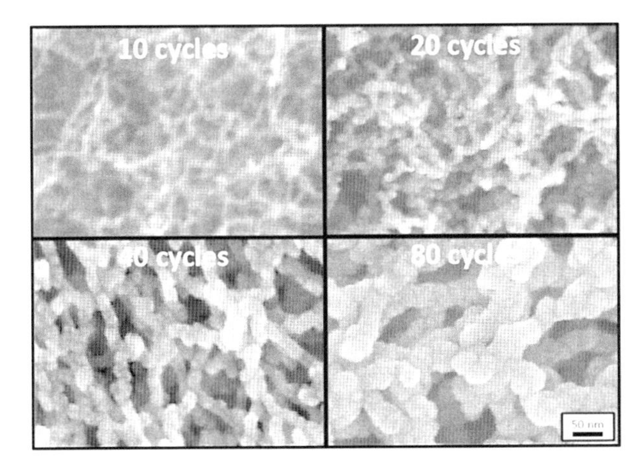

Figure 10.10 SEM images of silica aerogel versus ZnO ALD cycles illustrating progressive thickening of aerogel filaments. Reprinted with permission from Ref. [47]. Copyright 2006, J.W. Elam *et al.*

separated by 10 nm gaps defining an aspect ratio of $\sim 10^5$. The aerogel mass was found to increase quadratically after 3 cycles and then saturated beyond $\sim$10 cycles (Figure 10.12). In contrast, companion measurements performed on planar witness samples showed a linear increase in W thickness following a three-cycle nucleation period. This quadratic behavior is similar to ZnO ALD on silica aerogel [47], which

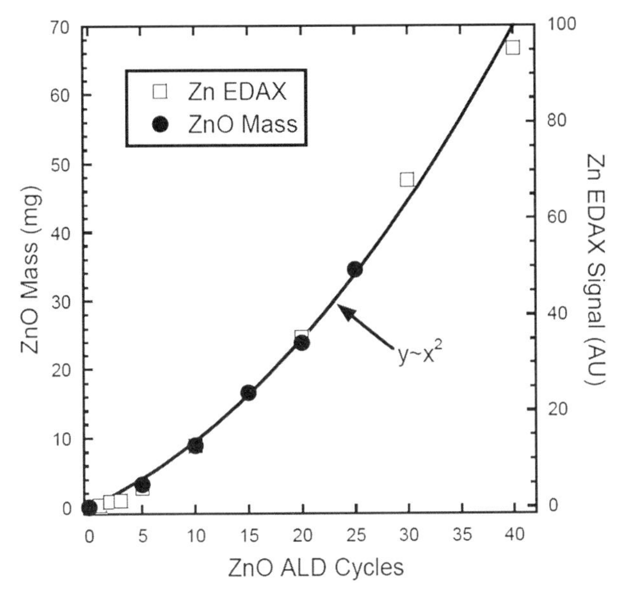

Figure 10.11 Mass gain and Zn EDAX signal versus ZnO ALD cycles on silica aerogel. Quadratic behavior results from linear increase in aerogel filament diameter with number of ZnO cycles. Reprinted with permission from Ref. [47]. Copyright 2006, J.W. Elam *et al.*

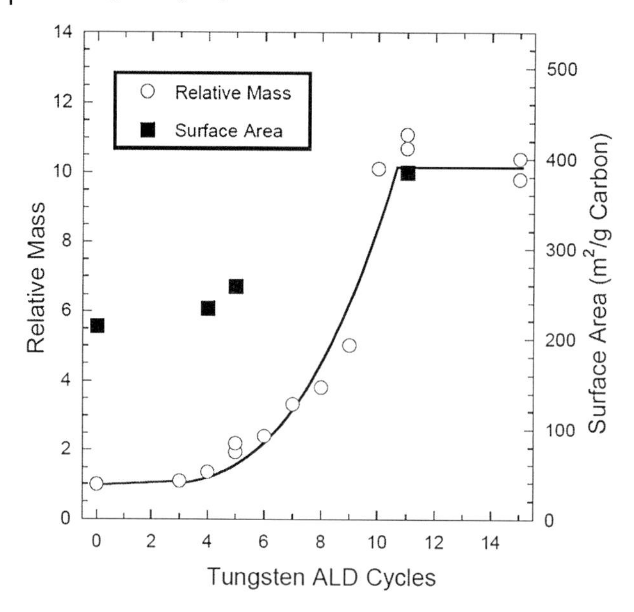

Figure 10.12 Relative mass and surface area of W-coated carbon aerogel versus number of W ALD cycles. Reprinted with permission from Ref. [50]. Copyright 2006, the American Institute of Physics.

resulted from an increase in aerogel filament diameter with ALD film thickness. In support of this mechanism, both the carbon aerogel surface area measured using nitrogen adsorption (Figure 10.12) and the filament diameter determined from TEM images (Figure 10.13) were found to increase with ALD W thickness. The saturation in aerogel mass beyond 10 ALD W cycles resulted from filling the 10 nm gaps with the ~5 nm thick ALD W films.

10.5 Nonideal Behavior during ALD in High Aspect Ratios

As described above, ALD excels at coating high aspect ratio materials. Nevertheless, there are instances where this method fails to produce conformal coatings as desired or where the thickness profile deviates significantly from that predicted by the models and it is worthwhile to review some of these examples of nonideal behavior. Roughly speaking, these examples can be divided into two categories: physical and chemical. One example of a physical nonideality leading to bad conformality is pore closure. If the ALD film thickness exceeds half the width or diameter of the pore, then the surface feature will become closed or clogged and this will prevent transport of the precursor vapors deeper into the pores [50]. In the case of a perfectly conformal coating in a uniform feature, closure will occur essentially simultaneously throughout the depth of the feature forming a solid plug and this tendency can be exploited to engineer nanomaterials [51]. Deliberate subsaturating ALD exposures can be used to seal the outer surfaces of a porous, low-*k* dielectric material prior to depositing a

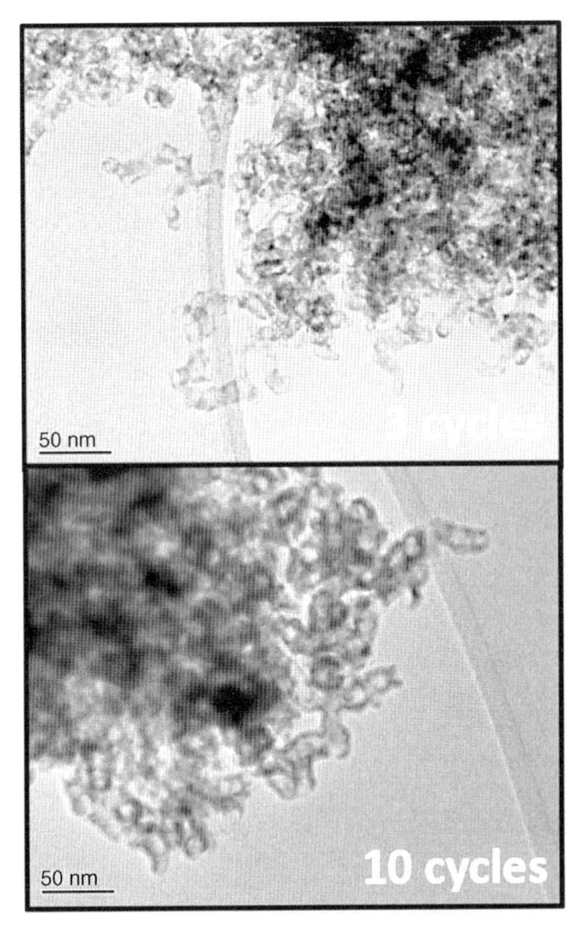

Figure 10.13 TEM images of carbon aerogel following three and seven cycles of W ALD. Reprinted with permission from Ref. [50]. Copyright 2006, the American Institute of Physics.

copper diffusion barrier such as ALD WN. This will prevent the ALD WN from infiltrating the low-*k* material and causing electrical shorts or increasing the dielectric constant [52]. Plasma treatment can also be used to densify and activate the exterior surfaces of the low-*k* material prior to ALD to facilitate intentional pore closure [53].

Another example of a physical nonideality during ALD in high aspect ratios is the precursor depletion (Section 10.2). Existing models for ALD in high aspect ratio structures assume that the precursor partial pressure remains constant so that saturation times can be predicted from Knudsen diffusion [5, 8]. But in the limit of a very high substrate surface area, the precursor is rapidly consumed through chemisorption on the reactive sites and the coating process will become transport limited. Conformal coatings can still be obtained if sufficient precursor is delivered to titrate all of the reactive sites, and the saturation exposure time will be independent of the aspect ratio and the reactive sticking coefficient [9].

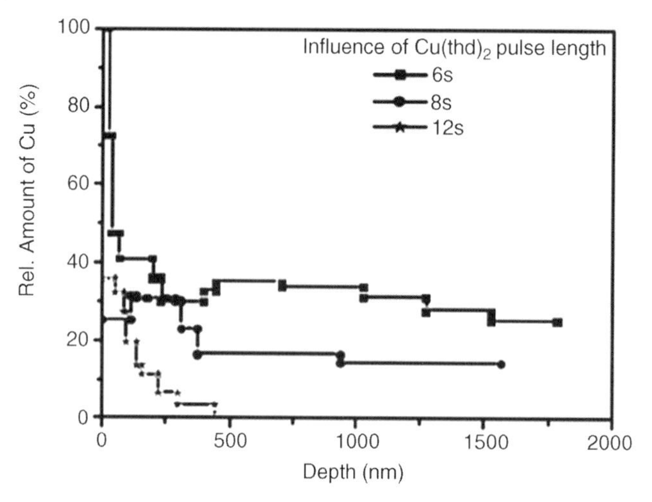

Figure 10.14 Influence of the Cu(thd)$_2$ pulse time on the infiltration depth of ALD Cu_2S in a TiO_2 matrix consisting of 25 nm particles as determined by RBS measurements. Reprinted with permission from Ref. [15]. Copyright 2004, the American Chemical Society.

Thermal decomposition is one example of a chemical nonideality that limits the uniformity of ALD films. Conformal ALD in high aspect ratios requires that the probability for thermal decomposition of the precursors must be negligible compared to the probability for diffusing to an empty site and reacting. In one study of Cu_2S ALD in a 4 μm porous film comprised of TiO_2 nanoparticles with sizes of 9–50 nm, RBS depth profile measurements yielded very nonuniform Cu thickness profiles (Figure 10.14) [15]. These experiments used alternating exposures to Cu(thd)$_2$ and H_2S at a temperature of 200 °C and the researchers found that as the Cu(thd)$_2$ exposure time was increased, the Cu_2S infiltration depth actually decreased. This behavior was attributed to thermal decomposition of the Cu(thd)$_2$ precursor and subsequent clogging of the TiO_2 nanopores.

The thermal decomposition of O_3 can reduce conformality during the ALD of metal oxides. Since some oxides will catalyze O_3 decomposition, the O_3 loss rate is highly dependent on both the temperature and composition of the substrate surface. For example, nonuniform thickness profiles generated during the ALD of In_2O_3 using cyclopentadienyl indium and O_3 were attributed to the catalytic destruction of the O_3 on the ALD In_2O_3 surface [54]. Nevertheless, reasonably uniform ALD ITO coatings could be prepared in high aspect ratio AAO membranes using this method by extending the O_3 exposure times to compensate for the reduced O_3 concentrations [55]. Similarly, the surface recombination of radicals such as O and H in plasma-enhanced ALD can profoundly affect the conformality of ALD films in high aspect ratio structures. As with O_3 decomposition, radical recombination is highly surface dependent so that film conformality will depend greatly on the material being deposited as well as the surface species involved [56].

Another chemical nonideality that can affect the conformality of ALD films is etching. Halogenated precursors such as $NbCl_5$ can react with metal oxide films to form volatile oxychloride species. This process can etch the surface to the point of complete removal of the film [57]. This problem will be particularly acute during ALD in porous supports where very large metal chloride exposures are needed. Metal chloride precursors can also form large, micron-sized agglomerates as a result of gaseous diffusion of hydroxychlorides and subsequent redeposition [58]. Another nonideality is blocking of adsorption sites by the gaseous by-product formed on neighboring sites. For instance, the HCl by-product of the $TiCl_4/H_2O$ process for TiO_2 ALD can readsorb onto the TiO_2 surface and prevent the adsorption and reaction of additional $TiCl_4$ molecules [59]. This problem is not unique to halogenated compounds since the isopropanol product of the $Ti(O^iPr)_4/H_2O$ process can also bind strongly to TiO_2 and limit additional $Ti(O^iPr)_4$ adsorption [60]. These phenomena may lead to nonconformal films in high aspect ratio substrates because the by-products generated on the outer regions of the pores will adsorb deeper within the pores and prevent or inhibit growth on these inner regions.

Nonideal behavior was observed during the ALD of Cu and Cu_3N films in aerogel substrates [61]. In these studies, the penetration depth of the films was found to be nearly independent of the precursor exposure time in complete contradiction of the expected diffusion-limited behavior. The authors postulated that this behavior resulted from thermal decomposition of the amidinate precursor, possibly catalyzed by metal nanoparticles, or from site blocking by the reaction by-products.

10.6 Conclusions and Future Outlook

One of the most useful attributes of ALD as it relates to nanomaterials synthesis is the capability to apply precise, conformal coatings on very high aspect ratio features including trenches, pores, and powders. This capability allows the surface properties to be tuned independently of surface structure and greatly expands the list of potential applications for ALD technology. Both analytical and numerical descriptions for the infiltration of high aspect ratio features during ALD provide useful descriptions and allow saturation conditions to be determined. This chapter has reviewed examples of ALD coatings on high aspect ratio substrates ranging from micromachined trenches with aspect ratios of $\sim$10 to aerogels with aspect ratios of 10^5. We expect that future developments will expand the list of available ALD materials for coating porous and high aspect ratio templates and also identify new, promising applications for this methodology. Continued growth in this field will require developing new processes and new precursors to overcome some of the nonidealities that are sometimes encountered, such as surface-catalyzed thermal decomposition. Due to the unique capability of ALD to apply conformal films on high aspect ratio structures, it is likely that this technology will play a central role in areas such as photovoltaics and batteries and help establish a new economy based on clean, renewable energy.

References

1 Ritala, M. and Leskelä, M. (2001) Atomic layer deposition, in *Handbook of Thin Film Materials*, vol. 1 (ed. H.S. Nalwa), Academic Press, San Diego, CA, pp. 103.

2 George, S.M. (2010) Atomic layer deposition: an overview. *Chem. Rev.*, **110** (1), 111–131.

3 Knez, M., Niesch, K., and Niinisto, L. (2007) Synthesis and surface engineering of complex nanostructures by atomic layer deposition. *Adv. Mater.*, **19** (21), 3425–3438.

4 Rolison, D.R., Long, R.W., Lytle, J.C., Fischer, A.E., Rhodes, C.P., McEvoy, T.M., Bourga, M.E., and Lubers, A.M. (2009) Multifunctional 3D nanoarchitectures for energy storage and conversion. *Chem. Soc. Rev.*, **38** (1), 226–252.

5 Gordon, R.G., Hausmann, D., Kim, E., and Shepard, J. (2003) A kinetic model for step coverage by atomic layer deposition in narrow holes or trenches. *Chem. Vapor Depos.*, **9** (2), 73–78.

6 Kim, J.Y., Ahn, J.H., Kang, S.W., and Kim, J.H. (2007) Step coverage modeling of thin films in atomic layer deposition. *J. Appl. Phys.*, **101** (7), 0735021–0735023.

7 Dendooven, J., Deduytsche, D., Musschoot, J., Vanmeirhaeghe, R.L., and Detavernier, C. (2009) Modeling the conformality of atomic layer deposition: the effect of sticking probability. *J. Electrochem. Soc.*, **156** (4), P63–P67.

8 Elam, J.W., Routkevitch, D., Mardilovich, P.P., and George, S.M. (2003) Conformal coating on ultrahigh-aspect-ratio nanopores of anodic alumina by atomic layer deposition. *Chem. Mater.*, **15** (18), 3507–3517.

9 Elam, J.W., Libera, J.A., Huynh, T.H., Feng, H., and Pellin, M.J. (2010) Atomic layer deposition of aluminum oxide in mesoporous silica gel. *J. Phys. Chem. C*, **114**, 17286–17292.

10 Libera, J.A., Elam, J.W., and Pellin, M.J. (2008) Conformal ZnO coatings on high surface area silica gel using atomic layer deposition. *Thin Solid Films*, **516** (18), 6158–6166.

11 Sneh, O., Clark-Phelps, R.B., Londergan, A.R., Winkler, J., and Seidel, T.E. (2002) Thin film atomic layer deposition equipment for semiconductor processing. *Thin Solid Films*, **402** (1–2), 248–261.

12 Perez, I., Robertson, E., Banerjee, P., Henn-Lecordier, L., Son, S.J., Lee, S.B., and Rubloff, G.W. (2008) TEM-based metrology for HfO_2 layers and nanotubes formed in anodic aluminum oxide nanopore structures. *Small*, **4** (8), 1223–1232.

13 Becker, J.S., Suh, S., Wang, S.L., and Gordon, R.G. (2003) Highly conformal thin films of tungsten nitride prepared by atomic layer deposition from a novel precursor. *Chem. Mater.*, **15** (15), 2969–2976.

14 Ducso, C., Khanh, N.Q., Horvath, Z., Barsony, I., Utriainen, M., Lehto, S., Nieminen, M., and Niinisto, L. (1996) Deposition of tin oxide into porous silicon by atomic layer epitaxy. *J. Electrochem. Soc.*, **143** (2), 683–687.

15 Reijnen, L., Feddes, B., Vredenberg, A.M., Schoonman, J., and Goossens, A. (2004) Rutherford backscattering spectroscopy study of $TiO_2/Cu_{1.8}S$ nanocomposites obtained by atomic layer deposition. *J. Phys. Chem. B*, **108** (26), 9133–9137.

16 Baumann, T.F., Biener, J., Wang, Y.M.M., Kucheyev, S.O., Nelson, E.J., Satcher, J.H., Elam, J.W., Pellin, M.J., and Hamza, A.V. (2006) Atomic layer deposition of uniform metal coatings on highly porous aerogel substrates. *Chem. Mater.*, **18** (26), 6106–6108.

17 Ferguson, J.D., Weimer, A.W., and George, S.M. (2000) ALD of Al_2O_3 and SiO_2 on BN particles using sequential surface reactions. *Appl. Surf. Sci.*, **162–163**, 280–292.

18 Pourret, A., Guyot-Sionnest, P., and Elam, J.W. (2009) Atomic layer deposition of ZnO in quantum dot thin films. *Adv. Mater.*, **21** (2), 232–235.

19 Gerritsen, E., Emonet, N., Caillat, C., Jourdan, N., Piazza, M., Fraboulet, D., Boeck, B., Berthelot, A., Smith, S., and Mazoyer, P. (2005) Evolution of materials technology for stacked-capacitors in 65 nm embedded-DRAM. *Solid-State Electron.*, **49** (11), 1767–1775.

20 Martinson, A.B.F., Elam, J.W., and Pellin, M.J. (2009) Atomic layer deposition of Cu_2S for future application in photovoltaics. *Appl. Phys. Lett.*, **94** (12), 1231071–1231073.

21 Herrmann, C.F., Delrio, F.W., Bright, V.M., and George, S.M. (2005) Conformal hydrophobic coatings prepared using atomic layer deposition seed layers and non-chlorinated hydrophobic precursors. *J. Micromech. Microeng.*, **15** (5), 984–992.

22 Scharf, T.W., Prasad, S.V., Dugger, M.T., Kotula, P.G., Goeke, R.S., and Grubbs, R.K. (2006) Growth, structure, and tribological behavior of atomic layer-deposited tungsten disulphide solid lubricant coatings with applications to MEMS. *Acta Mater.*, **54** (18), 4731–4743.

23 Mayer, T.M., Elam, J.W., George, S.M., Kotula, P.G., and Goeke, R.S. (2003) Atomic-layer deposition of wear-resistant coatings for microelectromechanical devices. *Appl. Phys. Lett.*, **82** (17), 2883–2885.

24 Hoivik, N.D., Elam, J.W., Linderman, R.J., Bright, V.M., George, S.M., and Lee, Y.C. (2003) Atomic layer deposited protective coatings for micro-electromechanical systems. *Sens. Actuators A*, **103** (1–2), 100–108.

25 Bae, C., Yoo, H., Kim, S., Lee, K., Kim, J., Sung, M.A., and Shin, H. (2008) Template-directed synthesis of oxide nanotubes: fabrication, characterization, and applications. *Chem. Mater.*, **20** (3), 756–767.

26 Banerjee, P., Perez, I., Henn-Lecordier, L., Lee, S.B., and Rubloff, G.W. (2009) Nanotubular metal–insulator–metal capacitor arrays for energy storage. *Nat. Nanotechnol.*, **4** (5), 292–296.

27 Miikkulainen, V., Suvanto, M., and Pakkanen, T.A. (2008) Molybdenum nitride nanotubes. *Thin Solid Films*, **516** (18), 6041–6047.

28 Daub, M., Knez, M., Goesele, U., and Nielsch, K. (2007) Ferromagnetic nanotubes by atomic layer deposition in anodic alumina membranes. *J. Appl. Phys.*, **101** (9), 09J1111–09J1113.

29 Comstock, D.J., Christensen, S.T., Elam, J.W., Pellin, M.J., and Hersam, M.C. (2010) Tuning the composition and nanostructure of Pt/Ir films via anodized aluminum oxide templated atomic layer deposition. *Adv. Funct. Mater.*, **20**, 3099–3105.

30 Kim, W.H., Park, S.J., Son, J.Y., and Kim, H. (2008) Ru nanostructure fabrication using an anodic aluminum oxide nanotemplate and highly conformal Ru atomic layer deposition. *Nanotechnology*, **19** (4), 1–8.

31 Lee, M., Kim, T., Bae, C., Shin, H., and Kim, J. (2010) Fabrication and applications of metal-oxide nano-tubes. *JOM*, **62** (4), 44–49.

32 Shin, H.J., Jeong, D.K., Lee, J.G., Sung, M.M., and Kim, J.Y. (2004) Formation of TiO_2 and ZrO_2 nanotubes using atomic layer deposition with ultraprecise control of the wall thickness. *Adv. Mater.*, **16** (14), 1197–1200.

33 Martinson, A.B.F., Elam, J.W., Hupp, J.T., and Pellin, M.J. (2007) ZnO nanotube based dye-sensitized solar cells. *Nano Lett.*, **7** (8), 2183–2187.

34 Stair, P.C., Marshall, C., Xiong, G., Feng, H., Pellin, M.J., Elam, J.W., Curtiss, L., Iton, L., Kung, H., Kung, M., and Wang, H.H. (2006) Novel, uniform nanostructured catalytic membranes. *Top. Catal.*, **39** (3–4), 181–186.

35 Elam, J.W., Libera, J.A., Pellin, M.J., and Stair, P.C. (2007) Spatially controlled atomic layer deposition in porous materials. *Appl. Phys. Lett.*, **91** (24), 2431051–2431053.

36 King, J.S., Graugnard, E., and Summers, C.J. (2005) TiO_2 inverse opals fabricated using low-temperature atomic layer deposition. *Adv. Mater.*, **17** (8), 1010–1013.

37 Scharrer, M., Wu, X., Yamilov, A., Cao, H., and Chang, R.P.H. (2005) Fabrication of inverted opal ZnO photonic crystals by atomic layer deposition. *Appl. Phys. Lett.*, **86** (15), 1511131–1511133.

38 Rugge, A., Becker, J.S., Gordon, R.G., and Tolbert, S.H. (2003) Tungsten nitride inverse opals by atomic layer deposition. *Nano Lett.*, **3** (9), 1293–1297.

39 Povey, I.M., Whitehead, D., Thomas, K., Pemble, M.E., Bardosova, M., and Renard, J. (2006) Photonic crystal thin films of GaAs prepared by atomic layer deposition. *Appl. Phys. Lett.*, **89** (10), 1041031–1041033.

40 King, J.S., Neff, C.W., Summers, C.J., Park, W., Blomquist, S., Forsythe, E., and Morton, D. (2003) High-filling-fraction inverted ZnS opals fabricated by atomic layer deposition. *Appl. Phys. Lett.*, **83** (13), 2566–2568.

41 King, J.S., Graugnard, E., and Summers, C.J. (2006) Photoluminescence modification by high-order photonic bands in TiO_2/ZnS: Mn multilayer inverse opals. *Appl. Phys. Lett.*, **88** (8), 0811091–0811093.

42 Baumann, T.F., Gash, A.E., Fox, G.A., Satcher, J.J.H., and Hrubesh, L.W. (2002) Oxidic aerogels, in *Handbook of Porous Solids* (eds F. Schuth, K.S.W. Sing, and J. Weitkamp), Wiley-VCH Verlag GmbH, Weinheim.

43 Rolison, D.R. (2003) Catalytic nanoarchitectures: the importance of nothing and the unimportance of periodicity. *Science*, **299** (5613), 1698–1701.

44 Kucheyev, S.O., Biener, J., Wang, Y.M., Baumann, T.F., Wu, K.J., van Buuren, T., Hamza, A.V., Satcher, J.H., Elam, J.W., and Pellin, M.J. (2005) Atomic layer deposition of ZnO on ultralow-density nanoporous silica aerogel monoliths. *Appl. Phys. Lett.*, **86** (8), 0831081–0831083.

45 Biener, J., Baumann, T.F., Wang, Y.M., Nelson, E.J., Kucheyev, S.O., Hamza, A.V., Kemell, M., Ritala, M., and Leskelä, M. (2007) Ruthenium/aerogel nanocomposites via atomic layer deposition. *Nanotechnology*, **18** (5), 1–4.

46 King, J.S., Wittstock, A., Biener, J., Kucheyev, S.O., Wang, Y.M., Baumann, T.F., Giri, S.K., Hamza, A.V., Baeumer, M., and Bent, S.F. (2008) Ultralow loading Pt nanocatalysts prepared by atomic layer deposition on carbon aerogels. *Nano Lett.*, **8** (8), 2405–2409.

47 Elam, J.W., Xiong, G., Han, C.Y., Wang, H.H., Birrell, J.P., Welp, U., Hryn, J.N., Pellin, M.J., Baumann, T.F., Poco, J.F., and Satcher, J.H. (2006) Atomic layer deposition for the conformal coating of nanoporous materials. *J. Nanomater.*, 1–5.

48 Hamann, T.W., Martinson, A.B.E., Elam, J.W., Pellin, M.J., and Hupp, J.T. (2008) Aerogel templated ZnO dye-sensitized solar cells. *Adv. Mater.*, **20** (8), 1560–1564.

49 Hamann, T.W., Martinson, A.B.F., Elam, J.W., Pellin, M.J., and Hupp, J.T. (2008) Atomic layer deposition of TiO_2 on aerogel templates: new photoanodes for dye-sensitized solar cells. *J. Phys. Chem. C*, **112** (27), 10303–10307.

50 Elam, J.W., Libera, J.A., Pellin, M.J., Zinovev, A.V., Greene, J.P., and Nolen, J.A. (2006) Atomic layer deposition of W on nanoporous carbon aerogels. *Appl. Phys. Lett.*, **89** (5), 0531241–0531243.

51 Wang, J.J., Deng, X.G., Varghese, R., Nikolov, A., Sciortino, P., Liu, F., Chen, L., and Liu, X.M. (2005) Filling high aspect-ratio nano-structures by atomic layer deposition and its applications in nano-optic devices and integrations. *J. Vac. Sci. Technol. B*, **23** (6), 3209–3213.

52 de Rouffignac, P., Li, Z.W., and Gordon, R.G. (2004) Sealing porous low-*k* dielectrics with silica. *Electrochem. Solid-State Lett.*, **7** (12), G306–G308.

53 Travaly, Y., Schuhmacher, J., Baklanov, M.R., Giangrandi, S., Richard, O., Brijs, B., Van Hove, M., Maex, K., Abell, T., Somers, K.R.F., Hendrickx, M.F.A., Vanquickenborne, L.G., Ceulemans, A., and Jonas, A.M. (2005) A theoretical and experimental study of atomic-layer-deposited films onto porous dielectric substrates. *J. Appl. Phys.*, **98** (8), 0835151–0835159.

54 Elam, J.W., Martinson, A.B.F., Pellin, M.J., and Hupp, J.T. (2006) Atomic layer deposition of In_2O_3 using cyclopentadienyl indium: a new synthetic route to transparent conducting oxide films. *Chem. Mater.*, **18** (15), 3571–3578.

55 Martinson, A.B.F., Elam, J.W., Liu, J., Pellin, M.J., Marks, T.J., and Hupp, J.T. (2008) Radial electron collection in dye-sensitized solar cells. *Nano Lett.*, **8** (9), 2862–2866.

56 Knoops, H.C.M., Langereis, E., van de Sanden, M.C.M., and Kessels, W.M.M. (2020) Conformality of plasma-assisted ALD: physical processes and modeling. *J. Electrochem. Soc.*, **157** (12), G241–G249.

57 Elers, K.-E., Ritala, M., Leskelä, M., and Rauhala, E. (1994) $NbCl_5$ as a precursor for ALE. *Appl. Surf. Sci.*, **82/83**, 468–474.

58 Puurunen, R.L. (2005) Formation of metal oxide particles in atomic layer deposition during the chemisorption of metal chlorides: a review. *Chem. Vapor Depos.*, **11** (2), 79–90.

59 Ritala, M., Leskelä, M., Nykanen, E., Soininen, P., and Niinisto, L. (1993) Growth of titanium dioxide thin films by atomic layer epitaxy. *Thin Solid Films*, **225**, 288–295.

60 Ritala, M., Leskelä, M., Niinisto, L., and Haussalo, P. (1993) Titanium isopropoxide as a precursor in atomic layer epitaxy of titanium-dioxide thin-films. *Chem. Mater.*, **5** (8), 1174–1181.

61 Kucheyev, S.O., Biener, J., Baumann, T.F., Wang, Y.M., Hamza, A.V., Li, Z., Lee, D.K., and Gordon, R.G. (2008) Mechanisms of atomic layer deposition on substrates with ultrahigh aspect ratios. *Langmuir*, **24** (3), 943–948.

11
Coatings of Nanoparticles and Nanowires

Hong Jin Fan and Kornelius Nielsch

11.1
ALD on Nanoparticles

Research of atomic layer deposition (ALD) on nanoparticles (e.g., metals, oxides, and polymers) has been mostly conducted by the Weiner and George groups in a scalable fluidized bed reactor or rotary reactor. The most demonstrated work is alumina ALD on bulk quantities of metal and ceramic micron-sized particles. For selected examples, see Refs [1–7]. Such conformal and extremely thin ALD coating is needed for modifying the surface properties of nanoparticles while keeping their bulk properties unchanged. Generally, for high aspect ratio surfaces, less steric effects occur and gas precursors can reach any available active sites more easily. Therefore, the ALD growth rates observed on nanoparticles are typically higher than those measured on planar substrates.

ALD oxide coating on nanoparticles provides surface passivation or surface functionalization. Several examples are given below. First, alumina coatings on titania particles can help prevent the potential skin damage in their cosmetic application [2]. Figure 11.1a shows the TEM image of the ALD alumina-coated titania nanoparticles. In order to test the photocatalytic activities of the uncoated and alumina-coated TiO_2 nanoparticles, the concentration of methylene blue was tracked over time to measure its degree of oxidation when exposed to different nanoparticle suspensions. As shown in Figure 11.1b, the coated titania does not show any measurable activity during the same time period. This is desirable for the real application of TiO_2 nanoparticles as sunscreens or cosmetics products, because an uncoated TiO_2 can potentially cause severe damage to the skin as a result of its catalytic oxidation effect to DNA *in vitro*. On the other hand, it is noted that the relative transmittance of titania nanoparticles remains unchanged over the entire UV spectrum. Therefore, it is demonstrated that ultrathin ALD alumina films can efficiently form a capping layer in sunscreen materials without altering the desired optical properties of the particle substrate.

Second, a conformal surface coating of a thin oxide layer can be used to avoid particle agglomeration when they are dispersed in liquid or polymer vehicles. As demonstrated by Wank *et al.* [6], the interactions between BN nanoparticles and the epoxy resin can be altered by an ultrathin Al_2O_3 coating, so that the adhesion between

Atomic Layer Deposition of Nanostructured Materials, First Edition. Edited by Nicola Pinna and Mato Knez.

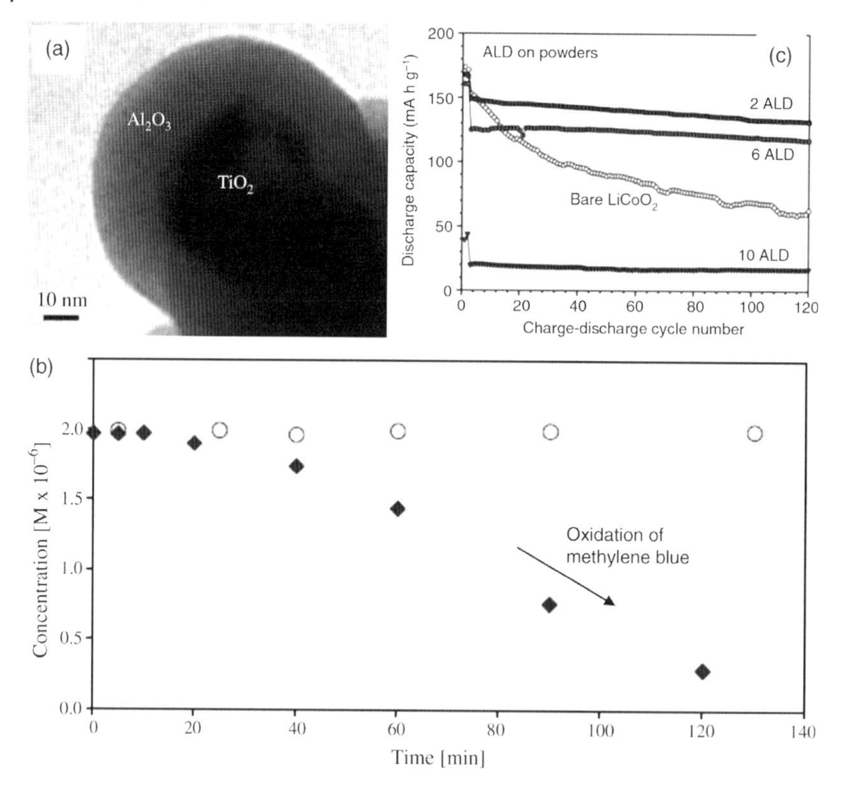

Figure 11.1 (a) TEM image of alumina-coated titania nanoparticles. (b) Comparison of the photocatalytic activity of uncoated (diamond) and alumina-coated (circle) titania nanoparticles. Reproduced with permission from Ref. [2]. (c) Charge–discharge cycle performance of lithium ion battery electrodes fabricated using the bare $LiCoO_2$ powder and the Al_2O_3 ALD-coated $LiCoO_2$ powder using 2, 6, and 10 ALD cycles. The thickness after two cycles is 3–4 Å. Reproduced with permission from Ref. [4].

the coated particles and a cured epoxy in a filled composite is strengthened by 25%. The George group also used ALD to enhance the performance of lithium ion batteries by coating a thin (3–4 Å) layer of Al_2O_3 on the $LiCoO_2$ cathode powder [4]. As shown in Figure 11.1c, the capacity retention for the 2 ALD cycle sample can be improved to 89% after 120 charge–discharge cycles, compared to 45% of the bare $LiCoO_2$ powder. Possible reason of the improvement is that the ultrathin Al_2O_3 film has acted to minimize Co dissolution or reduce surface electrolyte reactions. However, thicker coating reduces the capacity because of the restricted electron transport and possibly the slower Li^+ diffusion kinetics in the thick Al_2O_3 insulating layer.

Third, a conformal thin (<10 nm) ALD oxide layer can also be useful for functionalizing metal nanoparticles such as iron, nickel, cobalt, and copper [5, 7]. The ALD coating can provide effective protection for metal nanoparticles from oxidation and corrosion [3]. Figure 11.2a and b shows examples of Al_2O_3 ALD on Cu and Co nanoparticles with a film thickness of ~5.5 and 11 nm, respectively.

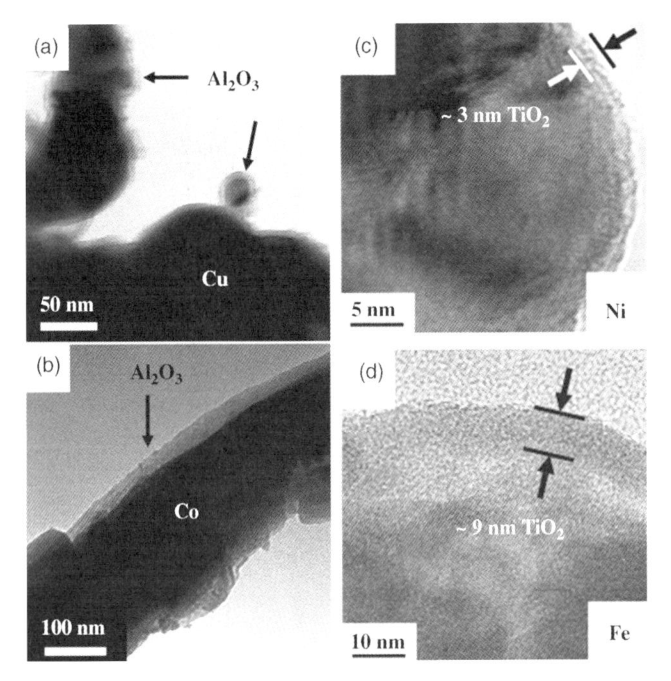

Figure 11.2 ALD coating on metal nanoparticles. TEM images of (a) ~5.5 nm Al_2O_3-coated copper and (b) 11 nm Al_2O_3-coated cobalt powders. TEM images of (c) 3 nm TiO_2-coated nickel and (d) 9 nm TiO_2-coated iron particles. Reproduced with permission from Ref. [5].

Such insulating oxide layer may alter the electrical, thermal, and magnetic properties of various nanometals to be superior to those of their bulk counterparts. Figure 11.2c and d shows amorphous TiO_2 coating on magnetic nanoparticles. It was observed that, during the early few cycles of TiO_2 ALD, the iron particle surface could be partially oxidized by the H_2O_2 precursor to Fe_3O_4 at 100 °C [5]. In order to circumvent this initial oxidation problem, prior to the TiO_2 coating, an ALD AlN layer can be deposited to passivate the Fe particle surface (using TMA and ammonia as precursor) as demonstrated by the same group [7]. Interestingly, such ALD TiO_2 films added an additional photoactive property to the magnetic metals.

ALD on polymer nanoparticles provides a straightforward method for forming hollow objects, where the nanoparticles serve as sacrificial template. One example was given by Ras *et al.* [8] who obtained both hollow nanospheres and nanotubes. As will be seen in other parts of the book, this is a generic process of ALD-assisted formation of hollow nanoobjects based on morphological transformation of polymer templates. Being a polymer, the ALD process needs low process temperature, which can be readily achieved by the Al_2O_3 recipe (80 °C).

ALD of TiO_2, Al_2O_3, or ZnS on closely packed monolayers of polystyrene (PS) and SiO_2 spherical particles has been conducted by the Summers group [9, 10]. An example is given in Figure 11.3, where a substrate patterned with a self-assembled monolayer of hexagonally arranged PS beads was processed by ALD to deposit

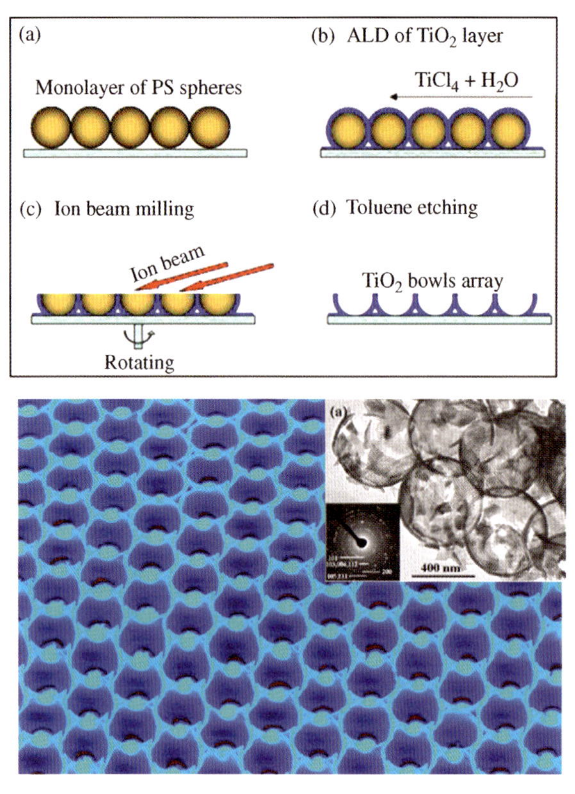

Figure 11.3 Nanobowls by ALD on polystyrene spheres. *Top*: Fabrication process. *Bottom*: SEM image of the nanobowl array. *Inset*: TEM image of some of the nanobowls showing their polycrystalline phase and spherical shape. Reproduced with permission from Ref. [10].

TiO_2 [10]. Then the top layer of TiO_2 was removed with an ion beam followed by removal of the polymer beads, and a micrometer-scale array of TiO_2 nanobowls was obtained (see Figure 11.3b). Compared to those high aspect ratio photonic crystals, condition of ALD on monolayer of spheres is less critical and it can be achieved in the continuous flow-type chamber. The precursors can easily access the "hidden" space between the nanobeads to achieve conformal coatings of the entire monolayer. Although other techniques such as pulsed laser deposition or electrochemical deposition are also capable of replicating the structure into nanobowl arrays, ALD proves itself to be the method of choice if high precision is desired or more complex templates (such as monolayers of spheres with two sizes or double-layer templates) are involved.

11.2
Vapor–Liquid–Solid Growth of Nanowires by ALD

Semiconductor nanowires (NWs), as a competing component for highly integrated photoelectronic device applications, have been under the research focus since

2001 [11]. The synthesis approaches of NWs include conventional top-down techniques such as lithography etching and bottom-up techniques such as vapor–liquid–-solid (VLS) or solution growth [12, 13]. In terms of perfect growth control, molecular beam epitaxy, chemical vapor deposition (CVD), and chemical beam epitaxy (CBE) are so far the most effective techniques for the growth of most semiconducting materials such as Si, Ge, ZnO, and III–V. The typical VLS is essentially a metal-catalyzed CVD process, in which metal nanoparticles (usually gold) catalyze the growth by forming eutectic alloys with the precursor molecules. For NWs of compound materials, both precursor molecules will alloy with the metal particles, forming ternary compound eutectic. There are cases where the ternary compounds are energetically unstable so that the resulting NWs can be nonhomogeneous in terms of composition or morphology. In this context, a separation of the precursor delivery, which is characteristic of ALD, might provide better control of the reaction. Nevertheless, probably because of its relative low growth rate, attempts of applying ALD for VLS growth of semiconductor NWs are few. The successful demonstration was first made by Huo and coworkers in 2002 [14], who conducted ALD growth of single crystalline superlattice NWs consisting of alternating layers of ZnSe and CdSe along the growth direction. Continuation of this work would be impactful since they are both currently important materials for nanostructured solar cell applications.

Recently, the Nielsch group at Hamburg demonstrated another VLS growth of binary semiconductor NWs Sb_2S_3 and Sb_2Se_3, as well as their heterogeneous NWs, using ALD [15]. Figure 11.4a illustrates schematically the pulse VLS reaction: upon exposure to the Sb precursor, the Se present in the Au catalyst reacts and the Sb_2Se_3 wire grows, after which excess Sb dissolves to saturation; similarly, subsequent reaction of Sb in Au with the Se precursor grows Sb_2Se_3 further and leaves Se-saturated Au ready for the next cycle. Exposure of Au nanoparticles prepared on a Si substrate to 300–400 alternating pulses of $Sb(NMe_2)_3$ and Et_2Se_2 vapors at 350 °C results in the growth of antimony sulfide and selenide NWs, as evidenced by SEM images in Figure 11.4b and c. Both reactions yield pure, stoichiometric, and crystalline material.

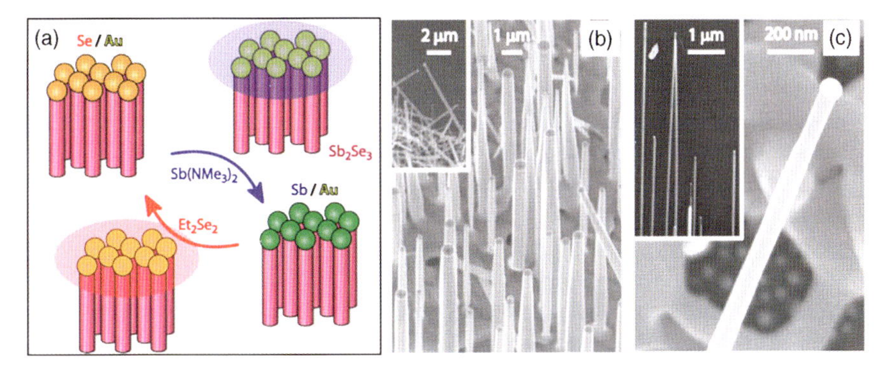

Figure 11.4 (a) Schematic of the ALD growth process of nanowires (Sb_2Se_3 as example) based on the VLS mechanism. (b) SEM image of Sb_2S_3 wires. (c) SEM image of Sb_2Se_3 wires. Reproduced with permission from Ref. [15].

Kipshidze *et al.* synthesized short light emitting GaN nanorods based on the so-called pulsed metal-organic CVD (MOCVD) process [16]. The precursor delivery of trimethylgallium and ammonia is strictly separated in a pulse mode, which makes this process exactly ALD in nature. Generally speaking, based on the similarity in the precursor types between MOCVD and ALD, it is expected that more semiconductor NWs that have been successfully synthesized by MOCVD (e.g., ZnO) can also be grown inside the ALD chamber. As commented by Knez *et al.* [17], ALD may provide a better control of VLS growth for binary semiconductor NWs for a broader range of materials. While a MOCVD system can have the pulse mode (as mentioned above for the growth of GaN nanorods), it is exciting to envisage that the ALD machine can also be modified to have MOCVD option.

11.3
Atomic Layer Epitaxy on Nanowires

In the pioneering time of ALD, epitaxial growth of semiconductor thin film was in the focus of the research community, but so far atomic layer epitaxy (ALE) has been applied very rarely for the nanostructure synthesis [14]. Recently, Nielsch and coworkers [18] have succeeded in coating single crystals of semiconductor nanowires using ALD at low temperatures (60–150 °C). Epitaxial growth means growth of the coating layer with the same orientation and atomic spacing as the crystal structure of the substrate. As shown in the enlarged TEM micrograph in Figure 11.5, a crystalline

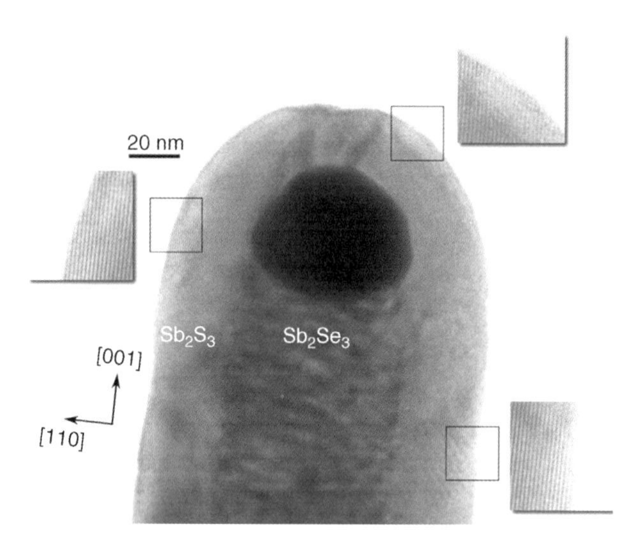

Figure 11.5 Conformal heteroepitaxial deposition of Sb_2S_3 at 90 °C onto Sb_2Se_3 nanowires. HRTEM image of the extremity of a wire, displaying the dark Au particle and identifying the crystallographic *c* axis parallel to the long axis of the nanowire. Reproduced with permission from Ref. [18].

layer of Sb_2S_3 has been homogenously grown without any lattice defects on top of a Sb_2Se_3 nanorod with a diameter of ~50 nm. Until now, comparable structures could only be realized by colloid chemical methods at moderate temperatures from liquids. The ALE route developed allows cost-effective production of three-dimensional semiconductor nanostructures in low vacuum or at high pressures. Thus, it is technologically very attractive for the development of solar cells, light emitting diodes, and thermoelectric components.

11.4
ALD on Semiconductor NWs for Surface Passivation

The initial idea of coating semiconductor NWs with an ALD dielectric layer was to provide surface passivation to reduce the unwanted surface traps. For example, two studies have shown that an Al_2O_3 cap layer efficiently lowers the defect-related deep level emissions of ZnO NWs [19, 20]. By dielectric capping, the separation of electrons and holes within the surface depletion region is reduced, which means that less holes are captured by surface traps and more electron–hole pairs are generated. As a result, the band gap emission can be enhanced while the green emission is suppressed.

Figure 11.6a shows energy diagrams of a fresh GaN NW and ALD Al_2O_3-coated GaN NW [21]. In the as-synthesized GaN NWs, the near band gap emission intensity is low. The accepted reason is that photogenerated excitons may reach the surface and recombine nonradiatively through the surface states, which are energetically located

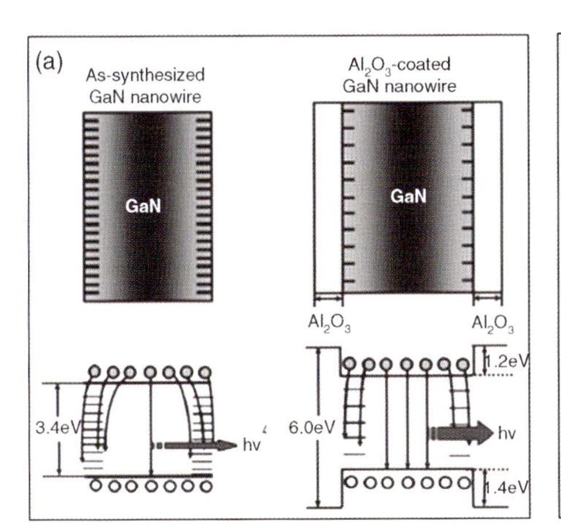

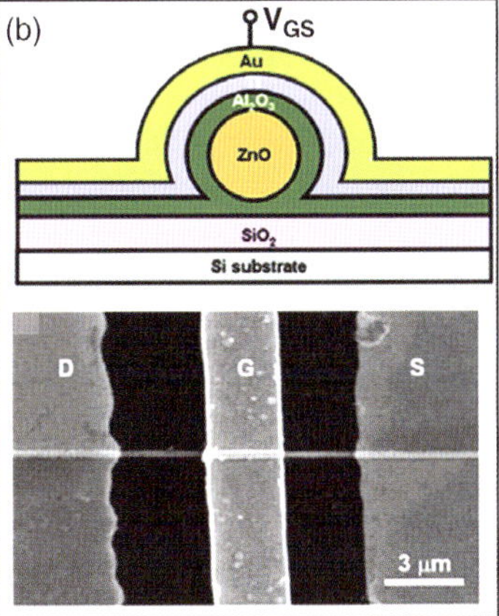

Figure 11.6 (a) *Top*: Energy diagrams of as-synthesized GaN nanowires and Al_2O_3-coated nanowires. *Bottom*: Illustrations showing the radiative and nonradiative electron–hole recombinations. (b) Cross-sectional schematic (*top*) and SEM image (*bottom*) of the omega-shaped gate ZnO nanowire field effect transistor device. Reproduced with permission from Ref. [21, 23].

within the band gap and structurally caused by dangling bonds or intrinsic defects. When the surface of the GaN NWs is coated with ALD Al_2O_3, the Al_2O_3 layer can passivate some of the surface states, causing stronger excitonic emissions than the as-synthesized GaN NWs. Also, Wang *et al.* [22] showed that an ALD Al_2O_3 layer on GaQ_3 NWs can reduce the NW degradation by oxygen and thus maintain the PL stability in air and water vapor.

The same argument can be applied to ZnO NWs as photoelectrode in dye-sensitized solar cells, as reported by the Yang group at UC Berkeley. ALD TiO_2 capping helps to reduce the recombination of photogenerated electron–hole pairs on the ZnO NW surface, so that the carriers can be more efficiently separated for improved efficiency [24]. It is noted that ALD has become a standard tool for surface passivation in crystalline Si photovoltaic devices. As Si NWs are now also being studied as the building block for miniaturized photovoltaics [25, 26], the application for Si NW surface passivation by ALD will be equally important.

In addition to optical properties, ALD coating of dielectric layer, Al_2O_3 or SiO_2, is also becoming a favorable technique for the application of semiconductor NWs as channel materials in field effect electronic devices. In these devices, the dielectric layer can have dual functions: first, it serves as the gate oxide, and second, it stabilizes the NW surface in air ambient (compared to the NW devices with planar back-gate oxide). For example, Keem *et al.* [23] fabricated omega-shaped gate field effect transistor (FET) devices using 17 nm thick ALD Al_2O_3 gate oxide surrounding the ZnO NW channel. Figure 11.6b shows such a transistor structure and the corresponding SEM of a real device. The authors found that the field effect mobility, peak transconductance, and I_{on}/I_{off} ratio for the omega-shaped gate FET are remarkably enhanced by nearly 3.5, 32, and 107 times, respectively, compared to the conventional back-gate FET. The enhancement is mostly attributed to the omega-shaped geometry and the surface passivation of the NW channel induced by the ALD Al_2O_3. For the gate oxide purpose, it is recently reported that the as-deposited ALD dielectric layer of SiO_2 or Al_2O_3 needs a post-annealing step. For example, in the case of ALD SiO_2, rapid thermal annealing at 1000 °C for 1 min can effectively reduce the number of –OH groups and the trap charge densities, which makes the ALD SiO_2 closer to thermal oxide in terms of dielectric quality.

11.5
ALD-Assisted Formation of Nanopeapods

Nanopeapod, with a chain of metal nanoparticles embedded within a solid or hollow oxide shell, is an emerging and fascinating type of nanostructure due to its potential optoelectronic applications. Various fabrication methods have been demonstrated in particular for Au@SiO_2 and Au@Ga_2O_3 systems, which are essentially a self-assembly of the catalyst metal particles during the VLS growth of NWs. First demonstrated nanopeapods using ALD are reported by the Knez group at MPI Halle (see Figure 11.7) [27]. In their experiments, CuO NWs were first coated with a 5 nm ALD layer of alumina or TiO_2. The NW core of CuO was then reduced to Cu by

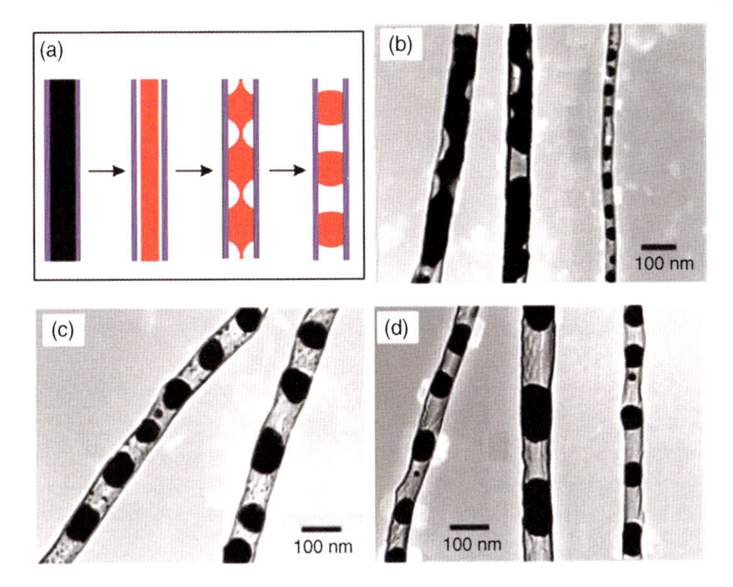

Figure 11.7 (a) Formation mechanism of Cu nanoparticle chains. CuO nanowires are first coated with ALD Al_2O_3, which are then subject to thermal reduction in H_2 ambient. The CuO core was reduced to Cu nanowires with a smaller diameter, during which process the surface diffusion of Cu atoms resulted in undulated Cu nanowires, and eventually nanoparticle chains. TEM images of Cu nanoparticle chains prepared by reduction of CuO nanowires with a 5 nm ALD Al_2O_3 shell in H_2 for 1 h at (b) 400 °C, (c) 500 °C, and (d) 600 °C. Reproduced with permission from Ref. [27].

H_2 annealing, causing a volume shrinkage by 43% according to the density ratio of Cu to CuO. Simultaneously, the thermal annealing breaks the continuous Cu into separated Cu nanoparticles, which move inside the hollow channel and reach an equilibrium chain profile under corresponding annealing temperatures and initial CuO NW diameter. Figure 11.7b–d shows the evolution of the nanopeapods after annealing at 400, 500, and 600 °C, respectively. In particular, the samples after 600 °C annealing exhibit a clear and clean nanopeapod structure. The NWs are perfectly fragmented into nanoparticle chains along the whole length of the tubes, without any remaining tiny particles attached to the inner wall surfaces of the tubes. The particles within one single tube have more uniform and regular shapes than those annealed at lower temperatures. Here the ALD layer preserves the morphological integrity during the phase transformation.

As copper is not an interesting enough material for plasmonics application, Au or Pt would be better candidates. In this case, a more straightforward approach is to have two ALD layers outside the metal NW; a middle sacrificial layer between inner metal and outer shell should be easily removed, leaving the space needed for the fragmentation of the metal wires. Indeed, such structures have been realized by the same group using porous alumina membrane templates, a slightly different strategy. In order to study the plasmon–exciton coupling, the outer shell can be light emitting materials such as ZnO or SnO_2 that are both ALD processable.

11.6
Photocorrosion of Semiconductor Nanowires Capped by ALD Shell

ZnO, as well as other wide band gap semiconductors, undergoes photolysis in the presence of moisture under strong UV irradiation. A bare ZnO NW will be seriously photoetched when it is put into neutral aqueous solutions under direct UV illumination, or even during aging under ambient conditions where trace UV photons from the daylight are unavoidable [28]. In this process, electron–hole pairs are generated by absorption of UV photons. When those unrecombined h^+ migrate to the NW surfaces, they react with water and produce H^+ ions through $H_2O + 2h^+ \rightarrow 2H^+ + \frac{1}{2}O_2$. The generated H^+ can then dissolve ZnO via $ZnO + 2H^+ \rightarrow Zn^+ + H_2O$. Such photoetching process might be slowed down by an inert ALD Al_2O_3, but it will be unexpectedly accelerated by an ALD TiO_2 coating. It is claimed to be related to the well-known photocatalytic effect of TiO_2, another semiconductor that can generate electron–hole pairs upon UV light. The reaction of water with TiO_2 under UV irradiation (i.e., water splitting) produces a higher density of H^+ ions, which may diffuse through the shell and reach the interface. Therefore, opposite to the initial belief that a TiO_2 coating can protect ZnO NWs from unwanted interface carrier recombination, it actually causes long-term instability of ZnO [28].

Such photocatalytic effect can be indirectly revealed from the difference in ZnO–Al_2O_3 and ZnO–TiO_2 wires. In the former case, the photoetching occurs only at the wire ends, similar to the uncoated ZnO NWs, whereas the etching of the ZnO–TiO_2 coaxial wires occurs both at the ends and at the core–shell interfaces.

Going one step further, the above-mentioned instability caused by the negative photocorrosion effect can be positively utilized for the fabrication of nanotubes. As demonstrated by the same group at MPI Halle [29], amorphous TiO_2 or Al_2O_3 nanotubes can be obtained through a self-induced photocorrosion of the ZnO or CuO core NWs whose surfaces are uniformly coated with ALD TiO_2 or Al_2O_3 layers. Also, again due to the unique photocatalytic effect of TiO_2, an ALD anatase phase TiO_2 shell can convert a core–shell NW into a nanotube of TiO_2. The results of a time-dependent experiment (see Figure 11.8) show that the voids start from the core–shell interface,

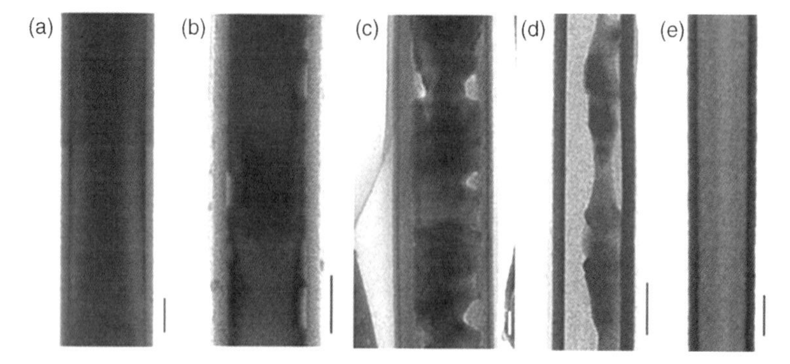

Figure 11.8 Formation of nanotubes by a photocatalytic effect. TEM images of ZnO–TiO_2 core–shell nanowires after irradiation with UV light for (a) 0 min, (b) 10 min, (c) 30 min, (d) 60 min, and (e) 120 min. Scale bar: 50 nm. Reproduced with permission from Ref. [29].

which enlarge over time and finally merge. A UV exposure for 2 h causes the initial core–shell NWs to be completely hollow (Figure 11.8e). Apart from the demonstrated ZnO and CuO, this new photocatalytic method for nanotube formation is expected to be applicable also for templates of other semiconductors such as ZnS, CdS, and CdSe, which are known to "suffer" from photocorrosion.

11.7 Interface Reaction of Nanowires with ALD Shell

Implementation of the solid-state reaction using NWs as one reactant provides a method for rational synthesis of compound oxide nanotubes. This method could be applicable to various ZnO- or MgO-based spinel oxides when the ZnO or MgO NWs are coated with a conformal layer of oxide (e.g., In_2O_3, SnO_2, TiO_2, Cr_2O_3), which is nowadays possible through ALD. Several examples of spinel 1D nanostructures are given below.

11.7.1 $ZnO–Al_2O_3$

When the ALD Al_2O_3-coated ZnO NWs are subject to high-temperature annealing, interface reaction occurs, leading to $ZnAl_2O_4$ spinel nanotubes (Figure 11.9) [30, 31]. This bilayer NW system provides a nice test bed of the nanoscale Kirkendall effect in nanotube formation. In this strategy, presynthesized ZnO NWs (10–30 nm thick and up to 20 μm long) were coated with a 10 nm ALD Al_2O_3. Thus formed $ZnO–Al_2O_3$ core–shell NWs were annealed in air at 700 °C for 3 h, causing an interfacial solid-state reaction and diffusion. Because the reaction is effectively a one-way transfer of ZnO into the alumina, it represents an extreme Kirkendall effect. Upon suitable matching of the thickness of the core and the shell, highly crystalline single-phase spinel nanotubes are obtained, as shown in Figure 11.9.

In addition, a suitable annealing temperature is very important [33]. The temperature window for an optimal solid-state reaction of $ZnO–Al_2O_3$ core–shell NWs is 700–800 °C. Tiny voids are generated at $ZnO–Al_2O_3$ interface near 600 °C, but do not grow into large voids due to kinetic reasons. A higher annealing temperature of about 800 °C results in hollow nanotubes even if the initial ZnO NW is thicker than necessary for a complete spinel formation of the former shell. Due to reactive desorption, the surface ZnO decomposes and evaporates in the presence of reactive Al_2O_3. However, when the temperature is further increased, the tube wall collapses due to the thermodynamic instability.

The conformity and uniformity characteristics of ALD are essential for the formation of smooth nanotubes. Particularly, the synthesis of complex tubular $ZnAl_2O_4$ nanostructures by such a shape-preserving transformation is possible if one starts with hierarchical 3D ZnO NWs [32]. Recent literature shows a large variety of ZnO 3D nanostructures including bridges, nails, springs, and stars. Figure 11.9 gives two examples: a Chinese firecracker-like 3D and a comb-like 2D hollow nanostructure $ZnAl_2O_4$ spinel. Such uniform-structured products cannot be obtained by other coating methods such as pulsed laser deposition, sputtering, or physical vapor deposition

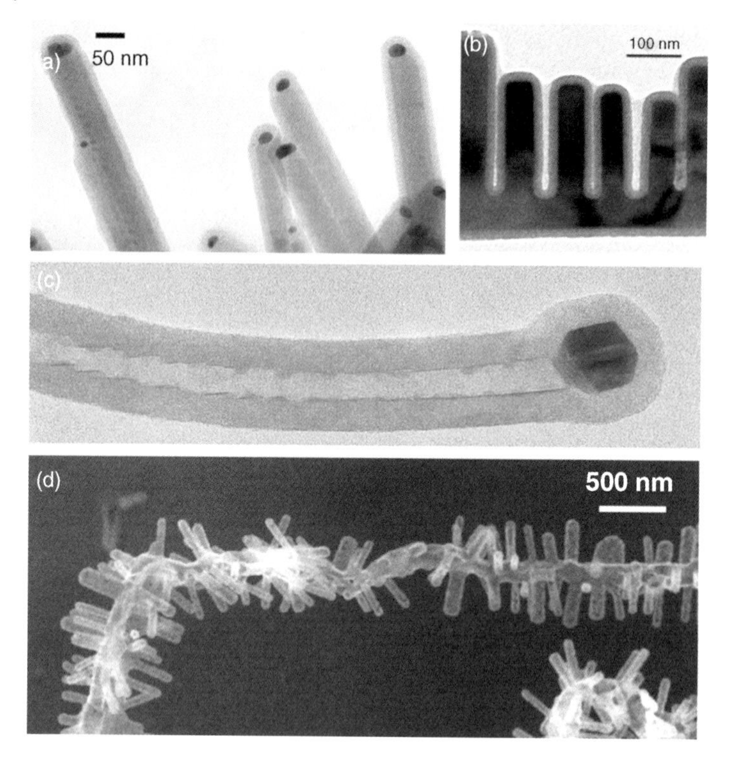

Figure 11.9 TEM images of (a) ALD Al_2O_3-coated ZnO nanorods and (b) comb-like ZnO structures. TEM images of (c) $ZnAl_2O_4$ nanotubes and (d) Chinese firecracker-like hollow $ZnAl_2O_4$ structures obtained after the temperature-induced reaction of the ZnO–Al_2O_3 core–shell nanowires. Reproduced with permission from Refs [31, 32].

because of their directionality-induced shadowing and poor step coverage. Same argument applies to other NW reactions in the following sections.

By using a slightly modified strategy, the Parsons group obtained a tube-in-tube structure consisting of electrospun polymer fibers that were consecutively ALD coated with three layers of Al_2O_3, ZnO, and Al_2O_3 [34]. Upon calcinations, the middle layer of ZnO diffuses both inward and outward (i.e., a bidirectional nanoscale Kirkendall effect) and reacts with Al_2O_3, resulting in ultralong double-tube hollow structures. Obviously, the layer uniformity provided by ALD is essential for the formation of such tube-in-tube nanostructures. Sputtering, for example, or CVD on the polymer fiber mesh would not be able to produce uniform nanotubes.

11.7.2
ZnO–TiO_2

The solid-state reaction of ZnO–TiO_2 core–shell NWs [35] is very different from that for the ZnO–Al_2O_3 system. The results are shown in Figure 11.10. When a thin (~5 nm) layer of ALD TiO_2 was deposited, the ZnO–TiO_2 core–shell NWs trans-

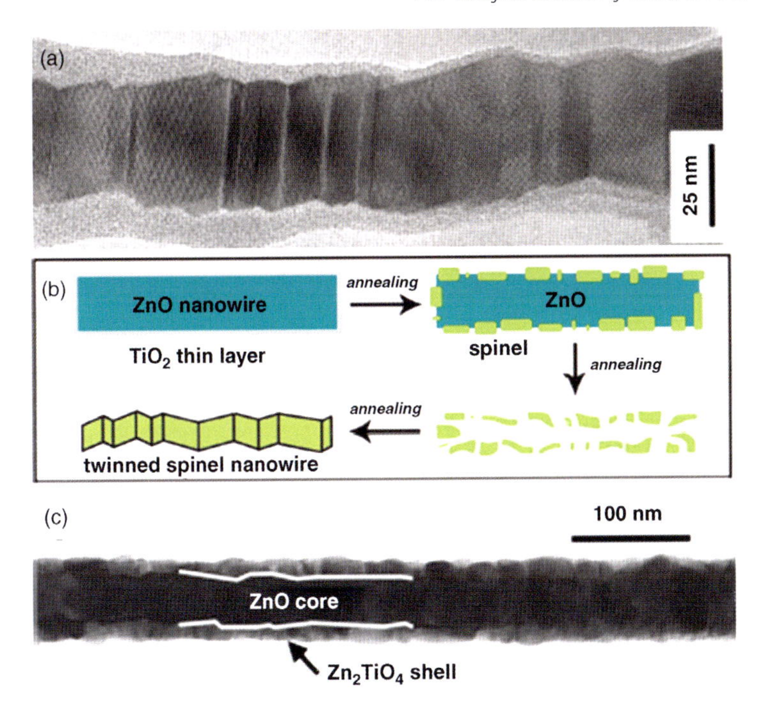

Figure 11.10 (a) TEM image of zigzag Zn_2TiO_4 multitwinned nanowires by solid-state reaction of ZnO–TiO_2 core–shell nanowires at 900 °C for 8 h. The ALD Al_2O_3 is 5 nm thick. (b) Schematic of the corresponding growth process for multitwinned Zn_2TiO_4 nanowires from ZnO–TiO_2 core–5 nm shell nanowires by annealing at high temperatures. (c) TEM image of ZnO nanowires shelled by a 17 nm TiO_2 layer after annealing at 700 °C for 4 h. Reproduced with permission from Ref. [35].

formed into a zigzag structure after annealing at 900 °C for 4 h in an open furnace (see the TEM image in Figure 11.10a). The twinned NW is composed of large parallelogram-shaped subcrystallites; the growth direction of each individual grain is along $\langle 111 \rangle$. No misfit dislocations were observed at the interface.

Formation of the multitwins depends on the thickness of the ALD TiO_2 layer. In the above experiment, the TiO_2 is a skinny layer relative to the diameter of the ZnO NW core (60–200 nm). When the ALD TiO_2 shell is thick enough (~17 nm), the resulting structure after solid-state reaction is a continuous Zn_2TiO_4 shell composed of nanocrystallites of a much larger size. Figure 11.10c shows a typical TEM image of the sample after annealing at 700 °C for 4 h. A core–shell NW with a rugged interface is now observed; no intervals exist between the nanocrystallites. When the core–shell NWs with a 17 nm TiO_2 layer are annealed at higher temperatures of 800 or 900 °C, no multitwinned spinel NWs are observed.

It is proposed that the formation of twinned Zn_2TiO_4 NWs out of the ~5 nm ALD TiO_2 shell in our case involves multiple stages (illustrated in Figure 11.10b) [35]. First, the initially 5 nm amorphous ALD TiO_2 shell transforms into anatase TiO_2 islands during thermal annealing, which are isolated from each other as a result of crystallization-induced volume shrinkage. Second, the TiO_2 reacts with ZnO core into individual spinel crystallites through lattice rearrangement. These spinel bricks are attached to the surface of the remaining ZnO stem. Finally, the unreacted ZnO is decomposed and evaporated through the gaps of the bricks, leaving a chain of loose-interconnected bricks. The bricks then assemble into solid multitwinned NWs via an oriented attachment and coalescence.

11.7.3 ZnO–SiO_2

Reaction of ZnO–SiO_2 is expected to be similar to ZnO–Al_2O_3. One unique experiment is the reaction of ZnO–Au–SiO_2 NWs [36], where the outer SiO_2 shell was coated by plasma ALD and the thin sandwich Au layer was coated by sputtering. Figure 11.11a presents a TEM image of the product after the trilayer NWs were

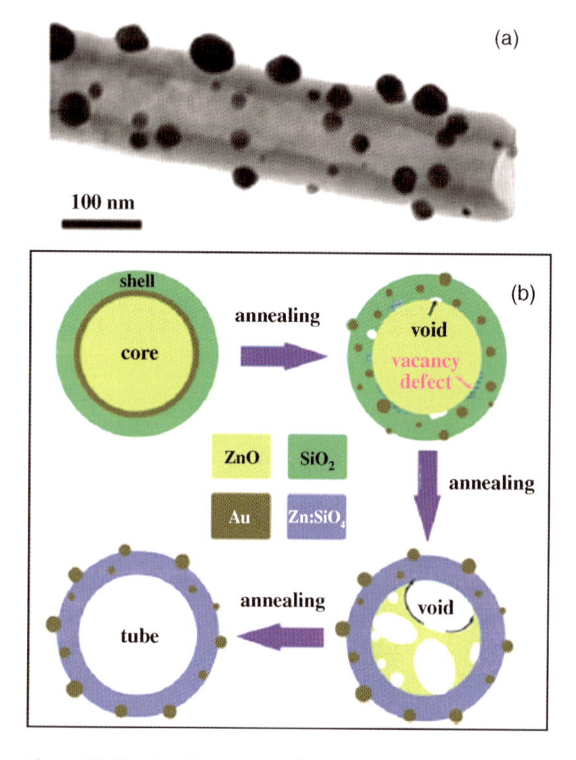

Figure 11.11 (a) TEM image of one Zn_2SiO_4 nanotube with Au nanocrystallites inlaided within the tube wall. The composite structure was obtained by annealing the ZnO–Au–SiO_2 trilayer nanowires at 900 °C for 3 h in air. (b) Schematic of the thermal-induced structural evolution of the ZnO–Au–SiO_2 trilayer nanowires into Zn_2SiO_4 nanotubes that are decorated with gold nanocrystallites. The ALD SiO_2 is 15 nm thick. Reproduced with permission from Ref. [36].

annealed at 900 °C for 3 h in air. The starting trilayer NWs transform into hollow or partially hollow Zn_2SiO_4 nanotubes decorated with ~50 nm gold particles. The formation process is illustrated in Figure 11.11b. The nanotubes are formed by one-way interfacial transfer of the ZnO core into the SiO_2 shell at high temperatures induced by the Kirkendall effect. The gold interlayer works as a gold precursor, which transforms into gold nanocrystallites and migrates to the outer surface of the SiO_2 shell during the annealing process. The key point here is that the motion of the gold interlayer facilitates the initial nucleation of the Kirkendall voids at the ZnO–SiO_2 interface and subsequently accelerates the formation of the Zn_2SiO_4 nanotubes. Such metal nanoparticle-decorated nanotubes might be useful for catalyst applications.

Similar experiments have also been conducted for the ZnO–Au–Al_2O_3 system (unpublished data). It is found that when the thickness of the ALD shell is increased from 10 to 20 nm, the aggregated Au particles can stay inside the tubes, rather than migrating toward the outer surface. This means that the outward migration of gold depends on the shell thickness. Another finding is that the sputtered metal particle prevents the formation of single crystalline tubes; the wall of the final $ZnAl_2O_4$ tubes is composed of spinel nanocrystallites. Such a striking difference from the above Zn_2SiO_4 is worth further investigation.

11.7.4 MgO–Al_2O_3

Spinel $MgAl_2O_4$ nanotubes have been obtained by a similar interface reaction between the ALD Al_2O_3 shell and the MgO core [37]. Figure 11.12a shows the fabrication process. Single-crystal MgO NWs were coated via ALD with a conformal layer of alumina. The interface solid-state reaction took place upon annealing the core–shell NWs at 700–800 °C under ambient conditions, during which cation pairs (Mg^{2+} and Al^{3+}, or Zn^{2+} and Fe^{3+}) diffuse in opposite directions while the oxygen sublattice is fixed. As the core NW has a larger diameter than necessary for a complete spinel reaction with the ALD shell, part of the core remains (Figure 11.12b) and can be etched in ammonia sulfuric solution. Spinel nanotubes of single crystalline walls can thus be obtained (Figure 11.12c). As MgO has an excellent lattice coherency with spinel $MgAl_2O_4$ and the ALD Al_2O_3 is uniform in thickness, a smooth interface will be formed between the spinel shell and the remaining MgO core, corresponding to a smooth inner wall of the eventual spinel nanotubes.

11.8 ALD ZnO on NWs/Tubes as Seed Layer for Growth of Hyperbranch

The ALD ZnO film can be used as seed layer for hydrothermal growth of ZnO NWs on various substrates. Using normal Si substrates, the optimum conditions for ALD ZnO seed layer are a deposition temperature of ≥95 °C and a thickness of about 10 nm. A conventional method for depositing the seed layer is dip coating or spin

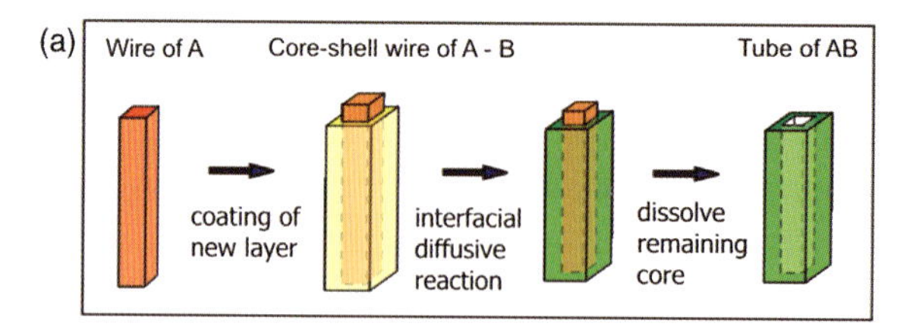

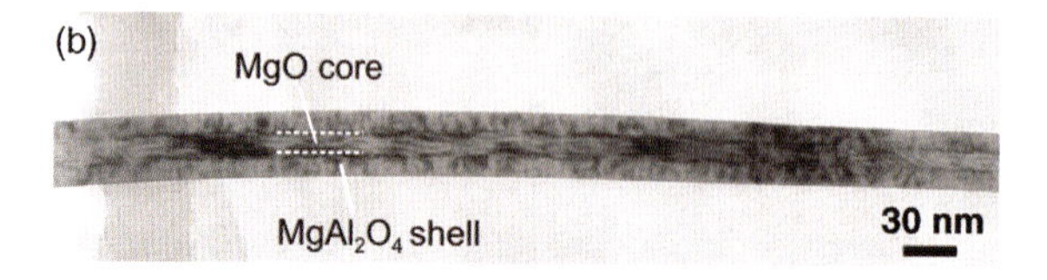

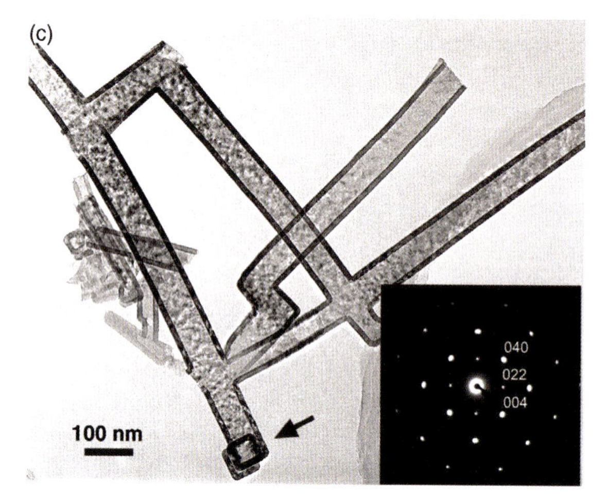

Figure 11.12 Formation of spinel $MgAl_2O_4$ nanotubes. (a) Schematic of the fabrication process involving Al_2O_3 ALD on MgO nanowires, interface reaction, and removal of remaining MgO. (b) TEM image of one MgO–$MgAl_2O_4$ core–shell nanowire. (c) TEM image of one branched tube of $MgAl_2O_4$. The arrow indicates the square cross section. Inset is the corresponding electron diffraction pattern revealing the single crystallinity. Reproduced with permission from Ref. [37].

coating of colloidal ZnO nanoparticles, followed by post-annealing at 300–500°C. This method works perfectly for planar and heat-resistant substrates such as Si and glass, but is not suitable for nonplanar 3D surfaces or polymers. In the following, examples of ALD ZnO seed layer are shown using CNT and Si nanopillar arrays as substrates [38, 39].

The schematic of the fabrication steps of the eventual CNT–ZnO 3D hybrid structure is shown in Figure 11.13a. The original MWCTNs or Si nanopillars are decorated with a shell of 80-cycle ZnO. The deposited ZnO shell is continuous and uniform along the tube. The ALD ZnO-coated substrate was immersed into a 35 ml aqueous solution of equimolar zinc nitrate $Zn(NO_3)_2{\cdot}6H_2O$ and hexamethylenetet-

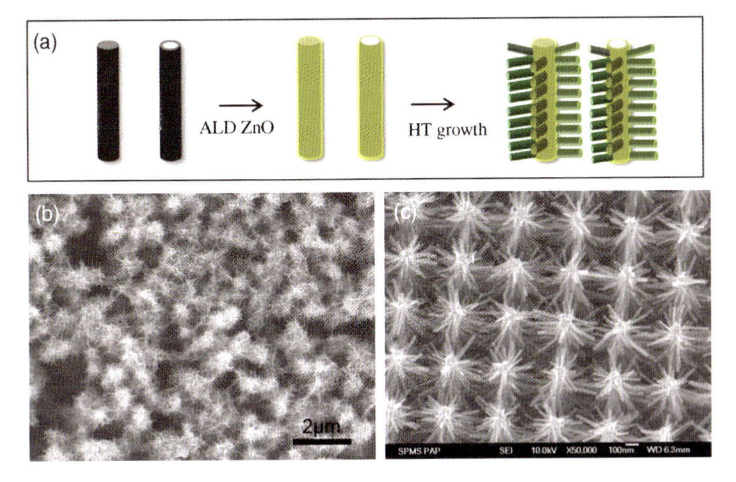

Figure 11.13 Fabrication of hyperbranched ZnO nanostructures based on hydrothermal growth using ALD ZnO seed layers. (a) Schematic of the fabrication process. The starting 1D backbone can be carbon nanotubes or any solid nanowires (e.g., Si, SnO_2). SEM images of two examples of the ZnO hyperbranches based on CNT backbones (b) and an ordered array of Si nanopillars (c). Reproduced with permission from Refs [38, 39].

ramine ($C_6H_{12}N_4$) in an autoclave. The reaction was conducted at 95 °C for 5 h. The resulting structures after the hydrothermal growth are 3D CNT–ZnO [39] or Si–ZnO hybrid nanotrees [38], with the overall alignment preserved (see Figure 11.13b and c). The densely packed ZnO nanorods are aligned roughly perpendicularly to the axis of the CNTs or Si pillars. Using the same technique, we have also achieved urchin-like ZnO nanostructures using monolayer of densely packed polystyrene spheres as template. Such ZnO-based 3D hybrids could be useful materials for energy-related applications.

11.9 Conclusions

When applied to inorganic nanowires and nanoparticles, ALD has proven useful for tuning the functional properties of the nanoobjects including magnetic properties, electronic transport, and optical absorption. The role of ALD in this case can be roughly categorized into three aspects: (i) It can provide a high-*k* dielectric surface passivation layer that works as a diffusion barrier for metal migration or charge transport. Similar dielectric wrapping has also been used as the gate dielectric for semiconductor nanowire transistor devices. (ii) For wide band gap semiconductor NWs, an ALD layer compensates the surface mobile charge traps and reduces the nonradiative exciton recombinations, thus enhancing the photoluminescence intensity. (iii) It allows new fabrication routes for various nanostructures, such as hollow nanoparticles, nanotubes, nanopeapods, core–shell nanowires, and nanowires of compounds, through interfacial solid-state reactions. In this application, the structure of the final product depends on several factors including the thickness and surface smoothness of the ALD layer.

On the other hand, different applications have different requirements for the ALD process. It is known that for substrates with deep trenches such as porous AAO and other framework structures, a closed chamber would be preferred that better facilitates pressure-induced infiltration. For most nanowires, particularly well-oriented arrays, this does not seem to be an issue and any ALD can deposit excellent coatings. However, for nanoparticles that are highly closely packed, the specially designed fluidized bed reactor or rotary reactor is regarded more efficient than conventional lamellar flow process in terms of 3D homogeneity.

References

1 Hakim, L.F., George, S.M., and Weimer, A.W. (2005) Conformal nanocoating of zirconia nanoparticles by atomic layer deposition in a fluidized bed reactor. *Nanotechnology*, **16**, S375–S381.

2 Hakim, L.F., King, D.M., Zhou, Y., Gump, C.J., George, S.M., and Weimer, A.W. (2007) Nanoparticle coating for advanced optical, mechanical and rheological properties. *Adv. Funct. Mater.*, **17**, 3175–3181.

3 Hakim, L.F., Vaughn, C.L., Dunsheath, H.J., Carney, C.S., Liang, X., Li, P., and Weimer, A.W. (2007) Synthesis of oxidation-resistant metal nanoparticles via atomic layer deposition. *Nanotechnology*, **18**, 345603.

4 Jung, Y.S., Cavanagh, A.S., Dillon, A.C., Groner, M.D., George, S.M., and Lee, S.H. (2010) Enhanced stability of $LiCoO_2$ cathodes in lithium-ion batteries using surface modification by atomic layer deposition. *J. Electrochem. Soc.*, **157**, A75–A81.

5 King, D.M., Zhou, Y., Hakim, L.F., Liang, X.H., Li, P., and Weimer, A.W. (2009) *In situ* synthesis of TiO_2-functionalized metal nanoparticles. *Ind. Eng. Chem. Res.*, **48**, 352–360.

6 Wank, J.R., George, S.M., and Weimer, A.W. (2004) Coating fine nickel particles with Al_2O_3 utilizing an atomic layer deposition-fluidized bed reactor (ALD-FBR). *J. Am. Ceram. Soc.*, **87**, 762–765.

7 Zhou, Y., King, D.M., Li, J., Barrett, K.S., Goldfarb, R.B., and Weimer, A.W. (2010) Synthesis of photoactive magnetic nanoparticles with atomic layer deposition. *Ind. Eng. Chem. Res.*, **49**, 6964–6971.

8 Ras, R.H.A., Kemell, M., de Wit, J., Ritala, M., ten Brinke, G., Leskelä, M., and Ikkala, O. (2007) Hollow inorganic nanospheres and nanotubes with tunable wall thicknesses by atomic layer deposition on self-assembled polymeric templates. *Adv. Mater.*, **19**, 102–106.

9 Graugnard, E., King, J.S., Gaillot, D.P., and Summers, C.J. (2006) Sacrificial-layer atomic layer deposition for fabrication of non-close-packed inverse-opal photonic crystals. *Adv. Funct. Mater.*, **16**, 1187–1196.

10 Wang, X.D., Graugnard, E., King, J.S., Wang, Z.L., and Summers, C.J. (2004) Large-scale fabrication of ordered nanobowl arrays. *Nano Lett.*, **4**, 2223–2226.

11 Fan, H.J., Werner, P., and Zacharias, M. (2006) Semiconductor nanowires: from self-organization to patterned growth. *Small*, **2**, 700–717.

12 Kuchibhatla, S., Karakoti, A.S., Bera, D., and Seal, S. (2007) One dimensional nanostructured materials. *Prog. Mater. Sci.*, **52**, 699–913.

13 Wang, N., Cai, Y., and Zhang, R.Q. (2008) Growth of nanowires. *Mater. Sci. Eng. R*, **60**, 1–51.

14 Solanki, R., Huo, J., Freeouf, J.L., and Miner, B. (2002) Atomic layer deposition of ZnSe/CdSe superlattice nanowires. *Appl. Phys. Lett.*, **81**, 3864–3866.

15 Yang, R.B., Bachmann, J., Pippel, E., Berger, A., Woltersdorf, J., Gosele, U., and Nielsch, K. (2009) Pulsed vapor–liquid–solid growth of antimony selenide and antimony sulfide nanowires. *Adv. Mater.*, **21**, 3170–3174.

16 Kipshidze, G., Yavich, B., Chandolu, A., Yun, J., Kuryatkov, V., Ahmad, I., Aurongzeb, D., Holtz, M., and Temkin, H. (2005) Controlled growth of GaN nanowires by pulsed metalorganic chemical vapor deposition. *Appl. Phys. Lett.*, **86**, 033104.

17 Knez, M., Niesch, K., and Niinisto, L. (2007) Synthesis and surface engineering of complex nanostructures by atomic layer deposition. *Adv. Mater.*, **19**, 3425–3438.

18 Yang, R.B., Zakharov, N., Moutanabbir, O., Scheerschmidt, K., Wu, L.-M., Gosele, U., Bachmann, J., and Nielsch, K. (2010) The transition between conformal atomic layer epitaxy and nanowire growth. *J. Am. Chem. Soc.*, **132**, 7592–7594.

19 Hui, K.C., Ong, H.C., Lee, P.F., and Dai, J.Y. (2005) Effects of AlO_x-cap layer on the luminescence and photoconductivity of ZnO thin films. *Appl. Phys. Lett.*, **86**.

20 Richters, J.P., Voss, T., Kim, D.S., Scholz, R., and Zacharias, M. (2008) Enhanced surface-excitonic emission in ZnO/Al_2O_3 core–shell nanowires. *Nanotechnology*, **19**, 305202.

21 Kang, M., Lee, J.S., Sim, S.K., Min, B., Cho, K., Kim, H., Sung, M.Y., Kim, S., Song, S.A., and Lee, M.S. (2004) Structural and optical properties of as-synthesized, Ga_2O_3-coated, and Al_2O_3-coated GaN nanowires. *Thin Solid Films*, **466**, 265–271.

22 Wang, C.C., Kei, C.C., Tao, Y., and Perng, T.P. (2009) Photoluminescence of GaQ_3–Al_2O_3 core–shell nanowires. *Electrochem. Solid-State Lett.*, **12**, K49–K52.

23 Keem, K., Jeong, D.Y., Kim, S., Lee, M.S., Yeo, I.S., Chung, U.I., and Moon, J.T. (2006) Fabrication and device characterization of omega-shaped-gate ZnO nanowire field-effect transistors. *Nano Lett.*, **6**, 1454–1458.

24 Law, M., Greene, L.E., Radenovic, A., Kuykendall, T., Liphardt, J., and Yang, P. (2006) ZnO–Al_2O_3 and ZnO–TiO_2 core–shell nanowire dye-sensitized solar cells. *J. Phys. Chem. B*, **110**, 22652–22663.

25 Hu, L. and Chen, G. (2007) Analysis of optical absorption in silicon nanowire arrays for photovoltaic applications. *Nano Lett.*, **7**, 3249–3252.

26 Tian, B., Zheng, X., Kempa, T.J., Fang, Y., Yu, N., Yu, G., Huang, J., and Lieber, C.M. (2007) Coaxial silicon nanowires as solar cells and nanoelectronic power sources. *Nature*, **449**, 885–889.

27 Qin, Y., Lee, S.M., Pan, A., Gosele, U., and Knez, M. (2008) Rayleigh-instability-induced metal nanoparticle chains encapsulated in nanotubes produced by atomic layer deposition. *Nano Lett.*, **8**, 114–118.

28 Yang, Y., Kim, D.S., Qin, Y., Berger, A., Scholz, R., Kim, H., Knez, M., and Gosele, U. (2009) Unexpected long-term instability of ZnO nanowires "protected" by a TiO_2 shell. *J. Am. Chem. Soc.*, **131**, 13920–13921.

29 Kim, D.S., Yang, Y., Kim, H., Berger, A., Knez, M., Gosele, U., and Schmidt, V. (2010) Formation of metal oxide nanotubes in neutral aqueous solution based on a photocatalytic effect. *Angew. Chem., Int. Ed.*, **49**, 210–212.

30 Fan, H.J., Gosele, U., and Zacharias, M. (2007) Formation of nanotubes and hollow nanoparticles based on Kirkendall and diffusion processes: a review. *Small*, **3**, 1660–1671.

31 Fan, H.J., Knez, M., Scholz, R., Nielsch, K., Pippel, E., Hesse, D., Zacharias, M., and Gosele, U. (2006) Monocrystalline spinel nanotube fabrication based on the Kirkendall effect. *Nat. Mater.*, **5**, 627–631.

32 Yang, Y., Kim, D.S., Scholz, R., Knez, M., Lee, S.M., Gosele, U., and Zacharias, M. (2008) Hierarchical three-dimensional ZnO and their shape-preserving transformation into hollow $ZnAl_2O_4$ nanostructures. *Chem. Mater.*, **20**, 3487–3494.

33 Yang, Y., Kim, D.S., Knez, M., Scholz, R., Berger, A., Pippel, E., Hesse, D., Gosele, U., and Zacharias, M. (2008) Influence of temperature on evolution of coaxial ZnO/

Al_2O_3 one-dimensional heterostructures: from core–shell nanowires to spinel nanotubes and porous nanowires. *J. Phys. Chem. C*, **112**, 4068–4074.

34 Peng, Q., Sun, X.Y., Spagnola, J.C., Saquing, C., Khan, S.A., Spontak, R.J., and Parsons, G.N. (2009) Bi-directional Kirkendall effect in coaxial microtuble nanolaminate assemblies fabricated by atomic layer deposition. *ACS Nano*, **3**, 546–554.

35 Yang, Y., Scholz, R., Fan, H.J., Hesse, D., Gosele, U., and Zacharias, M. (2009) Multitwinned spinel nanowires by assembly of nanobricks via oriented attachment: a case study of Zn_2TiO_4. *ACS Nano*, **3**, 555–562.

36 Yang, Y., Yang, R.B., Fan, H.J., Scholz, R., Huang, Z.P., Berger, A., Qin, Y., Knez, M., and Gosele, U. (2010) Diffusion-facilitated fabrication of gold-decorated Zn_2SiO_4 nanotubes by a one-step solid-state reaction. *Angew. Chem., Int. Ed.*, **49**, 1442–1446.

37 Fan, H.J., Knez, M., Scholz, R., Nielsch, K., Pippel, E., Hesse, D., Gosele, U., and Zacharias, M. (2006) Single-crystalline $MgAl_2O_4$ spinel nanotubes using a reactive and removable MgO nanowire template. *Nanotechnology*, **17**, 5157–5162.

38 Cheng, C., Yan, B., Wong, S.M., Li, X., Zhou, W., Yu, T., Shen, Z., Yu, H., and Fan, H.J. (2010) Fabrication and SERS performance of silver-nanoparticle-decorated Si/ZnO nanotrees in ordered arrays. *ACS Appl. Mater. Interfaces*, **2**, 1824–1828.

39 Li, X., Li, C., Zhang, Y., Chu, D., Milne, W., and Fan, H. (2010) Atomic layer deposition of ZnO on multi-walled carbon nanotubes and its use for synthesis of CNT–ZnO heterostructures. *Nanoscale Res. Lett.*, **5**, 1836–1840.

12
Atomic Layer Deposition on Soft Materials

Gregory N. Parsons

12.1
Introduction

Polymer materials find use in electronic fabrication and devices, biofunctional structures, energy generation and storage equipment, protective systems, textiles, and many more applications. Surface modification of polymers is typically desired and often necessary for most functional applications [1–6]. Strategies to modify natural polymer fibers and textiles can be traced to ancient times, and in the early nineteenth century, the textile industry developed mercerization and bleaching to modify natural polymer fibers, to improve color and luster, strength, and resistance to organic degradation [1, 6]. Currently, polymeric surface treatments find application in food packaging to improve encapsulation, in building and other engineering materials to improve surface adhesion, in medical devices and implants to produce biocidal finishes, to control surface wettability [3–5], and for a myriad of other practical uses. The fiber-forming nature and ability of polymers provides many opportunities for function and performance enhancement by surface coating and modification. Expanding applications for electrospun fibers [7–12] with diameter less than 200 nm is one example growing field. Most polymer surface modification is done using solution- or vapor-based methods, to coat the polymer with another organic or polymer layer, or to modify the surface termination. Sol–gel [6, 13] and other solution-based processes [14–17] are also performed to achieve inorganic coatings, including electroless deposition of conductive metals. Electroless copper plated onto a polypropylene (PP) fiber matrix [18] appears uniform in electron microscopy imaging, but it is not clear if the metal plating reaction proceeded uniformly throughout the thickness of the fiber mat structure, or if the growth was limited to the topmost visible fibers. One general downside of solvent-based processes is that solvents very often swell or degrade the polymer substrate. Vapor techniques often include plasmas [2, 19–22], chemical vapor deposition [23], or physical vapor deposition [24–31]. Plasmas can also cause damage to polymers, and most vapor deposition techniques require high temperature, or cause damage by impinging energetic species.

Atomic Layer Deposition of Nanostructured Materials, First Edition. Edited by Nicola Pinna and Mato Knez.

Compared to the more widespread use of plasma and physical vapor deposition, vapor-phase atomic layer deposition (ALD) may not be as readily implemented for high throughputs and extreme low costs that many large-area polymer structures need, but it offers some significant advantages over other techniques. The ALD process is typically thermally activated, and several metal oxide and metal coating processes are known that proceed readily at temperatures compatible with most polymer substrates. The exothermic reaction drives strong covalent bonding with the substrate, which can substantially improve adhesion, for example. In some cases, growth can proceed at room temperature, even if conditions are slightly outside the ALD process temperature "window" that defines well-behaved self-limiting growth. Also, the self-limiting binary reaction sequence that is well known in ALD allows the surface coating or modification process to be conformal and uniform across a large surface area substrate. It is the combination of thermal activation and self-limiting surface reactivity that allows highly uniform surface modification of high surface area polymer structures, such as textile or nonwoven polymer fiber media. As will be shown below, one must be careful, however, because even under conditions where the ALD precursor and reactant are known to undergo a controlled self-limited reaction, the precursor interaction with the polymer substrate may not be self-limiting, resulting in a polymer/inorganic interface that depends strongly on the polymer and the metal-organic precursors and other detailed deposition conditions.

Although no comprehensive review is found to date, several previous studies of ALD on polymers have summarized some past work [32–36]. This chapter attempts to accumulate and examine most of the studies published so far regarding ALD processes on polymers, including natural and synthetic materials. A summary of the articles collected is presented in Table 12.1. In assembling this article list, I found that the history and expanse of work in the area of ALD on polymers is much broader than originally anticipated, and it is therefore likely that some key research reports were not uncovered and not included in the table. The table shows that the field of ALD on soft materials is very broad, with many different polymers and deposited materials explored. However, there is still significant room to better understand the role precursor and polymer bonding structure plays in the nature of the ALD process on polymer and organic surfaces. The text therefore is presented from the point of view of what the previous work can teach us about ALD film growth initiation on soft materials. Since most published work to date on film nucleation addresses Al_2O_3 ALD from trimethylaluminum (TMA) and water, the text relates mostly to what we know about this growth reaction on polymers. Figure 12.1 presents a schematic diagram for the idealized TMA/water ALD processes on a reactive polymer surface. In this case, the polymer is shown as a fractional cross section of a polymer fiber, where the fiber surface presents functional groups (in this case hydroxyl groups) that readily react with the TMA. The TMA/OH surface reaction self-limits on the fiber surface, forming an aluminum-methyl surface prepared for subsequent water reaction. Continued film growth proceeds during subsequent cycles. A transmission electron microscopy (TEM) image of a cellulose cotton fiber exposed to many TMA/water cycles is also shown, giving a good view of a fairly abrupt organic/inorganic interface (within the resolution of the image), consistent with the ALD schematic

Table 12.1 Collection of research articles studying ALD and vapor infiltration processes on synthetic and natural polymers and other organic materials.

Polymer or molecule	ALD materials or precursors	References
Block copolymers: polystyrene-*block*-poly(4-vinylpyridine) and polystyrene-*block*-poly(2-vinylpyridine)	Al_2O_3	[38, 39]
Cellulose paper, cotton	TiO_2, Al_2O_3, Ir, TiO_xN_y	[37, 40–47]
Chitin (butterfly, cicada wings)	Al_2O_3, TiO_2	[43, 48–50]
Collagen	Al_2O_3, TiO_2, ZnO	[51, 52]
Ethylene tetrafluoroethylene	Al_2O_3, TiO_2	[34]
Ferritin and apoferritin proteins	TiO_2	[53, 54]
Keratin (wool)	Al_2O_3	[55]
Liquid crystal polymer membranes	Al_2O_3	[56]
Parylene	Al_2O_3	[57]
Poly(ether ether ketone)	Al_2O_3, TiO_2	[34]
Polytetrafluoroethylene	Al_2O_3, TiO_2	[34]
Poly(1-methoxy-4-(2′-ethyl-hexyloxy)-2,5-phenylene vinylene), other OLEDs	Al_2O_3/ZrO_2	[58–60]
Poly(*p*-xylylene)	Pd	[61]
Poly(styrene–divinylbenzene)	Al_2O_3	[62, 63]
Poly(3-hexylthiophene), other OPVs and OTFTs	Al_2O_3	[64–68]
Polyacrylonitrile	SnO_2	[69]
Polyamide (nylon 6)	Al_2O_3	[70]
Polybutylene terephthalate	Al_2O_3, ZnO, TiO_2	[71]
Polycarbonate	TiO_2, ZrO_2, W	[72–74]
Polydimethylsiloxane	Al_2O_3	[75]
Polyethersulfone	Al_2O_3	[57]
Polyethylene	Al_2O_3, W, TiO_2	[41, 55, 73, 76–80]
Polyethylene naphthalate	Al_2O_3, W, ZnO, TiO_2, SiO_2	[41, 71, 81–86]
Polyimide (Kapton)	Al_2O_3	[81–84, 87, 88]
Polyethylene terephthalate (PET, polyester)	Al_2O_3, ZnO, ZnO:Ga	[41, 43, 84, 89, 90]
Polylactic acid, polylactide	Al_2O_3, ZnO, TiO_2	[41, 71]
Poly(methacrylic acid)	Al_2O_3, TiO_2	[64, 91]
Poly(methyl methacrylate)	Al_2O_3, TiO_2, W, Ir, Pt, Ru	[34, 64, 73, 88, 92–94]
Polypropylene	Al_2O_3, W, TiO_2	[40, 55, 70, 73, 80, 92]
Polystyrene	Al_2O_3, W, TiO_2, SiO_2	[73, 92, 95–98]
Polyvinyl acetate	ZnO	[99]
Polyvinyl alcohol	Al_2O_3, ZnO	[70, 100, 101]
Polyvinyl chloride	Al_2O_3, W	[73, 92]
Polyvinylpyrrolidone	Al_2O_3, TiO_2, Ir, Pt, Ru, ZrO_2	[102–105]
Polyvinyl phenol	Al_2O_3	[64]
Porphyrins	Al_2O_3, ZnO, TiO_2	[106]
Self-assembled organic monolayers and surface oligomers	Al_2O_3, ZnO, HfO_2, ZrO_2, Ru, Rh, TiN, TaN, TiO_2, Cu	[107–129]
Si–O–C low-*k* films, SiLK™	TiN, TaN, Al_2O_3, WCN, Cu	[130–134]
Spider silk	Al_2O_3, TiO_2, ZnO	[135]
SU-8 photoresist	Al_2O_3, TiO_2	[136, 137]
Tobacco mosaic virus	Al_2O_3, TiO_2	[53]
Tris(8-hydroxyquinoline)gallium	Al_2O_3	[138]

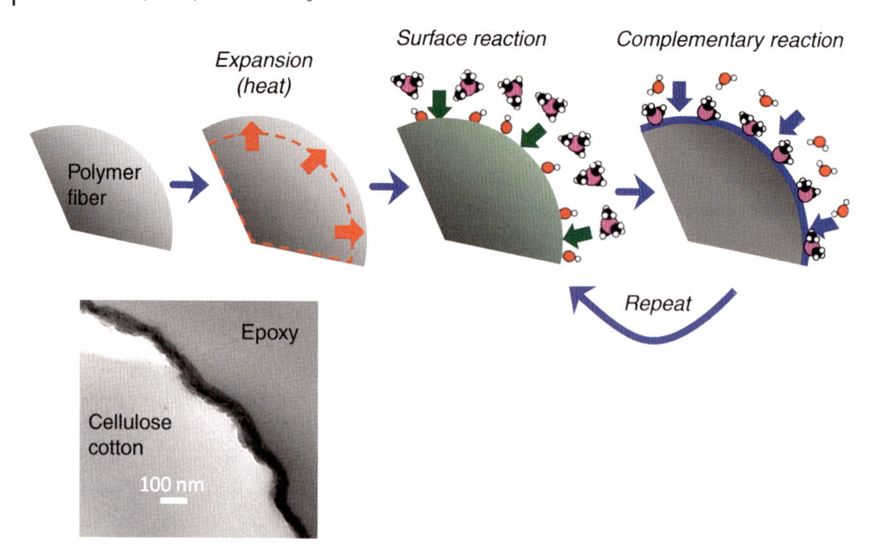

Figure 12.1 Idealized reaction scheme for atomic layer deposition of aluminum oxide from trimethylaluminum and water on a surface-functional polymer. In this example, the polymer surface displays OH groups (designated by the white/red circles) and these groups react readily with trimethylaluminum (designated by purple/black/white) to form surface aluminum-methyl groups. Reacting with water completes the ALD cycle, which is then repeated to form a well-defined overcoat on the polymer. The image is of an ALD coating on cellulose cotton polymer, showing a relatively abrupt polymer/oxide interface.

cartoon. Other images [37] show that ALD readily penetrates into the common convoluted tube structure of cotton fibers. Unfortunately, this well-defined growth is not typically observed for alumina deposition on polymers. A similar schematic diagram for TMA/water on a different polymer, now including subsurface nucleation, polymer swelling, and rough surface growth, is shown in Figure 12.2. A TEM image from a polypropylene fiber coated using TMA/water at 90 °C shows the nonuniform film structure developed. Toward the end of this chapter, we present data for TMA/water on three representative polymer materials: polypropylene, polyvinyl alcohol, and polyamide 6 (PA-6), and start to provide some possible guidelines to understand ALD reactions on various polymers. We hope that this collected overview will help push forward the fields of low-temperature ALD and inorganic/organic film integration, to enable researchers to make valuable advances for future use.

12.2
ALD on Polymers for Passivation, Encapsulation, and Surface Modification

Several groups have evaluated atomic layer deposition as an alternative low-temperature means to coat, passivate, and encapsulate polymer surfaces with inorganic thin films. While some of the earliest reports of ALD-based processes pointed to coating polymers (see reference 36 and citations therein), over the past 10 years or so the expansion of studies of ALD on polymers started with TiN, TaN, and WCN inorganic

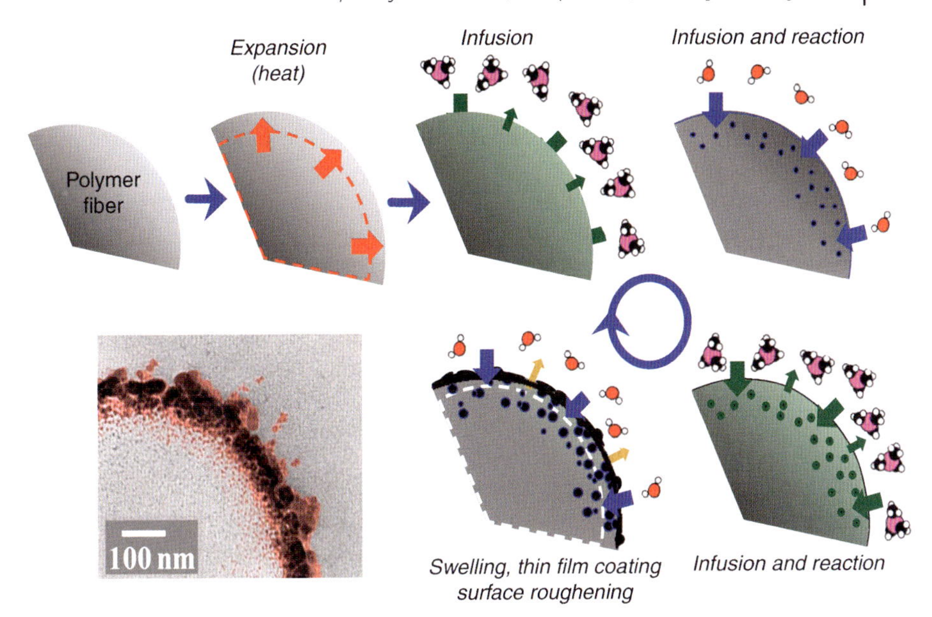

Figure 12.2 Likely reaction scheme for atomic layer deposition of aluminum oxide from trimethylaluminum and water on a surface-inert polymer, displaying no functional groups for reactions with ALD precursors. In this case, the polymer can absorb the metal-organic (TMA) and/or reactant (water) precursor resulting in subsurface reaction and nucleation. Subsequent ALD cycles result in further growth in subsurface particles, swelling of the substrate surface, and development of rough surface coverage. A TEM image shows a nonreactive polypropylene fiber after coating using TMA/water at 90 °C, where substantial subsurface nucleation and growth occur. A graded polymer/inorganic interface and rough surface texture are clearly observed.

diffusion barrier layers on carbon-containing low dielectric constant insulators [130–133]. The issue of TiN ALD layer nucleation was a primary concern, and it was immediately recognized that the chemically inert organic surface impeded the nucleation rate. It was also clear that the TiN ALD process on the untreated organic layer resulted in a graded organic/inorganic interface. Surface modification approaches including plasma exposure and aluminum oxide ALD coatings were identified as viable means to enhance nucleation. Exposing Si–O–C films to plasma produced a low carbon content silicon oxide layer that enabled abrupt oxide/TiN interfaces to form during ALD at 400 °C [130]. However, the conductivity was not as good as that for similar films deposited on oxidized silicon. Uniform TiN films were deposited on silicon low-*k* dielectrics when the dielectric was pretreated with ALD aluminum oxide [131]. However, the aluminum oxide ALD process still produced a graded organic/inorganic interface that could impede material performance. Precursor and/or water diffusion into the relatively nonreactive polymer was identified as a likely mechanism to account for the graded interface structure [131].

Work on ALD on polymers quickly expanded from metallic layers on low-*k* dielectrics to formation of dense oxides, predominantly Al_2O_3 films, as gas diffusion and protective layers on polymers. Some of the interesting polymers include polyimide (PI) [81–83, 87, 88], polyethylene terephthalate (PET) [89], polyethylene (PE) [76], polystyrene (PS) [92], polypropylene [40, 70, 77, 92], polyamide 6 [70], poly (methyl methacrylate) (PMMA) [34, 92, 93], polyvinyl chloride (PVC) [92], polyethylene naphthalate (PEN) [83], poly(ether ether ketone) (PEEK) [34], polytetrafluoroethylene (PTFE) [34], and ethylene tetrafluoroethylene (ETFE) [34]. These materials are interesting for ALD studies because they are widely used polymers, and they contain a variety of reactive and nonreactive groups for ALD nucleation, allowing researchers to explore and correlate nucleation mechanisms. The TMA and water process also allowed ALD less than 100 °C [57, 76, 92, 139]. Deposition of TiO_2 using $TiCl_4$ and water at 100 °C can also be performed on polymers [34, 72, 136]. Information regarding surface reaction mechanisms is probed using infrared (IR) transmission, quartz crystal microgravimetry (QCM), surface wetting analysis, and a variety of other techniques. The combination of IR and QCM confirm, for example, that subsurface nucleation and growth occur on several polymer materials during aluminum oxide ALD from TMA and water [92]. Surface wetting experiments by Kemell *et al.* [34] showed contact angle transitions consistent with surface film growth. Crystalline TiO_2 deposited at 250 °C on PMMA showed good UV-induced wetting transitions. Not all of these polymers showed the same trends in film growth rate and mass uptake, and the details were not always readily understood, but in general, most results could be understood in terms of subsurface reactant diffusion and excess subsurface growth, where the extent of reaction depends upon the precursors and polymer under study.

Metallic tungsten can also be formed by ALD at temperatures compatible with polymer substrates, and Wilson *et al.* [73] investigated W ALD using WF_6 and pure Si_2H_6 at 80 °C onto several polymer films, including polyethylene, polyvinyl chloride, polystyrene, polycarbonate, polypropylene, and PMMA. For the W ALD directly onto the polymers, many ALD cycles (>60) were needed to initiate the growth. By starting with 5–10 cycles of TMA/water on the polymers, the W films could nucleate and grow with an initiation period of ~25 Si_2H_6/WF_6 cycles. The authors hypothesized that the Si_2H_6 and WF_6 could not readily adsorb into the polymers, but the subsurface growth of small Al_2O_3 particles is sufficient to allow W ALD to begin, resulting in some subsurface growth with subsequent nuclei coalescence and film overgrowth, similar to an oxide coating.

A significant part of ALD process development on polymers addresses protective and oxygen barrier layers on plastic substrates for flexible electronic devices [57, 81, 82, 84–86, 90, 140, 141]. Park *et al.* [57] showed that 20–50 nm thick ALD Al_2O_3 on polyethersulfone or parylene polymers can significantly reduce the water vapor transmission rate (WVTR) through polymer sheets, although multiple layers, including plasma-deposited silicon nitride, were required to achieve performance approaching that of glass for functional OLED devices. Other work by Groner *et al.* on TMA/water ALD on polyethylene naphthalate and polyimide substrates [81] showed that films nucleated rapidly and film thickness increased linearly with time for deposition at 100–175 °C. The water vapor transmission rates for 5 nm thick Al_2O_3 films on PEN and polyimide were improved by a factor of 10 compared to the uncoated polymer, and alumina films 26 nm thick showed transmission performance better than plasma-deposited SiO_2 films that were >100 nm thick. In another study [82], Dameron *et al.* found that two bilayers of Al_2O_3/SiO_2 on the same polymers further improved barrier performance, but larger numbers of bilayers reduced performance, likely due to cracking. Similarly, Langereis *et al.* [86] found improvements in WVTR for alumina films deposited on PEN by plasma-enhanced ALD using TMA and a remote O_2 plasma. They also note improved performance at lower deposition temperatures, which they ascribe to an improved (i.e., more abrupt) interface between the polymer and inorganic using low-temperature plasma ALD. Aluminum oxide films ~35 Å thick formed by ALD onto polyimide foil also provide a good protective layer for high-energy oxygen atom erosion, as would be found in low Earth orbit [87, 88]. The thin alumina film performed better than a thicker (5 nm) silicon oxide film in oxygen atom erosion tests.

An interesting advantage of ALD is that in some cases it can be used to form nanoscale protective and encapsulating layers directly onto functional electronic devices, including organic thin film transistors (OTFTs) [64], organic photovoltaics (OPVs) [65–67], and organic light emitting diodes (OLEDs) [57–60]. Ferrari *et al.* [64] show that direct TMA/water exposure on poly(3-hexylthiophene) (P3HT) organic thin film transistors results in precursor penetration and oxide growth throughout the ~30 nm thick semiconductor layer, and a significant increase in surface roughness. Other studies of water-based ALD of Al_2O_3/HfO_2 encapsulation layers onto P3HT/[6,6]-phenyl-C_{61}-butyric acid methyl ester (PCBM) heterojunction solar cells also showed that the oxide films did not nucleate well on the organic [65]. In the OTFT study, Ferrari *et al.* hypothesized that reactive centers polyvinyl phenol (PVP) and poly (methyl methacrylate) coated onto P3HT would act as good nucleation sites for Al_2O_3 ALD. After spin coating PVP and PMMA interlayers on top of the P3HT and depositing ~50 nm of Al_2O_3 ALD at 125 °C, they found that the PVP produced a much sharper polymer/Al_2O_3 interface compared to P3HT/PMMA/Al_2O_3 stacks. This suggests that the carbonyl and ether groups in the polyester backbone are less reactive than the hydroxyl groups present in the alcohol polymer. The polyester therefore shows more subsurface reactant penetration and discontinuous film growth. However, the PVP layer is found to negatively affect the performance of the transistors, so more work is needed to define viable processes for direct transistor encapsulation.

Other studies also explore encapsulation of organic photovoltaics and OLEDs using ALD. Using TMA/water ALD at 100 °C on pentacene/C_{60} organic solar cells (capped with Al back contact) showed improved stability and performance compared to similar cells without the encapsulation layer [66]. Recently, Sarkar *et al.* [67] found that ALD Al_2O_3 using TMA/ozone performed much better than TMA/water for encapsulating and stabilizing bulk heterojunction organic PV devices under ambient exposure. Ghosh *et al.* [58] used ALD TMA/ozone with a top coating of parylene to encapsulate and stabilize OLEDs fabricated using tris(8-hydroxyquinolinato)aluminum (Alq_3). Other OLED devices using proprietary organics with top Al cathodes also showed improved performance when coated with Al_2O_3 or Al_2O_3/ZrO_2 multilayers by ALD [59]. In this case, the authors suggest that the Al contact material helped protect the active semiconductor from attack by the TMA.

Another application in OLEDs involves oxide barrier layers for reliable patterning of OLED active layers. In this case, poly(1-methoxy-4-(2′-ethyl-hexyloxy)-2,5-phenylene vinylene) (MEH-PPV) is used as the active layer, and ALD Al_2O_3 is meant to coat and protect the polymer during subsequent lithographic patterning and etching. However, infrared analysis shows that the vinylene groups present in the film readily react with TMA, resulting in significant polymer degradation. Surface pretreatment with isopropyl alcohol provides hydroxyl nucleation sites for ALD initiation, minimizing damage to the underlying polymer. Using the pretreatment step, the authors showed that an ultrathin (~1 nm) oxide film was sufficient to prevent MEH-PPV damage during lithographic patterning [60].

Researchers exploring hybrid organic/inorganic electronic devices, where the electronically active layers include both organic and inorganic materials, are beginning to use ALD for inorganic/inorganic integration. Katsia *et al.* [68], for example, showed good p/n junction performance using an n-type ZnO layer by ALD directly onto a spin-deposited p-type P3HT film. ALD ZnO has also been used as a selective electron transport layer in an inverted P3HT/PCBM solar cell [142].

In addition to electronic and optoelectronic barriers, ALD layers on polymers are also explored for other barrier systems. Paper and board materials covered with polyethylene and polylactide are often used for packaging food and other hermitic sealing applications, and ALD coatings can help improve material performance [41, 78, 79]. The ALD coatings help reduce transport of oxygen, water vapor, and aromatic organic compounds, and the thin inorganic layers have advantages over other common fiber coating methods in terms of environmental impact and recyclability. Plasma and corona pretreatments before ALD can also be used to improve performance and ALD layer adhesion [79].

Surface energy modification and surface wetting control is another area of potential interest for ALD coatings on polymers [34, 42, 48, 75, 143]. Hyde *et al.* [42] found that hydrophobic polypropylene fibers could be made hydrophilic by coating with ALD Al_2O_3. The transition from hydrophobic to hydrophilic occurred abruptly as the number of ALD cycles increased. The abrupt transition was related to the expected wetting behavior of nonplanar surfaces upon surface modification [42]. The naturally hydrophobic polymer starts out in a nonwetting stable Cassie state, which is highly nonwetting or even superhydrophobic. Upon coating with a more hydrophilic material, such as Al_2O_3, the surface transitions to a nonwetting metastable Cassie state, where the surface chemistry is more hydrophilic but the rough texture inhibits surface wetting. Upon further coating, the surface transforms to a wetting Wenzel state, where the hydrophilic nature of the surface finally drives the surface to fully wet. They also showed that the transformation from nonwetting to wetting depended on the temperature at which the coating was deposited, even though the surface energy of the ALD coating itself was not strongly temperature dependent. It was proposed [42], and later observed by TEM analysis [40, 70], that surface roughening on polypropylene was strongly dependent on the temperature for Al_2O_3 ALD. Higher deposition temperature resulted in more subsurface film growth, resulting in surface swelling and roughening. More roughening on surfaces coated at higher temperature then energetically favors the metastable Cassie state and inhibits transition to the readily wetting stable Wenzel surface. The surface wetting of cotton fibers was also strongly affected by ALD coating, where the cotton surface abruptly changed from hydrophilic to hydrophobic after a small number of ALD Al_2O_3 cycles, followed by a rapid change back to hydrophilic after further coating [42]. The initial transition to hydrophobicity is likely due to consumption of surface hydroxyl groups upon initial exposure to reactive metal-organics, or may involve a modification in nanoscale roughness visible in high-resolution TEM of native cotton [42]. Similar trends are observed after ALD coating of cotton/polyester blended fabrics, where the details of the transition depend on the macroscopic fiber alignment and density [43].

Several metal oxides are known to produce surface wetting transitions upon UV light exposure, and ALD coatings can work well to promote this effect. In addition to Kemell's results [34] mentioned above for UV wetting of crystalline TiO_2 formed at 250 °C on PMMA, Malm *et al.* [48] recently investigated photocontrolled wetting transitions using Al_2O_3/ZnO ALD onto natural nanotextured insect wings. They found that 20 cycles of ALD Al_2O_3 (to enable ZnO nucleation) followed by 10 cycles of ZnO (corresponding to a ~2 nm thick coating) did not produce any UV-induced wetting. However, 50 or 100 cycles of ZnO on Al_2O_3 showed good wetting/nonwetting transitions.

12.3
ALD for Bulk Modification of Natural and Synthetic Polymers and Molecules

Recent studies of ALD processes on natural polymers and small molecules, including porphyrins, collagen, and spider silk, have also led to interesting discoveries where ALD precursors can infiltrate [51, 52, 106, 135] and react to alter the material's optical [51, 106], chemical, and/or mechanical [51, 52, 135] properties. Lee *et al.* [51] exposed collagenous avian eggshell membranes to ZnO and TiO_2 ALD from diethylzinc (DEZ) and titanium isopropoxide (TIP), respectively, at ~100 °C and produced thin film coatings on the collagen fiber membranes that maintained good mechanical flexibility. The authors also discovered that the ALD process improved the collagen strength and toughness. Scanning electron micrograph images showed a sample with the film coating partially removed, supplying evidence for a well-defined film/fiber interface. However, they also hypothesized that metal-organic precursors could diffuse subsurface and react and bind at functional sites on the native polymer chain. The authors followed this lead and used long duration vapor exposures, referred to as multiple pulse vapor infiltration (MPI), pulsed vapor infiltration (PVI), or sequential vapor infiltration (SVI), to enhance subsurface diffusion into natural polymers and molecules. In some examples, subsequent exposure to water allows the reaction to go to completion. These process produce a hybrid organic/metal oxide network polymer with properties very different from the starting organic polymer. Zhang *et al.* [106] infiltrated porphyrin molecules with TMA, DEZ, and TIP and found that the diethylzinc exposure converted the free base to a Zn metalloporphyrin, resulting in a well-defined change in the optical absorption bands. In addition to the first work on collagen, Lee *et al.* further studied in two separate reports infiltration into collagen [52] and spider silk [135] protein, and found a substantial improvement in mechanical properties upon exposure to Al, Zn, and Ti precursors. Interestingly, SEM images of the ZnO-coated collagen [51] show that vapor infiltration and subsurface reaction can proceed along with formation of a distinct and intact film overcoat on the substrate. Cross-sectional TEM images of polyimide fibers coated with ALD alumina (discussed below and shown in Figure 12.5) also confirm cohesive film formation and substantial subsurface reaction. In other work, our group recently explored infiltration of synthetic polyesters [71], including polybutylene terephthalate, by TMA and other reactants and saw evidence in infrared transmission for disruption of the polymer bonding structure and formation of a hybrid organic/inorganic network structure. Calcination of the hybrid polymer produced a pure and porous metal oxide, and the porosity was controlled by the detailed conditions of precursor and reactant exposure and calcination [71]. All these results point to many new opportunities to expand understanding of starting materials and precursor reactions yet to be investigated, and suggest exciting new opportunities for natural and synthetic polymer modification.

12.4 ALD for Polymer Sacrificial Templating: Membranes, Fibers, and Biological and Optical Structures

In addition to depositing on planar polymer films, several groups evaluated ALD on polymer nanostructured templates to form inorganic nanotubes, spheres, and other shaped constructs. There is also growing interest in modifying surfaces of 3D polymer membranes and other materials for improved filtration, chemical separations, and bioadhesion [44, 56, 62, 63, 77, 144]. By performing TiO_2 and ZrO_2 ALD using titanium isopropoxide or zirconium *tert*-butoxide with water at 140 °C on commercial nanoporous polycarbonate membranes, Shin *et al.* [74] found that thin oxide nanotubes could be formed with length of ~12 μm, diameter of 30–200 nm, and wall thickness less than 5 nm. While some surface roughness is evident in the TEM images, they showed that film thickness scaled with ALD growth cycles from less than 1 nm, without significant growth incubation. Therefore, using their growth conditions and process, the polycarbonate substrate appears to readily promote ALD TiO_2 and ZrO_2 nucleation and growth.

ALD polymer templating also enables optically active 3D structures. For example, King *et al.* [136] deposited conformal TiO_2 using $TiCl_4$ and water at 100 °C onto SU-8 photoresist patterned as a porous 3D photonic crystal lattice. Scanning electron microscope images at the 200 nm scale showed smooth surface structure of the TiO_2, suggesting good nucleation of the TiO_2 on developed SU-8, although the authors did not present detailed analysis of the film nucleation. Graugnard *et al.* fabricated and characterized similar photonic crystals starting with Al_2O_3 ALD on patterned SU-8 [137]. Several groups also report structure reproduction using ALD on polystyrene templates [95–98]. Karuturi *et al.* [98] found that increasing the reactant pressure and extending the exposure time during TiO_2 ALD using $TiCl_4$ and water improved reactant transport into the carboxyl-modified polystyrene particle lattice. Interestingly, Hatton *et al.* showed that SiO_2 precursors could infiltrate and coat a polystyrene lattice by doing repeated exposures to reactants at atmospheric pressure in open-air test tubes [97]. For all these methods, dissolving the polymer produced inorganic coatings that faithfully retained the 3D template shape, and allowed analysis of photonic crystal performance.

During 2006–2008, several reports appeared using ALD as a means to template synthetic and natural polymer nanostructures [33]. Knez *et al.* [53] showed one of the first reports of ALD on biological materials, producing nanoscopic templates of tobacco mosaic virus and ferritin proteins using Al_2O_3 and TiO_2 deposition from trimethylaluminum and titanium isopropoxide and water, respectively, at 35 °C. Further exploration of titania coatings on ferritin and apoferritin showed conformal coating on the outside and inside of apoferritin cavities [54]. Another interesting example of ALD on natural polymers is the formation of photonic crystal and photonic band gap structures using butterfly wings as templates for Al_2O_3 or TiO_2 ALD at ~100 °C [49, 50]. Butterfly wings are comprised of the natural polymer chitin, which is a derivative of glucose and contains a considerable density of reactive pendant groups, including hydroxyl and carbonyl units. Microscopy images of butterfly wings after ALD coating show extremely fine reproduction of ~20 nm diameter ribs [50] and 90 nm long chambers [49], consistent with good nucleation and film growth on the chitin surface. The measured film thickness was similar to that expected from controlled growth on planar surfaces, also indicative of good nucleation. In addition, one of the wings studied contained sealed inner chambers [49], and when cracks in the chitin were not present, the ALD coating was not able to penetrate through the thin polymer wall to access the inside cavities. This also suggests that the initial growth was primarily confined to the outer surface of the polymer, without visible subsurface precursor diffusion. Because of the composition of the chitin, it is not too surprising that good ALD nucleation observed on this material is similar to that observed on cellulose cotton, as discussed below.

In contrast to nanostructured chitin butterfly wings, chitin-based cicada wings show very different behavior for ALD film nucleation and growth. The chitin cicada wings have a natural nanopillared texture, but Ras *et al.* [145] found that the wing surface termination chemistry impedes nucleation of ZnO ALD, allowing selective-area ALD on parts of the wings with different termination. Aluminum oxide ALD using TMA and water did produce growth, and allowed for a seed layer that could promote ZnO deposition [48]. Ras *et al.* also note that the cicada wing pillars have a natural waxy coating that promotes growth selectivity and permits unique control over lateral film propagation.

Natural polymer fibers provide another important template for ALD. In 2005, Kemell *et al.* showed that ALD TiO_2 could replicate the structure of cellulosic paper [45]. After ALD using $Ti(OMe)_4$ and water between 150 and 250 °C, the fibers were calcined resulting in hollow anatase TiO_2 tubules. The tubules showed photocatalytic activity for photodeposition of silver nanoparticles from $AgNO_3$. In a related study on cellulose fibers [46], ALD aluminum oxide or TiO_2 also provided a good barrier and nucleation layer for depositing catalytically active iridium from Ir(acac) and O_2 at 250 °C. Niskanen *et al.* examined alumina ALD using TMA and plasma-generated oxygen radicals on polyethylene, polypropylene, and keratin wool [55]. They found good replication of the wool structure by the deposited alumina. Even though the authors found that wool would combust readily upon exposure to oxygen radicals, the fact that it withstands the plasma ALD process suggests that the alumina layer begins to protect the fiber within the first ALD cycles.

Hyde *et al.* [37] also investigated ALD Al_2O_3 on cotton fibers and found that the growth rate as determined by transmission electron microscopy was somewhat larger than that on planar silicon wafers measured from the same runs. Water absorption into the fibers was one possible reason for the higher growth rate, although the film growth was confined to the fiber surface, creating good conformal films without significant subsurface nucleation and growth [37, 40]. By heating woven cotton samples in air after ALD, Hojo found that the woven structure was well preserved and replicated by the Al_2O_3 coating [47]. A mass analysis of the coatings after removal of the cotton substrate helped confirm linear growth, even when multiple layers of cotton fabric were stacked in the ALD reactor. Cellulose cotton fibers were also shown to be more biocompatible and bioadhesive after coating with ALD TiO_xN_y [44].

Groups are also studying ALD to modify surfaces for improved bioadhesion and to achieve more selective and efficient chemical separations [56, 62, 63, 77, 144, 146]. For example, Liang *et al.* [56] recently showed that for ALD using TMA and water, the effective pore size in a liquid crystal polymer membrane could be reduced from 0.75 to ~0.55 nm, producing a significant improvement in selectivity for H_2/N_2 separation. Size selectivity is a critical feature in membrane and filtration separation systems, and the facility of ALD to control film thickness and conformality, with simultaneous capacity to adjust surface functionalization and chemical termination, will likely have significant impact in these fields moving forward.

Block copolymers also offer unique nanoscale morphologies for templating ALD films and coatings. Ras *et al.* [38] examined ALD Al_2O_3 on polystyrene-*block*-poly(4-vinylpyridine) (PS-*b*-P4VP) spheres and spun nanorods. In this work, many of the resulting coated nanorods appeared with a diffuse, low-density Al_2O_3 coating, which the authors ascribed to water penetration and retention in the hygroscopic P4VP polymer block. The water uptake then led to reaction with TMA during the subsequent pulse producing a graded composition on the nanostructure surface. Wang *et al.* [39] found that a polystyrene-*block*-poly(2-vinylpyridine) (PS-*b*-P2VP) copolymer geometrically constrained in anodic aluminum oxide (AAO) nanopores produces block copolymer nanorods containing patterned nanoscopic domains. Selectively swelling the minority P2VP domains using ethanol followed by solvent evaporation produced mesoporous rods consisting of the more solid PS matrix, maintaining the initial morphological shape of the constrained copolymer. Exposing the polymer nanorods to 200 ALD cycles of diethylzinc and water at 60 °C resulted in film coating on internal walls of the mesopores, as well as on the outer surface of the nanorods. Transmission electron microscope images showed good replication of the structure by the ALD coating. Energy-dispersive X-ray mapping of the Zn concentration may indicate some precursor penetration into the PS core. The authors observed thinner layers on the inner sides of the mesopores, indicating that the growth on the outer surface impeded precursor transport into the internal surfaces. The work on ALD on block copolymers, along with work on ALD on tobacco mosaic virus [53], shows the extraordinary ability for ALD to penetrate into nanometer-scale organic contours and profile the shape with well-defined thin film coatings.

Electrospun synthetic polymer nanofibers also provide an interesting template for ALD coatings. The size of electrospun nanofibers presents unprecedented surface area to mass ratio. Leskelä *et al.* [102] showed good conformal coverage of ALD Al_2O_3 on ~400 nm diameter electrospun polyvinylpyrrolidone fibers, producing well-defined nanotubes after polymer calcination in air at 500 °C. The resulting tube walls appear smooth on the outside and inside, suggesting good nucleation of the oxide film on the PVP polymer substrate. In addition to Al_2O_3, Kim *et al.* [103] showed that titanium oxide ALD at 70 °C from titanium isopropoxide and water on ~200 nm electrospun polyvinylpyrrolidone could produce anatase TiO_2 after calcination at 500 °C. Again, the tubules showed smooth conformal growth, indicative of good nucleation and growth on the polymer. Work reported by Santala *et al.* [104] showed how ALD titania coatings on electrospun PVP and PVP/$NiFe_2O_4$ nanocomposite fibers could produce inorganic magnetic and photocatalytic tubules after calcination. The magnetic properties of these materials allowed them to be readily collected from solution after photocatalytic performance. Peng *et al.* [100] quantified the ALD Al_2O_3 film growth rate on electrospun polyvinyl alcohol fibers at 45 °C and compared the growth rate on the polymer to that on planar silicon. The growth rate was the same on the PVA and silicon, except for a slight excess growth on PVA during the first ALD cycles. They showed good conformal growth and fiber structure replication, as well as smooth inner and outer tube surfaces after calcination in air [100, 101]. Similar alumina nanotubes were also demonstrated by Wang *et al.* [138] who used sub-100 nm tris(8-hydroxyquinoline) gallium (GaQ_3) nanowires as a template for deposition from TMA and water at 25 °C. Electrospun fibers also provide a good ALD substrate to produce high surface area oxide nanotubes for gas sensors. Recent reports [69, 99] show good sensor devices operating at high temperature using ALD ZnO and SnO_2 nanotubes formed using electrospun nanofiber templates. The high surface area of polymer nanofibers can also help with material characterization, for example, enabling direct analysis of the polymer/reactant interactions by *in situ* infrared transmission analysis [70]. Results from some of these analyses are presented below. Generally, many results confirm that nanostructured polymer materials can be good templates for ALD film growth, producing a wide variety of unique materials and shapes.

12.5 ALD Nucleation on Patterned and Planar SAMs and Surface Oligomers

Interactions between ALD precursors and polymers are also highly interesting for thin film patterning applications where the receptivity of polymer surfaces is designed to minimize nucleation, to allow, for example, for selective-area thin film deposition. Physically, patterns of polymer films can be used to impede or inhibit

ALD film nucleation on the substrate [94, 105, 145, 147–149], or allow liftoff of any resulting ALD coating. In other cases, the polymers are developed in the form of patterned self-assembled monolayers (SAMs), where the surface functionality impedes ALD nucleation [107–120]. Chemical patterns on SAMs or polymer films, formed through UV irradiation, can permit ALD on some parts of the polymer while not allowing growth on other surface regions. Sinha *et al.* [91] used this approach to integrate TiO_2 ALD into a bilayer resist scheme for semiconductor patterning.

There is also interest in preparing tailored organic films and SAMs with terminal groups designed to be chemically receptive to ALD [121–125, 143]. Seitz *et al.* [125] explored surface reactions during ALD of Cu to form a conducting top contact to a molecular monolayer for molecular electronic applications. Chen *et al.* [126] used ALD to form hafnium oxide coatings onto organic films embedded with redox-active small molecules to produce organic–inorganic bilayer structures that permitted electrochemical charge transfer from molecules to electrolyte, while controlling charge leakage to the underlying substrate [126]. Many other schemes involving ALD for organic/inorganic integration in electronic systems can also be imagined.

Dube *et al.* [121, 127] and Sharma *et al.* [124] used molecular beam techniques to explore surface reactions important for ALD on SAMs and branched dendritic oligomer interface layers. They analyzed tetrakis(dimethylamido)titanium, $Ti[N(CH_3)_2]_4$, adsorption on OH- and NH_2-terminated SAMs as well as pentakis(dimethylamido)tantalum adsorption on surface-branched polyglycidol and amino-modified polyglycidol films. Haran *et al.* [128] explored atomistic mechanisms for growth reactions and modeled them using *ab initio* methods. A review of the nucleation studies on interfacial organic layers was also recently presented by Hughes and Engstrom [129]. Results show that the organic termination impedes the initial rate of film growth compared to deposition on oxidized silicon, even when the organic is terminated with reactive amine or hydroxyl groups on straight-chained or branched molecules. The metal-organic precursors react more as temperature is increased from 160 to $\sim$200 °C, suggesting degradation of the organic layer at higher temperatures, leading to enhanced roughening. The precursors are generally able to penetrate into the organic layer, and the reaction between the Ta precursor and the hydroxyl group was more facile than for amine termination, resulting in a larger extent of dimethyl amine ligand exchange. Reaction models [128] indicate that precursor adsorption proceeds through dative bonding complexes, similar to those for other metal-organics, as well as through hydrogen-bonded complexes. Rapid surface roughening was also noted during initial growth on all the organics, relative to that on oxide, indicating more 3D growth on the organics.

Kobayashi *et al.* [143] also evaluated ALD reactions on SAMs and showed that the TMA/water reaction on methyl-terminated monolayers produced rough, uneven film growth. This is consistent with results discussed above and those from Chen *et al.* [112] who showed that very high density methyl-SAMs are needed to inhibit the metal oxide precursor from penetrating through the SAM and nucleating on the underlying substrate surface. Kobayashi also devised an alcohol treatment as a means to promote smooth Al_2O_3 ALD on hydrophobic polymers, where physisorbed alcohol presents hydroxyl groups that enable ready film nucleation [122]. In another study of

ALD on SAMs, Li *et al.* [123] used *in situ* infrared analysis to characterize surface reactions during TMA/D_2O exposures on a carboxylic acid-terminated monolayer on silicon. Their results suggest that the TMA will react with accessible COOH groups forming a complete layer that can help block subsequent TMA or oxygen from reaching the underlying substrate. A high-density monolayer is required in order to impede TMA diffusion into the layer.

Of course, organic layers can be readily deposited using vapor sources, so mechanisms to integrate vapor-phase organic monomers into ALD processes are also gaining interest. These molecular layer deposition (MLD) sequences are being explored by several groups to produce surface polymers as well as new hybrid organic/inorganic thin film materials.

For all of these material systems, a good understanding of the fundamental interactions between organic functional groups and inorganic precursors is essential to achieve new and innovative applications. Toward this end, Xu and Musgrave undertook a density functional theory study of Al_2O_3 ALD on self-assembled monolayers with various end terminations. Another study by Derk *et al.* [150] addresses reactions between TMA and a range of functional groups present in organic systems. The work shows that reaction between TMA and most organic functional groups, including amine, carbonyl, cyano, ether, and others, is generally exothermic to form Al–O or Al–N bonds. Reactions proceed by formation of a stable Lewis acid/base adduct, resulting in a kinetic barrier to the final product state. Often, the barrier between the stable adduct and the transition state is relatively small (i.e., less than 25–30 kcal/mol), so the reaction will proceed at low temperature (although it may proceed slowly). The barriers for reaction between TMA and a methyl amine or a methyl acetate group, for example, are in the range of 22–30 kcal/mol. Other functional groups, such as dimethyl ether, are predicted to have a much larger barrier for reaction (>70 kcal/mol). These reaction schemes are calculated for reaction between one functional group and one TMA molecule. However, TMA is known to be a strong Lewis acid catalyst, so TMA bound in an adduct state may enhance reactivity for a second TMA with the organic [151, 152]. These calculated reaction schemes provide guidelines to understand ALD processes on polymers and organic surfaces.

The work described above shows that ALD on some polymers can produce good nucleation and conformal growth, while similar processes on other polymers result in poor surface nucleation and a graded interface. Studies of ALD on organic monolayer support the view that the density and surface termination of the polymer as well as the precursor size, ligand structure, and reactivity with the polymer surface all contribute to the subsequent structure and nature of the film deposition. We can begin to understand the correlation between polymer structure and nucleation behavior by exploring specific mechanisms associated with film nucleation on various polymers with and without functional groups for film nucleation. As shown below, we find that polymers with functional groups considered amenable to film nucleation do not always permit smooth conformal film growth. The goal of these studies is to better understand the relationship between the polymer structure, ALD precursor, and detailed reaction conditions that control conformal film nucleation and growth.

12.6 Reactions during Al_2O_3 ALD on Representative Polymer Materials

In this section, we discuss and summarize recent reports of Al_2O_3 ALD on three polymer materials: polypropylene, polyvinyl alcohol, and polyamide 6. These materials were chosen because they are common materials that present three different bonding structures for ALD film nucleation and growth. The polypropylene does not contain reactive groups, whereas the PVA contains hydroxyl pendant groups off the polymer chain. The PA-6 contains carbonyl and amine groups within the polymer backbone. These different groups result in very different response to TMA/water exposure, and may provide a starting point to better understand the relations between polymer structure and ALD film growth habit on soft organic materials in general.

12.6.1 Al_2O_3 ALD on Polypropylene

Polypropylene is a common semicrystalline thermoplastic with nonpolar structure and highly nonreactive surface. It has a high chemical resistance to organic acids, alcohols, esters, ketones, inorganic acids, and alkalis, and it does not easily absorb water. The coefficient of linear expansion is larger than most common polymers, with a value of $\sim(0.6\text{–}1.1) \times 10^{-4}\,°C^{-1}$ between 60 and 90 °C. The crystalline melting temperature is 165–176 °C and the glass transition temperature is approximately −10 °C. The softening point is ~90 °C, which limits the material functional operating temperature, although materials were sufficiently stable to allow thin film coating at 120 °C.

As discussed above, several studies report results of ALD on PP [55, 73, 80, 92]. Wilson *et al.* [92] cast thin polypropylene films onto QCM crystals, and used QCM to analyze TMA/H_2O ALD onto the film surface. Results showed mass uptake consistent with TMA diffusing into the polymer bulk. Spagnola *et al.* [70] and Jur *et al.* [40] also investigated TMA/water reaction on polypropylene at 60 and 90 °C. Those studies utilized polypropylene films cast onto QCM crystals, as well as melt-blown polypropylene nonwoven fiber mats, where the fibers have a radius between 500 nm and 5 μm, and the overall mat thickness is ~0.3 mm. The QCM results showed that during the first few (5–10) ALD cycles, the mass uptake on polypropylene was independent of temperature, but as growth continued, significantly larger mass uptake was observed at 90 °C compared to that at 60 °C. The QCM trace data also showed evidence for TMA absorption during ALD. We expect some differences between the reports from different groups because polypropylene films formed in different labs likely contain different crystallinity fraction or different tacticity that will affect the reaction details.

Representative *in situ* FTIR results collected from melt-blown polypropylene fibers during TMA/water exposure are plotted in Figure 12.3. For the samples used here, the fibers have fairly large diameter (1–10 μm), and using BET surface analysis data, we estimate that the surface area of the fiber mat is ~50× enhanced over a planar film with the same rectangular dimensions. The IR beam passes directly through the fiber mat without any beam focusing, allowing a large surface area to be characterized. To begin the experiments, the fiber mat substrate was placed in the reactor and held

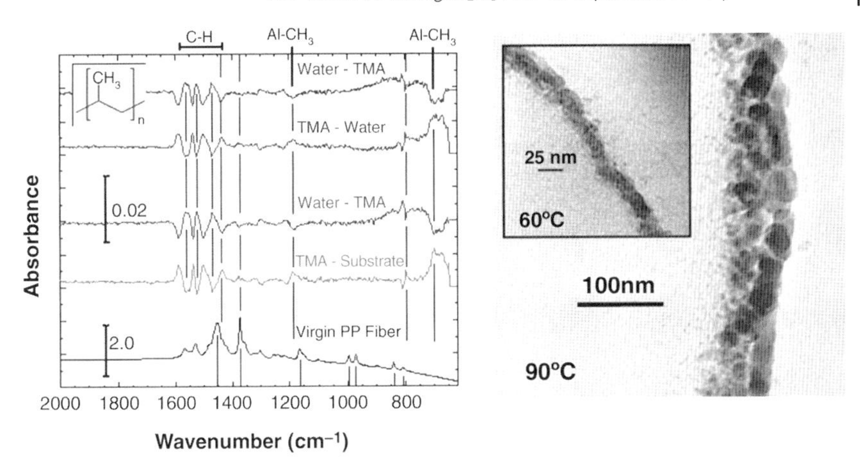

Figure 12.3 Differential infrared transmission data and TEM micrographs of polypropylene after coating using TMA/water at 60 and 90 °C. The IR data, collected at 90 °C, show changes consistent with diffusion of precursor reactants into the polymer without substantial reaction. The TEM images show subsurface growth at 90 °C, with more conformal coverage and a more abrupt organic/inorganic interface for deposition at 60 °C.

under vacuum overnight to remove any adsorbed water, and we then collected the transmission spectrum of the untreated fiber at 90 °C as a background reference. Spectra were collected again during inert gas flow after each TMA and water exposure, respectively, for several ALD cycles holding the reactor temperature fixed. The starting background spectrum is shown, along with the difference spectra collected after each exposure, using the previous collected spectrum as the background reference. After the first TMA pulse, the infrared transmission spectrum showed little change relative to the background. (The scale for the difference spectra collected upon precursor exposure is a factor of 100 larger than that for the background transmission spectrum.) A small increase in the CH_3 deformation peak ($\sim$1200 cm^{-1}) indicated the adsorption of small amount of TMA, with no obvious change of the IR peaks related to PP. A change in the modes between 1450 and 1600 cm^{-1} produces an oscillatory shape in the difference spectra. The oscillatory shape is consistent with a shift of CH_2 bending modes due to a local change in the polarization when the electrophilic TMA penetrates into the polymer network. The mode shifts could be due to charge transfer effects or due to deformation of the crystallite regions within the polymer. We note that upon the subsequent water dose, oscillatory features appear again, consistent with the CH_2 modes shifting back toward their original position. We ascribe this to TMA out-diffusion from the polymer bulk. Some TMA will remain in the polymer and react with water during the reactant pulse, producing subsurface film nucleation. Figure 12.3 also shows transmission electron micrographs of polypropylene after 100 cycles of TMA/water exposure at 90 and 60 °C. We see that ALD at 90 °C produces significant subsurface nucleation and growth. At lower deposition temperatures, the coating is more uniform and conformal, consistent with strong temperature dependence in the subsurface diffusion. *In situ* QCM analysis [40] also shows larger mass uptake at higher temperature, consistent with these TEM images. To examine the extent of TMA adsorption in polypropylene, a polypropylene film spun cast on silicon was characterized using X-ray photoelectron spectroscopy (XPS) before and after one long TMA exposure step and one long water exposure. Interestingly, the XPS data showed no aluminum after TMA/water soaking, and the C 1s peak detail scan showed almost no change in shape. This further confirms that the TMA reacts very little, if at all, with the PP polymer matrix.

12.6.2 Al_2O_3 ALD on Polyvinyl Alcohol

Polyvinyl alcohol is widely used for films and fibers, and has good adhesive properties. It is water soluble, but resistant to many solvents. The melting point depends on the amount of water absorbed, but generally ranges from 230 °C for fully hydrolyzed to ~180 °C for less hydrolyzed material. Its glass transition temperature is ~85 °C. The polymer is atactic (i.e., the hydroxyl groups are positioned randomly along the polymer chain) and it shows crystalline properties. Its density ranges from ~1.2 to 1.3 g/cm^3. For the ALD studies [100], we used electrospun PVA nanofibers (200–300 nm in diameter) generated from starting material with a molecular weight of 127 kDa and a degree of hydrolysis of 88%.

In situ FTIR spectra collected from electrospun PVA fibers are plotted in Figure 12.4. The IR setup allowed us to monitor the amount of water present in

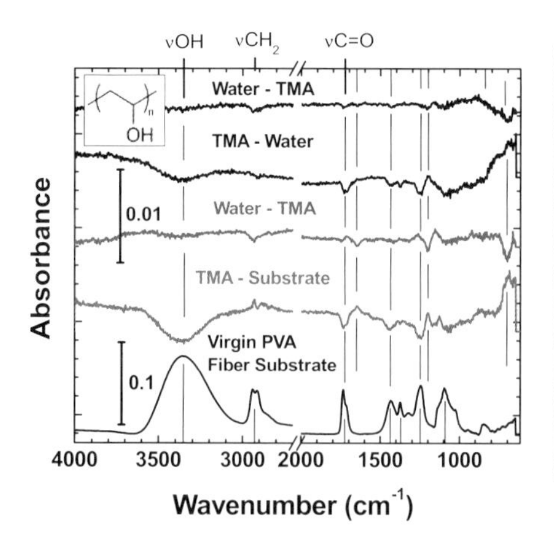

Figure 12.4 Infrared data and TEM images for Al_2O_3 ALD onto polyvinyl alcohol fibers at 45 °C. The PVA contains a large density of hydroxyl groups for ALD nucleation. Exposing the PVA to TMA results in a marked decrease in OH absorption and onset of modes for Al–CH_3 deformation, consistent with the ALD schematic shown in Figure 12.1. The TEM micrographs show smooth surface texture on the inside and outside of Al_2O_3 tubules formed after ALD and removing the PVA polymer, also consistent with an abrupt organic/inorganic interface for this system.

the PVA fiber matrix, and the concentration of hydroxyl groups decreases over time under vacuum exposure. In principle, the much smaller fiber diameter for the PVA versus PP fibers described above can produce a much larger surface area enhancement. However, the electrospun fiber mats are overall much thinner than the meltblown PP mats, so the surface area increase is also ~50× compared to a planar film. After ~24 h in vacuum, the hydroxyl concentration stabilizes, and the ALD experiments can begin. The first TMA dose showed a dramatic decrease of OH absorbance at 3700 cm^{-1} and increase of CH_3 deformation at 1200 cm^{-1}, indicating reaction between the TMA and the OH groups on the polymer chain. The second TMA and water doses show similar, but smaller changes in the IR modes. After six cycles of TMA/water, the IR data (not shown) are consistent with an Al_2O_3 layer on the PVA surface. After 100 ALD cycles, the coated PVA fibers were heated in air to 400 °C for 24 h to allow the polymer to be removed, and TEM images of the materials remaining are also shown in Figure 12.4. The images show well-shaped tubes with very smooth inner and outer walls. An inset shows another sample at higher magnification, prepared with fewer ALD cycles. The Al_2O_3 layer less than 10 nm thick is produced, with surface roughness less than ~1 nm. Other images collected after only a few ALD cycles (not included here) also show well-defined structures, but thinner layers do not have the mechanical rigidity to maintain the tubular shape shown in Figure 12.4. The IR and TEM data all indicate that ALD proceeds readily on PVA polymer, where there are a large number of reactive groups ready to promote Al_2O_3 growth initiation.

12.6.3
Al_2O_3 ALD on Polyamide 6

Polyamide 6, or nylon 6, is a variation of the well-known nylon 6,6. It is semicrystalline and is used for its mechanical strength, elasticity, and luster. It is widely available in fiber form in ropes, filaments, and nets. It has a melting point of ~220 °C and a glass transition temperature of ~50 °C. The density ranges from ~1.1 to 1.2 g/cm^3, and it readily absorbs water.

For ALD experiments [70], we spun coated a silicon wafer with a nylon 6 film and placed it in the reactor under vacuum, and collected the starting transmission spectrum. We then exposed the film to Al_2O_3 ALD cycles at 90 °C and collected differential absorbance spectra after each exposure step. Figure 12.5 displays the results. The absorbance of an untreated nylon 6 film shows a hydrogen-bonded N–H stretching mode at 3304 cm^{-1}, the C–H_2 asymmetric and symmetric stretching features at 2930 and 2860 cm^{-1}, respectively, and the amide I (C=O stretch) feature at 1640 cm^{-1}. Also visible are the amide II (N–H bend/C–N stretch) and amide III (N–C=O skeletal vibration) bands at 1541 and 1280 cm^{-1}, respectively. Exposing the nylon 6 film to TMA produced significant changes in the IR modes, even though the surface area of the film is much smaller compared to the fiber substrates shown above. Upon TMA exposure, the N–H feature at 3304 cm^{-1} and amide I feature at 1640 cm^{-1} decreased, and some change in the C–H stretching modes is also visible. The spectrum after TMA exposure also shows visible methyl rocking and deformation modes at 1437, 1190, and 690 cm^{-1}. Water exposure removed the Al–CH_3

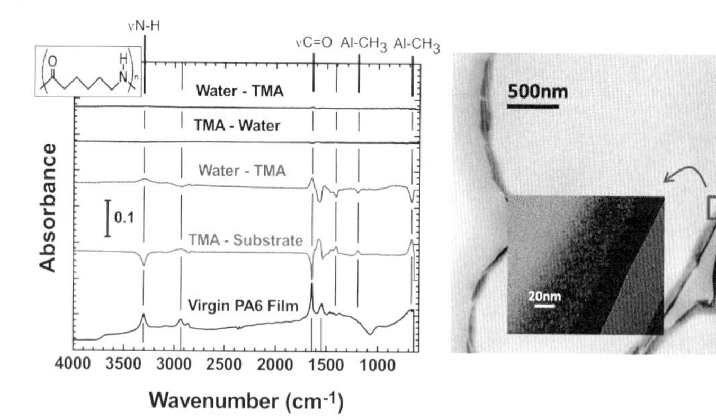

Figure 12.5 Similar infrared and TEM results for Al_2O_3 ALD onto polyamide nylon 6 at 90 °C. Even though the polyamide contains amine and carbonyl functional groups, substantial subsurface reaction is observed in both the IR and TEM data. A higher resolution TEM image in the inset shows that the ALD process results in conformal uniform film on the PA-6 fibers, even with the subsurface diffusion and nucleation.

features and produced an increase in Al−O stretching modes in the 700 cm^{-1} range. Interestingly, the N−H stretching mode appears to have increased after water exposure. Figure 12.5 also shows TEM images of nylon 6 fibers after exposure to 100 Al_2O_3 ALD cycles. A graded interface ~100 nm thick is clearly visible, and a higher resolution image from the same sample (in the figure inset) shows a ~15 nm layer on the diffuse interface. We see from this image that the ALD process can produce significant subsurface diffusion even when it creates a well-defined conformal thin film coating. QCM data also show that the initial TMA exposure on PA-6 results in significant mass uptake, more than 10× that for a similar exposure on PVA [70].

12.6.4 Mechanisms for Al_2O_3 ALD on PP, PVA, and PA-6

Polypropylene is relatively inert to reaction with trimethylaluminum. However, during the ALD exposure sequence, TMA does appear to readily diffuse into the near-surface region where it can get kinetically trapped and react with water during the second half-cycle step. The extent of reactant diffusion and subsurface growth can be controlled by temperature. We expect that other process conditions, such as the magnitude of the TMA and water exposure, the duration of the purge cycle times, and the purge gas flow velocity near the growth surface, will also affect the structure of the polymer/film interface.

The mechanisms for ALD nucleation are very different on polymers such as PVA and PA-6 with reactive groups present. Even for films with reactive groups, the mechanisms depend strongly on the polymer structure. Polyamide 6 has reactive carbonyl and amine groups in the polymer chain, and the IR data clearly show that the C=O and N−H stretching modes decrease upon TMA exposure. Similarly, the hydroxyl groups in PVA readily react with TMA. However, under the conditions studied, the reaction on PVA seems to be limited mostly to the polymer surface, whereas on PA-6 significant subsurface reaction is visible. Several mechanisms could account for this difference. One possibility is that the PVA may display a higher density of reactive groups on the surface, so that when the TMA reacts at the surface, it is able to form a relatively dense O–Al–O layer that effectively blocks subsurface TMA diffusion. A lower density of reactive groups on the PA-6 may therefore not react so readily to form a blocking layer, permitting more subsurface reaction. Another possibility has to do with reaction kinetics. *Ab initio* studies [150] indicate that TMA forms a stable Lewis acid/base adduct with hydroxyl, carbonyl, and amine groups, and a thermodynamic barrier is present between the adduct and the final product state. The barrier is relatively small for Al–O and methane formation from the dative-bonded TMA/hydroxyl unit, but it is larger for TMA bound to the carbonyl and amine groups, so a slower reaction rate is expected. A slower rate for TMA reaction with the carbonyl and amine groups on the surface could give the TMA more time to diffuse and get trapped below the surface. This mechanism, however, does not take into account the expected catalytic effect of the TMA. Organoaluminum compounds are known to accelerate reactions by binding reactants that can readily donate electron lone pairs [151, 152]. With many TMA molecules present, a dative-bonded TMA can increase the rate of insertion of a second TMA into a functional group. Perhaps if the density of functional groups on the PA-6 surface is sufficiently low, TMA can diffuse subsurface and bond to create active centers for reaction with other TMA molecules. It may also be important to note that the PA-6 is bound together to a large extent by hydrogen bonds between the carbonyl and amine groups. Since the IR data show reaction at the carbonyl and amine sites, it is reasonable to expect that the TMA will disrupt the hydrogen bonding environment. This may "open up" the polymer chain network, promoting further TMA diffusion into the subsurface region. In any case, it is clear that the chemical mechanisms associated with ALD on polymers are strongly influenced by the nature of the polymer and ALD precursor, and the broad knowledge and understanding of ALD nucleation on hard surfaces does not translate readily to ALD on soft materials and interfaces.

12.7 Summary

The benefits and capabilities of atomic layer deposition technology are becoming better known, and many researchers are using ALD to solve many new and interesting problems. New exploration will certainly push ALD coatings and surface modification onto polymers and biological structures that we currently do not consider. Expanding uses will require a more full and basic understanding of the chemical mechanisms associated with precursor and reactant species interactions with organic functional groups and polymer chains. This chapter highlights the tremendous amount of work that has already been done to show the applications, capabilities, and limits for ALD on polymers, and points out some of the problems, challenges, and accessible opportunities. The new application areas are open, and tools and capabilities to expand our fundamental knowledge in this field are readily available.

Acknowledgment

Time to write this chapter was made available through funding from the US National Science Foundation under grant CBET-1034374.

References

1 Adanur, S. (1995) *Wellington Sears Handbook of Industrial Textiles*, Technomic Publishing Co., Inc., Lancaster, PA.

2 Lewis, J.S. and Weaver, M.S. (2004) Thin-film permeation-barrier technology for flexible organic light-emitting devices. *IEEE J. Sel. Top. Quant. Electron.*, **10**, 45–57.

3 Hersel, U., Dahmen, C., and Kessler, H. (2003) RGD modified polymers: biomaterials for stimulated cell adhesion and beyond. *Biomaterials*, **24**, 4385–4415.

4 Gowri, S., Almeida, L., Amorim, T., Carneiro, N., Souto, A.P., and Esteves, M.F. (2010) Polymer nanocomposites for multifunctional finishing of textiles: a review. *Text. Res. J.*, **80**, 1290–1306.

5 Genzer, J. and Efimenko, K. (2000) Creating long-lived superhydrophobic polymer surfaces through mechanically assembled monolayers. *Science*, **290**, 2130–2133.

6 Mahltig, B., Haufe, H., and Bottcher, H. (2005) Functionalisation of textiles by inorganic sol–gel coatings. *J. Mater. Chem.*, **15**, 4385–4398.

7 Reneker, D.H., Yarin, A.L., Fong, H., and Koombhongse, S. (2000) Bending instability of electrically charged liquid jets of polymer solutions in electrospinning. *J. Appl. Phys.*, **87**, 4531–4547.

8 Theron, A., Zussman, E., and Yarin, A.L. (2001) Electrostatic field-assisted alignment of electrospun nanofibres. *Nanotechnology*, **12**, 384–390.

9 Burger, C., Hsiao, B.S., and Chu, B. (2006) Nanofibrous materials and their applications. *Annu. Rev. Mater. Res.*, **36**, 333–368.

10 Ma, Z.W., Kotaki, M., Inai, R., and Ramakrishna, S. (2005) Potential of nanofiber matrix as tissue-engineering scaffolds. *Tissue Eng.*, **11**, 101–109.

11 Ramakrishna, S., Fujihara, K., Teo, W.E., Yong, T., Ma, Z.W., and Ramaseshan, R. (2006) Electrospun nanofibers: solving global issues. *Mater. Today*, **9**, 40–50.

12 Ma, Z.W., Kotaki, M., Yong, T., He, W., and Ramakrishna, S. (2005) Surface engineering of electrospun polyethylene terephthalate (PET) nanofibers towards development of a new material for blood vessel engineering. *Biomaterials*, **26**, 2527–2536.

13 Xing, Y.J., Yang, X.J., and Dai, J.J. (2007) Antimicrobial finishing of cotton textile based on water glass by sol–gel method. *J. Sol-Gel Sci. Technol.*, **43**, 187–192.

14 Shimizu, K., Imai, H., Hirashima, H., and Tsukuma, K. (1999) Low-temperature synthesis of anatase thin films on glass and organic substrates by direct deposition from aqueous solutions. *Thin Solid Films*, **351**, 220–224.

15 Drew, C., Liu, X., Ziegler, D., Wang, X.Y., Bruno, F.F., Whitten, J., Samuelson, L.A., and Kumar, J. (2003) Metal oxide-coated polymer nanofibers. *Nano Lett.*, **3**, 143–147.

16 Charbonnier, M., Alami, M., and Romand, M. (1998) Electroless plating of polymers: XPS study of the initiation

mechanisms. *J. Appl. Electrochem.*, **28**, 449–453.

17 Wu, S.Y., Kang, E.T., Neoh, K.G., and Tan, K.L. (2000) Electroless deposition of copper on surface modified poly (tetrafluoroethylene) films from graft copolymerization and silanization. *Langmuir*, **16**, 5192–5198.

18 Stefecka, M., Kando, M., Matsuo, H., Nakashima, Y., Koyanagi, M., Kamiya, T., and Cernak, M. (2004) Electromagnetic shielding efficiency of plasma treated and electroless metal plated polypropylene nonwoven fabrics. *J. Mater. Sci.*, **39**, 2215–2217.

19 Chatham, H. (1996) Oxygen diffusion barrier properties of transparent oxide coatings on polymeric substrates. *Surf. Coat. Technol.*, **78**, 1–9.

20 Sobrinho, A.S.D., Latreche, M., Czeremuszkin, G., Klemberg-Sapieha, J.E., and Wertheimer, M.R. (1998) Transparent barrier coatings on polyethylene terephthalate by single- and dual-frequency plasma-enhanced chemical vapor deposition. *J. Vac. Sci. Technol. A*, **16**, 3190–3198.

21 Liston, E.M. (1989) Plasma treatment for improved bonding: a review. *J. Adhes.*, **30**, 199–218.

22 Inagaki, N., Tasaka, S., and Makino, M. (1997) Plasma polymer deposition from mixture of tetramethoxysilane and oxygen on PET films and their oxygen gas barrier properties. *J. Appl. Polym. Sci.*, **64**, 1031–1039.

23 Ma, M.L., Mao, Y., Gupta, M., Gleason, K.K., and Rutledge, G.C. (2005) Superhydrophobic fabrics produced by electrospinning and chemical vapor deposition. *Macromolecules*, **38**, 9742–9748.

24 Leterrier, Y. (2003) Durability of nanosized oxygen-barrier coatings on polymers. *Prog. Mater. Sci.*, **48**, 1–55.

25 Deng, C.S., Assender, H.E., Dinelli, F., Kolosov, O.V., Briggs, G.A.D., Miyamoto, T., and Tsukahara, Y. (2000) Nucleation and growth of gas barrier aluminium oxide on surfaces of poly(ethylene terephthalate) and polypropylene: effects of the polymer surface properties. *J. Polym. Sci. Polym. Phys.*, **38**, 3151–3162.

26 Rotger, J.C., Pireaux, J.J., Caudano, R., Thorne, N.A., Dunlop, H.M., and Benmalek, M. (1995) Deposition of silicon oxide onto polyethylene and polyethyleneterephthalate: an X-ray photoelectron spectroscopy interfacial study. *J. Vac. Sci. Technol. A*, **13**, 260–267.

27 Bichler, C., Kerbstadt, T., Langowski, H.C., and Moosheimer, U. (1997) The substrate-barrier film interface in thin barrier film coating. *Surf. Coat. Technol.*, **97**, 299–307.

28 Hooper, A., Fisher, G.L., Konstadinidis, K., Jung, D., Nguyen, H., Opila, R., Collins, R.W., Winograd, N., and Allara, D.L. (1999) Chemical effects of methyl and methyl ester groups on the nucleation and growth of vapor-deposited aluminum films. *J. Am. Chem. Soc.*, **121**, 8052–8064.

29 Cueff, R., Baud, G., Benmalek, M., Besse, J.P., Butruille, J.R., Dunlop, H.M., and Jacquet, M. (1995) Characterization and adhesion study of thin alumina coatings sputtered on PFT. *Thin Solid Films*, **270**, 230–236.

30 Bodino, F., Baud, G., Benmalek, M., Besse, J.P., Dunlop, H.M., and Jacquet, M. (1994) Alumina coating on polyethylene terephthalate. *Thin Solid Films*, **241**, 21–24.

31 Vallon, S., Hofrichter, A., Guyot, L., Drevillon, B., Klemberg-Sapieha, J.E., Martinu, L., and Poncin-Epaillard, F. (1996) Adhesion mechanisms of silica layers on plasma-treated polymers. 1. Polycarbonate. *J. Adhes. Sci. Technol.*, **10**, 1287–1311.

32 Trifonov, S.A., Lapikov, V.A., and Malygin, A.A. (2002) Reactivity of phenol–formaldehyde microspheres toward PCl_3, $VOCl_3$, and CrO_2Cl_2 vapors. *Russ. J. Appl. Chem.*, **75**, 969–973.

33 Knez, M., Niesch, K., and Niinisto, L. (2007) Synthesis and surface engineering of complex nanostructures by atomic layer deposition. *Adv. Mater.*, **19**, 3425–3438.

34 Kemell, M., Farm, E., Ritala, M., and Leskelä, M. (2008) Surface modification of thermoplastics by atomic layer deposition of Al_2O_3 and TiO_2 thin films. *Eur. Polym. J.*, **44**, 3564–3570.

35 Jur, J.S., Spagnola, J.C., Lee, K., Gong, B., Peng, Q., and Parsons, G.N. (2010) Temperature-dependent subsurface growth during atomic layer deposition on polypropylene and cellulose fibers. *Langmuir*, **26**, 8239–8244.

36 Malygin, A.A. (2006) The molecular layering nanotechnology: basis and application. *J. Ind. Eng. Chem.*, **12**, 1–11.

37 Hyde, G.K., Park, K.J., Stewart, S.M., Hinestroza, J.P., and Parsons, G.N. (2007) Atomic layer deposition of conformal inorganic nanoscale coatings on three-dimensional natural fiber systems: effect of surface topology on film growth characteristics. *Langmuir*, **23**, 9844–9849.

38 Ras, R.H.A., Kemell, M., de Wit, J., Ritala, M., ten Brinke, G., Leskelä, M., and Ikkala, O. (2007) Hollow inorganic nanospheres by nanotubes with tunable wall thickness by atomic layer deposition on self-assembled polymer templates. *Adv. Mater.*, **19**, 102–106.

39 Wang, Y., Qin, Y., Berger, A., Yau, E., He, C.C., Zhang, L.B., Gosele, U., Knez, M., and Steinhart, M. (2009) Nanoscopic morphologies in block copolymer nanorods as templates for atomic-layer deposition of semiconductors. *Adv. Mater.*, **21**, 2763.

40 Jur, J.S., Spagnola, J.C., Lee, K., Gong, B., Peng, Q., and Parsons, G.N. (2010) Temperature-dependent subsurface growth during atomic layer deposition on polypropylene and cellulose fibers. *Langmuir*, **26**, 8239–8244.

41 Hirvikorpi, T., Vaha-Nissi, M., Mustonen, T., Iiskola, E., and Karppinen, M. (2010) Atomic layer deposited aluminum oxide barrier coatings for packaging materials. *Thin Solid Films*, **518**, 2654–2658.

42 Hyde, G.K., Scarel, G., Spagnola, J.C., Peng, Q., Lee, K., Gong, B., Roberts, K.G., Roth, K.M., Hanson, C.A., Devine, C.K., Stewart, S.M., Hojo, D., Na, J.S., Jur, J.S., and Parsons, G.N. (2010). Atomic layer deposition and abrupt wetting transitions on nonwoven polypropylene and woven cotton fabrics. *Langmuir*, **26**, 2550–2558.

43 Roth, K.M., Roberts, K.G., and Hyde, G.K. (2010) Effect of weave geometry on surface energy modification of textile materials via atomic layer deposition. *Text. Res. J.*, **80**, 1970–1981.

44 Hyde, G.K., McCullen, S.D., Jeon, S., Stewart, S.M., Jeon, H., Loboa, E.G., and Parsons, G.N. (2009) Atomic layer deposition and biocompatibility of titanium nitride nano-coatings on cellulose fiber substrates. *Biomed. Mater.*, **4**, 10.

45 Kemell, M., Pore, V., Ritala, M., Leskelä, M., and Linden, M. (2005) Atomic layer deposition in nanometer-level replication of cellulosic substances and preparation of photocatalytic TiO_2/cellulose composites. *J. Am. Chem. Soc.*, **127**, 14178–14179.

46 Kemell, M., Pore, V., Ritala, M., and Leskelä, M. (2006) Ir/oxide/cellulose composites for catalytic purposes prepared by atomic layer deposition. *Chem. Vapor Depos.*, **12**, 419–422.

47 Hojo, D. and Adschiri, T. (2010) Mass analysis of growth of Al_2O_3 thin films from low-temperature atomic layer deposition on woven cotton. *Chem. Vapor Depos.*, **16**, 248–253.

48 Malm, J., Sahramo, E., Karppinen, M., and Ras, R.H.A. (2010) Photo-controlled wettability switching by conformal coating of nanoscale topographies with ultrathin oxide films. *Chem. Mater.*, **22**, 3349–3352.

49 Gaillot, D.P., Deparis, O., Welch, V., Wagner, B.K., Vigneron, J.P., and Summers, C.J. (2008) Composite organic–inorganic butterfly scales: production of photonic structures with atomic layer deposition. *Phys. Rev. E*, **78**, 6.

50 Huang, J.Y., Wang, X.D., and Wang, Z.L. (2006) Controlled replication of butterfly wings for achieving tunable photonic properties. *Nano Lett.*, **6**, 2325–2331.

51 Lee, S.M., Grass, G., Kim, G.M., Dresbach, C., Zhang, L.B., Gosele, U., and Knez, M. (2009) Low-temperature ZnO atomic layer deposition on biotemplates: flexible photocatalytic ZnO structures from eggshell membranes. *Phys. Chem. Chem. Phys.*, **11**, 3608–3614.

52 Lee, S.M., Pippel, E., Moutanabbir, O., Gunkel, I., Thurn-Albrecht, T., and Knez, M. (2010) Improved mechanical stability

of dried collagen membrane after metal infiltration. *ACS Appl. Mater. Interfaces*, **2**, 2436–2441.

53 Knez, M., Kadri, A., Wege, C., Gosele, U., Jeske, H., and Nielsch, K. (2006) Atomic layer deposition on biological macromolecules: metal oxide coating of tobacco mosaic virus and ferritin. *Nano Lett.*, **6**, 1172–1177.

54 Kim, H., Pippel, E., Gosele, U., and Knez, M. (2009) Titania nanostructures fabricated by atomic layer deposition using spherical protein cages. *Langmuir*, **25**, 13284–13289.

55 Niskanen, A., Arstila, K., Ritala, M., and Leskelä, M. (2005) Low-temperature deposition of aluminum oxide by radical enhanced atomic layer deposition. *J. Electrochem. Soc.*, **152**, F90–F93.

56 Liang, X.H., Lu, X.Y., Yu, M., Cavanagh, A.S., Gin, D.L., and Weimer, A.W. (2010) Modification of nanoporous supported lyotropic liquid crystal polymer membranes by atomic layer deposition. *J. Membr. Sci.*, **349**, 1–5.

57 Park, S.H.K., Oh, J., Hwang, C.S., Lee, J.I., Yang, Y.S., and Chu, H.Y. (2005) Ultrathin film encapsulation of an OLED by ALD. *Electrochem. Solid-State Lett.*, **8**, H21–H23.

58 Ghosh, A.P., Gerenser, L.J., Jarman, C.M., and Fornalik, J.E. (2005) Thin-film encapsulation of organic light-emitting devices. *Appl. Phys. Lett.*, **86**, 223503-1–223503-3.

59 Meyer, J., Schneidenbach, D., Winkler, T., Hamwi, S., Weimann, T., Hinze, P., Ammermann, S., Johannes, H.H., Riedl, T., and Kowalsky, W. (2009) Reliable thin film encapsulation for organic light emitting diodes grown by low-temperature atomic layer deposition. *Appl. Phys. Lett.*, **94**, 233305-1–233305-3.

60 Chang, C.Y., Tsai, F.Y., Jhuo, S.J., and Chen, M.J. (2008) Enhanced OLED performance upon photolithographic patterning by using an atomic-layer-deposited buffer layer. *Org. Electron.*, **9**, 667–672.

61 Ten Eyck, G.A., Pimanpang, S., Juneja, J.S., Bakhru, H., Lu, T.M., and Wang, G.C. (2007) Plasma-enhanced atomic layer deposition of palladium on a polymer substrate. *Chem. Vapor Depos.*, **13**, 307–311.

62 Liang, X.H., George, S.M., and Weimer, A.W. (2007) Synthesis of a novel porous polymer/ceramic composite material by low-temperature atomic layer deposition. *Chem. Mater.*, **19**, 5388–5394.

63 Liang, X.H., Lynn, A.D., King, D.M., Bryant, S.J., and Weimer, A.W. (2009) Biocompatible interface films deposited within porous polymers by atomic layer deposition (ALD). *ACS Appl. Mater. Interfaces*, **1**, 1988–1995.

64 Ferrari, S., Perissinotti, F., Peron, E., Fumagalli, L., Natali, D., and Sampietro, M. (2007) Atomic layer deposited Al_2O_3 as a capping layer for polymer based transistors. *Org. Electron.*, **8**, 407–414.

65 Chang, C.Y., Chou, C.T., Lee, Y.J., Chen, M.J., and Tsai, F.Y. (2009) Thin-film encapsulation of polymer-based bulk-heterojunction photovoltaic cells by atomic layer deposition. *Org. Electron.*, **10**, 1300–1306.

66 Potscavage, W.J., Yoo, S., Domercq, B., and Kippelen, B. (2007) Encapsulation of pentacene/C-60 organic solar cells with Al_2O_3 deposited by atomic layer deposition. *Appl. Phys. Lett.*, **90**, 3.

67 Sarkar, S., Culp, J.H., Whyland, J.T., Garvan, M., and Misra, V. (2010) Encapsulation of organic solar cells with ultrathin barrier layers deposited by ozone-based atomic layer deposition. *Org. Electron.*, **11**, 1896–1900.

68 Katsia, E., Huby, N., Tallarida, G., Kutrzeba-Kotowska, B., Perego, M., Ferrari, S., Krebs, F.C., Guziewicz, E., Godlewski, M., Osinniy, V., and Luka, G. (2009) Poly(3-hexylthiophene)/ZnO hybrid pn junctions for microelectronics applications. *Appl. Phys. Lett.*, **94**, 143501-1–143501-3.

69 Kim, W.S., Lee, B.S., Kim, D.H., Kim, H.C., Yu, W.R., and Hong, S.H. (2010) SnO_2 nanotubes fabricated using electrospinning and atomic layer deposition and their gas sensing performance. *Nanotechnology*, **21**, 7.

70 Spagnola, J.C., Gong, B., Arvidson, S.A., Jur, J.S., Khan, S.A., and Parsons, G.N. (2010) Surface and sub-surface reactions during low temperature aluminium oxide atomic layer deposition on fiber-

forming polymers. *J. Mater. Chem.*, **20**, 4213–4222.
71 Gong, B., Peng, Q., Jur, J.S., Devine, C.K., Lee, K., and Parsons, G.N. (2011) Sequential Vapor Infiltration of Metal Oxides into Sacrificial Polyester Fibers: Shape Replication and Controlled porosity of Microporous/Mesoporous Oxide Monoliths. *Chem. Mater.*, (in press) doi: 10.1021/cm200694w.
72 Triani, G., Campbell, J.A., Evans, P.J., Davis, J., Latella, B.A., and Burford, R.P. (2010) Low temperature atomic layer deposition of titania thin films. *Thin Solid Films*, **518**, 3182–3189.
73 Wilson, C.A., McCormick, J.A., Cavanagh, A.S., Goldstein, D.N., Weimer, A.W., and George, S.M. (2008) Tungsten atomic layer deposition on polymers. *Thin Solid Films*, **516**, 6175–6185.
74 Shin, H., Jeong, D.-K., Lee, J., Sung, M.M., and Kim, J. (2004) Formation of TiO_2 and ZrO_2 nanotubes using atomic layer deposition with ultraprecise control of the wall thickness. *Adv. Mater.*, **16**, 1197–1200.
75 Spagnola, J.C., Gong, B., and Parsons, G.N. (2010) Surface texture and wetting stability of polydimethylsiloxane coated with aluminum oxide at low temperature by atomic layer deposition. *J. Vac. Sci. Technol. A*, **28**, 1330–1337.
76 Ferguson, J.D., Weimer, A.W., and George, S.M. (2004) Atomic layer deposition of Al_2O_3 films on polyethylene particles. *Chem. Mater.*, **16**, 5602–5609.
77 Liang, X., King, D.M., Groner, M.D., Blackson, J.H., Harris, J.D., George, S.M., and Weimer, A.W. (2008) Barrier properties of polymer/alumina nanocomposite membranes fabricated by atomic layer deposition. *J. Membr. Sci.*, **322**, 105–112.
78 Hirvikorpi, T., Vaha-Nissi, M., Harlin, A., and Karppinen, M. (2010) Comparison of some coating techniques to fabricate barrier layers on packaging materials. *Thin Solid Films*, **518**, 5463–5466.
79 Hirvikorpi, T., Vaha-Nissi, M., Harlin, A., Marles, J., Miikkulainen, V., and Karppinen, M. (2010) Effect of corona pre-treatment on the performance of gas barrier layers applied by atomic layer deposition onto polymer-coated paperboard. *Appl. Surf. Sci.*, **257**, 736–740.
80 Niskanen, A., Arstila, K., Leskelä, M., and Ritala, M. (2007) Radical enhanced atomic layer deposition of titanium dioxide. *Chem. Vapor Depos.*, **13**, 152–157.
81 Groner, M.D., George, S.M., McLean, R.S., and Carcia, P.F. (2006) Gas diffusion barriers on polymers using Al_2O_3 atomic layer deposition. *Appl. Phys. Lett.*, **88**, 051907.
82 Dameron, A.A., Davidson, S.D., Burton, B.B., Carcia, P.F., McLean, R.S., and George, S.M. (2008) Gas diffusion barriers on polymers using multilayers fabricated by Al_2O_3 and rapid SiO_2 atomic layer deposition. *J. Phys. Chem. C*, **112**, 4573–4580.
83 Fabreguette, F.H. and George, S.M. (2007) X-ray mirrors on flexible polymer substrates fabricated by atomic layer deposition. *Thin Solid Films*, **515**, 7177–7180.
84 Carcia, P.F., McLean, R.S., Groner, M.D., Dameron, A.A., and George, S.M. (2009) Gas diffusion ultrabarriers on polymer substrates using Al_2O_3 atomic layer deposition and SiN plasma-enhanced chemical vapor deposition. *J. Appl. Phys.*, **106**, 6.
85 Carcia, P.F., McLean, R.S., Reilly, M.H., Groner, M.D., and George, S.M. (2006) Ca test of Al_2O_3 gas diffusion barriers grown by atomic layer deposition on polymers. *Appl. Phys. Lett.*, **89**, 3.
86 Langereis, E., Creatore, M., Heil, S.B.S., van de Sanden, M.C.M., and Kessels, W.M.M. (2006) Plasma-assisted atomic layer deposition of Al_2O_3 moisture permeation barriers on polymers. *Appl. Phys. Lett.*, **89**, 3.
87 Cooper, R., Upadhyaya, H.P., Minton, T.K., Berman, M.R., Du, X.H., and George, S.M. (2008) Protection of polymer from atomic-oxygen erosion using Al_2O_3 atomic layer deposition coatings. *Thin Solid Films*, **516**, 4036–4039.
88 Minton, T.K., Wu, B.H., Zhang, J.M., Lindholm, N.F., Abdulagatov, A.I., O'Patchen, J., George, S.M., and Groner, M.D. (2010) Protecting polymers in space

with atomic layer deposition coatings. *ACS Appl. Mater. Interfaces*, **2**, 2515–2520.

89 Ott, A.W. and Chang, R.P.H. (1999) Atomic layer-controlled growth of transparent conducting ZnO on plastic substrates. *Mater. Chem. Phys.*, **58**, 132–138.

90 Carcia, P.F., McLean, R.S., and Reilly, M.H. (2010) Permeation measurements and modeling of highly defective Al_2O_3 thin films grown by atomic layer deposition on polymers. *Appl. Phys. Lett.*, **97**, 3.

91 Sinha, A., Hess, D.W., and Henderson, C.L. (2006) A top surface imaging method using area selective ALD on chemically amplified polymer photoresist films. *Electrochem. Solid-State Lett.*, **9**, G330–G333.

92 Wilson, C.A., Grubbs, R.K., and George, S.M. (2005) Nucleation and growth during Al_2O_3 atomic layer deposition on polymers. *Chem. Mater.*, **17**, 5625–5634.

93 Kaariainen, T.O., Cameron, D.C., and Tanttari, M. (2009) Adhesion of Ti and TiC coatings on PMMA subject to plasma treatment: effect of intermediate layers of Al_2O_3 and TiO_2 deposited by atomic layer deposition. *Plasma Process. Polym.*, **6**, 631–641.

94 Farm, E., Kemell, M., Ritala, M., and Leskelä, M. (2008) Selective-area atomic layer deposition using poly(methyl methacrylate) films as mask. *J. Phys. Chem. C*, **112**, 15791–15795.

95 Xu, D.W., Graugnard, E., King, J.S., Zhong, L.W., and Summers, C.J. (2004) Large-scale fabrication of ordered nanobowl arrays. *Nano Lett.*, **4**, 2223–2226.

96 Graugnard, E., King, J.S., Jain, S., Summers, C.J., Zhang-Williams, Y., and Khoo, I.C. (2005) Electric-field tuning of the Bragg peak in large-pore TiO_2 inverse shell opals. *Phys. Rev. B*, **72**, 4.

97 Hatton, B., Kitaev, V., Perovic, D., Ozin, G., and Aizenberg, J. (2010) Low-temperature synthesis of nanoscale silica multilayers: atomic layer deposition in a test tube. *J. Mater. Chem.*, **20**, 6009–6013.

98 Karuturi, S.K., Liu, L.J., Su, L.T., Zhao, Y., Fan, H.J., Ge, X.C., He, S.L., and Yoong, A.T.I. (2010) Kinetics of stop-flow atomic layer deposition for high aspect ratio template filling through photonic band gap measurements. *J. Phys. Chem. C*, **114**, 14843–14848.

99 Park, J.Y., Choi, S.W., and Kim, S.S. (2010) A synthesis and sensing application of hollow ZnO nanofibers with uniform wall thicknesses grown using polymer templates. *Nanotechnology*, **21**, 9.

100 Peng, Q., Sun, X.Y., Spagnola, J.C., Hyde, G.K., Spontak, R.J., and Parsons, G.N. (2007) Atomic layer deposition on electrospun polymer fibers as a direct route to Al_2O_3 microtubes with precise wall thickness control. *Nano Lett.*, **7**, 719–722.

101 Peng, Q., Sun, X.-Y., Spagnola, J.C., Saquing, C., Khan, S.A., Spontak, R.J., and Parsons, G.N. (2009) Bi-directional Kirkendall effect in coaxial microtube nanolaminate assemblies fabricated by atomic layer deposition. *ACS Nano*, **3**, 546–554.

102 Leskelä, M., Kemell, M., Kukli, K., Pore, V., Santala, E., Ritala, M., and Lu, J. (2007) Exploitation of atomic layer deposition for nanostructured materials. *Mater. Sci. Eng. C*, **27**, 1504–1508.

103 Kim, G.M., Lee, S.M., Michler, G.H., Roggendorf, H., Gosele, U., and Knez, M. (2008) Nanostructured pure anatase titania tubes replicated from electrospun polymer fiber templates by atomic layer deposition. *Chem. Mater.*, **20**, 3085–3091.

104 Santala, E., Kemell, M., Leskelä, M., and Ritala, M. (2009) The preparation of reusable magnetic and photocatalytic composite nanofibers by electrospinning and atomic layer deposition. *Nanotechnology*, **20**, 5.

105 Farm, E., Kemell, M., Santala, E., Ritala, M., and Leskelä, M. (2010) Selective-area atomic layer deposition using poly(vinyl pyrrolidone) as a passivation layer. *J. Electrochem. Soc.*, **157**, K10–K14.

106 Zhang, L.B., Patil, A.J., Li, L., Schierhorn, A., Mann, S., Gosele, U., and Knez, M. (2009) Chemical infiltration during atomic layer deposition: metalation of porphyrins as model substrates. *Angew. Chem., Int. Ed.*, **48**, 4982–4985.

107 Bae, C., Shin, H.J., Moon, J., and Sung, M.M. (2006) Contact area lithography (CAL): a new approach to direct formation

of nanometric chemical patterns. *Chem. Mater.*, **18**, 1085–1088.

108 Park, M.H., Jang, Y.J., Sung-Suh, H.M., and Sung, M.M. (2004) Selective atomic layer deposition of titanium oxide on patterned self-assembled monolayers formed by microcontact printing. *Langmuir*, **20**, 2257–2260.

109 Yan, M., Koide, Y., Babcock, J.R., Markworth, P.R., Belot, J.A., Marks, T.J., and Chang, R.P.H. (2001) Selective-area atomic layer epitaxy growth of ZnO features on soft lithography-patterned substrates. *Appl. Phys. Lett.*, **79**, 1709–1711.

110 Xu, Y. and Musgrave, C.B. (2004) A DFT study of the Al_2O_3 atomic layer deposition on SAMs: effect of SAM termination. *Chem. Mater.*, **16**, 646–653.

111 Seo, E.K., Lee, J.W., Sung-Suh, H.M., and Sung, M.M. (2004) Atomic layer deposition of titanium oxide on self-assembled-monolayer-coated gold. *Chem. Mater.*, **16**, 1878–1883.

112 Chen, R., Kim, H., McIntyre, P.C., and Bent, S.F. (2004) Self-assembled monolayer resist for atomic layer deposition of HfO_2 and ZrO_2 high-kappa gate dielectrics. *Appl. Phys. Lett.*, **84**, 4017–4019.

113 Park, K.J., Doub, J.M., Gougousi, T., and Parsons, G.N. (2005) Microcontact patterning of ruthenium gate electrodes by selective area atomic layer deposition. *Appl. Phys. Lett.*, **86**, 3.

114 Chen, R., Kim, H., McIntyre, P.C., and Bent, S.F. (2005) Investigation of self-assembled monolayer resists for hafnium dioxide atomic layer deposition. *Chem. Mater.*, **17**, 536–544.

115 Chen, R., Kim, H., McIntyre, P.C., Porter, D.W., and Bent, S.F. (2005) Achieving area-selective atomic layer deposition on patterned substrates by selective surface modification. *Appl. Phys. Lett.*, **86**, 3.

116 Park, K.J. and Parsons, G.N. (2006) Selective area atomic layer deposition of rhodium and effective work function characterization in capacitor structures. *Appl. Phys. Lett.*, **89**, 3.

117 Chen, R. and Bent, S.F. (2006) Highly stable monolayer resists for atomic layer deposition on germanium and silicon. *Chem. Mater.*, **18**, 3733–3741.

118 Chen, R. and Bent, S.F. (2006) Chemistry for positive pattern transfer using area-selective atomic layer deposition. *Adv. Mater.*, **18**, 1086.

119 Jiang, X.R., Chen, R., and Bent, S.F. (2007) Spatial control over atomic layer deposition using microcontact-printed resists. *Surf. Coat. Technol.*, **201**, 8799–8807.

120 Preiner, M.J. and Melosh, N.A. (2009) Identification and passivation of defects in self-assembled monolayers. *Langmuir*, **25**, 2585–2587.

121 Dube, A., Sharma, M., Ma, P.F., and Engstrom, J.R. (2006) Effects of interfacial organic layers on thin film nucleation in atomic layer deposition. *Appl. Phys. Lett.*, **89**, 3.

122 Kobayashi, N.P. and Williams, R.S. (2008) Two-stage atomic layer deposition of aluminum oxide on alkanethiolate self-assembled monolayers using *n*-propanol and water as oxygen sources. *Chem. Mater.*, **20**, 5356–5360.

123 Li, M., Dai, M., and Chabal, Y.J. (2009) Atomic layer deposition of aluminum oxide on carboxylic acid-terminated self-assembled monolayers. *Langmuir*, **25**, 1911–1914.

124 Sharma, M., Dube, A., Hughes, K.J., and Engstrom, J.R. (2008) Gas-surface reactions between pentakis (dimethylamido)tantalum and surface grown hyperbranched polyglycidol films. *Langmuir*, **24**, 8610–8619.

125 Seitz, O., Dai, M., Aguirre-Tostado, F.S., Wallace, R.M., and Chabal, Y.J. (2009) Copper-metal deposition on self assembled monolayer for making top contacts in molecular electronic devices. *J. Am. Chem. Soc.*, **131**, 18159–18167.

126 Chen, Z., Sarkar, S., Biswas, N., and Misra, V. (2009) Atomic layer deposition of hafnium dioxide on TiN and self-assembled monolayer molecular film. *J. Electrochem. Soc.*, **156**, H561–H566.

127 Dube, A., Sharma, M., Ma, P.F., Ercius, P.A., Muller, D.A., and Engstrom, J.R. (2007) Effects of interfacial organic layers on nucleation, growth, and morphological evolution in atomic layer thin film deposition. *J. Phys. Chem. C*, **111**, 11045–11058.

128 Haran, M., Engstrom, J.R., and Clancy, P. (2006) *Ab initio* calculations of the reaction mechanisms for metal-nitride deposition from organo-metallic precursors onto functionalized self-assembled monolayers. *J. Am. Chem. Soc.*, **128**, 836–847.

129 Hughes, K.J. and Engstrom, J.R. (2010) Interfacial organic layers: tailored surface chemistry for nucleation and growth. *J. Vac. Sci. Technol. A*, **28**, 1033–1059.

130 Satta, A., Baklanov, M., Richard, O., Vantomme, A., Bender, H., Conard, T., Maex, K., Li, W.M., Elers, K.E., and Haukka, S. (2002) Enhancement of ALCVD (TM) TiN growth on Si–O–C and alpha-SiC:H films by O_2-based plasma treatments. *Microelectron. Eng.*, **60**, 59–69.

131 Elam, J.W., Wilson, C.A., Schuisky, M., Sechrist, Z.A., and George, S.M. (2003) Improved nucleation of TiN atomic layer deposition films on SILK low-*k* polymer dielectric using an Al_2O_3 atomic layer deposition adhesion layer. *J. Vac. Sci. Technol. B*, **21**, 1099–1107.

132 Travaly, Y., Schuhmacher, J., Baklanov, M.R., Giangrandi, S., Richard, O., Brijs, B., Van Hove, M., Maex, K., Abell, T., Somers, K.R.F., Hendrickx, M.F.A., Vanquickenborne, L.G., Ceulemans, A., and Jonas, A.M. (2005) A theoretical and experimental study of atomic-layer-deposited films onto porous dielectric substrates. *J. Appl. Phys.*, **98**, 9.

133 Hoyas, A.M., Schuhmacher, J., Shamiryan, D., Waeterloos, J., Besling, W., Celis, J.P., and Maex, K. (2004) Growth and characterization of atomic layer deposited $WC_{0.7}N_{0.3}$ on polymer films. *J. Appl. Phys.*, **95**, 381–388.

134 Niskanen, A., Rahtu, A., Sajavaara, T., Arstila, K., Ritala, M., and Leskelä, M. (2005) Radical-enhanced atomic layer deposition of metallic copper thin films. *J. Electrochem. Soc.*, **152**, G25–G28.

135 Lee, S.M., Pippel, E., Gosele, U., Dresbach, C., Qin, Y., Chandran, C.V., Brauniger, T., Hause, G., and Knez, M. (2009) Greatly increased toughness of infiltrated spider silk. *Science*, **324**, 488–492.

136 King, J.S., Graugnard, E., Roche, O.M., Sharp, D.N., Scrimgeour, J., Denning, R.G., Turberfield, A.J., and Summers, C.J. (2006) Infiltration and inversion of holographically defined polymer photonic crystal templates by atomic layer deposition. *Adv. Mater.*, **18**, 1561.

137 Graugnard, E., Roche, O.M., Dunham, S.N., King, J.S., Sharp, D.N., Denning, R.G., Turberfield, A.J., and Summers, C.J. (2009) Replicated photonic crystals by atomic layer deposition within holographically defined polymer templates. *Appl. Phys. Lett.*, **94**, 3.

138 Wang, C.C., Kei, C.C., Yu, Y.W., and Perng, T.P. (2007) Organic nanowire-templated fabrication of alumina nanotubes by atomic layer deposition. *Nano Lett.*, **7**, 1566–1569.

139 Groner, M.D., Fabreguette, F.H., Elam, J.W., and George, S.M. (2004) Low-temperature Al_2O_3 atomic layer deposition. *Chem. Mater.*, **16**, 639–645.

140 Carcia, P.F., McLean, R.S., and Reilly, M.H. (2005) Oxide engineering of ZnO thin-film transistors for flexible electronics. *J. Soc. Inform. Display*, **13**, 547–554.

141 Meyer, J., Gorrn, P., Bertram, F., Hamwi, S., Winkler, T., Johannes, H.H., Weimann, T., Hinze, P., Riedl, T., and Kowalsky, W. (2009) Al_2O_3/ZrO_2 nanolaminates as ultrahigh gas-diffusion barriers: a strategy for reliable encapsulation of organic electronics. *Adv. Mater.*, **21**, 1845.

142 Wang, J.C., Weng, W.T., Tsai, M.Y., Lee, M.K., Horng, S.F., Perng, T.P., Kei, C.C., Yu, C.C., and Meng, H.F. (2010) Highly efficient flexible inverted organic solar cells using atomic layer deposited ZnO as electron selective layer. *J. Mater. Chem.*, **20**, 862–866.

143 Kobayashi, N.P., Donley, C.L., Wang, S.Y., and Williams, R.S. (2007) Atomic layer deposition of aluminum oxide on hydrophobic and hydrophilic surfaces. *J. Cryst. Growth*, **299**, 218–222.

144 Liang, X.H., Hakim, L.F., Zhan, G.D., McCormick, J.A., George, S.M., Weimer, A.W., Spencer, J.A., Buechler, K.J., Blackson, J., Wood, C.J., and Dorgan, J.R. (2007) Novel processing to produce polymer/ceramic nanocomposites by atomic layer deposition. *J. Am. Ceram. Soc.*, **90**, 57–63.

145 Ras, R.H.A., Sahramo, E., Malm, J., Raula, J., and Karppinen, M. (2008) Blocking the lateral film growth at the nanoscale in area-selective atomic layer deposition. *J. Am. Chem. Soc.*, **130**, 11252.

146 Adiga, S.P., Curtiss, L.A., Elam, J.W., Pellin, M.J., Shih, C.C., Shih, C.M., Lin, S.J., Su, Y.Y., Gittard, S.A., Zhang, J., and Narayan, R.J. (2008) Nanoporous materials for biomedical devices. *JOM*, **60**, 26–32.

147 Sinha, A., Hess, D.W., and Henderson, C.L. (2006) Area-selective ALD of titanium dioxide using lithographically defined poly(methyl methacrylate) films. *J. Electrochem. Soc.*, **153**, G465–G469.

148 Sinha, A., Hess, D.W., and Henderson, C.L. (2006) Area selective atomic layer deposition of titanium dioxide: effect of precursor chemistry. *J. Vac. Sci. Technol. B*, **24**, 2523–2532.

149 Sinha, A., Hess, D.W., and Henderson, C.L. (2007) Transport behavior of atomic layer deposition precursors through polymer masking layers: influence on area selective atomic layer deposition. *J. Vac. Sci. Technol. B*, **25**, 1721–1728.

150 Derk, A.R., Zimmerman, P., and Musgrave, C.B. (2010) Reactivity of trimethylaluminum towards organic functional groups for the deposition of hybrid organic–inorganic films by molecular and atomic layer deposition, submitted.

151 Sugimoto, H., Kawamura, C., Kuroki, M., Aida, T., and Inoue, S. (1994) Lewis acid-assisted anionic ring-opening polymerization of epoxide by the aluminum complexes of porphyrin, phthalocyanine, tetraazaannulene, and Schiff-base as initiators. *Macromolecules*, **27**, 2013–2018.

152 Suzuki, T., Saimoto, H., Tomioka, H., Oshima, K., and Nozaki, H. (1982) Regioselective and stereoselective ring-opening of epoxy alcohols with organo-aluminum compounds leading to 1,2-diols. *Tetrahedron Lett.*, **23**, 3597–3600.

13
Application of ALD to Biomaterials and Biocompatible Coatings

Mato Knez

The previous chapters showed that atomic layer deposition (ALD) has many facets for fundamental and applied research in nanoscience and nanotechnology. Aside from the molecular layer deposition (MLD), all the processes were applied with inorganic materials on inorganic substrates. However, one type of substrate was not yet considered, namely, biomaterials. Nature offers an enormous pool of structures and materials on the microscale as well as nanoscale that potentially can act as template for ALD, thus widening the range of available structures and materials. Biomaterials are sometimes not easy to be classified. Depending on the scientific discipline, biologists, chemists, physicists, and materials scientists often define biomaterials to be strictly protein-based or biomineralized and naturally available, while others also include non-protein-based materials such as polysaccharides (e.g., cellulose) and materials that are artificially obtained from peptides or other bioorganic molecules through (self-)assembly. Here, we will not distinguish between those cases, but will rather cover all the biomaterials in the broader sense. The main difficulty with those substrates (especially the protein- or peptide-based substrates) was that due to the thermal sensitivity of those materials, the ALD seemed to be incompatible and thus the application of ALD to biomaterials was not considered for a long time. After recognizing that some ALD processes also work at lower temperatures, first experiments using biological templates for ALD coatings have been performed. This chapter will give an overview of the biotemplates that were used as substrates for ALD coatings and the resulting novel nanostructures and functional materials. Most of the approaches are rather curiosity-driven. In contrast to those, applications of inorganic ALD coatings for biocompatible surfaces, which will be covered in the second part of this chapter, are of great interest for biology or biomedicine as they offer the possibility to induce or enhance biocompatibility of technical materials for possible use as biological implants or as substrates for *in vitro* tissue growth.

Atomic Layer Deposition of Nanostructured Materials, First Edition. Edited by Nicola Pinna and Mato Knez.

13.1 Application of ALD to Biomaterials

Biomaterials are usually very robust, however, primarily in their natural environment, which means that an appropriate temperature, an aqueous environment, and so on should be preserved. Exceeding those conditions can easily lead to denaturation or destruction of the material if, for example, the temperature is too high. Presumably, this is the reason for the application of ALD to biomaterials not being considered before 2006, as the early ALD tools did not easily enable reasonable deposition processes significantly below 100 °C, which for many proteins is far too high. Therefore, an important factor for applying ALD deposition to real (protein-based) biomaterials was finding processes and precursors that allow coating temperatures significantly below 100 °C. Of particular importance for this step was the work from the group of Steve George [1], coating a polymer bottle with Al_2O_3 at temperatures as low as 33 °C. At least from the thermal point of view, this process opted for successful coating of biomaterials. Of course, there are several further factors that play a role, such as drying the biomaterial in vacuum, which could easily lead to destruction, but also the exposure of the biomaterial to highly reactive precursors such as TMA, which could potentially destructively interact with the proteins. The best way to find out was simply trying some biological substrates in combination with promising ALD processes.

13.1.1 Protein-Based Nanostructures

13.1.1.1 Tobacco Mosaic Virus

The first protein-based substrate that was coated by ALD was a plant virus called tobacco mosaic virus (TMV). The virus is a protein tube with a length of 300 nm and a diameter of 18 nm, comprising a hollow central channel with a diameter of 4 nm. The TMV is the first known virus [2] and the initiator of the whole research field of modern virology. This particular virus is robust (can resist temperatures up to 80 °C), not harmful to mammals, and easy to handle in laboratories, and thus the ideal biological nanostructure for various investigations, starting from the development of electron microscopy in 1939 [3], nanostructure fabrication [4–9], and even going toward technological aspects [10–12]. It is not surprising that this virus was selected as the first biomaterial for an ALD coating. The most important question was whether or not the vacuum treatment and the highly reactive precursors will destroy the virus. As only few processes were known to work successfully at the required low temperatures, trimethylaluminum (TMA) and titanium tetraisopropoxide (TIP) were used for deposition of Al_2O_3 and TiO_2, respectively. The counterpart in both cases was water and the process temperatures were 35 °C. Being adsorbed on a polymer-coated TEM grid, the viruses became coated with the corresponding metal oxides [13, 14]. Interestingly, the viruses did not encounter destruction, at least not morphologically. The diameters remained 18 nm, indicating that the shape was preserved. The metal oxide was deposited on

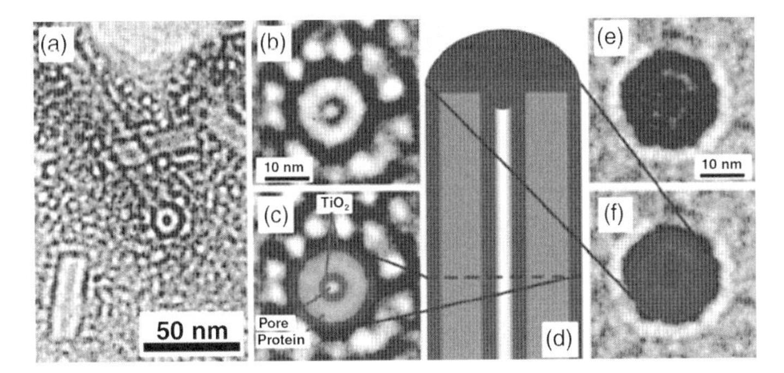

Figure 13.1 (a) TEM (200 kV) image of TMV treated with TiO_2 by ALD. A disk from a broken TMV (circular particle) embedded in an amorphous TiO_2 film can be seen in the cross-sectional view. The TiO_2 covering the interior channel appears hollow with a pore diameter of 1–1.5 nm and a wall thickness of 1 nm. The covered inner channel of the viruses appears brighter along the axis, indicating a hollow TiO_2 nanotube. (b) Magnification of a further TiO_2-covered disk showing a hollow area inside the TiO_2-coated interior channel of the virus. (c) Optically enhanced image (b). The bright circle represents the viral protein sheath. Darker circles show the TiO_2 coating of the viral surface (outer surface and channel surface). The surrounding gray area is the embedding amorphous TiO_2 film. (d) Sketch of the cross section of a TiO_2-covered TMV. In the top part of the virus, no pore is visible in the center. This part represents the assumed clogged area of the inner viral channel. (e) Magnification of a further TiO_2-covered disk showing a clogged interior channel of the virus. (f) Optically enhanced image (e). Reprinted with permission from Ref. [13]. Copyright 2006, the American Chemical Society.

the viruses as well as in the space between them. Importantly, the deposition also occurred within the central channel of the virions, which is another indication of the shape preservation of the TMV, since the channel would not be accessible for ALD deposition in the case of a collapse of the structure. In some cases, on small subunits of the virus standing upright at the grid, indications could be found that the coating of the central channel takes place on the channel wall only, leading to metal oxide nanotubes, templated by the walls of the hollow channel (see Figure 13.1). Due to the problematic handling for electron microscopy, as the substrate (polymer-coated TEM grid) is an amorphous film, and the small size of the features, more clear TEM micrographs could not be obtained.

The TMV, being a well-investigated virus, offers a wide spectrum of possibilities for manipulation. On the one hand, the amino acid sequence of the protein tube can be modified and the virus can be coated and filled with inorganic materials [5–9]. On the other hand, adsorption of TMV on various substrates can be achieved, even in a patterned way [15, 16]. A combination of metallization, patterning, and ALD was recently published by Gerasopoulos *et al.*, presenting a strategy towards hierarchical patterning with TMV as template [17]. In this work, surfaces were structured with uncoated or Ni-coated TMV by means of modified liftoff techniques and the Ni-TMV structured patterns were coated with Al_2O_3 or TiO_2. Due to the stabilizing effect of the Ni coating on TMV, the process for the ALD coating did not require explicitly low temperatures, but was performed at 220 °C for Al_2O_3 or 150 °C for TiO_2. The resulting patterns have a large surface area and could be of great interest for catalysis, dye-sensitized solar cells (DSSCs), and so on.

13.1.1.2 Ferritin and Apoferritin

The next type of biomaterial that was processed by ALD was ferritin. Ferritin is a globular protein containing 24 subunits. The diameter is 12 nm and it contains a cavity with a diameter of about 7 nm. In nature, ferritin acts as a storage container and transport vehicle for iron; thus, the ferritin contains a core of ferrihydrite. Removal of the iron-containing core leads to a hollow ferritin, the so-called apoferritin. In aqueous conditions, the iron-containing core is removed or formed through small, 3–4 Å channels that are located at the boundaries of the protein subunits. The ALD deposition on ferritin leads to freestanding metal oxide (Al_2O_3 or TiO_2) films with embedded ferritins [13, 14]. The films were flexible and robust enough to be spread over holes of holey carbon TEM grids without significant destruction. Upon electron beam irradiation during the TEM investigation, the films started to deform. The significance of this work lies not in any kind of application, but rather in the fact that the ferritin proteins appear to be robust enough to withstand the process conditions that are considered to be rather harsh for biomaterials. A follow-up work in 2009 [18] showed that indeed the protein sphere (this time apoferritin) is not destroyed during the process, but it is rather the case that the ALD process can be used to fill the hollow cavity with titania. Depending on the pretreatment (drying with and without the presence of NaCl), the deposition of titania could be directed either exclusively towards the outer surface of the apoferritin or towards both the outer surface and inner cavity. In both cases, the sphere appeared intact and the ALD deposition was influenced by the presence of water molecules within the channels. If present, the water hydrolyzes the precursor already at the entrance of the channel and further diffusion of precursor molecules is hindered due to clogging.

It should be mentioned at this point that water plays an important role for the proteins and is usually omnipresent. Even after evacuation for a long time, it is not expected that all the water molecules are removed from the protein. Therefore, at least for the first ALD cycles, the deposition behavior should be rather CVD-like as precursors instantaneously react with water molecules bound to the proteins.

13.1.1.3 S-Layers

Further protein assemblies that were investigated as substrates for ALD deposition were S-layers. These self-assembled sheets consist of proteins of cell envelopes of some archaea and bacteria and can after proper treatment reassemble to form structured protein films in 2D with a very regular distribution of nanosized pores. Liu *et al.* [19] made use of the S-layers to produce regularly distributed nanopatterns of HfO_2. The critical step in this work is the functionalization of the S-layer by means of attaching octadecyltrichlorosilane (ODTS), which was already successfully used for surface passivation and area-selective ALD of HfO_2 [20]. A pretreated Si wafer (by HF etching) was covered with an S-layer and subsequently the S-layer was

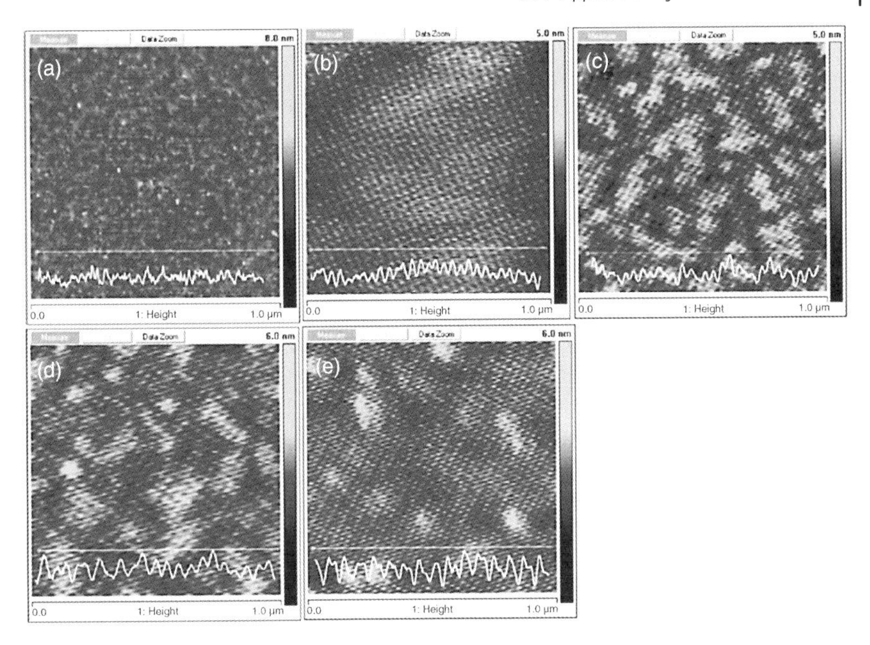

Figure 13.2 AFM images of (a) a HF-cleaned Si substrate; (b) HF-cleaned Si substrate reassembled with S-layer proteins from supernatant with 10 mM $CaCl_2$ for 1 h; (c) sample (b) subsequently modified by 10 mM ODTS for 2 h; (d) after 30 Å HfO_2 deposited on sample (c); and (e) HfO_2 nanopatterns on a Si substrate generated by annealing sample (d) in air at 600 °C for 2 h. The line scans in (a)–(e) show the AFM cross-sectional profiles along the lines shown in each image. Reprinted with permission from Ref. [19]. Copyright 2008, the American Chemical Society.

terminated with ODTS. The subsequent HfO_2 coating led to deposition only within the pores of the S-layer that could be removed after the process by thermal treatment at 600 °C in air. The resulting structure was a Si wafer surface with regularly arranged HfO_2 nanodots with feature sizes of about 9 nm (Figure 13.2 shows such nanopatterns). The S-layers showed remarkable thermal stability as the ALD process was performed with hafnium tetra-*tert*-butoxide (HTB) and water at 200 °C. Presumably, the Ca^{2+} treatment, which is necessary for the formation of the regular S-layer structure and/or the ODTS treatment, had a positive effect on the thermal stability of the structure forming additional electrostatic or van der Waals interactions.

13.1.2 Peptide Assemblies

Peptides are small subunits of proteins, consisting of at least two amino acids, connected via a peptide bond. Through assembly, some peptides can form nanostructure networks. If one, for example, uses diphenylalanine, after electrospinning even thousands of micrometer long fibers can be obtained [21]. After organogel formation, ribbons with lengths of hundreds of micrometers, widths of some hundreds of nanometers, and thicknesses of some tens of nanometers are obtained. Those organogels can be easily transformed into xerogels once the solvent (e.g., chloroform) is evaporated.

Such ribbons, formed after gelation of diphenylalanine, were coated with TiO_2 by ALD in a series of works by Kim and coworkers [22–24]. The TiO_2 coating was performed at around 140 °C from TIP and ammonia gas, and the structures were subsequently calcinated at temperatures above 300 °C. In this way, the peptides are thermally degraded and the TiO_2 coating recrystallized to obtain hollow structures of anatase, just in a similar way as shown before with electrospun polymer fibers as templates [25]. Various characteristics of those coated ribbons were investigated, namely, the wetting properties [22], electrochemical properties [23], and the application potential for dye-sensitized solar cells [24]. In the first case, the wetting properties of the anatase phase TiO_2 were altered by switching a UV light source off and on. While the switchable wetting effect is not new [26], the structure and the ease of synthesis might show some technological relevance for the synthesis of hydrophobic structures. A comparison of the efficiency to further synthetic methods (e.g., titania powder), however, needs to be performed first.

More interesting application schemes were shown in follow-up papers. The hollow structure of the replicated nanoribbons was proposed as potential electrode material for Li ion batteries [23]. The thin TiO_2 layer (only 15 nm) and the structural parameters allow an easy diffusion of the electrolyte into the structure and thus the Li ions to diffuse into the TiO_2 matrix. The nanoribbons show good performance in terms of capacity and rate capability compared to TiO_2 nanopowders. The specific capacity of the hollow nanoribbons reaches the fivefold capacity of TiO_2 powders with 100 nm particle size.

Another proposed application with the same nanoribbons is related to dye-sensitized solar cells [24]. These commonly consist of titania, an organic dye, and an electrolyte sandwiched between two electrodes [27]. Usually transparent conductive oxides (TCOs) are used as electrode. In the present example, instead of titania powders, the nanoribbons were used for the anode and the dye was *cis*-bis(isothiocyanato)bis(2,2′-bipyridi-4,4′-dicarboxylato)ruthenium(II) bis-tetrabutylammonium (Ru535-bisTBA). The titania ribbons were calcinated at two different temperatures, namely, 400 and 500 °C, which resulted in pure anatase structures and mixed anatase–rutile structures, respectively. While the anatase DSSC showed a power conversion efficiency of 2.4%, the mixed anatase–rutile DSSC had an enhanced efficiency of around 3.8%, being similar or even slightly higher than a DSSC constructed with conventional titania nanoparticles (3.6%). These particles, similar to the ribbons calcinated at 500 °C, also showed a mixed phase of anatase and rutile. The large overall surface of the ribbons, together with a large crystallite size, reduced grain boundaries, and densely packed crystallites apparently allowed a comparatively high power conversion efficiency.

13.1.3
Natural Fibers

13.1.3.1 DNA

Fibers are common features found in nature. The probably widest spread biological fiber is a nanowire, the DNA, which was (and still is) one of the most prominent biomolecules that scientists would like to implement into technology. DNA already acted as template for the synthesis of metallic nanowires [28], however, without any technological application up to now. DNA was also coated by ALD, once for curiosity [14] and once as functional substrate [29]. The latter case is an interesting approach towards the production of nanotube-based transistors. The transistor is constructed of a carbon nanotube (CNT) that is known to be a difficult substrate for an ALD coating. Usually, ALD films without functionalization after ALD coating show cluster formation and a surface roughening. The deposited films also usually do not stick very well to the nanotubes, but are rather attached by physical interactions. The work by Lu *et al.* shows an approach where the carbon nanotube is wrapped with a DNA molecule [29]. The DNA molecule attaches to the CNT via π-stacking and provides a new surface for the ALD precursors to bind. The result is an effective and uniform coating of the CNT with HfO_2, which is not easily possible without the wrapped DNA molecules (see also Figure 13.3).

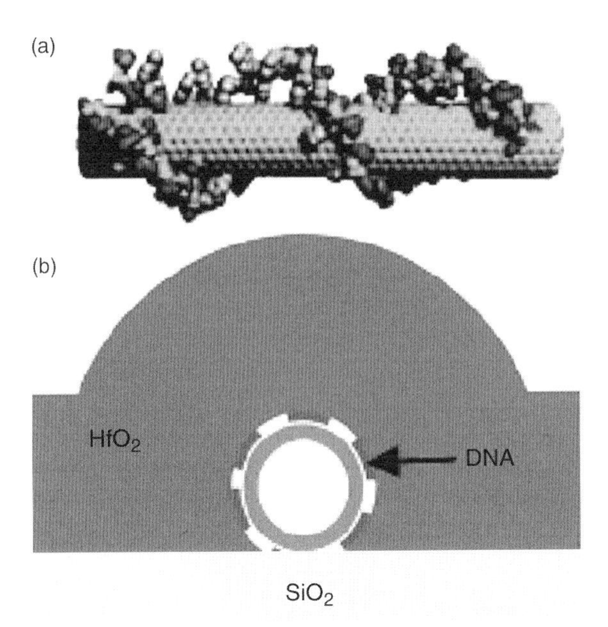

Figure 13.3 DNA functionalization for nanotube electronics. (a) Schematic of a DNA-coated SWNT. (b) Cross-sectional view of HfO_2 (~3 nm by ALD) conformally deposited on a DNA functionalized nanotube lying on a SiO_2 substrate. Reprinted with permission from Ref. [29]. Copyright 2006, the American Chemical Society.

13.1.3.2 Collagen

Biological materials such as muscles, tissues, and so on are assemblies of protein fibers in various molecular compositions, microscopic shapes, and macroscopic organizations. Many of those fibers are composed of collagen, which is a protein frequently showing the repeating amino acid motif glycine–proline–hydroxyproline [30, 31]. The latter two amino acids are sometimes substituted with further amino acids. One simple example for such collagen fiber networks can be obtained when peeling a breakfast egg, namely, the soft tissue that is located between the rigid exterior calcite shell and the interior egg. This soft membrane has the function to protect the embryo from bacterial invasion, still allowing an effective gas exchange. Such a membrane was matter of investigation by Lee *et al.* [32]. The initial aim was to learn whether an ALD coating can be applied to collagen without destruction of its morphology. The coating was performed with TiO_2 or ZnO at temperatures ranging from 70 to 300 °C. In particular for higher temperatures, an initial coating at 70 °C was applied in order to stabilize the structure, since otherwise denaturation occurred. The XRD investigations indicated that the ZnO coatings deposited at temperatures of 70 or 100 °C readily exhibit photocatalytic effects as crystalline features of wurtzite type were observed at those deposition temperatures, while the TiO_2 coatings only showed reasonable reflections of anatase at deposition temperatures above 160 °C. The photocatalytic effect was tested with the survival rate of *Escherichia coli* bacteria upon illumination of the membrane with UV light. Indeed, the ZnO deposited at 100 °C already showed a good bactericidal effect, which was obtained from TiO_2 only if the deposition temperature exceeded 160 °C. This approach shows a way to produce photocatalytic coatings on temperature-sensitive materials just by switching the photocatalyst from TiO_2 to ZnO. The efficiency is lower in direct comparison, but one benefits from the lower deposition temperature, thus being less harmful to the substrate. An interesting additional feature that was observed was a change in the mechanical behavior of the membrane after the ALD coating. The membrane increased its toughness by simultaneously increasing the strength and ductility, which is remarkable and unusual in physics [33]. A detailed investigation showed that the core of the collagen contains Zn or Ti, which appears to be the source for the change of the mechanical properties of the collagen. It is, however, not comparable to the naturally occurring insertion of metals into a collagen structure, the so-called biomineralization, which results in very strong materials such as bones and tendons. In contrast to biomineralized collagen where calcite is formed locally between the collagen helices, after ALD the metals seem to be very well dispersed without any observable clusters or aggregates.

13.1.3.3 Spider Silk

The insertion of metals into the biomaterial was considered to be an effect of the extended exposure time during the ALD process. In this case, the precursor is injected into the ALD chamber simultaneously closing the vacuum valve. The residual time for the precursor could in this way be extended to several tens of seconds. In order to verify this effect, another type of biological fiber, a spider silk, was subjected to an analogous process. Spider silks are remarkably tough materials,

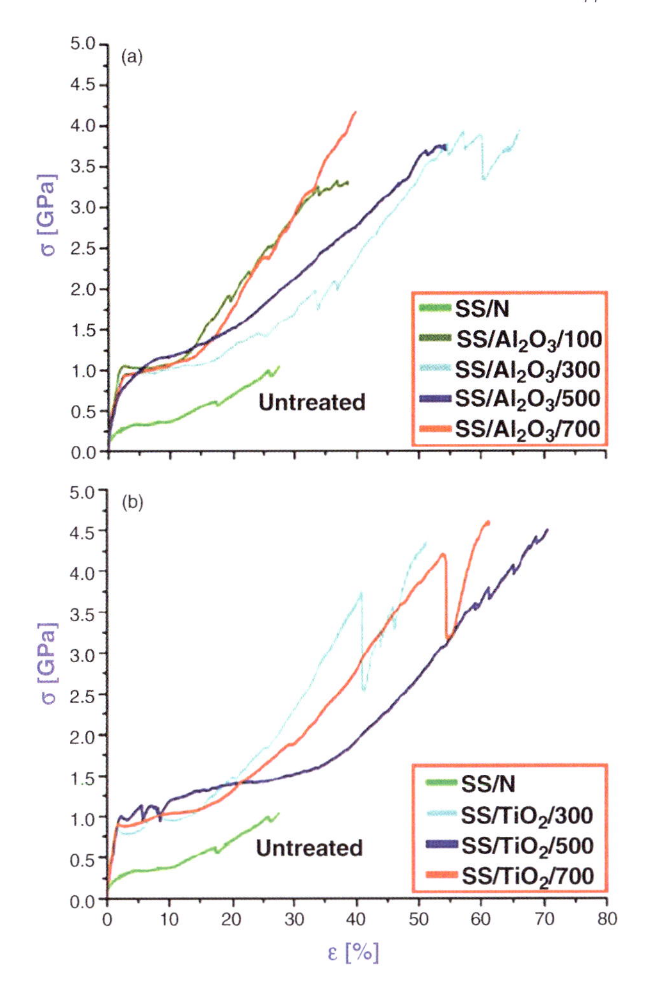

Figure 13.4 Tensile test curves of silk fiber samples treated by TMA/H_2O and TIP/H_2O pulse exposure pairs and comparison to untreated samples and samples treated with various numbers of cycles. (a and b) Stress (σ) and strain (ε) curves of silk fibers treated with TMA/H_2O and TIP/H_2O pulse pairs with various numbers of cycles ranging from 100 to 700. Reprinted with permission from Ref. [34]. Copyright 2009, the American Association for the Advancement of Science.

outperforming most of the man-made materials, such as carbon fibers, Kevlar, nylon, and so on. After an ALD process of TiO_2, Al_2O_3, or ZnO with extended exposure time of the substrates to the precursors, the toughness of the resulting fibers could be increased up to 10-fold [34] (see Figure 13.4). A series of experiments, including NMR, XRD, TEM-EDX, TOF-SIMS, and so on, showed that the ALD-deposited metal indeed is also found in the bulk of the spider silk, which points towards an infiltration of the protein from the gas phase. The proposed mechanism is somewhat different from a simple coating as it relies on diffusion of the precursor into the protein and reaction with it, which requires a certain residence time of the precursor in the reaction chamber. In fact, the same process was performed in a regular coating mode (exposure time equal to the pulsing time) with almost no influence to the mechanical properties of the fiber [34]. With this respect, the process was given another name in order to distinguish between the coating and the infiltration mode, namely, multiple pulsed vapor phase infiltration (MPI).

The infiltration mechanism is a very interesting topic to be considered if soft materials are processed by ALD. Indications for such infiltration were already observed before [35] as a hindrance for the precise coating of polymers. Additional studies have shown that the infiltration, particularly with diethylzinc (DEZ), can be used to controllably produce nanostructures from block copolymers [36] and also to synthesize Zn-porphyrin through infiltration of porphyrin aggregates, the so-called J-aggregates [37].

13.1.3.4 Cellulose Fibers from Paper

First attempts to coat natural or nature-derived materials by ALD were performed by the group of Ritala [38, 39], coating cellulose fibers with TiO_2 or bilayers of Ir/Al_2O_3 or Ir/TiO_2. Cellulose does not consist of proteins but is a polysaccharide consisting of several hundred or thousand linked D-glucose units and shows much higher resistance to thermal treatment than most protein-based materials. In the mentioned works, the ALD coating of the cellulose fibers was conducted at temperatures of 150 or 250 °C for the metal oxides and Ir was deposited at 250 °C, which was still below the decomposition temperature of the substrates. Titania coatings obtained at 250 °C showed crystalline anatase and thus were expectedly photocatalytically active. An additional Ir coating even enhanced the photocatalytic activity of the coated fibers. Interestingly, a protective effect of the metal oxide coating was observed. Namely, the by-product during the Ir deposition from $Ir(acac)_3$ and O_2 is atomic oxygen, which is expected to decompose the cellulose at 250 °C. In the present case, a degradation of the cellulose was not observed, indicating that the initially deposited metal oxide acts as protective layer.

13.1.3.5 Cotton Fibers

A similar substrate for ALD deposition was also used by the group of Parsons. The substrates in this work were cotton fibers that primarily consist of highly crystalline cellulose. Those cotton fibers should be functionalized by ALD to produce fabrics with enhanced wear resistance, promising a large application potential in the textile industry. ALD was performed by depositing alumina at 100 °C with a coating thickness of 50 nm. The obtained coating was uniform throughout the fibers with growth rates initially being 0.5 nm per cycle and decreasing to 0.3 nm per cycle during the process. The enhanced growth rate is related to the surface morphology of the fibers and water trapped within the cotton, eventually reacting with TMA in a subsequent cycle [40]. A follow-up work showed that the wetting behavior of the fabrics can be easily affected with the thickness of the ALD coating [41]. Only one ALD

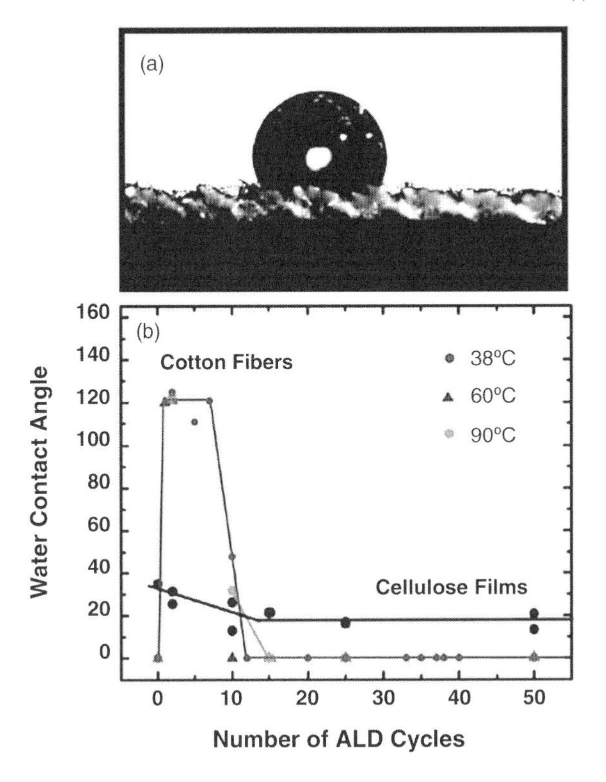

Figure 13.5 (a) Image of a water droplet on a nonwetting cotton surface after two A/B cycles of aluminum oxide. (b) Transition of water contact angle on cotton fibers and cellulose films as a function of ALD cycles at various deposition temperatures. One cycle consisted of TMA/Ar/H_2O/Ar 1/30/1/60 s. Reprinted with permission from Ref. [41]. Copyright 2010, the American Chemical Society.

cycle of alumina abruptly switched the wetting cotton to nonwetting (see Figure 13.5 for the contact angle as a function of the number of ALD cycles). An increasing number of ALD cycles caused a reverse switch, from nonwetting to wetting. The effect could be explained with the nanoscale fibrils present at the native cotton fibers. First ALD cycles might increase the surface roughening by nucleation on those fibrils, which eventually become completely coated and smoothened after 50 or more cycles. The explanation appears reasonable as the surface morphology plays an important role for the wetting properties, which will be discussed further in Section 13.1.4.

13.1.3.6 **Sea Mouse Bristles**

A very recent approach to apply ALD to high aspect ratio biomaterials is shown with the coating of bristles of a sea mouse, a species of Polychaeta [42]. The bristles are hair and spines, both of which exhibit centimeter long and parallel hollow channels with diameters of around 200 nm. The structure is composed of chitin and proteins and is continuously formed by the sea mouse. Being very regular, the structure shows photonic effects, effectively reflecting light at certain wavelengths resulting in blue/green iridescence of the hair and red/orange iridescence of the spines. The structures were coated with Al_2O_3 to obtain nanotubes templated by the pores. The wall thickness of the resulting nanotubes was around 20 nm, but the total length of the nanotubes could not be measured, since it was difficult to completely release the nanotubes from the template. As this publication shows very early work, it remains for the future to see how efficient the ALD can be in coating such extremely high aspect ratio structures.

13.1.4 Patterned Biomaterials

13.1.4.1 Butterfly Wings

Another interesting biotemplate approach was published in 2006 by the group of Wang [43]. Instead of applying ALD to individual biological nanostructures, they concentrated on structured surfaces, namely, butterfly wings. Those frequently show interesting coloration, which is significantly influenced by photonic effects as a result of an overlay of microstructures and nanostructures. In nature, colors appear either as "chemical color," based on pigmentation, or as "structural or physical color," based on structural effects and the interaction of structures with light [44]. In most cases, a mixture of both effects appears. If no pigments are present, the coloration relies purely on the structural coloration, which is, for example, found in photonic structures. In the work of Huang *et al.* [43], the wings of a blue colored butterfly (*Morpho peleides*) were used as templates for ALD deposition of Al_2O_3. The wings of butterflies are composed of chitin that is also found in, for example, exoskeletons of insects and arthropods, cell walls of fungi, and so on. The chitin is chemically similar to cellulose (see Section 13.1.3.4) with *N*-acetylglucosamine being the constituting unit for the polymer. The alumina coatings that were deposited on the butterfly wings had thicknesses of 10–40 nm in steps of 10 nm. The coating induced a color change of the wings, which depending on the thickness shifted from the originally found blue color towards pink with a 40 nm coating. The color change is physically reasonable as with coating one adds a thin film of a material with a different refractive index from the underlying chitin of the butterfly wing. After the coating, the template was removed by thermal treatment and the resulting replicas were examined. The TEM micrographs showed a conformal coating of the fine structures of the wing, which proves that also the chitin-based biomaterials can act as substrate for ALD deposition. The optical properties of the structure after removal of the organic part were similar to the optical properties of the coated structure, however, showing a shift of the main reflectance peak from 550 towards 420 nm (see reflectance spectra in Figure 13.6). Given the structural properties, even a sort of optical waveguide with beam splitters was proposed that the authors consider a potential route for mass production of such structures in future.

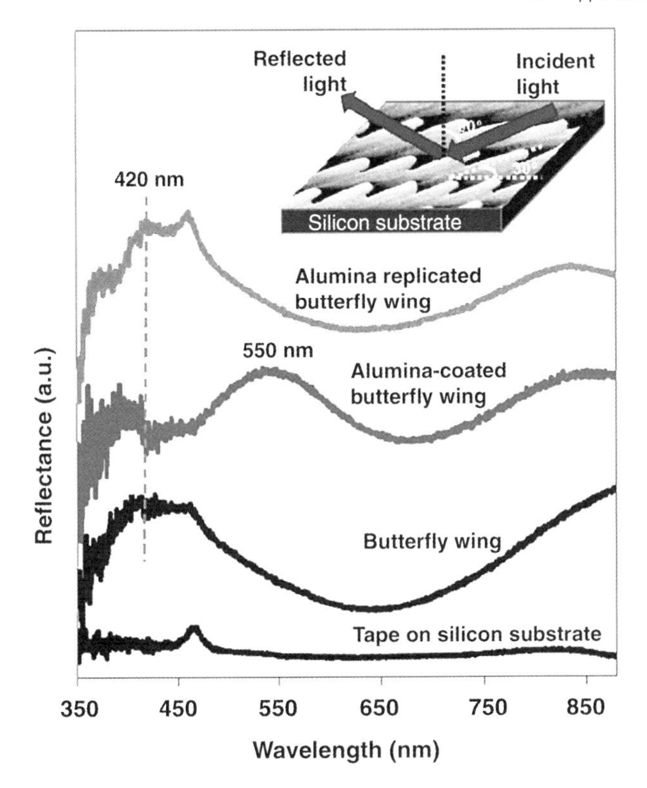

Figure 13.6 Reflectance spectra of original butterfly wings, alumina-coated wings, alumina replicas, and the carbon tape background in UV–visible light range. The inset is the schematic of reflection measurement setup. Figure courtesy of Z.L. Wang, Georgia Tech. Reprinted with permission from Ref. [43]. Copyright 2006, the American Chemical Society.

A similar approach was shown by Gaillot *et al.* [45]. The investigation was performed on another type of butterfly, the *Papilio blumei*, which shows green iridescence on the wings with blue colored tails. Also here, the wings resemble natural photonic structures with embedded air cavities. The main scientific question was the accessibility of the air pockets to ALD precursors during a coating process. The coating was performed with TiO_2, which has higher refractive index than the chitin-based substrate. A model was constructed for simulating the photonic properties both for deposition on the top and bottom surfaces of the wing only and for a coating of the surfaces of the embedded air pockets, which are the two possible models for coatings. The simulations and the measured reflectance spectra for both cases were in good agreement, showing that both of the schemes for the deposition are possible. Coating of the buried air pockets presumably becomes possible once the structure has defects in the form of cracks or from the edges of the prepared samples if the pockets are interconnected. The significant optical difference for those two cases is the shift of the main reflection peak at 524 nm, which proceeds towards larger wavelengths with about 1 nm per nm TiO_2 coating if the structure is coated only externally, while it shifts by 12 nm per nm TiO_2 after coating the air pockets, which is a difference of more than one order of magnitude. Both cases were experimentally observed, but it is unclear whether the two modes can be controllably produced.

13.1.4.2 Fly Eyes

A further investigated biological structure with optical function is an eye of a household fly [46]. Those structures are very complicated and a result of nature's adaptation necessary for the survival of the fly. The eyes of a household fly consist of smaller subunits, the so-called ommatidia, which again contain many nanostructured protuberances. The native eye shows antireflective properties above a wavelength of 400 nm with a reflection peak at a wavelength of 330 nm. In a similar manner to the above-mentioned butterfly wing replication by the same group, the fly eye was coated with Al_2O_3 at 100 °C. After removal of the template by annealing at 500 °C in air, the alumina replica showed very similar antireflection behavior to the original structure, with a slight shift of the reflection peak from 330 to 375 nm (see Figure 13.7). Taking into account that the source of the master material, the fly, is

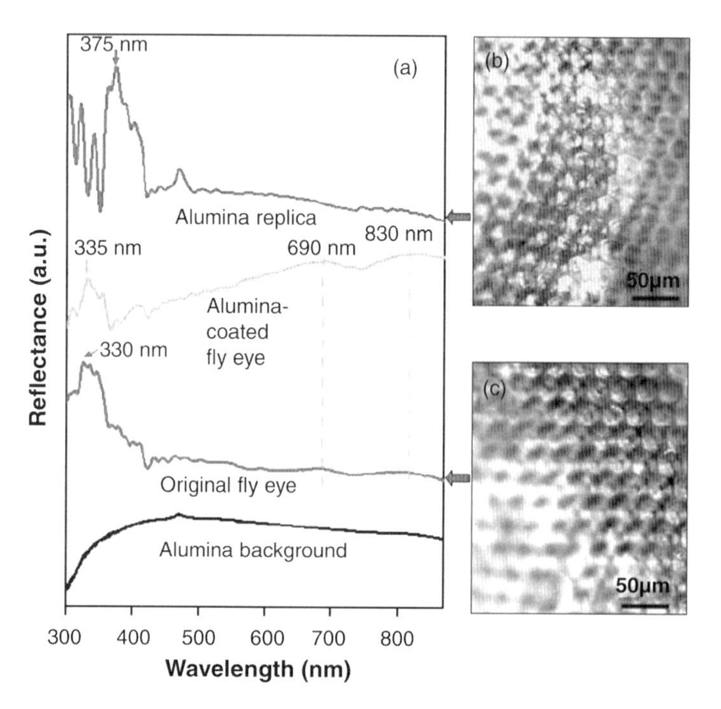

Figure 13.7 (a) Reflectance spectra recorded on a 100 nm thick alumina replication of a fly eye, a fly eye coated with 100 nm alumina thin film, the original fly eye, and a glass substrate coated with 100 nm alumina thin film as the background signal. The reflectance was measured at 30° to the normal direction. Optical microscopy images of the alumina replica of fly eyes (b) and the original fly eyes (c). Figure courtesy of Z.L. Wang, Georgia Tech. Reprinted with permission from Ref. [46]. Copyright 2008, IOP Publishing Ltd.

unlimited, this approach could indeed be a means to produce antireflective structures in masses. However, the small size of the eyes limits the practical applicability of this approach for antireflective structures as competing strategies, for example, by imprinting or micelle patterning [47], are more cost effective at the present.

13.1.4.3 **Legumes**

Photocatalytically active structures were produced from pods (legumes) as templates [48]. The used legume, a fruit of *Wisteria* from the pea family, contains microscaled tubular trichomes, which were coated with about 60 nm thick Al_2O_3 films. The alumina coating acts as a support for further material growth. A ZnAl layered double hydroxide (LDH) film was grown on top, which after calcination resulted in a ZnAl mixed metal oxide (MMO) framework, which showed crystalline phases of ZnO and $ZnAl_2O_4$ upon XRD examination. The spinel formation during calcination is expected as the systems Al_2O_3 and ZnO have a great tendency for interfacial diffusion, which was already shown for the formation of Kirkendall voids [49]. The nanocrystals formed on the structure after calcination showed a greatly enhanced surface area in comparison to a powder sample, with the specific surface being 134.32 m^2/g for the MMO versus 55.44 m^2/g for the powder. Thus, it is not astonishing that the enhanced specific surface leads to a higher photocatalytic activity, which was confirmed with investigations of the degradation of two dyes, poly {1-4[4-(3-carboxy-4-hydroxyphenylazo)benzenesulfonamido]-1,2-ethanediyl} sodium salt (PAZO) and sulforhodamine B, upon UV irradiation. In both cases, the biotemplated structure showed much higher activity for the photodegradation of the dyes than the corresponding powder sample.

13.1.4.4 **Water Strider Legs**

Ordered structures do not only show optical properties that were discussed above. The very delicate structure of butterfly wings, for example, also has water-repelling properties, which is important for the survival of the insect, as small amounts of water on the wing would readily make flying impossible. The hydrophobic butterfly wings (here again *M. peleides* was used) were coated with hydrophilic amorphous Al_2O_3 and compared to coated water strider legs, which are known to be superhydrophobic [50]. Generally, the hydrophobic nature of structured surfaces is characterized by two model states, the so-called Cassie state [51] and the Wenzel state [52], which differ in their contact angle hysteresis [53]. For antiwetting behavior and self-cleaning features, the Cassie state is the favored state, which is applicable to both the structures, the butterfly wing and the water strider leg in their native form. Once coated with a hydrophilic film (in this case 30 nm of Al_2O_3), the water strider leg persists in the Cassie state while the butterfly wing changes to the Wenzel state. The influence of the native wax coating appears to be stronger in the case of the butterfly wing, while structural effects dominate the wetting behavior in the water strider case. This is explained by the fact that the structure of the butterfly wing is dominantly in plane, which means in 2D. The water strider, however, has round shaped features inclining from the surface, thus providing the possibility to trap air that, regardless of the hydrophilic or hydrophobic nature of the surface, will favor the Cassie state and let the whole structure remain superhydrophobic.

13.1.5
Biomineralized Structures

Biominerals are special types of materials found in nature. Although these are rather inorganic materials, they are produced by a number of organisms in a very complicated synthetic process. Prominent examples of such organisms are the unicellular algae called diatoms. Those algae produce a patterned exoskeleton of porous silica, the frustule, which can be found in nature in various shapes and sizes.

Such frustules were also coated by ALD. This approach is not too demanding, as the frustules consist of inorganic material, which in contrast to proteins or other biological soft materials is not sensitive to thermal decomposition. In the first example, the frustule was coated with titania at 100 °C, using $TiCl_4$ and water as precursors [54] (see Figure 13.8). The initial pore size of the frustule valves was about 40 nm and could be conformally coated to reduce the pore size to about 5 nm. The two resulting effects, the pore size reduction and the modification of the surface chemistry with a new material (here TiO_2 instead of SiO_2), envisage an application

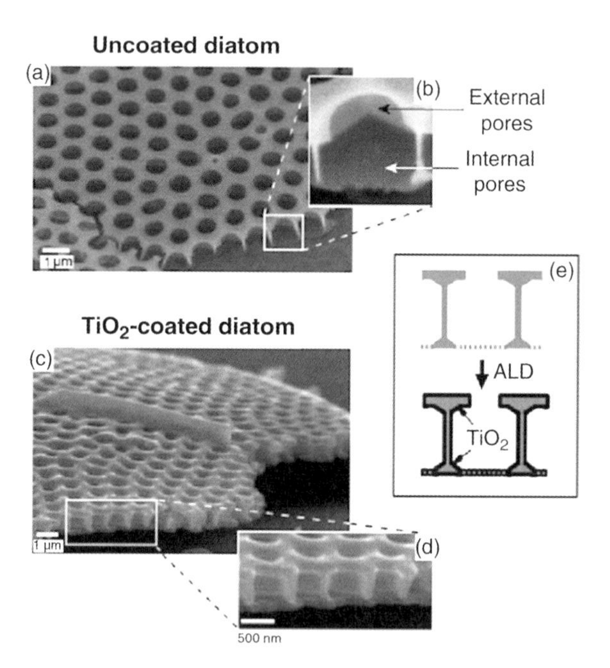

Figure 13.8 (a and b) SEM images of uncoated frustules of *T. eccentrica* with profiles of the pore structure; (c and d) SEM images of fractured frustule after TiO_2 coating (500 cycles); and (e) scheme of formation of conformal TiO_2 inside pore. Reprinted with permission from Ref. [54]. Copyright 2006, the Royal Society of Chemistry.

potential for molecular separation by size and affinity. The latter one can potentially be tuned with coating materials other than TiO_2.

In the second example, germanium was metabolically inserted into the frustule biosilica first, resulting in a concentration of about 1.6 wt% Ge in the frustule [55]. The frustule acts as a 2D photonic crystal slab, which was incorporated into an electroluminescent device. The coating of the frustules, which were dispersed on a TCO electrode, was performed with 400 nm $HfSiO_4$ by ALD. Photoluminescence at a wavelength of 450 nm was observed after irradiation with a 337 nm laser. Electroluminescence was observed with sharp lines below 400 nm and above 640 nm with a band gap between 500 and 640 nm after excitation at 150 V and 10 kHz. The results appear promising for future investigation and optimization, particularly with respect to emission color and brightness.

13.2
Biocompatible Coatings

Biocompatible coatings are of great interest in the field of biology and medicine. In particular for implants, such coatings are highly desired as they can turn an incompatible material biocompatible with only a thin layer, thus providing the possibility to use cheaper sources for the construction of the implant. ALD, being a highly precise deposition strategy, is considered most powerful for performing such coating and it is astonishing that so far investigations in this area are sparsely found in the literature.

13.2.1
Biocompatibility of Alumina

Most of the investigations concentrated on the biocompatibility of materials that are routinely deposited by ALD. The goal is to prove that a coating is at least not less biocompatible than common materials such as glass. The most common material for ALD deposition is Al_2O_3 that was also the first ALD material tested for biocompatibility. In the work by Finch *et al.* [56], the effects of proliferation of coronary artery smooth muscle cells on 60 nm alumina-coated glass were compared to those for uncoated glass and a silane-terminated hydrophobic surface. The cells proliferate very similarly or even more rapidly on the alumina coating than on the glass cover slips during the first 24 h. The similarity in proliferation is explained with the hydrophilic nature of the alumina coating, which is due to the large concentration of surface-exposed hydroxyl groups. The glass slips have a similar wetting behavior (the contact angles are close with 55° for the alumina versus 61° for the glass), indicating the reason for the similar proliferation. Accordingly, tests on surfaces modified with a hydrophobic silane (contact angle of 108°) showed a significant decrease of the biocompatibility of the surface. Apparently, an alumina coating already shows biocompatibility that is at least as good as common glass slips.

13.2.2
Biocompatibility of Titania

Alumina coatings were also compared to titania coatings in terms of bioactivity. The work of Liang *et al.* [57] describes a biomimetic process where after ALD deposition of alumina or titania on polymer particles a biocompatible film of hydroxyapatite (HA) is grown after immersion of the coated particles into a simulated body fluid (SBF). Compared to uncoated polymer particles, the ALD-coated particles showed enhanced nucleation and growth of HA in the SBF. After 2 weeks of immersion, the surfaces showed virtually complete coverage with a high surface roughness. The nucleation and formation of HA is believed to rely on the surface charge of the ceramic coating. Namely, after immersion into the SBF, Ti–OH or Al–OH groups, respectively, may dissociate leading to a negative surface charge, which is beneficial for the HA growth. In terms of biocompatibility, however, no significant improvement of the coatings could be observed. Fibroblast cells (NIH/3T3) were used to investigate the cell attachment, leading to similar results in density and morphology for uncoated as well as coated polymer particles.

13.2.3
Biocompatibility of Hydroxyapatite

A very direct approach to obtain biocompatible coatings by ALD was shown by Putkonen *et al.* [58]. The main objective is to directly access HA by an ALD process, which is chemically very demanding, being a ternary compound with many variation possibilities in stoichiometry. A four-precursor ALD process consisting of $Ca(thd)_2/O_3$ and $(CH_3O)_3PO/H_2O$ at 300 °C was proposed. After the initial formation of a calcium carbonate film by the first precursor pair, the carbonate groups are subsequently exchanged with phosphate groups from the second precursor pair. Thermal annealing in dry or moist N_2 was applied to further reduce the carbon content, eventually leading to HA films if the annealing temperature was above 500 °C. As-deposited films and films annealed at 800 °C were tested for biocompatibility by attaching preosteoblast MC 3T3-E1 cells (see Figure 13.9). Particularly, the annealed films showed good biocompatibility as the cells started spreading and forming lamellipodia. Although this ALD process is not as straightforward as the binary processes for the deposition of alumina or titania, this approach looks very promising for the direct synthesis of biocompatible coatings by ALD.

13.2.4
Biocompatibility of Pt-, TiO_2-, or ZnO-Coated Porous Alumina

As already mentioned above, the wetting properties of surfaces apparently have a strong influence on the biocompatibility. The wetting properties not only can be modified by coatings exhibiting various wettabilities, but also strongly depend on the structural features of the surface, as already described in Section 13.1.4. One type of structures exhibiting nanoscale features, which might be beneficial for biomedical

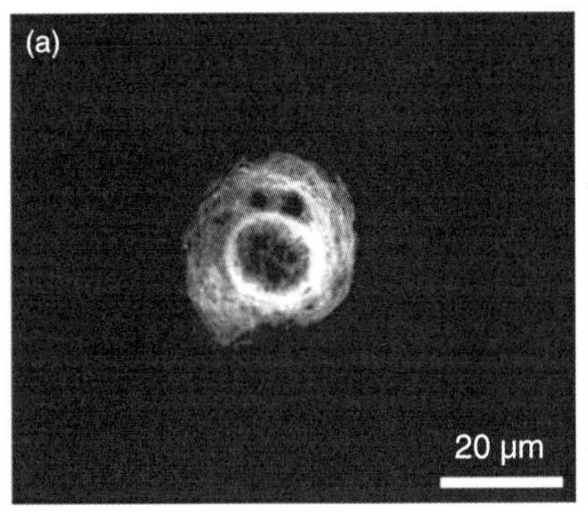

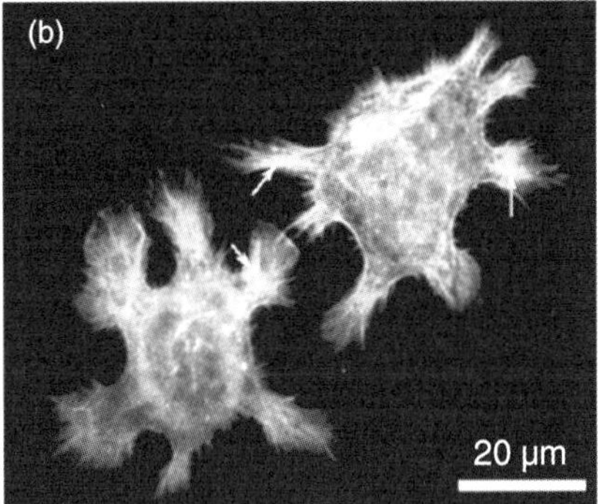

Figure 13.9 MC 3T3 E1 cells were grown for 3 h on as-deposited (a) and on annealed (b) hydroxyapatite films formed on Si(100) by ALD. On the annealed, crystallized film, the cells show clear lamellipodia and filopodia structures. Reprinted with permission from Ref. [58]. Copyright 2009, Elsevier.

applications, is porous anodic aluminum oxide (AAO). In a series of works, AAO was coated with Pt, TiO_2, or ZnO by ALD and the proliferation of neonatal human epidermal keratinocytes was examined [59–61]. In the case of TiO_2 and ZnO, the viability slightly increased in direct comparison to uncoated AAO, while Pt-coated AAO showed a slightly lower viability than the uncoated AAO. Aside from the biocompatibility, the metal oxide-coated AAO also showed bactericidal effects, which was demonstrated with the survival rate of *E. coli* or *Staphylococcus aureus* bacteria upon irradiation with UV light, in a similar way to the work performed by Lee *et al.* [32].

13.2.5
Biocompatibility of TiN-Coated Cotton Fabrics

A further study on the biocompatibility of rough surfaces, namely, ALD-coated cotton fibers, was performed by Hyde *et al.* [62]. Similar to the investigations of the wetting properties (see also Section 13.1.3), the cotton fabrics were coated by ALD, however, at this time with TiN. The TiN coating was obtained from tetrakis(dimethylamido) titanium (TDMAT) and ammonia at process temperatures of around 150 °C. The initiation of the growth as well as the covalent attachment of the coating to the cotton fibers is believed to rely on the hydroxyl groups that the cotton already provides naturally. The TiN coating showed some oxidation, presumably on the surface, after storage in air for 10–15 days. Subsequently, the biocompatibility of the coated fabrics was investigated by the cell adherence of human adipose-derived adult stem cells (hADSCs), taking the coating thickness and the wetting properties of the surface into consideration. The biocompatibility was good for all coating thicknesses and a DNA quantification of the attached hADSCs showed that the cell adhesion was maximized on the most hydrophobic surfaces with only about 2 nm thickness of TiN (see Figure 13.10). Further examination and optimization might lead to even enhanced biocompatibility of the materials, which is, with respect to the unlimited natural

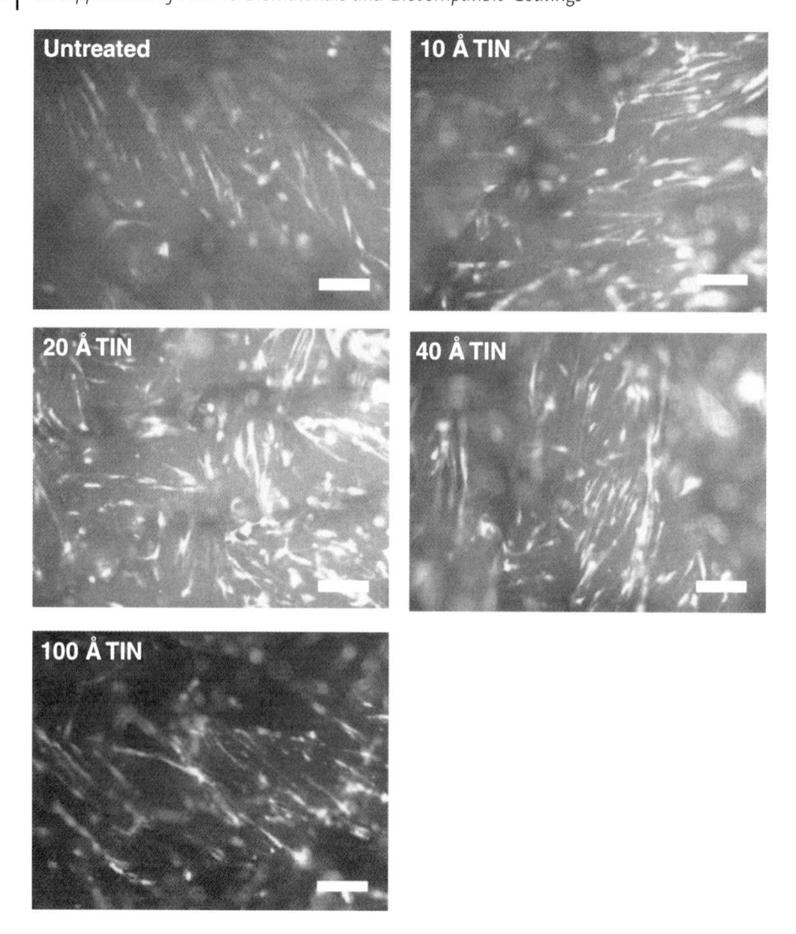

Figure 13.10 Viability images of hADSCs showing adhesion on TiN-coated substrates after 24 h of incubation. hADSCs are well spread and elongated on all surfaces showing enhanced adhesion on the hydrophobic coatings. TiN coating thickness is labeled. A scale bar = 200 μm is included for reference in each panel image. Reprinted with permission from Ref. [62]. Copyright 2009, IOP Publishing Ltd.

resources of cotton and the easy handling of the material, very promising for an application in biology or medicine.

13.3 Summary

This chapter showed that ALD in combination with biomaterials can indeed be an interesting strategy to either produce novel (nano)structures or modify existing structures and materials. Several technical limitations (e.g., the process temperature, vacuum, etc.) either are overcome in modern ALD tools or will be overcome in near future. The increased interest of researchers to combine the fields of biology and the ALD process technology is obvious from the increasing number of papers published in the field. One can expect that the number of publications will steadily increase as ALD proceeds to become a routine tool in many laboratories. Novel ideas will appear in near future and the tools will be adapted for the specific requirements of various types of biomaterials.

Biocompatible coatings are presumably the most promising and also most important field for an ALD application in (bio)medicine. Although until now only few groups stepped into this area, this is presumably just the beginning. Intense investigations might quickly lead to scientifically and technologically important advances in the fields of artificial tissues, implant coatings, and cell cultivation.

Acknowledgments

The author greatly acknowledges the financial support by the BMBF in the frame of the "Nanofutur" program (FKZ 03X5507). This chapter is dedicated to the late Ulrich Gösele whom the author is thankful for his everlasting support and inspiring discussions.

References

1 Groner, M.D., Fabreguette, F.H., Elam, J.W., and George, S.M. (2004) Low-temperature Al_2O_3 atomic layer deposition. *Chem. Mater.*, **16** (4), 639–645.

2 Beijerinck, M.J. (1898) Ueber ein contagium vivum fuidum als Ursache der Fleckenkrankheit der Tabaksblätter. *Verh. Kon. Akad. Wetensch.*, **65**, 3–21.

3 Kausche, G.A. and Ruska, H. (1939) The visualisation of adsorption of metal colloids on protein bodies: the reaction between colloidal gold and tobacco mosaic virus. *Kolloid Z.*, **89**, 21–26.

4 Shenton, W., Douglas, T., Young, M., Stubbs, G., and Mann, S. (1999) Inorganic–organic nanotube composites from template mineralization of tobacco mosaic virus. *Adv. Mater.*, **11** (3), 253–256.

5 Fowler, C.E., Shenton, W., Stubbs, G., and Mann, S. (2001) Tobacco mosaic virus liquid crystals as templates for the interior design of silica mesophases and nanoparticles. *Adv. Mater.*, **13** (16), 1266–1269; Dujardin, E., Peet, C., Stubbs, G., Culver, J.N., and Mann, S. (2003) Organization of metallic nanoparticles using tobacco mosaic virus templates. *Nano Lett.*, **3** (3), 413–417.

6 Knez, M., Sumser, M., Bittner, A.M., Wege, C., Jeske, H., Kooi, S., Burghard, M., and Kern, K. (2002) Electrochemical modification of individual nano-objects. *J. Electroanal. Chem.*, **522** (1), 70–74.

7 Knez, M., Bittner, A.M., Boes, F., Wege, C., Jeske, H., Maiss, E., and Kern, K. (2003) Biotemplate synthesis of 3-nm nickel and cobalt nanowires. *Nano Lett.*, **3** (8), 1079–1082.

8 Knez, M., Sumser, M., Bittner, A.M., Wege, C., Jeske, H., Martin, T.P., and Kern, K. (2004) Spatially selective nucleation of metal clusters on the tobacco mosaic virus. *Adv. Funct. Mater.*, **14** (2), 116–124.

9 Balci, S., Bittner, A.M., Hahn, K., Scheu, C., Knez, M., Kadri, A., Wege, C., Jeske, H., and Kern, K. (2006) Copper nanowires within the central channel of tobacco mosaic virus particles. *Electrochim. Acta*, **51** (28), 6251–6257.
10 Gerasopoulos, K., McCarthy, M., Royston, E., Culver, J.N., and Ghodssi, R. (2008) Nanostructured nickel electrodes using the tobacco mosaic virus for microbattery applications. *J. Micromech. Microeng.*, **18** (10), 104003.
11 Royston, E., Ghosh, A., Kofinas, P., Harris, M.T., and Culver, J.N. (2008) Self-assembly of virus-structured high surface area nanomaterials and their application as battery electrodes. *Langmuir*, **24** (3), 906–912.
12 Tseng, R.J., Tsai, C.L., Ma, L.P., and Ouyang, J.Y. (2006) Digital memory device based on tobacco mosaic virus conjugated with nanoparticles. *Nat. Nanotechnol.*, **1** (1), 72–77.
13 Knez, M., Kadri, A., Wege, C., Goesele, U., Jeske, H., and Nielsch, K. (2006) Atomic layer deposition on biological macromolecules: metal oxide coating of tobacco mosaic virus and ferritin. *Nano Lett.*, **6** (6), 1172–1177.
14 Knez, M., Nielsch, K., Patil, A.J., Mann, S., and Goesele, U. (2007) Atomic layer deposition on biological macromolecules. *ECS Trans.*, **3** (15), 219–225.
15 Knez, M., Sumser, M.P., Bittner, A.M., Wege, C., Jeske, H., Hoffmann, D.M.P., Kuhnke, K., and Kern, K. (2004) Binding the tobacco mosaic virus to inorganic surfaces. *Langmuir*, **20** (2), 441–447.
16 Balci, S., Leinberger, D.M., Knez, M., Bittner, A.M., Boes, F., Kadri, A., Wege, C., Jeske, H., and Kern, K. (2008) Printing and aligning mesoscale patterns of tobacco mosaic virus on surfaces. *Adv. Mater.*, **20** (11), 2195–2200.
17 Gerasopoulos, K., McCarthy, M., Banerjee, P., Fan, X., Culver, J.N., and Ghodssi, R. (2010) Biofabrication methods for the patterned assembly and synthesis of viral nanotemplates. *Nanotechnology*, **21** (5), 055304.
18 Kim, H., Pippel, E., Goesele, U., and Knez, M. (2009) Titania nanostructures fabricated by atomic layer deposition using spherical protein cages. *Langmuir*, **25** (23), 13284–13289.
19 Liu, J., Mao, Y., Lan, E., Banatao, D.R., Forse, G.J., Lu, J., Blom, H.-O., Yeates, T.O., Dunn, B., and Chang, J.P. (2008) Generation of oxide nanopatterns by combining self-assembly of S-layer proteins and area-selective atomic layer deposition. *J. Am. Chem. Soc.*, **130** (50), 16908–16913.
20 Chen, R., Kim, H., McIntyre, P.C., and Bent, S.F. (2004) Self-assembled monolayer resist for atomic layer deposition of HfO_2 and ZrO_2 high-kappa gate dielectrics. *Appl. Phys. Lett.*, **84** (20), 4017–4019.
21 Singh, G., Bittner, A.M., Loscher, S., Malinowski, N., and Kern, K. (2008) Electrospinning of diphenylalanine nanotubes. *Adv. Mater.*, **20**, 2332–2336.
22 Han, T.H., Oh, J.K., Park, J.S., Kwon, S.-H., Kim, S.-W., and Kim, S.O. (2009) Highly entangled hollow TiO_2 nanoribbons templating diphenylalanine assembly. *J. Mater. Chem.*, **19**, 3512–3516.
23 Kim, S.-W., Han, T.H., Kim, J., Gwon, H., Moon, H.-S., Kang, S.-W., Kim, S.O., and Kang, K. (2009) Fabrication and electrochemical characterization of TiO_2 three-dimensional nanonetwork based on peptide assembly. *ACS Nano*, **3** (5), 1085–1090.
24 Han, T.H., Moon, H.-S., Hwang, J.O., Seok, S.I., Im, S.H., and Kim, S.O. (2010) Peptide-templating dye-sensitized solar cells. *Nanotechnology*, **21**, 185601.
25 Kim, G.-M., Lee, S.-M., Michler, G.H., Roggendorf, H., Gösele, U., and Knez, M. (2008) Nanostructured pure anatase titania tubes replicated from electrospun polymer fiber templates by atomic layer deposition. *Chem. Mater.*, **20** (9), 3085–3091.
26 Wang, R., Hashimoto, K., Fujishima, A., Chikuni, M., Kojima, E., Kitamura, A., Shimohigoshi, M., and Watanabe, T. (1997) Light-induced amphiphilic surfaces. *Nature*, **388** (6641), 431–432.

27 Bach, U., Lupo, D., Comte, P., Moser, J.E., Weissortel, F., Salbeck, J., Spreitzer, H., and Grätzel, M. (1998) Solid-state dye-sensitized mesoporous TiO_2 solar cells with high photon-to-electron conversion efficiencies. *Nature*, **395** (6702), 583–585.
28 Richter, J., Seidel, R., Kirsch, R., Mertig, M., Pompe, W., Plaschke, J., and Schackert, H.K. (2000) Nanoscale palladium metallization of DNA. *Adv. Mater.*, **12** (7), 507–510.
29 Lu, Y., Bangsarutip, S., Wang, X., Zhang, L., Nishi, Y., and Dai, H. (2006) DNA functionalization of carbon nanotubes for ultrathin atomic layer deposition of high *k* dielectrics for nanotubes transistors with 60 mV/decade switching. *J. Am. Chem. Soc.*, **128** (11), 3518–3519.
30 Ramachandran, G.N. and Kartha, G. (1955) Structure of collagen. *Nature*, **176** (4482), 593–595.
31 Rich, A. and Crick, F.H.C. (1961) The molecular structure of collagen. *J. Mol. Biol.*, **3** (5), 483–506.
32 Lee, S.-M., Grass, G., Kim, G.-M., Dresbach, C., Zhang, L., Gösele, U., and Knez, M. (2009) Low-temperature ZnO atomic layer deposition on biotemplates: flexible photocatalytic ZnO structures from eggshell membranes. *Phys. Chem. Chem. Phys.*, **11**, 3608–3614.
33 Lee, S.-M., Pippel, E., Moutannabir, O., Gunkel, I., Thurn-Albrecht, T., and Knez, M. (2010) Improved mechanical stability of dried collagen membrane after metal infiltration. *ACS Appl. Mater. Interfaces*, **2** (5), 2436–2441.
34 Lee, S.-M., Pippel, E., Gösele, U., Dresbach, C., Qin, Y., Chandran, C.V., Bräuniger, T., Hause, G., and Knez, M. (2009) Greatly increased toughness of infiltrated spider silk. *Science*, **324** (5926), 488–492; Lee, S.-M., Pippel, E., and Knez, M. (2011) Metal infiltration into biomaterials by ALD and CVD: A comparative study. *Chem. Phys. Chem.*, **12**, 791–798.
35 Wilson, C.A., Grubbs, R.K., and George, S.M. (2005) Nucleation and growth during Al_2O_3 atomic layer deposition on polymers. *Chem. Mater.*, **17** (23), 5625–5634.
36 Wang, Y., Qin, Y., Berger, A., Yau, E., He, C., Zhang, L., Gösele, U., Knez, M., and Steinhart, M. (2009) Nanoscopic morphologies in block copolymer nanorods as templates for atomic-layer deposition of semiconductors. *Adv. Mater.*, **21** (27), 2763–2766.
37 Zhang, L., Patil, A.J., Li, L., Schierhorn, A., Mann, S., Gösele, U., and Knez, M. (2009) Chemical infiltration during atomic layer deposition: metalation of porphyrins as model substrates. *Angew. Chem., Int. Ed.*, **48** (27), 4982–4985.
38 Kemell, M., Pore, V., Ritala, M., Leskelä, M., and Linden, M. (2005) Atomic layer deposition in nanometer-level replication of cellulosic substances and preparation of photocatalytic TiO_2/cellulose composites. *J. Am. Chem. Soc.*, **127** (41), 14178–14179.
39 Kemell, M., Pore, V., Ritala, M., and Leskelä, M. (2006) Ir/oxide/cellulose composites for catalytic purposes prepared by atomic layer deposition. *Chem. Vapor Depos.*, **12** (7), 419–422.
40 Hyde, D.K., Park, K.J., Stewart, S.M., Hinestroza, J.P., and Parsons, G.N. (2007) Atomic layer deposition of conformal inorganic nanoscale coatings on three-dimensional natural fiber systems: effect of surface topology on film growth characteristics. *Langmuir*, **23** (19), 9844–9849.
41 Hyde, G.K., Scarel, G., Spagnola, J.C., Peng, Q., Lee, K., Gong, B., Roberts, K.G., Roth, K.M., Hanson, C.A., Devine, C.K., Stewart, S.M., Hojo, D., Na, J.-S., Jur, J.S., and Parsons, G.N. (2010) Atomic layer deposition and abrupt wetting transitions on nonwoven polypropylene and woven cotton fabrics. *Langmuir*, **26** (4), 2550–2558.
42 Mumm, F., Kemell, M., Leskelä, M., and Sikorski, P. (2010) A bio-originated porous template for the fabrication of very long, inorganic nanotubes and nanowires. *Bioinspir. Biomim.*, **5**, 026005.
43 Huang, J., Wang, X., and Wang, Z.L. (2006) Controlled replication of butterfly wings for achieving tunable

photonic properties. *Nano Lett.*, **6** (10), 2325–2331.

44 Lee, S.-M., Üpping, J., Bielawny, A., and Knez, M. (2011) The structure based color of natural petals discriminated by polymer replication. *ACS Appl. Mater. Interfaces*, **3** (1), 30–34.

45 Gaillot, D.P., Deparis, O., Welch, V., Wagner, B.K., Vigneron, J.P., and Summers, C.J. (2008) Composite organic–inorganic butterfly scales: production of photonic structures with atomic layer deposition. *Phys. Rev. E*, **78** (3), 031922.

46 Huang, J., Wang, X., and Wang, Z.L. (2008) Bio-inspired fabrication of antireflection nanostructures by replicating fly eyes. *Nanotechnology*, **19** (2), 025602.

47 Brunner, R., Deparnay, A., Helgert, M., Burkhardt, M., Lohmuller, T., and Spatz, J.P. (2008) Product piracy from nature: biomimetic microstructures and interfaces for high-performance optics. *Proc. SPIE*, **7057**, 705705.

48 Zhao, Y., Wei, M., Wang, Z.L., and Duan, X. (2009) Biotemplated hierarchical nanostructure of layered double hydroxides with improved photocatalysis performance. *ACS Nano*, **3** (12), 4009–4016.

49 Fan, H.J., Knez, M., Scholz, R., Nielsch, K., Pippel, E., Hesse, D., Zacharias, M., and Gösele, U. (2006) Monocrystalline spinel nanotube fabrication based on the Kirkendall effect. *Nat. Mater.*, **5** (8), 627–631.

50 Ding, Y., Xu, S., Zhang, Y., Wang, A.C., Wang, M.H., Xiu, Y., Wong, C.P., and Wang, Z.L. (2008) Modifying the anti-wetting property of butterfly wings and water strider legs by atomic layer deposition coating: surface materials versus geometry. *Nanotechnology*, **19** (35), 355708.

51 Cassie, A.B.D. and Baxter, S. (1944) Wettability of porous surfaces. *Trans. Faraday Soc.*, **40**, 546–551.

52 Wenzel, R.N. (1936) Resistance of solid surfaces to wetting by water. *Ind. Eng. Chem.*, **28** (8), 988–994.

53 Quere, D. (2005) Non-sticking drops. *Rep. Prog. Phys.*, **68** (11), 2495–2532.

54 Losic, D., Triani, G., Evans, P.J., Atanacio, A., Mitchell, J.G., and Voelcker, N.H. (2006) Controlled pore structure modification of diatoms by atomic layer deposition. *J. Mater. Chem.*, **16** (41), 4029–4034.

55 Rorrer, G.L., Jeffryes, C., Chang, C.-H., Lee, D.-H., Gutu, T., Jiao, J., and Solanki, R. (2007) Biological fabrication of nanostructured silicon–germanium photonic crystals possessing unique photoluminescent and electroluminescent properties. *Proc. SPIE*, **6645**, 66450A.

56 Finch, D.S., Oreskovic, T., Ramadurai, K., Herrmann, C.F., George, S.M., and Mahajan, R.L. (2008) Biocompatibility of atomic layer-deposited alumina thin films. *J. Biomed. Mater. Res. A*, **87A** (1), 100–106.

57 Liang, X., Lynn, A.D., King, D.M., Bryant, S.J., and Weimer, A.W. (2009) Biocompatible interface films deposited within porous polymers by atomic layer deposition (ALD). *ACS Appl. Mater. Interfaces*, **1** (9), 1988–1995.

58 Putkonen, M., Sajavaara, T., Rahkila, P., Xu, L., Cheng, S., Niinistö, L., and Whitlow, H.J. (2009) Atomic layer deposition and characterization of biocompatible hydroxyapatite thin films. *Thin Solid Films*, **517** (20), 5819–5824.

59 Narayan, R.J., Monteiro-Riviere, N.A., Brigmon, R.L., Pellin, M.J., and Elam, J.W. (2009) Atomic layer deposition of TiO_2 thin films on nanoporous alumina templates: medical applications. *JOM*, **61** (6), 12–16.

60 Narayan, R.J., Adiga, S.P., Pellin, M.J., Curtiss, L.A., Stafslien, S., Chisholm, B., Monteiro-Riviere, N.A., Brigmon, R.L., and Elam, J.W. (2010) Atomic layer deposition of nanoporous biomaterials. *Mater. Today*, **13** (3), 60–64.

61 Narayan, R.J., Adiga, S.P., Pellin, M.J., Curtiss, L.A., Hryn, A.J., Stafslien, S., Chisholm, B., Shih, C.-C., Shih, C.-M.,

Lin, S.-J., Su, Y.-Y., Jin, C., Zhang, J., Monteiro-Riviere, N.A., and Elarn, J.W. (2010) Atomic layer deposition-based functionalization of materials for medical and environmental health applications. *Philos. Trans. R. Soc. A*, **368** (1917), 2033–2064.

62 Hyde, G.K., McCullen, S.D., Jeon, S., Stewart, S.M., Jeon, H., Loboa, E.G., and Parsons, G.N. (2009) Atomic layer deposition and biocompatibility of titanium nitride nano-coatings on cellulose fiber substrates. *Biomed. Mater.*, **4** (2), 025001.

14
Coating of Carbon Nanotubes

Catherine Marichy, Andrea Pucci, Marc-Georg Willinger, and Nicola Pinna

14.1
Introduction

Carbon nanotubes (CNTs) offer a high surface area, good thermal and electrical conductivity, and mechanical as well as chemical stability. As such, they are ideally suited as support for a second material that can be deposited onto their surface either as particles or as a thin film. Such heterostructures find applications in catalysis, energy storage, or gas sensing, where it is essential to expose the active phase on a large surface area. Due to the small dimensions, interactions between the deposited material and the tubes at the interface can significantly alter the properties of the composite. This is specifically the case for semiconducting materials when the dimensions are in the range of the Debye length. Therefore, the modification of CNTs is of outmost scientific and technological importance. Many reports dealing with the synthesis, characterization, and properties of such hybrid materials can be found in the recent literature [1]. By using different material combinations and synthesis procedures, MO_x/CNT composites, MO_x-coated CNTs, or MO_x-filled CNTs have been prepared and their electrical, electrochemical, (photo)catalytic, field emission, or gas sensing properties investigated [1]. Studying such synergetic phenomena and size-dependent properties in a systematic way relies on a method that allows a precise control of the particle or film growth during the deposition. Nevertheless, the common features of almost all the techniques used are the nonhomogeneity of the metal oxide coating and often only partial covering of the entire surface of the tubes, resulting in discontinuous films or islands of inorganic coatings (cf. Sections 14.3 and 14.4). Moreover, control and reproducibility of the film thickness are often difficult or impossible to obtain. Among the techniques available, atomic layer deposition (ALD) is specifically suited for the production of such heterostructures. It is based on subsequent gas–surface reactions and therefore allows the coating of flat surfaces as well as complex and high surface area nanostructures in a conformal and homogeneous manner with a precise control over the thickness of the deposited film in the range of a few angstroms (cf. Section 14.5). The main drawback of ALD is the difficulty to process a large amount of CNTs. However, this

Atomic Layer Deposition of Nanostructured Materials, First Edition. Edited by Nicola Pinna and Mato Knez.

problem was recently addressed by making use of fluidized bed or rotary ALD reactors (cf. Section 14.6).

The objective of this chapter is to review the decoration/coating of CNTs by ALD and to discuss it in comparison to the chemical and physical techniques commonly employed. The chapter starts with a short description of the purification and surface functionalization of CNTs (Section 14.2). Indeed, their understanding and application is fundamental to the decoration/coating process and ultimately to achieve the desired nanostructure. Then, a short overview of solution- (Section 14.3) and gas-phase techniques (Section 14.4) precedes the main part dealing with ALD (Section 14.5). Before concluding, the ALD on graphene and graphite (Section 14.7) is also discussed as the challenges that have to be faced for their coating are very similar to those found for CNTs.

14.2
Purification and Surface Functionalization of Carbon Nanotubes

In order to decorate and coat CNTs, it is primordial to characterize and modify their surface chemistry. Although CNTs can be synthesized using several techniques and routes, in all the cases, the pristine CNTs contain a variety of impurities ranging from other forms of carbon (i.e., fullerenes, graphite, and amorphous carbon) to metal nanoparticles used as catalyst to promote the CNT growth. Numerous procedures have been used to purify CNTs [2]. Two types of functionalizations can be distinguished: covalent and noncovalent. The most common ways involve covalent functionalization by gas- and liquid-phase oxidative treatments. Even if the chemical functionalization permits to obtain a fine dispersion and a relatively selective separation, no way to purify the product without damaging the nanotubes or changing the original morphology can be unfortunately achieved. Here, an overview of the two classes of approaches for the surface functionalization of CNTs is presented. More details can be found in the review of Tasis *et al.* [3].

On the one hand, covalent functionalization by treatment of pristine CNTs with strong acids such as HNO_3 or other strong oxidizing agents (e.g., O_3) permits to generate oxygenated functional groups such as alcohol, ketone, ether, carboxylic acid, and ester [4, 5]. This first chemical modification of the CNT surface paves the way to an endless possibility of further attachments. Esterification and amidation, together with the generation of a zwitterionic linkage, are the most studied and developed approaches. As will be demonstrated in the next sections, the presence of these oxygenated functional groups is needed for the initiation of the ALD growth onto CNTs. As a matter of fact, no growth is observed when ALD is attempted on highly graphitized and unfunctionalized or defect-free CNT walls.

On the other hand, noncovalent functionalization is mainly based on supramolecular complexation using various adsorptive and anchoring forces, such as van der Waals and π–π interactions, hydrogen bonding, and electrostatic forces. One of the main advantages of the noncovalent functionalization, besides the use of mild conditions, is that no serious changes in the electrical and mechanical properties of the pristine CNTs occur. Indeed, the carbon sp^2 structure and the conjugation of carbon atoms are preserved. This functionalization can be achieved, for instance, by the adsorption or anchoring of various functional molecules such as polymers or surfactants [6]. As a prominent example, the adsorption of sodium dodecyl sulfate (SDS) was used prior to Al_2O_3 ALD for the homogeneous coating of single-wall CNTs (SWCNTs) [7] as discussed in Section 14.5. Another particular noncovalent surface modification of CNTs was introduced by Farmer and Gordon [8]. In order to coat freestanding SWCNTs by Al_2O_3 ALD, the authors showed that the *in situ* physisorption of NO_2 on SWCNTs permitted to initiate the metal oxide growth. Because CNTs were already proposed as NO_2 gas sensors, NO_2 adsorption is relatively well documented as well as its influence on the CNT electrical response. NO_2 has an unpaired electron and it is known as a strong oxidizer. Therefore, upon adsorption, a charge transfer is likely to occur from the CNTs to adsorbed NO_2, resulting in the formation of electronic levels within the gap of the semiconducting tubes. These states are located close to the Fermi level and give rise to an increased conductivity [9]. Due to the reversible and nondissociative adsorption of NO_2, the surface modification had to be performed *in situ* and it is not applicable to liquid phase.

14.3 Decoration/Coating of Carbon Nanotubes by Solution Routes

The decoration and the coating of CNTs by solution routes can be realized in two ways: (i) the *in situ* synthesis of the coating compound directly at the surface of the nanotubes, or (ii) the attachment of preformed nanobuilding blocks at the nanotube surface. A large amount of work was published in the past decade on the subject; therefore, here we are only describing few prominent results. Deeper discussion of the subject can be found in the very recent review of Eder [1].

14.3.1 *In Situ* Coating

The *in situ* formation of materials directly at the surface of the CNTs not only has the clear advantage to be a one-step process, but also permits the formation of continuous and homogeneous coatings, instead of only attached nanoparticles at the surface of the CNTs.

As an example, the preparation of a vanadium oxide–CNT composite based on hydrolysis of NH_4VO_3 solution in the presence of acid-treated multiwalled CNTs (MWCNTs) was recently demonstrated in view of applications in heterogeneous catalysis [10].

The homogeneous coating can also be achieved on noncovalently functionalized MWCNTs. For example, benzyl alcohol adsorbs onto the surface of the CNTs via π–π interactions with the phenyl rings. The hydroxyl group of benzyl alcohol promotes the TiO_2 growth at the CNT surface via a classical sol–gel process [11].

Electrochemical deposition permits the growth of metal and metal oxide nanoparticles on CNT-modified electrodes. As a typical example, Pt electrodeposition was applied to produce CNT-modified electrodes for proton exchange membrane fuel cells [12].

Thermal decomposition under supercritical CO_2 conditions also allows the decoration/coating of CNTs as demonstrated by Peng *et al.* [13]. Indeed, MWCNT forest was coated with Ga_2O_3 from gallium acetylacetonate. The obtained film acted as a base for Ru ALD in order to synthesize heterostructures.

14.3.2
Attachment of Preformed Nanobuilding Blocks

The attachment of preformed nanobuilding blocks permits to fabricate complex architecture by grafting various kinds of materials and compounds to the surface of pristine or chemically functionalized CNTs.

A typical example showing the versatility of the approach is given by Correa-Duarte *et al.* [14]. The authors reported a multistep approach for the controlled deposition of $Au@SiO_2$ nanoparticles at the surface of MWCNTs. They used the polymer wrapping technique with an anionic polyelectrolyte (poly(sodium 4-styrenesulfonate)) to obtain a stable colloid suspension of the MWCNTs in water. In the second step, the MWCNTs were reacted with a cationic polyelectrolyte (poly(diallyldimethylammonium chloride)) to produce a second shell of polymer. Finally, the twice-wrapped MWCNTs were exposed to $Au@SiO_2$ colloids to form the final complex hybrid nanostructure.

Another interesting example is given by Li *et al.* [15]. Their approach consists of impregnation and sol–gel processes. CNTs were decorated with preformed nanoparticles by sonicating and mixing tubes with decorating materials. The sonication permitted to create additional defects making the CNT pretreatment nonmandatory. Using the same process, an additional material could be added onto the decorated CNTs. Via a step-by-step self-assembly approach, binary, ternary, and quaternary CNT-based composites have been achieved with various materials such as TiO_2, Co_3O_4, Au, Co, and CoO.

The importance of surface functional groups of CNT walls was highlighted by Pan *et al.* [16]. Three different functionalized CNTs (CNT-COOH, CNT-OH, and CNT-NH_2) were mixed with carboxyl-capped CdSe quantum dots (QDs). The strong electrostatic interaction between the positively charged amine-modified CNTs and the negatively charged capped QDs led to densely coated CNTs, contrary to the case of hydroxyl- and carboxyl-modified CNTs.

14.4
Decoration/Coating of Carbon Nanotubes by Gas-Phase Techniques

Various physical and chemical gas-phase deposition techniques have been used as well for the decoration and/or coating of CNTs.

Electron beam evaporation of Ti, Ni, Pd, Au, Al, and Fe onto freestanding SWCNTs led to pearl necklace-like morphology [17]. The size of the attached nanoparticles strongly depends on the deposition conditions and on the evaporated metal. As a matter of fact, the electron beam deposition of Au, Al, and Fe produced large particles disconnected from each other, while Ti, Ni, and Pd form a nanoparticle-like continuous coating. The pretreatment of the SWCNTs with a surfactant permitted to obtain a more continuous and homogeneous layer.

Radio-frequency plasma-assisted sputtering was used to deposit Zn onto MWCNTs. Oxidization of the coated sample in air led to ZnO nanoparticles forming a pearl necklace-like structure. The size of the particles could be simply controlled by the sputtering time [18].

Pulsed laser deposition permitted the coaxial coating of MWCNTs with various inorganic materials (i.e., ZrO_x, HfO_x, AlO_x, ZnO_x, and Au) [19]. On the one hand, the high-κ oxide layers were found to be amorphous and conformal. On the other hand, gold and zinc oxide deposition led to a coating made of nanocrystalline particles.

Chemical vapor deposition from tin hydride (SnH_4) in nitrogen at 550 °C was employed for the coating of MWCNTs with SnO_2 nanoparticles. The oxidation of Sn^0 to the oxide was attributed to residual oxygen in the reaction chamber. The degree of coverage of the CNTs with SnO_2 nanoparticles and the particle size can be adjusted by simply controlling the deposition time and the flow rate of the precursor mixture during the CVD procedure [20].

Cycled vacuum feeding CVD, by the *in situ* decomposition of SiH_4, has been used by Wang *et al.* to deposit silicon on MWCNTs [21]. This process is comparable to ALD in terms of intermittent feeding of the gaseous precursor and the controlled growth mechanism, but leads to a higher growth rate. A continuous and dense layer of Si particles has been achieved by this technique.

These typical examples prove that different gas-phase techniques can be employed for the decoration and/or the coating of carbon nanotubes. However, a conformal and continuous thin film is still difficult to achieve with these techniques. As will be described in the next section, ALD permits to obtain a more conformal and controlled coating of the CNTs, which is, for example, a prerequisite when a continuous high-κ oxide coating is required as gate dielectric in nanotube field effect transistors.

14.5 Atomic Layer Deposition on Carbon Nanotubes

The coating of MWCNTs with Al_2O_3, by ALD, from trimethylaluminum (TMA) and water was already reported in 2003 [22]. Multilayer coatings on MWCNTs could be obtained by the sequential Al_2O_3 and W ALD processes as described by Herrmann *et al.* [23]. TMA/H_2O and WF_6/Si_2H_6 cycles were used for the Al_2O_3 and tungsten deposition, respectively. The obtained "coaxial cables" are constituted by rather conformal ALD layers. This multistep approach permits to produce multifunctional nanostructures with interesting flexibility in terms of type of materials that can be deposited and their combination at the nanoscale.

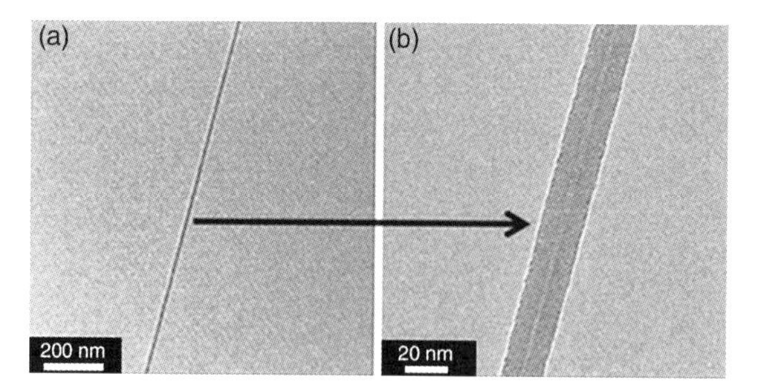

Figure 14.1 ALD coating behavior of suspended NO_2/TMA functionalized SWCNTs. (a) TEM micrograph of an Al_2O_3-coated SWCNT. (b) Higher magnification of the ~2 nm nanotube reveals the 10 nm coating to be uniform and continuous. Reproduced with permission from Ref. [8].

Suspended SWCNTs could be conformally coated with Al_2O_3, after a NO_2 pretreatment [8]. NO_2 reversibly physisorbs on the SWCNTs acting as anchoring site for further oxide growth (cf. also Section 14.2). One of the most relevant advantages of this approach is that the SWCNTs are not modified with chemical groups that would alter their physical properties. This is particularly important for the application of the freestanding coated SWCNTs as field effect transistors. The noncovalent functionalization was achieved by several NO_2 and TMA cycles prior to TMA and water cycles. Such a pretreatment produces a stable complex that does not desorb from the nanotube sidewalls at room temperature and allows the coating of SWCNTs with ALD material. TEM micrographs (Figure 14.1a and b) revealed the uniform and continuous resulting coating. The oxide layer thickness of 10 nm was in perfect agreement with the expected one from the number of ALD cycles carried out. Therefore, there is no evidence of inhibition or delay in nucleation of ALD Al_2O_3 onto the NO_2–TMA pretreated SWCNTs.

The idea of the alumina growth onto NO_2-treated SWCNTs arose from a previous paper of the same group, in which Al_2O_3 and HfO_2 were grown onto SWCNTs functionalized with aniline and nitroaniline [24]. The treated SWCNTs presented surface phenyl and nitrophenyl groups, respectively, as depicted in Figure 14.2 (left). Depending on the surface species, aluminum and hafnium oxide depositions onto the SWCNTs exhibited either localized nucleation or continuous coating, as clearly shown in the TEM images (Figure 14.2, right). The Al_2O_3 spheres measured between 15 and 20 nm and the diameter of the Al_2O_3 wire was 22 nm corresponding to a typical growth per cycle of 0.1 nm for the TMA/H_2O process at 225 °C. This finding once again proved the absence of evident inhibition or delay in the nucleation of Al_2O_3 onto surface nitryl species.

Min *et al.* [25] used MWCNT arrays grown on porous anodic alumina as templates for the growth of Ru thin films. The deposition took place after dissolution of the alumina matrix by phosphoric acid (Figure 14.3a and d), which also permitted to open the CNT tips, leading to the coating of the inside and outside (Figure 14.3b and e). After thermal treatment in oxygen at 500 °C, ruthenium was oxidized into RuO_2 and the MWCNTs burned off producing freestanding RuO_2 nanotube array (Figure 14.3c and f).

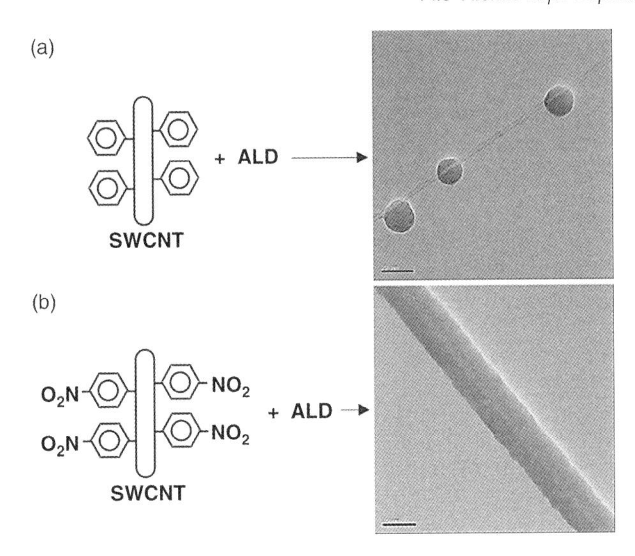

Figure 14.2 Reaction schematic with TEM results of functionalized suspended SWCNTs exposed to 100 ALD cycles of Al_2O_3. (a) SWCNTs functionalized using aniline. (b) SWCNTs functionalized using nitroaniline. Scale bar: 20 nm. Reproduced with permission from Ref. [24].

Gomathi *et al.* [26] coated HNO_3-treated MWCNTs with SiO_2, TiO_2, and Al_2O_3 from the respective metal chloride and water at 80 or 150 °C. In this work, an unconventional glass reactor was used. It consisted of two glass chambers for the precursors connected to the reaction chamber by high-vacuum stopcocks, which were manually operated. Despite the simplicity of the setup, the quality and conformality of the coatings were surprisingly good. The as-deposited hydroxides needed to be annealed to 350 °C to form the oxide.

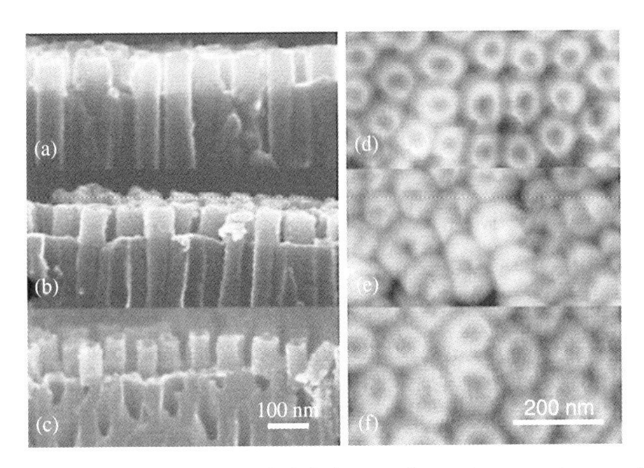

Figure 14.3 (a–c) SEM and (d–f) dynamic force microscopy images of (a and d) MWCNT array, (b and e) Ru-coated MWCNT array, and (c and f) RuO_2 nanotube array. Reproduced with permission from Ref. [25].

Javey *et al.* [27] deposited zirconium oxide thin films (~8 nm), as gate dielectrics for nanotube field effect transistors, on top of individual single-walled carbon nanotubes. Individual semiconducting SWCNTs, bridging metal source and drain electrodes (of spacing ~3 μm), were lying onto a SiO_2/Si substrate. Helbling *et al.* [28] also investigated SWCNT-based field effect transistors and showed that their encapsulation by approximately 100 nm Al_2O_3 ALD leads to stable device operation for 260 days and reduces the sensitivity to the environment. Recently, Nakashima *et al.* and Shen *et al.* coated CNTs with HfO_2 for CNT field effect transistors applied to biosensors and for nanoelectrodes based on a single CNT, respectively [29, 30].

Noncovalent coating of SWCNTs with poly-T DNA improved the nucleation of HfO_2 ALD [31]. Indeed, for SWCNTs on SiO_2 substrates with DNA functionalization, concurrent ALD growth on both nanotubes and SiO_2 surfaces was observed. Actually, a quasicontinuous and conformal layer was obtained. On the contrary, for SWCNTs without DNA functionalization, no HfO_2 nucleation and growth directly on the "defect-free" tube surface took place.

Kim *et al.* [32] fabricated coaxial nanotubes with highly conformal ZnO film, by employing an intermediate Al_2O_3 coating on the nanotubes. Indeed, the direct coating with ZnO led to highly irregular ZnO shells, but the Al_2O_3 shell formed a uniform coat on the nanotubes allowing deposition of a conformal ZnO layer.

In a recent study, Al_2O_3 ALD (from TMA and H_2O) on SWCNTs was studied in a viscous flow reactor and in fluidized bed reactor [7] (cf. also Section 14.6). The SWCNTs were pretreated with ethanol or with SDS prior to deposition. As expected, neither continuous nor conformal ALD films could be achieved on unfunctionalized SWCNTs. Only nucleation at defective sites was observed, leading to a pearl necklace-like morphology. However, after a simple ethanol pretreatment, SWCNT ropes could be coated with continuous, although not conformal, Al_2O_3 layers (Figure 14.4), proving that physisorbed ethanol promotes Al_2O_3 nucleation and that the density of nucleation sites is much higher than in the case of the unfunctionalized SWCNTs. The SDS-based surfactant dispersion technique permitted to successfully promote a conformal Al_2O_3 shell that contained no visible nodules or particles. The advantage of

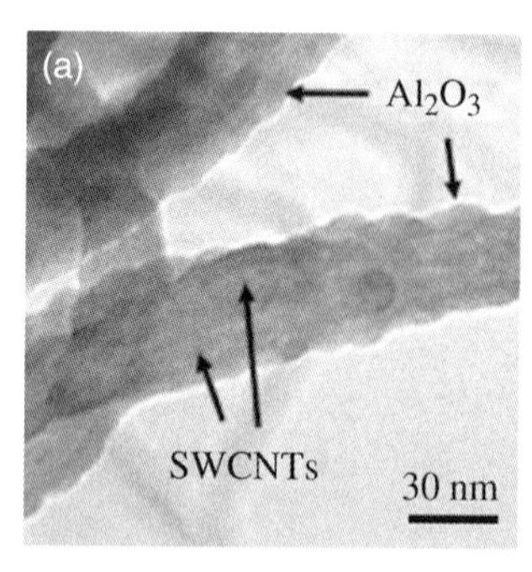

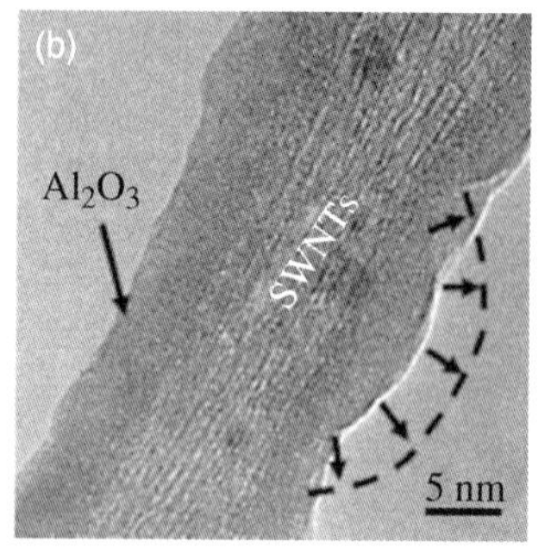

Figure 14.4 (a) TEM image of Al_2O_3 ALD on ethanol-dispersed SWCNT bundles, detailing continuous Al_2O_3 growth on bulk quantities of SWCNTs. (b) High-resolution TEM image of the same, showing radial growth from closely packed nodules that form a continuous film. Reproduced with permission from Ref. [7].

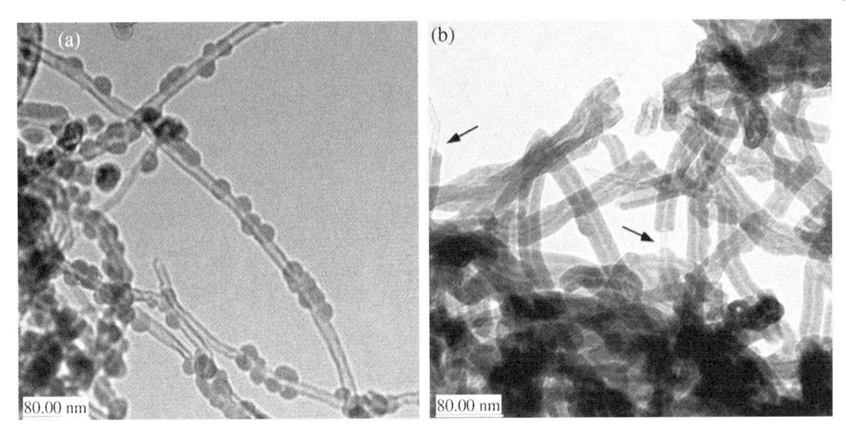

Figure 14.5 (a) TEM image of untreated MWCNTs after 50 ALD cycles of Al_2O_3. (b) TEM image of NO_2/TMA functionalized MWCNTs after 50 ALD cycles. Reproduced with permission from Ref. [33].

the functionalization with SDS is that no covalent chemical modification is required, which, as already discussed, considerably alters the physical properties of the pristine SWCNT.

ALD of Al_2O_3 (from TMA and H_2O) on untreated NC7000 MWCNTs from Nanocyl in a rotary reactor (cf. also Section 14.6) resulted in the growth of Al_2O_3 nanospheres forming a pearl necklace-like morphology. A TEM micrograph of these spherical and monodisperse nanospheres after 50 ALD cycles is shown in Figure 14.5a [33]. The TEM images suggested that Al_2O_3 nucleated at defects on the MWCNT surface during the initial Al_2O_3 ALD cycles. The nanospheres then grew isotropically with the number of ALD cycles to yield fairly monodisperse sphere diameters.

To grow conformal Al_2O_3, prefunctionalization of the MWCNT, similar to that proposed previously [8], from NO_2 and TMA was carried out at room temperature prior to the Al_2O_3 deposition at 180 °C. The TEM image after 50 ALD cycles proved that a smooth and conformal film can be grown under these conditions (Figure 14.5b).

Recently, CNTs have been coated via a nonhydrolytic approach applied to ALD, by using metal alkoxides and carboxylic acids as reactants [34, 35]. Based on this procedure, hafnium, titanium, vanadium, and tin oxide thin films of controlled thickness have been deposited onto the surface of functionalized CNTs. The obtained heterostructures were characterized by analytical electron microscopy using high-resolution imaging, electron energy loss and energy-dispersive X-ray spectroscopy, and elemental mapping. The morphology of the obtained hybrid materials with TiO_2, HfO_2, and V_2O_4 was nicely revealed by scanning electron microscope (SEM) images recorded using the secondary (SE) or backscattering electron (BSE) detector (not shown). In the TEM (Figure 14.6a and b), the contrast-rich, darker regions on the outer and inner walls of the CNTs correspond to the metal oxide layers deposited by the ALD process. In the energy filtered TEM images, the lighter regions correspond to the metal oxide (Figure 14.6e and f). The coating, only a few nanometers thick, was

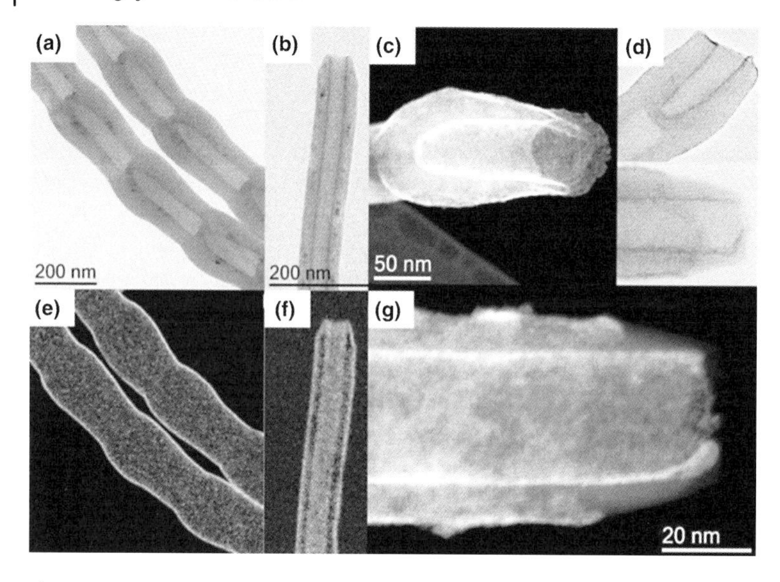

Figure 14.6 TEM images recorded from CNTs coated with V_2O_4 (a and b) and corresponding energy filtered images showing the distribution of vanadium on the surface of the tubes (e and f). High-angle annular dark field STEM images of CNTs coated with SnO_2 and corresponding TEM images are shown in (c), (g), and (d), respectively [9, 36, 37].

uniform along the whole surface of the tubes and presented approximately the same thickness on the inner and outer surfaces. Only the inner cavities in the bamboo-like tubes remain inaccessible for the precursor vapor and hence uncoated as can be clearly seen in the energy filtered images (Figure 14.6e). The film thickness abstracted from TEM measurements on the coated tubes was in agreement with reflectometry measurements performed on silicon wafers coated during the same deposition experiments. High-resolution TEM and electron diffraction experiments showed that the as-deposited films are amorphous and can be directly grown on the graphitic surface of the nitric acid-treated CNTs [9, 36]. The coating of CNTs with these metal oxides by the carboxylic acid route revealed well-defined and homogenous coatings of controllable thickness on the outer and inner parts of the walls.

The SnO_2-coated CNTs present a different growth behavior [37]. While the dark contrast due to the coating of the surface is relatively homogeneous in the high-resolution TEM images, a granular and particulate aspect of the deposited film can be seen in the dark field STEM images (Figure 14.6c, d, and g). The particulate structure of the film was confirmed by high-resolution TEM images. Close inspection revealed small contrast variations due to lattice fringes of nanometer-sized crystalline SnO_2 particles. Contrary to TiO_2, V_2O_4, and HfO_2 depositions [36], different growths per cycle were obtained for the coating of CNTs and Si wafer. Moreover, in the case of the CNTs used in this work, the walls are made up of stacked graphite cones and then the surface of the tubular structures terminates with graphitic planes only at the beginning and ending (Figure 14.6). As evidenced by the figure, the coating was homogeneous on the inner and outer walls of the tubes, whereas the exposed graphitic planes often remained uncoated [37]. This nicely demonstrates that initiation of film growth requires a certain degree of functionalization or the presence of defects in order to create anchoring points, as previously also noticed by Cavanagh *et al.* [33].

Finally, these various examples show that ALD permits, by controlling the surface chemistry and the density of nucleation sites, to produce complex nanostructures (e.g., conformal coating or perl necklace-like) and therefore to tune the functionalities and/or the properties of the final material.

14.6
Coating of Large Quantity of CNTs by ALD

A general drawback of the ALD processing of nanoparticles or CNTs is the low amount of material that can be homogeneously coated. As a matter of fact, the studies discussed in Section 14.5 were made on single CNT or just few milligrams deposited by drop coating on a substrate. For the production of commercially viable CNT-based composites, bulk quantities of CNTs per batch must be used [38]. For this purpose, fluidized bed reactors [39–42] or rotary reactors [43–45] can be used as in the case of nanoparticles.

Zhan *et al.* [7] used an ALD fluidized bed reactor composed of a small fluidized bed holding (4 cm in diameter) for the Al_2O_3 ALD coating of SWCNTs. The schematic diagram of this fluidized bed reactor is shown in Figure 14.7 [39]. The bed was fluidized using N_2 as the gas source at a pressure around 100 Pa. The Al_2O_3 deposition was carried out at 450 K using TMA and H_2O pulses of 750 s separated by 120 s purge times. By using a "large" reactor (15 cm in diameter), 100 g of SWCNTs per batch could be processed. The main drawback of fluidized bed reactors is that long precursor pulses are required leading to a large amount of reactants lost to the vacuum pump. Rotary reactors can overcome this drawback due to their possibility of static reactant exposures [43]. Cavanagh *et al.* [33] used a rotary reactor, recently used for the ALD on nanoparticles, for the coating of gram quantity of MWCNTs with Al_2O_3 and WO_3.

A schematic of the rotary reactor is shown in Figure 14.8. Each reactant entered the reaction chamber (A) through a needle and pneumatic valves (B) that were attached to 0.25″ welded ports on a custom 6.0″ conflat cap (C). During the reaction, the MWCNTs were mechanically agitated in a rotating porous cylinder (D). The rotation was achieved via a magnetically coupled rotary feedthrough (E). The pressure was monitored with 10 and 1000 Torr capacitance manometers (F). A gate valve (G) was opened to exhaust product gases and any excess precursor to the pump (H) [33].

The development of these reactors and the extension of their use from the coating of nanopowders to CNTs permit the fabrication of large quantity of multifunctional nanostructures at a quality not achievable by any other technique so far.

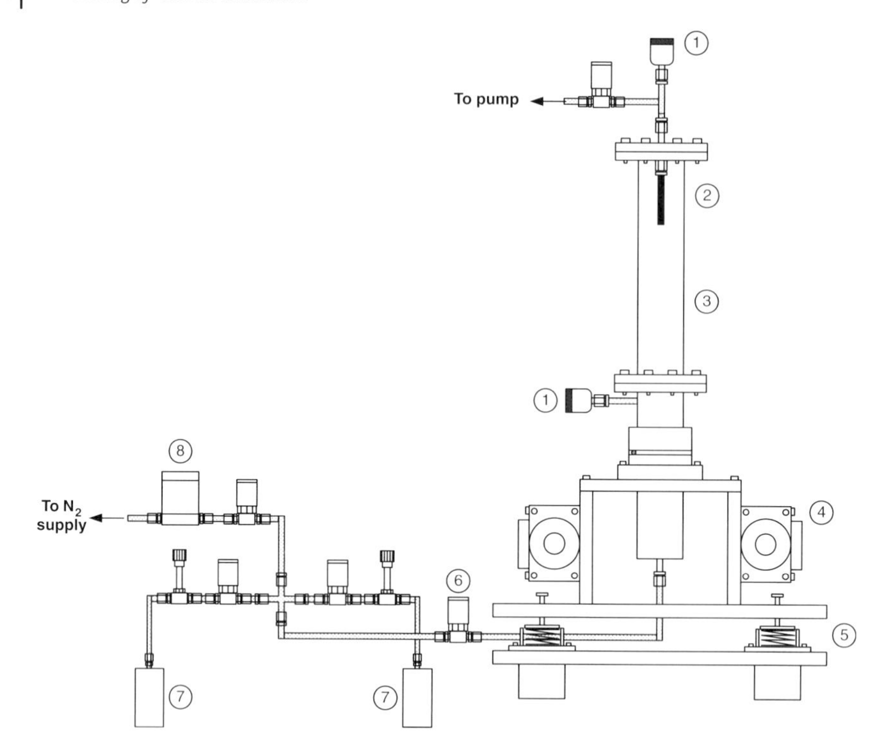

Figure 14.7 Schematic diagram of the fluidized bed reactor apparatus: (1) pressure transducers, (2) sintered metal filter, (3) reaction column, (4) vibro motors, (5) spring supports, (6) pneumatic valve, (7) reactant containers, and (8) mass flow controller. Reproduced with permission from Ref. [39].

14.7
ALD Coating of Other sp^2-Bonded Carbon Materials

Since the surface chemistry of graphene is analogous to that of CNTs, it is interesting to shortly describe ALD growth behavior onto these nanostructures as well.

Williams *et al.* [46] reported the realization of a single-layer graphene p–n junction, where the top gate was made by Al_2O_3 ALD onto graphene by a noncovalent functionalization. Indeed, several cycles of NO_2 and TMA have been realized prior to TMA and water cycles, as was reported for ALD on CNTs [8], in order to functionalize the surface without affecting the electronic structure and the degree of disorder of the graphene.

The noncovalent functionalization of pristine graphene with 3,4,9,10-perylene tetracarboxylic acid was used to promote the Al_2O_3 ALD [47]. The carboxylate functional groups on the perylene molecules serve as uniformly distributed sites for nucleation of ALD.

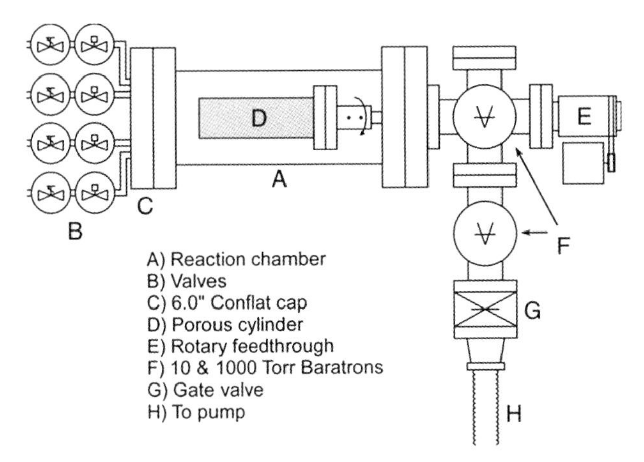

Figure 14.8 Schematic of the rotary reactor used to coat gram quantities of MWCNTs. Reproduced with permission from Ref. [33].

The growth of Al_2O_3 top gate dielectric onto graphene for field effect transistor fabrication was also achieved by deposition of a thin nucleation layer of oxidized Al between the graphene layer and the dielectric [48]. Prior to the Al_2O_3 layer growth by ALD, a 1–2 nm thick aluminum layer was deposited on the graphene surface by electron beam evaporation. After the Al deposition, the samples were exposed to air and transferred to the ALD chamber for the deposition of Al_2O_3 using TMA and water.

Lee *et al.* [49] compared the growth behavior of Al_2O_3 ALD onto highly oriented pyrolytic graphite (HOPG) as a function of the oxygen source and surface pretreatment. As expected, the TMA/water process onto clean HOPG substrates led to Al_2O_3 growth only along the sharp step edges and not on the chemically inert basal planes. When ozone was used as oxygen source, the growth occurred not only at the step edges but also on the basal planes. However, in order to reach a conformal growth onto the whole substrate, a pretreatment of the HOPG with ozone prior to the Al_2O_3 ALD was required. This functionalization increases the concentration of nucleation sites. Indeed, ozonolysis leads to the cleavage of C—C double bonds and to the formation of carboxylic acid/ester, ketone/aldehyde, and alcohol surface groups [4].

Finally, Speck *et al.* [50] studied the growth behavior of Al_2O_3 top gate dielectric onto epitaxial graphene and HOPG. When water was used as oxygen source, an inhomogeneous nucleation was observed, whereas ozone led to the formation of continuous oxide layers. However, significant degradation of the epitaxial graphene took place at high temperatures and ozone exposure. At lower temperature (i.e., 250 °C), the authors found the best deposition conditions, which do not damage too severely the epitaxial graphene but provide enough nucleation centers for the growth of pinhole-free ALD films.

14.8 Conclusions

The combination of different materials, especially when it is done at the nanometer scale, is the key to advanced functional materials. In this objective, due to their intrinsic properties, CNTs are ideally suited as support for other materials. In this chapter, an overview of the challenge of combining CNTs with other materials in order to couple and/or enhance the properties of both components in a symbiotic relation was presented. By traditional solution and physical routes, conformality, homogeneity, and control of the CNT decoration/coating appear still challenging. Moreover, a chemical functionalization is often required, leading to a modification of the intrinsic CNT physical properties, and a full coverage or a defect-free coating is rarely obtained.

The atomic layer deposition technique emerges as a powerful alternative to this challenge. Indeed, it allows the coating of flat surfaces as well as complex and high surface area nanostructures in a conformal and homogeneous manner with a precise control over the thickness of the deposited film in the range of a few angstroms. An *in situ* noncovalent functionalization, prior the ALD deposition, is achievable, permitting the formation of a conformal film as in the case of Al_2O_3 and HfO_2. Moreover, due to the versatility of the ALD processes, various materials can be deposited with different grades of homogeneity. Some of them lead only to the decoration of the CNTs, or to films made of particles. On the other hand, additional routes such as that involving nonhydrolytic chemistry allow the homogenous and defect-free coating of the inner and outer walls of the CNTs with a controlled thickness.

Furthermore, the coating of graphene and graphite appears as challenging as the one of the CNTs. As a matter of fact, only few ALD processes have been used so far. Similarly to CNTs, also the *in situ* functionalization and the type of the oxygen source are key parameters for the success of the deposition.

Finally, the most relevant challenge to make these CNT-based heterostructures appealing for industrial-scale applications is the scaling up of the process. Indeed, the ALD does not enable the coating of a large amount of material. However, new perspectives are opened by the use of rotary or fluidized bed reactors, which recently allowed the gram-scale ALD coating of CNTs.

References

1 Eder, D. (2010) Carbon nanotube–inorganic hybrids. *Chem. Rev.*, **110**, 1348–1385.

2 Park, T.-J., Banerjee, S., Hemraj-Benny, T., and Wong, S.S. (2006) Purification strategies and purity visualization techniques for single-walled carbon nanotubes. *J. Mater. Chem.*, **16**, 141–154.

3 Tasis, D., Tagmatarchis, N., Bianco, A., and Prato, M. (2006) Chemistry of carbon nanotubes. *Chem. Rev.*, **106**, 1105–1136.

4 Banerjee, S., Hemraj-Benny, T., and Wong, S.S. (2005) Covalent surface chemistry of single-walled carbon nanotubes. *Adv. Mater.*, **17**, 17–29.

5 Hirsch, A. (2002) Functionalization of single-walled carbon nanotubes. *Angew. Chem., Int. Ed.*, **41**, 1853–1859.

6 O'Connell, M.J., Boul, P., Ericson, L.M., Huffman, C., Wang, Y., Haroz, E., Kuper, C., Tour, J., Ausman, K.D., and Smalley, R.E. (2001) Reversible water-solubilization of single-walled carbon nanotubes by polymer wrapping. *Chem. Phys. Lett.*, **342**, 265–271.

7 Zhan, G.-D., Du, X., King, D.M., Hakim, L.F., Liang, X., McCormick, J.A., and Weimer, A.W. (2008) Atomic layer deposition on bulk quantities of surfactant-modified single-walled carbon nanotubes. *J. Am. Ceram. Soc.*, **91**, 831–835.

8 Farmer, D.B. and Gordon, R.G. (2006) Atomic layer deposition on suspended single-walled carbon nanotubes via gas-phase noncovalent functionalization. *Nano Lett.*, **6**, 699–703.

9 Willinger, M., Neri, G., Rauwel, E., Bonavita, A., Micali, G., and Pinna, N. (2008) Vanadium oxide sensing layer grown on carbon nanotubes by a new atomic layer deposition process. *Nano Lett.*, **8**, 4201–4204.

10 Chen, X.-W., Zhu, Z., Hävecker, M., Su, D.S., and Schlögl, R. (2007) Carbon nanotube-induced preparation of vanadium oxide nanorods: application as a catalyst for the partial oxidation of *n*-butane. *Mater. Res. Bull.*, **42**, 354–361.

11 Eder, D. and Windle, A.H. (2008) Carbon–inorganic hybrid materials: the carbon-nanotube/TiO_2 interface. *Adv. Mater.*, **20**, 1787–1793.

12 Wang, C., Waje, M., Wang, X., Tang, J.M., Haddon, R.C., and Yan, Y. (2004) Proton exchange membrane fuel cells with carbon nanotube based electrodes. *Nano Lett.*, **4**, 345–348.

13 Peng, Q., Spagnola, J.C., Daisuke, H., Park, K.J., and Parsons, G.N. (2008) Conformal metal oxide coatings on nanotubes by direct low temperature metal-organic pyrolysis in supercritical carbon dioxide. *J. Vac. Sci. Technol. B*, **26**, 978–982.

14 Correa-Duarte, M.A., Sobal, N., Liz-Marzán, L.M., and Giersig, M. (2004) Linear assemblies of silica-coated gold nanoparticles using carbon nanotubes as templates. *Adv. Mater.*, **16**, 2179–2184.

15 Li, J., Tang, S., Lu, L., and Zeng, H.C. (2007) Preparation of nanocomposites of metals, metal oxides, and carbon nanotubes via self-assembly. *J. Am. Chem. Soc.*, **129**, 9401–9409.

16 Pan, B., Cui, D., Ozkan, C.S., Ozkan, M., Xu, P., Huang, T., Liu, F., Chen, H., Li, Q., He, R., and Gao, F. (2008) Effects of carbon nanotubes on photoluminescence properties of quantum dots. *J. Phys. Chem. C*, **112**, 939–944.

17 Zhang, Y., Franklin, N.W., Chen, R.J., and Dai, H. (2000) Metal coating on suspended carbon nanotubes and its implication to metal–tube interaction. *Chem. Phys. Lett.*, **331**, 35–41.

18 Zhu, Y., Elim, H.I., Foo, Y.-L., Yu, T., Liu, Y., Ji, W., Lee, J.-Y., Shen, Z., Wee, A.T.S., Thong, J.T.L., and Sow, C.H. (2006) Multiwalled carbon nanotubes beaded with ZnO nanoparticles for ultrafast nonlinear optical switching. *Adv. Mater.*, **18**, 587–592.

19 Ikuno, T., Yasuda, T., Honda, S.-I., Oura, K., Katayama, M., Lee, J.-G., and Mori, H. (2005) Coating carbon nanotubes with inorganic materials by pulsed laser deposition. *J. Appl. Phys.*, **98**, 114305.

20 Kuang, Q., Li, S.-F., Xie, Z.-X., Lin, S.-C., Zhang, X.-H., Xie, S.-Y., Huang, R.-B., and Zheng, L.-S. (2006) Controllable fabrication of SnO_2-coated multiwalled carbon nanotubes by chemical vapor deposition. *Carbon*, **44**, 1166–1172.

21 Wang, Y.H., Li, Y.N., Lu, J., Zang, J.B., and Huang, H. (2006) Microstructure and thermal characteristic of Si-coated multi-walled carbon nanotubes. *Nanotechnology*, **17**, 3817.

22 Lee, J.S., Min, B., Cho, K., Kim, S., Park, J., Lee, Y.T., Kim, N.S., Lee, M.S., Park, S.O., and Moon, J.T. (2003) Al_2O_3 nanotubes and nanorods fabricated by coating and filling of carbon nanotubes with atomic-layer deposition. *J. Cryst. Growth*, **254**, 443–448.

23 Herrmann, C.F., Fabreguette, F.H., Finch, D.S., Geiss, R., and George, S.M. (2005) Multilayer and functional coatings on carbon nanotubes using atomic layer deposition. *Appl. Phys. Lett.*, **87**, 123110.

24 Farmer, D.B. and Gordon, R.G. (2005) ALD of high-kappa dielectrics on suspended functionalized SWNTs. *Electrochem. Solid-State Lett.*, **8**, G89–G91.

25 Min, Y.-S., Bae, E.J., Jeong, K.S., Cho, Y.J., Lee, J.-H., Choi, W.B., and Park, G.-S. (2003) Ruthenium oxide nanotube arrays fabricated by atomic layer deposition using a carbon nanotube template. *Adv. Mater.*, **15**, 1019–1022.

26 Gomathi, A., Vivekchand, S.R.C., Govindaraj, A., and Rao, C.N.R. (2005) Chemically bonded ceramic oxide coatings on carbon nanotubes and inorganic nanowires. *Adv. Mater.*, **17**, 2757–2761.

27 Javey, A., Kim, H., Brink, M., Wang, Q., Ural, A., Guo, J., McIntyre, P., McEuen, P., Lundstrom, M., and Dai, H. (2002) High-κ dielectrics for advanced carbon-nanotube transistors and logic gates. *Nat. Mater.*, **1**, 241–246.

28 Helbling, T., Hierold, C., Roman, C., Durrer, L., Mattmann, M., and Bright, V.M. (2009) Long term investigations of carbon nanotube transistors encapsulated by atomic-layer-deposited Al_2O_3 for sensor applications. *Nanotechnology*, **20**, 434010.

29 Shen, J., Wang, W., Chen, Q., Wang, M., Xu, S., Zhou, Y., and Zhang, X.-X. (2009) The fabrication of nanoelectrodes based on a single carbon nanotube. *Nanotechnology*, **20**, 245307.

30 Nakashima, Y., Ohno, Y., Kishimoto, S., Okochi, M., Honda, H., and Mizutani, T. (2010) Fabrication process of carbon nanotube field effect transistors using atomic layer deposition passivation for biosensors. *J. Nanosci. Nanotechnol.*, **10**, 3805–3809.

31 Lu, Y., Bangsaruntip, S., Wang, X., Zhang, L., Nishi, Y., and Dai, H. (2006) DNA functionalization of carbon nanotubes for ultrathin atomic layer deposition of high-κ dielectrics for nanotube transistors with 60 mV/decade switching. *J. Am. Chem. Soc.*, **128**, 3518–3519.

32 Kim, D.S., Lee, S.-M., Scholz, R., Knez, M., Gösele, U., Fallert, J., Kalt, H., and Zacharias, M. (2008) Synthesis and optical properties of ZnO and carbon nanotube based coaxial heterostructures. *Appl. Phys. Lett.*, **93**, 103108.

33 Cavanagh, A.S., Wilson, C.A., Weimer, A.W., and George, S.M. (2009) Atomic layer deposition on gram quantities of multi-walled carbon nanotubes. *Nanotechnology*, **20**, 255602.

34 Rauwel, E., Clavel, G., Willinger, M.-G., Rauwel, P., and Pinna, N. (2008) Non-aqueous routes to metal oxide thin films by atomic layer deposition. *Angew. Chem., Int. Ed.*, **47**, 3592–3595.

35 Rauwel, E., Willinger, M.-G., Ducroquet, F., Rauwel, P., Matko, I., Kiselev, D., and Pinna, N. (2008) Carboxylic acids as oxygen sources for the atomic layer deposition of high-κ metal oxides. *J. Phys. Chem. C*, **112**, 12754–12759.

36 Willinger, M., Neri, G., Bonavita, A., Micali, G., Rauwel, E., Herntrich, T., and Pinna, N. (2009) The controlled deposition of metal oxides onto carbon nanotubes by atomic layer deposition: examples and a case study on the application of V_2O_4 coated nanotubes in gas sensing. *Phys. Chem. Chem. Phys.*, **11**, 3615–3622.

37 Marichy, C., Donato, N., Willinger, M.-G., Latino, M., Karpinsky, D., Yu, S.-H., Neri, G., and Pinna, N. (2011) Tin dioxide sensing layer grown on tubular nanostructures by a non-aqueous atomic layer deposition process. *Adv. Funct. Mater.*, **21**, 658–666.

38 King, D.M., Liang, X., and Weimer, A.W. (2009) Functionalization of fine particles using atomic and molecular layer deposition. *ECS Trans.*, **25**, 163–190.

39 Hakim, L.F., George, S.M., and Weimer, A.W. (2005) Conformal nanocoating of zirconia nanoparticles by atomic layer deposition in a fluidized bed reactor. *Nanotechnology*, **16**, S375.

40 King, D.M., Spencer, J.A., II, Liang, X., Hakim, L.F., and Weimer, A.W. (2007) Atomic layer deposition on particles using a fluidized bed reactor with *in situ* mass spectrometry. *Surf. Coat. Technol.*, **201**, 9163–9171.

41 Liang, X., King, D.M., Li, P., and Weimer, A.W. (2009) Low-temperature atomic layer-deposited TiO_2 films with low photoactivity. *J. Am. Ceram. Soc.*, **92**, 649–654.

42 Wank, J.R., George, S.M., and Weimer, A.W. (2004) Nanocoating individual cohesive boron nitride particles in a fluidized bed by ALD. *Powder Technol.*, **142**, 59–69.

43 McCormick, J.A., Cloutier, B.L., Weimer, A.W., and George, S.M. (2007) Rotary reactor for atomic layer deposition on large quantities of nanoparticles. *J. Vac. Sci. Technol. A*, **25**, 67–74.

44 McCormick, J.A., Rice, K.P., Pau, D.F., Weimer, A.W., and George, S.M. (2007) Analysis of Al_2O_3 atomic layer deposition on ZrO_2 nanoparticles in a rotary reactor. *Chem. Vapor Depos.*, **13**, 491–498.

45 Wilson, C.A., McCormick, J.A., Cavanagh, A.S., Goldstein, D.N., Weimer, A.W., and George, S.M. (2008) Tungsten atomic layer deposition on polymers. *Thin Solid Films*, **516**, 6175–6185.

46 Williams, J.R., DiCarlo, L., and Marcus, C.M. (2007) Quantum hall effect in a gate-controlled p–n junction of graphene. *Science*, **317**, 638–641.

47 Wang, X., Tabakman, S.M., and Dai, H. (2008) Atomic layer deposition of metal oxides on pristine and functionalized graphene. *J. Am. Chem. Soc.*, **130**, 8152–8153.

48 Kim, S., Nah, J., Jo, I., Shahrjerdi, D., Colombo, L., Yao, Z., Tutuc, E., and Banerjee, S.K. (2009) Realization of a high mobility dual-gated graphene field-effect transistor with Al_2O_3 dielectric. *Appl. Phys. Lett.*, **94**, 062107.

49 Lee, B., Park, S.-Y., Kim, H.-C., Cho, K., Vogel, E.M., Kim, M.J., Wallace, R.M., and Kim, J. (2008) Conformal Al_2O_3 dielectric layer deposited by atomic layer deposition for graphene-based nanoelectronics. *Appl. Phys. Lett.*, **92**, 203102.

50 Speck, F., Ostler, M., Röhrl, J., Emtsev, K.V., Hundhausen, M., Ley, L., and Seyller, T. (2010) Atomic layer deposited aluminum oxide films on graphite and graphene studied by XPS and AFM. *Phys. Status Solidi C*, **7**, 398–401.

15
Inverse Opal Photonics

Davy P. Gaillot and Christopher J. Summers

15.1
Introduction and Background

Opals have long been admired as gemstones that exhibit beautiful colors spanning the entire visible spectrum. In opals, a slight deviation of the angle of observation or ambient light suffices to change its color, as shown in Figure 15.1a. Abundant in various countries, the universal admiration for this intriguing optical phenomenon made opals remarkable fashion accessories [1–3] such that these colorful jewels have been sought for centuries due to their obvious decoration and market value. Later, opals were scientifically studied for their unique optical properties and the insight they give into the formation of naturally occurring periodic structures. The iridescence properties of opals (Figure 15.1b) were related to Bragg diffractions, and thus, they are one of the first examples of the impact of three-dimensional periodic materials on optical properties. Also, early mineral analysis showed that opals are merely composed of silicon dioxide (silica) beads arrayed in a face-centered cubic (fcc) close-packed lattice and cemented together by hydrolyzed amorphous silica, as shown in Figure 15.1c.

Following the introduction of the "photonic crystal" (PC) concept by Yablonovitch and John [4, 5], it was very soon realized that opal-based structures provided an excellent test bed for experimentally investigating optical phenomena in 3D periodic media. Thus, they and their inverse structures have been responsible for many significant advances. Essentially, PCs are dielectric heterostructures whose properties can be described by the subtle marriage between solid-state physics and the electromagnetic theory of photons. By using architectures based on one-, two-, or three-dimensional periodic arrangements of conventional dielectric materials, unique photonic effects can be obtained. Surprisingly, although 1D stacks of alternating dielectric layers have been used for almost a century as highly efficient mirrors (>99%) for filtering or blocking a wide range of optical frequencies [6], it was many years before this technology was revisited and extended to more complex structures. This occurred in 1987 when Sajeev John and Eli Yablonovitch simultaneously reported that 3D periodic structures scaled at the wavelength of interest

Atomic Layer Deposition of Nanostructured Materials, First Edition. Edited by Nicola Pinna and Mato Knez.

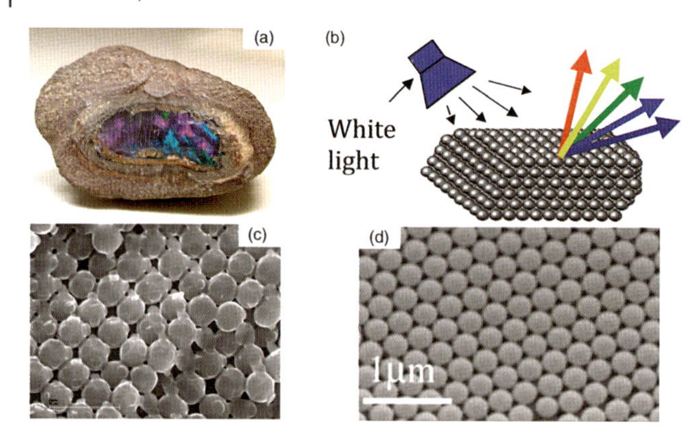

Figure 15.1 (a) Natural opal gemstone crystallized in rock. (b) Representation of colors produced by Bragg diffraction of white light from an opal. (c) High-magnification SEM image of (a) showing the arrangement of silica beads in a highly organized fcc lattice. The naturally formed silica spheres are cemented with hydrolyzed amorphous silica [3]. (d) SEM image of a synthetic opal obtained by self-assembly of ~250 nm silica spheres in an fcc lattice. Top view presents the densest close-packed direction of the crystal [111] direction. Synthetic opals exhibit a sharp Bragg peak along this direction, which can be predicted using Eq. (15.4).

exhibit unexpected optical properties. They showed that not only the Bragg stack diffraction mechanism can be extended to two- or three-dimensional dielectric scatters, as shown in Figure 15.1, to inhibit the spontaneous emission rate, but light can also be strongly localized as a means to enhance the spontaneous emission rate [4, 5]. Moreover, they predicted the formation of photonic band gaps (PBGs) that have become a cornerstone feature in PC technology. These arise because of the interaction between light and periodic low and high dielectric regions, which exactly mirrors the interaction between electrons and the periodic modulation of the atoms' potential in a crystalline lattice (Kronig–Penny model) that gives rise to energy band gaps. Therefore, the classical optical equivalent to the PBG is the Bragg diffraction mechanism. Light propagating through the crystal along symmetry directions experiences the periodicity of the corresponding planes and Bragg diffraction occurs as in X-ray diffraction. Hence, the Bragg peak is responsible for the sharp and vibrant color that makes opals so attractive. As in X-ray diffraction theory, the Bragg peak can be analytically derived from structure parameters, but with the additional consideration of material properties. For instance, opals exhibit a strong Bragg peak along the densest [111] directions (Figure 15.1d), which is given by the Bragg–Snell relationship

$$\lambda_{\mathrm{Bragg}} = 2d_{111}\sqrt{n_{\mathrm{eff}}^2 - (\sin\theta)^2}. \tag{15.1}$$

Equation (15.1) shows that the Bragg peak, λ_{Bragg}, strongly depends on the distance separating the (111) planes, d_{111}, the effective index (dielectric constant) of the material, n_{eff}, and the angle of incident of light, θ. In addition, d_{111} is related to the dielectric sphere diameter, D, by

$$d_{111} = \sqrt{(2/3)}D = 0.816D, \tag{15.2}$$

and the effective index, n_{eff}, is related to the fcc nature of the network by

$$n_{\text{eff}}^2 = n_{\text{silica}}^2 f_{\text{opal}} + n_{\text{air}}^2 f_{\text{air}} = 1.3475, \tag{15.3}$$

where the volume fractions of the silica opal, f_{opal}, and air, f_{air}, are 0.74 and 0.26, with refractive indices of 1.45 and 1, respectively. Finally, by combining Eqs. (15.2) and (15.3) for these conditions, Eq. (15.1) becomes (for silica spheres) at normal incidence

$$\lambda_{\text{Bragg}} \approx 2.2D. \tag{15.4}$$

A significant change in opal properties can be realized by noting that the air voids, which represent ~26% of the opal volume (Figure 15.2a), can be homogeneously or partially filled to create an infiltrated opal as shown in Figure 15.2b. Figure 15.2c and d shows the inverted structure that can be formed by selectively etching the dielectric spheres to leave air spheres within a high dielectric constant backbone and smaller tetrahedral and octahedral sites.

It should be noted that infiltrated and inverted structures still obey the Bragg–Snell law but the high-index material and air now occupy 26 and 74% of the total volume, respectively. Also, conformal infiltration of an opal forms a more complicated structure, basically an inverse shell or a highly conformal inverse opal in which the

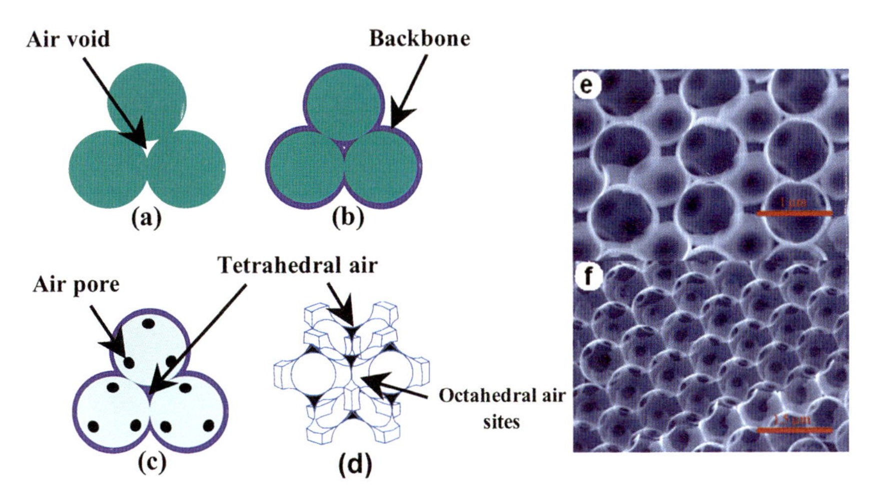

Figure 15.2 Sequential experimental steps to fabricate inverse shell opals. (a) Bare opal. The dielectric spheres account for 74% of the total dielectric volume, whereas the air interstitials account for the remaining 26%. (b) Partially infiltrated opal. High-index dielectric layers are conformally deposited onto the low-index spheres. (c) Inverse shell opal. The original template is selectively etched resulting in a high-index backbone in air. (d) Air network within the backbone. Small tetrahedral- and octahedral-shaped air pockets remain trapped within the backbone. Inverse shell opals exhibit a full PBG provided the dielectric contrast between the backbone and air is sufficiently high [7]. (e and f) SEM images from Ref. [8] of a cleaved silicon inverse shell opal along the [110] and [111] facets, respectively.

coating conforms to the topography of the spheres and consequently leaves air pockets at the interstitials. This is because in conformal deposition methods the fcc geometry hinders complete infiltration because the air channels eventually clog with increasing thickness. As a consequence, the maximum point for conformal infiltration is limited to 86% of the air void volume or 22.4% of the total opal volume fraction. Figure 15.2d shows that the interstitials of an inverse shell opal form a network composed of air channels connecting octahedral and tetrahedral air sites that comprise 3.6% of the pore volume. The octahedral air sites form an fcc air lattice shifted from the main fcc air lattice by half a unit cell, while the tetrahedral air sites form a simple cubic (sc) air lattice with a smaller lattice constant. The consequence of this inversion of the architecture was shown to allow the formation of wider Bragg peaks with respect to the original opal, which provided a very strong impetus to more fully exploit the optical properties of opal-based structures and to expand their range of applications. Both the conventional opal and inverse opal structures and their respective photonic band structures and properties, and dependence on refractive index contrast, are shown in Figure 15.3.

As shown in Figure 15.3, the optical properties of a PC are characterized by their normalized photonic band diagrams, which define the allowed photon frequencies (ω_n) to photon wave vector ($ka/2\pi$) relationships within the crystal [9, 10]. This approach was borrowed from solid-state physics in which electronic band structures

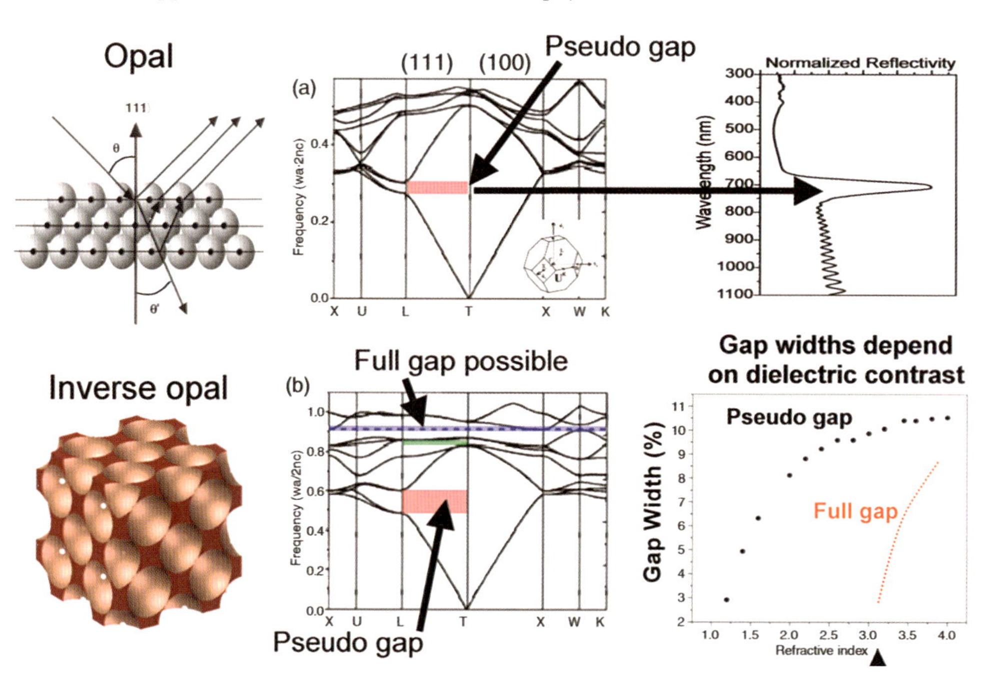

Figure 15.3 Schematic showing the Bragg diffraction properties of opals and inverse opals, their photonic band diagrams, and their resulting spectral and optical properties. (a) For an opal note the correspondence between the Bragg diffraction and photonic band gap along the L–Γ (111) direction. (b) Schematic of an inverse opal and corresponding photonic band diagram and presence of both directional and full photonic band gaps are also shown with their dependence on refractive index contrast.

are used to define the properties of electrons within the bulk matrix. The normalization scheme is powerful because it gives directly the unscaled information of the PBG properties: PBG width and spectral position. The midpoint frequency of the PBG ω_{PBG} is inversely proportional to the structure periodicity, whereas the PBG width is given by the ratio of the PBG frequency width $\Delta\omega_{PBG}$ to the midpoint frequency ω_{PBG}. Many simulation techniques such as the finite-difference time-domain (FDTD) or plane wave expansion (PWE) have been developed over the past 15 years to compute these band diagrams and predict the PBG properties. A comprehensive summary of these techniques is given in recent texts by Joannopoulos *et al.* [9, 10].

15.2 Properties of Three-Dimensional Photonic Band Structures

In 1987, Sajeev John and Eli Yablonovitch proposed that omnidirectional PBGs or stop bands would enable modifications to the spontaneous emission rate and that strong Anderson localization would occur in periodically modulated dielectric materials (PCs). In certain cases, the density of states (DOS) goes to zero within the PBG and photons in this gap are not allowed either to propagate through or to be emitted from the crystal. A wide PBG makes it feasible to engineer localized electromagnetic states within the PBG to obtain far more control over the optical properties [11]. This can be achieved by locally removing (donor) or adding (acceptor) dielectric regions. Conceptually, dielectric doping is the same as the doping typically performed in semiconductors. Figure 15.4 shows the potential of a titania/air inverse opal to enhance or inhibit the spontaneous emission rate of embedded CdSe quantum dots with various diameters [12]. Although the prepared inverse structure only exhibits a partial photonic gap, changes in lifetime by up to a factor of 2, compared to a homogeneous medium baseline, were clearly observed over a broad range of frequencies.

Therefore, it can be immediately seen that most advantages and impact will be obtained with PCs exhibiting a full and large 3D PBG, and that ultimate control of light can only be achieved in 3D PBG PCs in which photons generated from embedded light emitters are completely trapped or light rays externally interacting with the PC are totally reflected. Consequently, many techniques have been developed or invented to fabricate inorganic periodic networks or suitable organic networks that can then be further processed with higher index materials.

15.2.1 Synthetic Opals and Inverse Opals

In 1990, Ho and coworkers theoretically investigated the dependence of 3D PC properties on structure for the diamond, diamond-like, and fcc lattices [13]. They predicted a 28% full photonic band gap (FPBG) in the diamond structure and that a refractive index contrast as low as 2.0 was sufficient to open a FPBG. No gap was predicted in the fcc lattice, but later in 1992 Sozuer *et al.* repeated these calculations

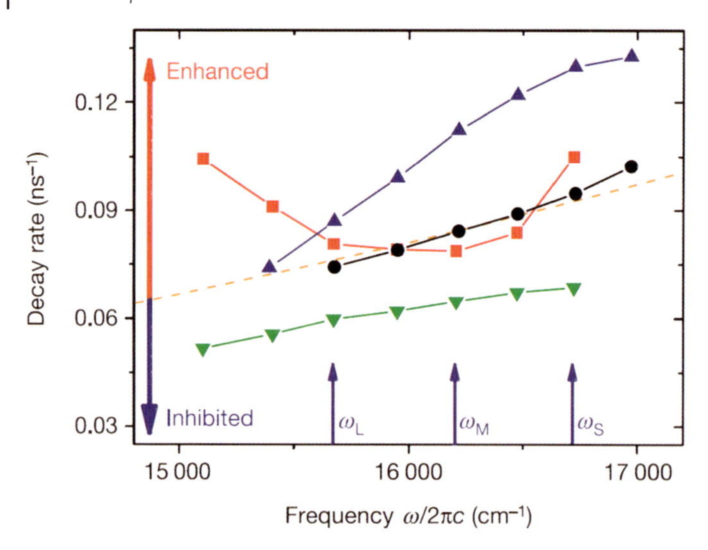

Figure 15.4 Measured decay rates of the excited states of quantum dots in photonic crystals with different lattice parameters. Lattice parameters are $a = 370$ nm (black circles), $a = 420$ nm (red squares), $a = 500$ nm (green upside-down triangles), and $a = 580$ nm (blue triangles). The dashed curve is the calculated decay rate for an electric dipole in a homogeneous medium. Three fixed emission frequencies are specified corresponding to large (ω_L, diameter 6.0 nm), medium (ω_M, diameter 4.5 nm), and small (ω_S, diameter 3.8 nm) sized quantum dots. After Ref. [12].

and identified a direct full PBG of 5.1% between the eighth and ninth bands in inverse opal structures for refractive index values >3.5 [14]. Thus, Ho *et al.*'s work inspired the fabrication of the diamond-like structure and the first experimental demonstration of a FPBG material. This was reported in the microwave regime in 1991 by Yablonovitch *et al.* [15]. The "Yablonovite" structure was fabricated by mechanical drilling along three of the diamond lattice axes in a dielectric slab at predetermined angles. The long-range periodicity of the crystal was then obtained by repetitively applying this technique to a mask array of triangular holes much as in a television set. Later, the repetitive application of the traditional "top-down" 2D semiconductor fabrication was used to form diamond-like structures as reported by Lin *et al.*, Ozbay *et al.*, and Noda *et al.* [16–18]. Figure 15.5 shows 3D PCs fabricated by electron beam lithography [16], holographic lithography [19], two-photon absorption [20], glancing angle deposition (GLAD) [21], robotic manipulation [22], and self-assembly [17]. Apart from the electron beam and GLAD techniques, the importance of all the other methods is in forming high-precision templates that can then be infiltrated and converted to the inverse structure. (*Note*: Unfortunately, the fabrication of a true diamond structure requires overcoming enormous technological issues since it possesses two "atoms" or scattered elements per unit cell and it is only recently that effective methods have been suggested for the micron range [24].) Applying the scalability rule to these calculations shows that a ~150–300 nm periodicity is required for operation at optical wavelengths, which coupled with

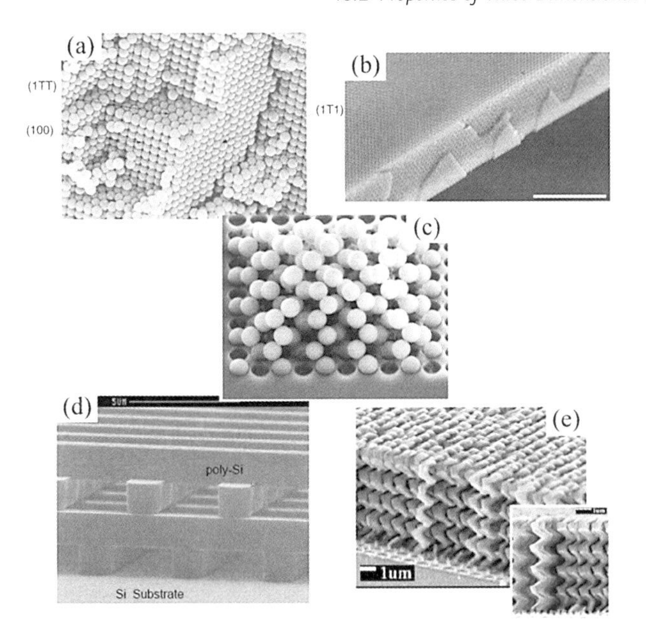

Figure 15.5 Montage of 3D PCs fabricated with various processes. (a) Self-assembly of colloidal particles in an fcc lattice [23]. (b) Synthetic fcc lattice realized by holographic lithography of photosensitive SU-8 polymer [19]. (c) Diamond lattice obtained by nanorobotic manipulation of silica and polystyrene spheres [22]. (d) Woodpile structure fabricated by electron beam lithography and MOCVD deposition of polysilicon [16]. (e) Square spiral lattices fabricated by the GLAD method [21].

inevitable dielectric material processing constraints presents a most challenging technological barrier that during the past decade has promoted many studies.

However, despite the fact that synthetic opals were predicted to be good candidates for achieving 3D light confinement, the low refractive index n of a silica opal ($n \sim 1.45$) opens only directional or pseudo-PBGs (PPBGs) between the second and third bands in the photonic band diagram [11, 25, 26]. In contrast, the work of Sozuer *et al.* suggested that inverse opal structures could provide a very effective route to obtaining structures with a full PBG and provide a direct experimental path to explore photonic effects in 3D structures. Consequently, one of Nature's amazing recipes was investigated to see if it could be emulated to form high-quality opals that could then be inverted to form viable photonic crystal structures. This triggered studies to understand the origin of optical effects in conventional opals, as well as methods to enable the rapid formation of opals and to fabricate inverse structures.

In 1979, Iler had already demonstrated "bottom-up" self-assembly and cementing techniques to close-pack and bond presynthesized silica spherical particles into an opal network [3] and this technique was first adopted by Van Blaaderen *et al.* to form a synthetic optical opal structure [27]. Iler had increased the sedimentation process by a factor of a million (growth rate of 1 year), but faster techniques were obviously desirable to fabricate synthetic opals. Nevertheless, this work provided a major fabrication breakthrough and was therefore revisited to build fcc structures with long-range order for PBG applications. Typically, colloidal particles or dielectric spherical particles diluted in a viscous solvent are ordered by van der Waals forces into a self-assembled close-packed array [1]. Simultaneously, as the solvent is removed, crystallization into an fcc lattice (or a hexagonal closed-packed (hcp) lattice) occurs under thermodynamic constraints that only allow certain phases to form [28, 29]. Figure 15.1d shows a scanning electron microscope (SEM) image of a synthetic opal fabricated by sedimentation. A number of advanced techniques now support the fabrication of high-quality opals, which enable uniform PBG properties over a large scale [27, 30, 31]. These techniques offer effective and efficient approaches to build large-scale 3D periodic architectures and achieve robust FPBGs with large bandwidth [8, 32–34], by supplying the templates for inverse opal growth. For example, Ozin's group has developed the isothermal heating evaporation-induced self-assembly (IHEISA) technique, which is a very fast method for the controlled deposition of small to large, 200–1100 nm diameter, spheres into large-area colloids with a very high degree of perfection [31]. This approach also involves a temperature treatment of SiO_2, to improve their homogeneity. In addition, the Pemble and Clays groups have reported extremely uniform large-area opals by the Langmuir–Blodgett (LB) technique [35, 36]. Also, the IHEISA and LB [37] approaches are amenable to the deposition of hetero- and superlattice structures.

15.3
Large-Pore and Non-Close-Packed Inverse Opals

In 1998 and 1999, Busch and John studied the dependence of PBG properties on the inverse opal architecture fabricated by conformal deposition or by other total infiltration means [11]. They showed that the sintering process that is typically performed to mechanically strengthen the opal films also favored the formation of wider PBGs. As shown in Figure 15.6a by slightly sintering an opal template just below its melting point, the dielectric spheres coalesce at their close-packed neck points, as depicted in Figure 15.6b. Geometrically, this thermal process results in overlapping spheres and consequently a smaller lattice constant with subsequent short- and long-range structural disorders [30]. After conformal or complete infiltration with the desired material (Figure 15.6c) and removal of the original template, the structure consists of a non-close-packed (NCP) array of air spheres interconnected by large pores, as shown in Figure 15.6d. Note that in this chapter, we identify all structures in which the air cavity radii are greater than the original close-packed opal radii as large-pore (LP) architectures. The NCP terminology will be used for a second set of structures introduced in the next paragraph. Busch and John predicted wider gaps in LP inverse opals fabricated by total infiltration techniques, as shown in Figure 15.6e. On the other hand, partially infiltrated LP inverse shell opals fabricated by conformal means using such techniques as atomic layer deposition (ALD) do not necessarily exhibit a full photonic band gap unless very high-index (>3.0) materials are used. Nevertheless, these architectures have many structural advantages. For

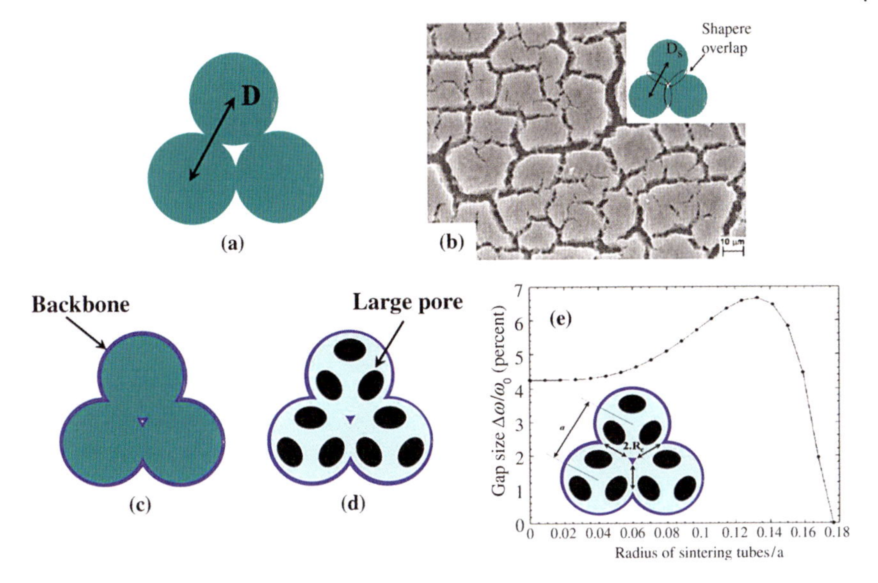

Figure 15.6 (a) Schematic of a close-packed opal, where the lattice constant D is also the diameter of the component spheres. (b) Low-magnification SEM image of a heavily sintered synthetic silica opal (after Ref. [38]). As shown in the inset, the spheres strongly overlap such that large necks are formed at the close-packing locations (solid ellipse), resulting in lattice shrinkage. The observed film cracking is an attribute of the long- and short-range structural disorders introduced during the heat treatment at 1000 °C. (c) Schematic of a heavily sintered synthetic opal after conformal infiltration. (d) Removal of the original sintered template results in an inverse shell opal with large pores. The size of the pores depends on the sintering temperature and process time. (e) Gap size $\Delta\omega/\omega_o$ dependence on the radius of sintering tubes R_c normalized to the fcc lattice constant a (after Ref. [11]). The inset depicts the geometry of the simulated LP inverse silicon structure ($n_{Si} = 3.45$). A maximum gap of ~6.8% is predicted for a normalized radius value of 0.133D, where a is the lattice constant.

example, the surface area of the interior of the backbone is quite large and can be functionalized. Also, the available volume (>80%) makes their infiltration with different materials (liquid, gas, solid) easier. This is an attractive approach for designing sensors or filters, or more advanced architectures as discussed later.

15.4 Experimental Studies

15.4.1 Inversion of Opal Structures

As a consequence of the prediction and realization that inverse opals could exhibit a FPBG, many experimental investigations were initiated on how to convert an opal structure into an inverse structure. In the first series of studies, attempts were made to fill the interstitial air voids with a high-index material using electrochemical

[39, 40], chemical conversion [41, 42], or sol–gel techniques [43–46]. However, these techniques produce a high degree of porosity. Nevertheless, several groups have successfully fabricated inverse opals with FPBGs ranging from the near-visible to the microwave range. A second approach was driven by Wijnhoven and Vos [47] who reported the fabrication of partially filled inverted opals where a titanium oxide (TiO_2) backbone was obtained by precipitation from a liquid-phase chemical reaction. The most successful was a cooperation between the Lopez and Ozin groups [8] in which high-index Si ($n = 3.45$) was conformally deposited into the pore structure. This growth was a result of using silane (Si_2H_6) as the precursor. At an ampoule pressure of 200 Torr and for temperatures between 200 and 400 °C, a layer-by-layer growth mode can be achieved for silicon resulting in the conformal deposition within the silica template. Following growth and annealing, the original silica opal was then removed by etching in HF, thus producing an inverted opal with a dielectric contrast close to 1 : 3.45. As calculated by Busch and John, this structure produces additional (tetrahedral and octahedral) scattering sites (Figure 15.2d) that were predicted to enhance the magnitude of the PBG. By using a larger lattice (sphere) diameter, the PBG was calculated to occur at 1460 nm where Si is transparent and at which many optical processing functions are performed. Figure 15.7 shows reflectivity and transmission data, where the large peak at 2.5 μm was attributed to the Bragg diffraction or PPBG along the Γ–L direction and the peak at 1.46 μm to the full PBG. From reflectivity measurements taken as a function of angle, a full photonic band gap of 6% was determined. A similar result was also reported by Vlasov *et al.* [46], in which a low-temperature (550 °C), low-pressure chemical vapor deposition reaction was used to conformally coat the opal structure. This occurs because in this regime Si is also deposited by a layer-by-layer surface-limited reaction process.

In a following paper, Míguez *et al.* reported a similar study using germane (GeH_4) and obtained a FPBG of 6.1% at ~2 μm [32]. In an attempt to obtain a full PBG in the visible region, additional studies were reported for Sb_2S_3 [48], which in the

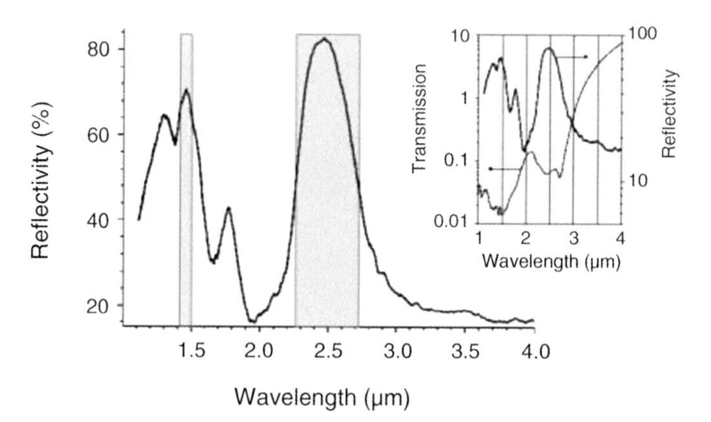

Figure 15.7 Reflection spectrum of an inverse Si opal. The shaded regions at 1.5 and 2.5 μm indicate the calculated positions of a complete 3D PBG and the first (111) stop band. The inset shows the same spectrum plotted logarithmically. After Ref. [8].

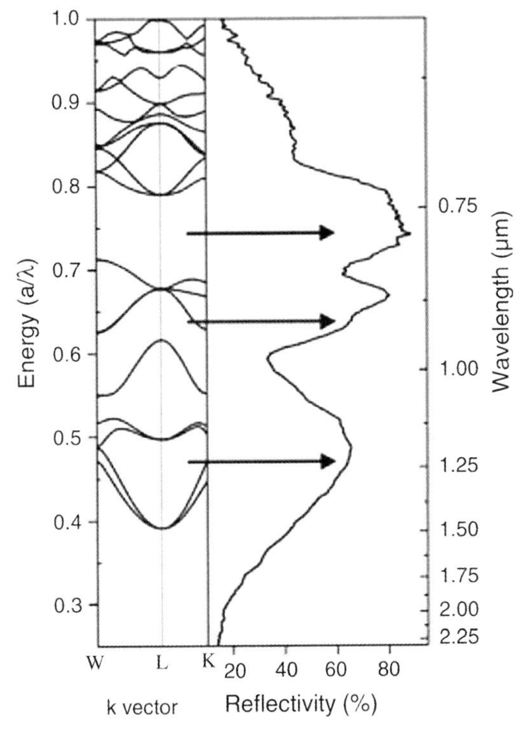

Figure 15.8 Comparison between the photonic band diagram and reflectance spectra of a 410 nm Sb_2S_3 inverted opal showing a wide complete gap at ~0.75 μm. After Ref. [48].

amorphous and crystalline phases has energy gaps with corresponding wavelengths and dielectric constants of 564 nm, 10.9 and 697 nm, 14.4, respectively. A two-step reaction process was used in which a silica opal was first infiltrated with a liquid antimony alkoxide precursor to form Sb_2O_3. The silica template was then removed by a chemical etch and the Sb_2O_3 converted to Sb_2S_3, by sulfidation by placing the sample in flowing H_2S in a reactor held at atmospheric pressure and 300 °C. In Figure 15.8, the comparison between the photonic band diagram and reflectivity shows very good agreement and the presence of a full PGB centered at 750 nm. Thus, the potential to form a FPBG was demonstrated, but will require an opal template formed from smaller spheres to move the PBG into the visible region.

In addition to increasing the selection of high-index materials and spectral range of operation, investigations were also initiated into engineering the photonic band structure of opal and inverse opals by the growth of multilayered shells. This has application to controlling the dispersion properties that affect refraction, diffraction (superprism) [49], and slow to superluminal light effects in 3D structures [50]. Specifically, Garcia-Santamaria *et al.* demonstrated that significant changes in the photonic band structure were obtained by conformally filling opals with silicon and germanium and using different etchants to remove one of the layers [33]. Thus, by first depositing germanium and then a silicon layer, to thicknesses representing 30 and 80% of the pore space volume, respectively, and then removing the Ge by an aqua regia etch, a "freestanding" Si shell was obtained within the original silica opal template. Using these techniques, a structure with enhanced PBG and a full PBG between lower bands was predicted.

From this discussion, inverse shell opals fabricated by conformal deposition have been demonstrated as promising candidates to obtain a FPBG in the visible region or at longer wavelengths. On the other hand, the intrinsic limitations of the inverse shell opal architecture, the infiltrated material properties, and the process itself impede the engineering of wide FPBGs in the visible regime. Thus, the investigation of more advanced architectures was necessary to enable the use of desirable optically transparent materials with reduced refractive index.

Hence, Fenollosa and Meseguer demonstrated selective chemical etching to achieve templates for non-close-packed (NCP) inverse structures [51]. In 2003, Míguez *et al.* fabricated NCP inverse structures by first depositing silica, at room temperature, into a polymer opal and then heating to produce a silica inverse opal. Following this step, the conformal backfilling of silicon by CVD can be achieved into the silica micromold as illustrated in Figure 15.9 to increase the effective index and achieve non-close-packed topologies. This technique of micromolding in silica inverse opals (MISO) was demonstrated to produce a wider range of structures that exhibit a complete PBG [52]. Indeed, the final structure depends not only on the micromold geometry but also on the thickness of the silicon film, which can be adjusted precisely.

On the other hand, the structural and optical qualities of the structures obtained with the MISO approach are limited by the sol–gel step performed to obtain the silica micromold. The porosity and degree of infiltration are poorly controlled in this initial step despite being the most critical process. Blanco and López [53] later improved the routine by substituting the sol–gel step by a CVD step that offers additional control over the quality and geometry of the silica micromold. Hence, the silica–silicon inverted structure that presents an onion layer structure was shown to gain in structural and optical quality. A variety of inverse structures taken from the literature are presented in Figure 15.10.

In 2000, Doosje *et al.* theoretically investigated a second class of NCP inverse structures [55]. In the first class discussed earlier, the air spheres overlap creating large pores. In the second class, the air spheres no longer overlap but are now connected by tubular air channels with a dumbbell-like shape, as shown in Figure 15.11a. A $\sim$10%

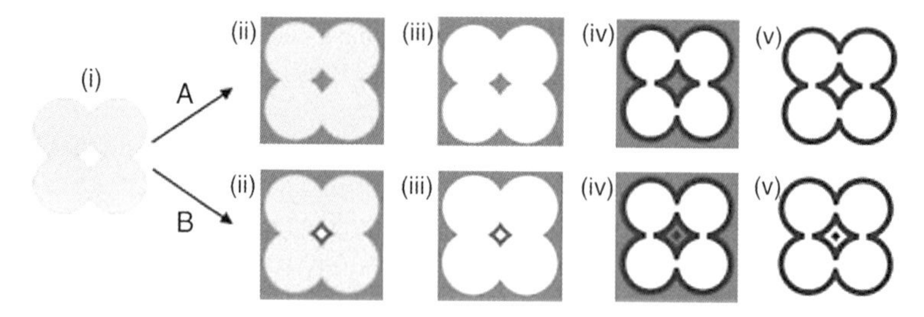

Figure 15.9 MISO process steps. (i) Latex colloidal crystal. (ii) Silica infiltrated latex colloidal crystal via sol–gel. (iii) Silica micromold after removal of the latex template. (iv) Conformal backfill infiltration of silicon using CVD. (v) Silicon inverted structure after removal of the silica micromold. After Ref. [52].

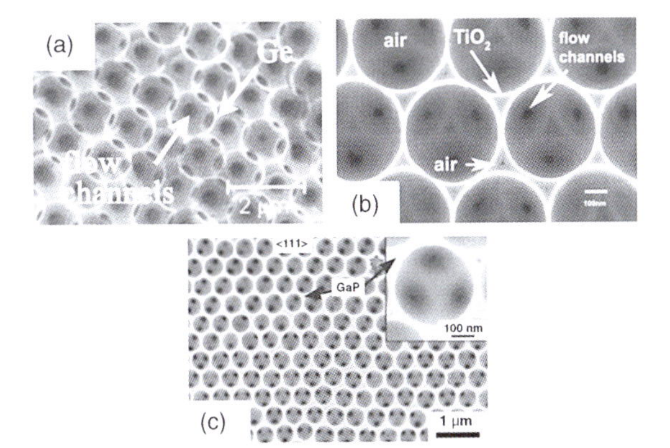

Figure 15.10 Examples of high-index inverse shell opals obtained with two different infiltration techniques. (a) CVD of amorphous Ge at 300 °C followed by annealing at 950 °C ($n_{Ge} \sim 3.9$ at 1.5 μm) (after Ref. [32]). A 6.5% full PBG at 2.1 μm was experimentally observed. (b) ALD of amorphous TiO_2 at 100 °C followed by phase transformation into anatase TiO_2 at 400 °C for 2 h ($n_{an} \sim 2.65$ at 500 nm) (after Ref. [54]). This structure does not exhibit a full PBG. (c) ALD of crystalline GaP deposited at 450 °C ($n_{GaP} \sim 3.34$ at 600 nm) (after Ref. [34]). The structure was predicted to exhibit a 3% full PBG at 736 nm.

FPBG was predicted for a fully infiltrated silicon ($n_{Si} = 3.45$) NCP inverse opal (Figure 15.11b). These results opened up this technology for practical applications and most importantly demonstrated the significance of the lattice structure properties: lattice constant, lattice basis, and the placement of dielectric material(s) as well as the magnitude of the refractive index on the magnitude of the photonic band gap and the shape of the photonic band structure. As a consequence, new infiltration techniques, such as atomic layer deposition, were vigorously investigated.

15.4.2
Atomic Layer Deposition

As discussed in the introductory chapter, the ALD technique originated at Helsinki University when in 1977 Tuomo Suntola was investigating the growth of yellow-

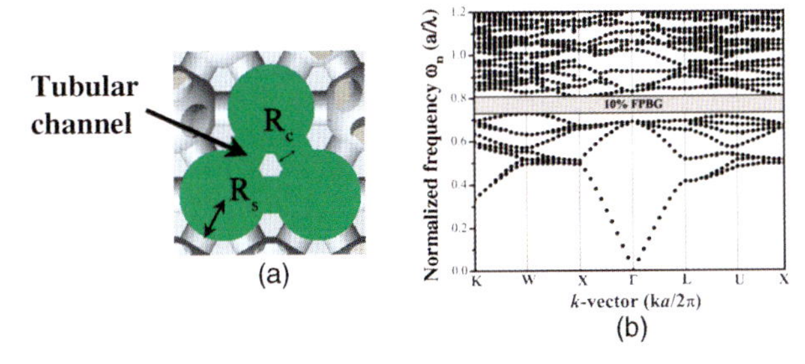

Figure 15.11 (a) Schematic of a "tube-like" NCP inverse opal studied by Doosje *et al.* in which the spherical air chambers are connected by tubular channels [55]. (b) Photonic band structure of the optimized Doosje silicon structure ($n_{Si} = 3.45$) performed by 3D FDTD computations. A full PBG of 10% was predicted between the eighth and ninth bands (gray region). $R_s = 0.3201a$ and $R_c = 0.398R_s$, where a is the cubic lattice constant ($a = D/\sqrt{2}$).

emitting zinc sulfide (ZnS : Mn) films for thin film electroluminescent (TFEL) displays [56]. Controlling the thickness of these thin multilayered ATO/ZnS:Mn/ATO (ATO represents a Al_2O_3/TiO_2 sandwich) devices was well beyond the capabilities of current epitaxial techniques in which large growth rates were commonly encountered. In exploring the influence of self-limiting surface chemical reactions, Suntola [57] realized that these provided a mechanism for controlled layer-by-layer growth with subnanometer precision over the deposition rate because film thickness is directly proportional to the number of precursor cycles. Thus, the crystal lattice structure achieved is thin, uniform, and exactly conforms to the topology of the substrate, irrespective of its three-dimensional complexity.

First named atomic layer epitaxy (ALE), the technique is now generally called atomic layer deposition, and frequently presented as a two-cycle variant of CVD or metal-organic CVD (MOCVD). ALD excels in the deposition of thin films with thickness values ranging from a few angstroms to a few hundred nanometers and has garnered renewed interest because there is a strong demand for electronic and electrooptic nanosized devices composed of alternating layers of different materials (i.e., five atomic layers of silica are needed as a gate material in 65 nm transistor technology) and various groups have studied and successfully achieved high-quality conformal growth of metals [58–61], oxides [62–65], semiconductors [34, 66, 67], and nitride materials [68–70]. Growing a wide variety of materials is an unprecedented advantage compared to conventional epitaxial methods, which are limited to a precise class of materials and growth temperatures. In addition, ALD enables conformal deposition in porous networks, provided the pulse/purge times are sufficiently large to saturate/evacuate the whole surface. This attribute is exploited, for example, in the deposition of conformal layers in opals, 2D PC silicon waveguides [71, 72], and biological scaffolds [73].

The potential of ALD as a means to conformally deposit dielectric layers of high-index materials into highly porous structures, such as opal photonic crystal structures, was pioneered by King *et al.* in 2003 [74] and Rugge *et al.* [75] who used atomic layer deposition to, respectively, infiltrate ZnS:Mn and Ta_3N_5 into silica opals. In the approach reported by Rugge *et al.*, a silica opal was infiltrated at 250 °C to form Ta_2O_5 using pentakis(dimethylamino)tantalum and water precursors and the silica opal subsequently etched out. The Ta_2O_5 overlayer was then placed in a reactor and converted into Ta_3N_5 by a nitridation reaction initiated by flowing ammonia at 1 atm held at 800 °C for 2 h. Inverse opal formation was confirmed by SEM measurements and the observation of a reflectance peak at ~600 nm as expected from theoretical simulations. The inverse ZnS:Mn opals were formed in a conventional ALD reactor using $ZnCl_4$ and H_2S precursors at a temperature of 450 °C. To dope with Mn, one in a hundred pulses of $ZnCl_4$ were substituted by a pulse of $MnCl_4$. SEM, reflectance, and photoluminescence studies confirmed the formation of an inverse opal.

Subsequently, King *et al.* further demonstrated the robustness and flexibility of ALD for fabricating inverse shell opals with an unprecedented degree of control and repeatability by using anatase TiO_2 ($n \sim 2.65$) [54, 76–78]. This exploited another key advantage of ALD: the capability to grow high-index materials at low temperature, typically below 100 °C (Figure 15.10b). In these works, the smoothness of deposited TiO_2 films (roughness <2 Å rms) provided extreme control over the dielectric

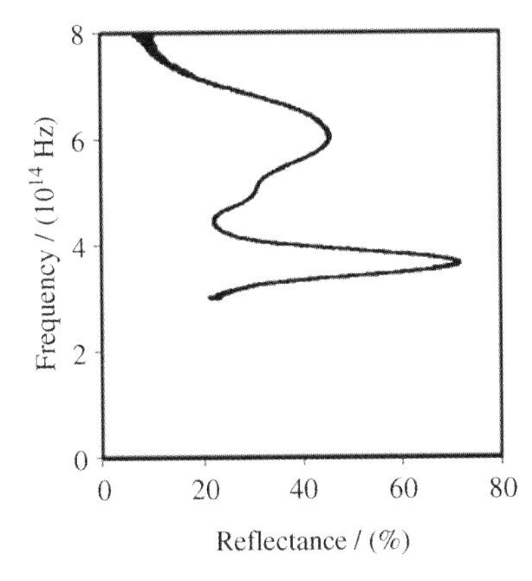

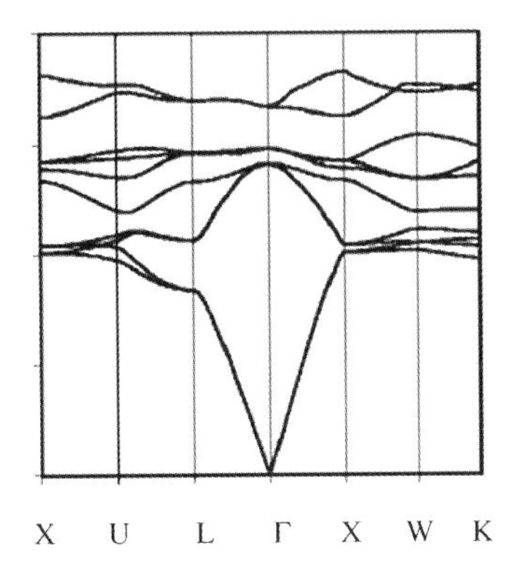

Figure 15.12 Reflectance of an inverted GaAs opal compared to the corresponding calculated Brillouin zone. After Ref. [79].

architecture and therefore optical properties. In addition, Graugnard *et al.* reported similar capabilities using gallium phosphide ($n \sim 3.3$) [34] as shown in Figure 15.10c. GaP is an optoelectronic III–V compound that is partially transparent at optical wavelengths (cutoff wavelength ~550 nm). A FPBG of 3% was predicted in GaP inverse shell opals [53, 61]. Unfortunately, the growth of III–V semiconductors is performed at 400–500 °C resulting in a polycrystalline morphology with rougher surfaces (~5 nm rms). Previously, Povey *et al.* [79] had demonstrated the same result in GaAs as shown in Figure 15.12, where the broad reflective band observed at 6×10^{14} Hz is well fitted by the calculated photonic band structure. This frequency corresponds to a wavelength of 0.58 μm, which is above the band gap of GaAs. Thus, in their analysis the authors included a complex refractive index component.

More recently, the ALD technique was further developed by King *et al.*, whereby a silica template was presintered, infiltrated with TiO_2, and after removal of the sintered template conformally backfilled with TiO_2, resulting in an NCP inverse opal [80]. However, despite the titania structures exhibiting larger pseudo-PBGs, sintering provided insufficient modification and control over the original template to further increase the PBG and resulted in inevitable disorder such as thin film cracking.

Also, several other functional materials have been infiltrated into opals: for example, Scharrer *et al.* have fabricated three-dimensional optically active ZnO photonic crystals by infiltrating polystyrene opal templates (whose surfaces were carboxylated to enhance attachment) using low-temperature atomic layer deposition [81]. ZnO was deposited using diethylzinc and water precursors in a continuous flow ALD reactor at growth temperatures below 85 °C to avoid deformation or melting of the PS structures. The ALD cycle consisted of a 2 s exposure to DEtZn, followed by a 30 s N_2 purge, and then a 4 s exposure to H_2O, and a similar N_2 purge. After deposition, the polystyrene was removed by firing at elevated temperatures, and reactive ion etching was used to remove the top layer of ZnO and expose the (111) photonic crystal surface. The resulting structures were reported to have high filling fractions and photonic band gaps in the near-UV to visible spectrum. The samples

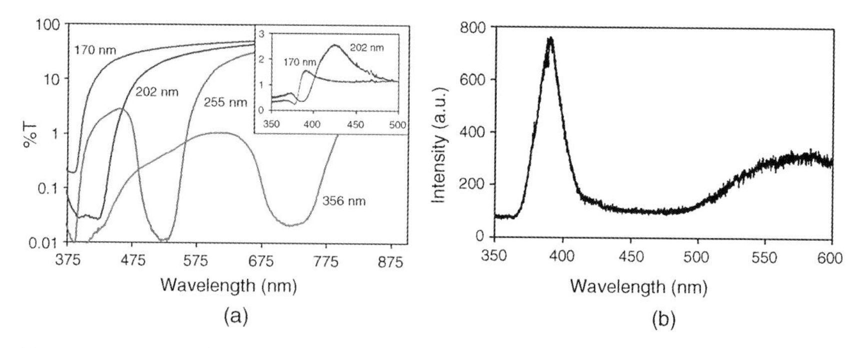

Figure 15.13 Optical properties of ZnO inverse opals (after Ref. [81]). (a) Transmission spectra showing dependence on sphere diameter and (b) photoluminescence showing near band edge emission from ZnO and a longer (lower energy) defect emission band at ~550 nm.

also exhibited efficient photoluminescence as shown in Figure 15.13, which shows the dependence of transmission on sphere diameter and the "bulk-like" photoluminescence spectra typical of high-quality crystalline structures. In addition, efficient UV lasing was recently reported in ZnO inverse opals by optical pumping in high order photonic bands with small group velocity [82]. These and other inverse opals made from optoelectronic materials, such as GaAs and GaP, are of interest for studying and modifying spontaneous and stimulated emission characteristics in active photonic crystals.

We also note that luminescent inverse opals structures have been reported by Withall *et al.* in which Y_2O_3:Tb^{3+} was infiltrated into a silica opal. Evidence was presented to show a modification of the optical and emission characteristics in the green region of the spectra [83].

In addition, a wide range of materials has also been used to infiltrate synthetic opal templates, including oxides [57, 84, 85], metals [86, 87], glasses [88], and luminescent materials [46, 56, 89, 90].

15.4.3
Multilayer Fabrication Steps for Advanced Photonic Crystals

The ALD tool has also demonstrated a unique approach to fabricate high-quality multilayered inverse structures that exhibit additional optical functionalities. An example of multilayer growth with ALD is shown in Figure 15.14 for enhanced luminescence applications [89]. First, a silica opal was partially infiltrated with ZnS:Mn and then the infiltration was completed with a second layer of TiO_2. After removal of the silica spheres, additional layers of TiO_2 were sequentially deposited on the inside of the inverse opal structure. This composite TiO_2/ZnS:Mn/TiO_2 film allows independent control over refractive index and luminescence. The SEM images clearly show the high definition obtained between each layer and the photoluminescence spectrum of Mn ion recorded for different thicknesses of the "backfilled" third layer, from 1 to 5 nm. Note that the peak intensity of the Mn ion at 585 nm increases by a factor of 2 as the reflection spectrum (photonic band gap) moves to longer wavelengths.

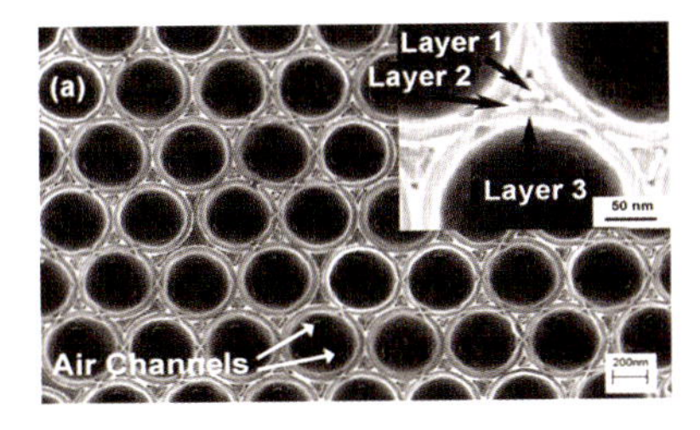

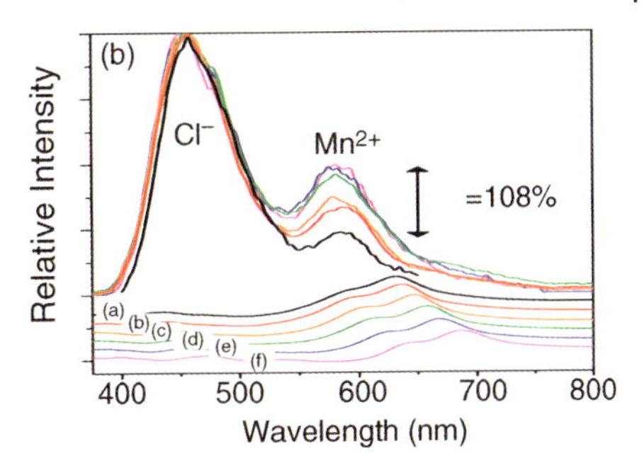

Figure 15.14 Multilayered inverse opal. (a) SEM image of structure showing conformal deposition of each layer. (b) Photoluminescence spectrum of Mn ion recorded for different thicknesses of the "backfilled" third layer. Note that the peak position of the Mn ion at 585 nm increases by a factor of 2 as the reflection spectrum moves to longer wavelengths. This is because the photonic band gap moves to longer wavelengths off the Mn emission. The peak at shorter wavelengths (~460 nm) is due to the presence of Cl impurities. After Ref. [89].

The conformal multilayer approach has been the key to achieve high-quality inverse structures with advanced geometrical control and enhanced optical properties. In particular, to overcome the limitations of the sintering technique to fabricate uncracked thin films, a unique modified NCP fabrication scheme using ALD was developed by Graugnard *et al.* involving the deposition of conformal multilayered materials [38]. With this technique, structural modifications to the original template and inverse shell opal backbone can be made with precise and independent control of the tube and sphere radii, adding additional geometrical flexibility and PBG adjustment. In this technique (Figure 15.15), a synthetic opal is conformally and sequentially infiltrated with a buffer material (that can be selectively etched and therefore acts as a sacrificial buffer layer), resulting in an overlapping template, as shown in Figure 15.15b. A high refractive index material is then deposited into the remaining air void up to the maximum conformal radii, as depicted in Figure 15.15c. After selective removal of the modified template (original opal: Figure 15.15d; sacrificial shell: Figure 15.15e), a low filling fraction inverse shell structure is formed (Figure 15.15e), which becomes a backbone template. Finally, this backbone template can be further modified by conformally coating dielectric layers onto its interior, as reported by Míguez *et al.* [32], King *et al.* [80], and Graugnard *et al.* [38], and shown in Figure 15.15f. The introduction of a sacrificial layer (SL), which was designed to mimic the sintering process, coupled with conformal backfilling adds significantly more flexibility and precision to geometrical manipulation allowing fabrication of many inverse structures as shown in Figure 15.16.

Figure 15.16 shows cross-sectional SEM images of a non-close-packed opal at various stages during the fabrication process and accompanying schematics of each stage. The fabricated and predicted structures match very well. Note as shown in the final SEM image of a completely backfilled structure, a hollow sphere is unavoidably left at the center of each lattice point. This unfortunately does limit the optical density of these structures. The effects of these structural changes in the optical properties

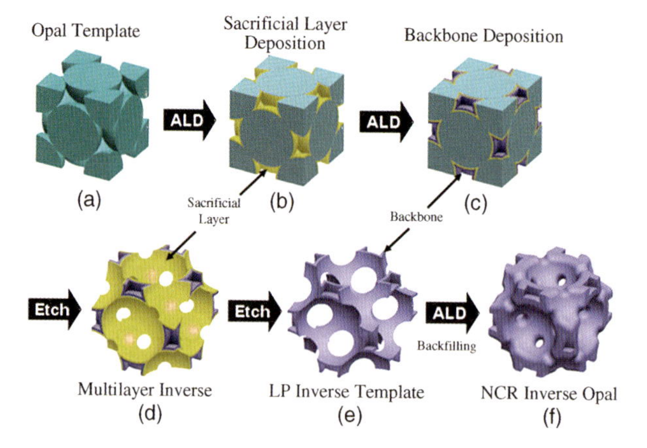

Figure 15.15 Template-patterned sacrificial layer ALD applied to create an NCP inverse opal [38]. The original opal template in (a) is coated with a conformal thin film, resulting in the new template shown in (b). (c) A second ALD infiltration is performed in the remaining air volume to the maximum opal infiltration of 86%. (d) Removal of the original template creates a lattice of close-packed interconnected air spheres in a multilayered dielectric matrix. (e) A lattice of larger diameter overlapping air spheres forms by removing the sacrificial buffer layer. As a consequence, larger air channels now connect the air spheres. (f) Backfilling with additional dielectric forms an NCP lattice of separated air spheres, which are connected by small hyperboloid-shaped air tubes.

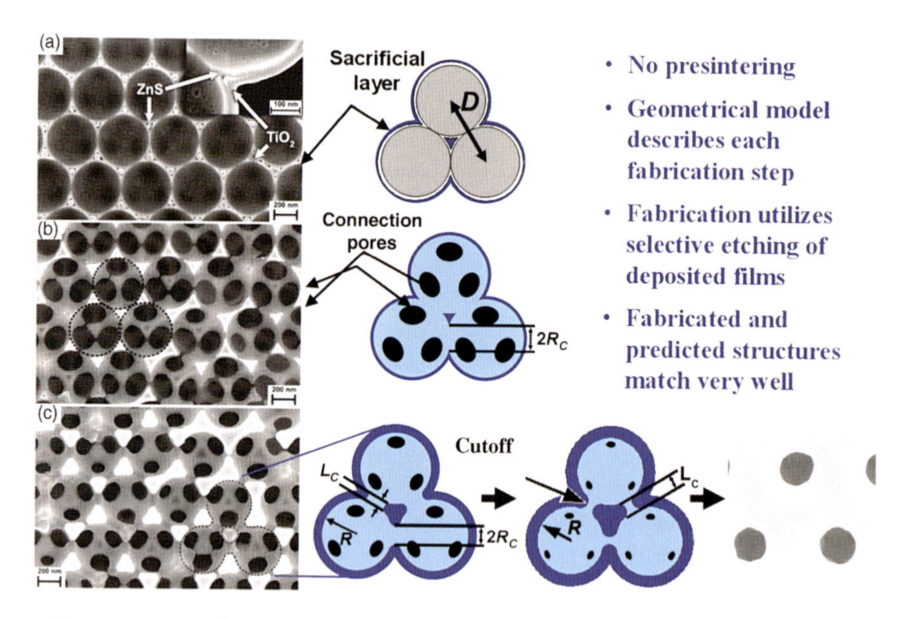

Figure 15.16 Sacrificial layer techniques for the fabrication of an NCP opal. Note that there is no presintering and the fabrication utilizes selective etching of deposited films and/or the original template. A geometrical model accurately describes each fabrication stage. The final SEM image shows a cross section through a completely backfilled structure.

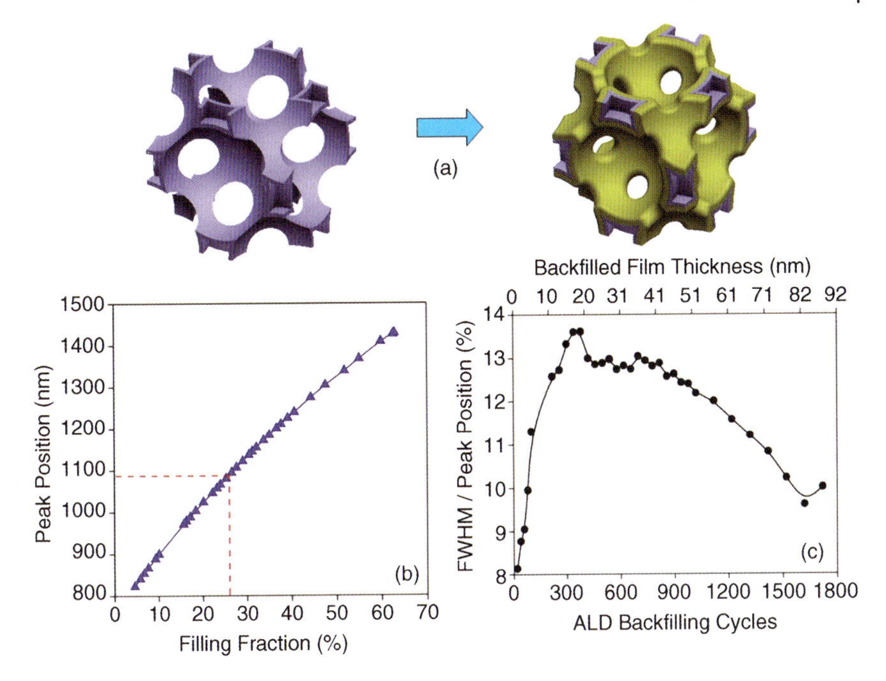

Figure 15.17 (a) Schematic of change in NCP opal with backfilling and (b) the dependence of the peak reflectance on filling fraction. (c) Corresponding dependence of the full-width at half-maximum (FWHM) as a function of number for ALD cycles and backfilled film thickness. After Ref. [38].

are shown in Figure 15.17, which demonstrates how the conformal backfilling of an NCP inverse shell opal statically tunes the position of the photonic bang gap and increases the width of the photonic band gap. As can be seen, ALD provides precise filling fraction control and a tuning range of over 600 nm with a precision in wavelength of 0.35 nm. Commensurate with this change in wavelength, a 67% increase in the width of the PBG was observed.

Figure 15.18 shows a comparison between the experimentally measured and calculated dependence of backfilling on the thickness of the backfilled TiO_2 layer. Excellent agreement was found for both a simple geometric model and a full computer simulation. In Figure 15.18b, the peak wavelength position is shown as a function of backfill thickness. The model again gives very good agreement with experiment and is also used to calculate the effects that can potentially be obtained using materials of lower or higher index. We note that these calculations can also be used to calculate the degree of dynamical tunability that could be realized by infiltrating with a liquid crystal or ferroelectric material. This is discussed in the next section.

Following this study, Gaillot *et al.* reported in 2006 a comprehensive study of the optical properties of inverse NCP and LP opals. In this paper, they explored the photonic band gap potential of silicon inverted structures fabricated via the sacrificial layer fabrication route developed by Graugnard *et al.* [91]. Their results clearly highlighted the photonic band gap width dependence on the inverted backbone

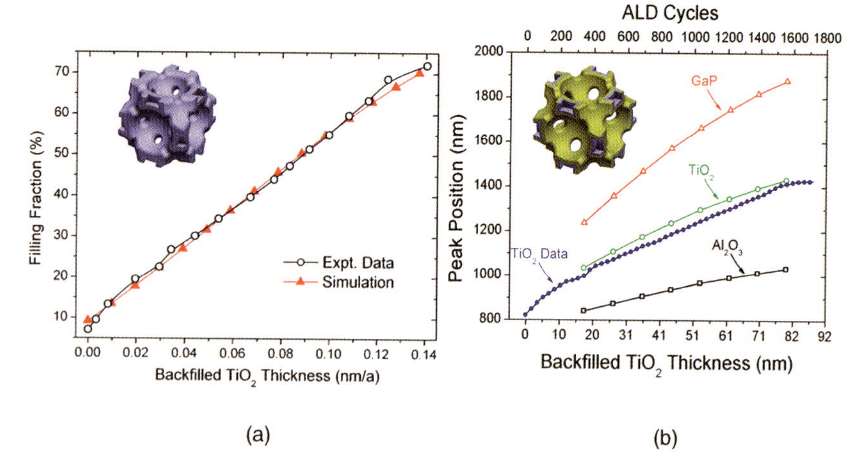

Figure 15.18 Static tunability versus backfilling in sacrificial layer NCP inverse opals. (a) Numerical simulation algorithm accurately models ALD growth in complex 3D geometries. (b) Tuning ranges achieved by combining filling fraction control with higher index or lower index materials. After Ref. [38].

geometry. In addition, partially filled air voids at the tetrahedral locations (i.e., beyond the conformal infiltration limit) were shown to increase the optical gap width and reduce the refractive index requirement to open a gap down to 2.7. This latter parameter is critical to synthesize inverted structures for visible applications with low refractive index materials and further highlights the potential of ALD to achieve the desired structures.

In addition to the SL technique, double infiltration techniques have also been explored to convert polymer inverse structures fabricated via holographic lithography back into another material [92]. Indeed, in addition to opal formation using silica or polymer spheres, a promising technique being developed for photonic crystal fabrication is holographic lithography that uses a three-dimensional 3D interference pattern to define periodic microstructures in a photosensitive polymer film (see Figure 15.5b) [19]. Holographic lithography may also be combined with direct two-photon laser writing to embed waveguides directly into the 3D PC [20]. However, polymer spheres and polymeric resists (SU-8) suffer from a low refractive index $n = 1.63$, which greatly restricts their usefulness as a PC matrix. Nevertheless, this limitation can be overcome by infilling the air space with high-index material, thus inverting the index contrast, which can then be enhanced by etching out the polymer. For example, we have reported the infiltration of TiO_2 into holographically defined opal-like polymer templates by low-temperature atomic layer deposition [92].

As shown in Figure 15.19, the initial inversion phase can be accomplished by using ALD of either TiO_2 or Al_2O_3. However, Al_2O_3 was preferred to avoid the conversion of TiO_2 into a rougher anatase phase during GaP growth at higher temperatures of $\sim 450\,^{\circ}C$. However, to initiate the growth of GaP, a very thin ~ 5 nm film of TiO_2 was found to be necessary to seed nucleation and achieve growth.

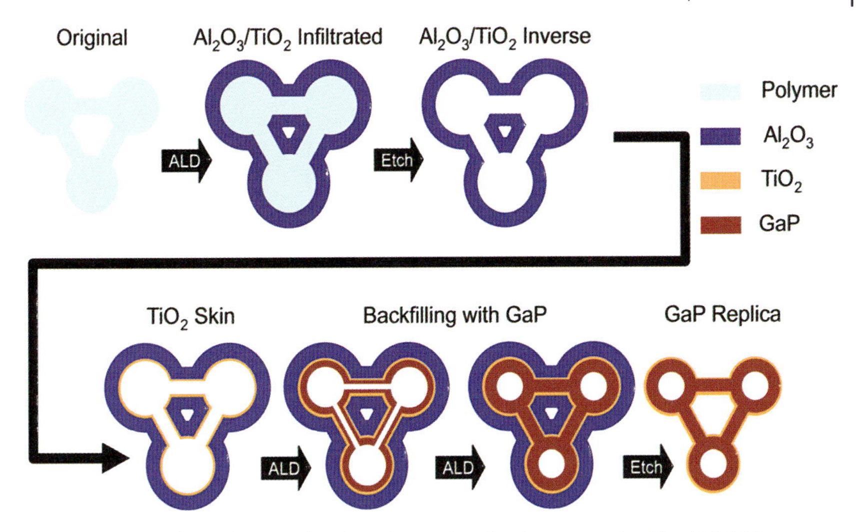

Figure 15.19 Holographic template inversion and replication protocols. After Ref. [92].

Reviewing these facts, we can then summarize the requirements for photonic crystal fabrication and materials as the following. Structural requirements are high-quality templates, deposition on/within the template lattice, suitable materials for conversion to an inverse structure, precise control of conformal deposition within the porous template lattice, and finally etching selectivity between the deposited material and template. The complementary optical material requirements are high refractive index ($n > 3.0$) required for full PBG in opals, optically transparency in the visible region, and smooth surfaces to reduce scattering. The advantages of ALD are therefore digital layer control, highly conformal growth, smooth surfaces (rms < 0.2 nm), low-temperature growth, and the capability to grow over a wide temperature range from room temperature up to 900 °C. In addition, as mentioned earlier, ALD is applicable to many materials: oxides, semiconductors, and metals. Finally, the growth conditions (pulse/purge times) must be optimized for highly porous structures.

The typical growth and processing flow path for forming an inverse opal is as follows:

1) Prepare template.
   - Opal self-assembly holographic lithography.
   - Two-photon polymerization, inkjet direct writing.
2) Infiltrate structure by ALD.
   - Optimize deposition protocols for infiltrating highly porous structures by taking into consideration the greater surface area and complexity of pore structure in opals.
     - Surface area of opal film >200 times larger than the planar area.
     - Longer pulse and purge times ~10 and 60 s (for a 10 mm thick opal).

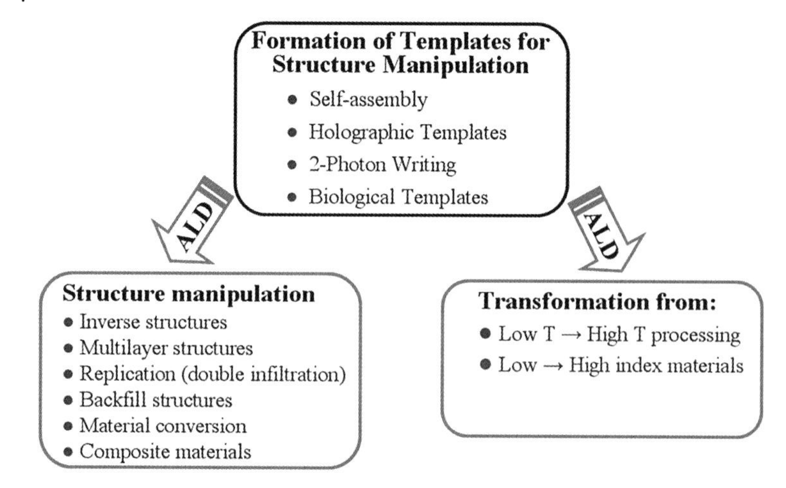

Figure 15.20 Summary of processes and capabilities of atomic layer deposition.

3) Surface preparation/functionalization.
   - Ion mill "top surface" to expose original template.
4) Selective etching process to remove template or initial layer.
   - HF, H_2SO_4, plasma etch, burn out.
5) For double inversion:
   - Growth/ion milling/etching, and so on.
6) Characterization: SEM, XRD, reflectance, PL.

The advantages and capabilities of ALD are succinctly summarized in Figure 15.20.

15.5 Tunable PC Structures

The development of optically reconfigurable 3D photonic structures is certainly the cornerstone to provide high-end applications to the photonics market but is not recent. The idea of tuning the PBG position and width was first introduced by John and Busch in 1999 [93, 94] and has been experimentally verified with relative success by many groups [95–98]. The investigation of electrooptical (EO) materials infiltrated into 3D PC is critical since it potentially offers a wealth of functionalities and device applications. For example, the effective refractive index, n_{eff}, of the opal structure can be controlled by the infiltration of a liquid crystal (LC) to tune the Bragg wavelength, as initially reported by Busch and John and Yoshino *et al.* [98], since LCs possess a birefringence, Δn_{LC}. When the orientation of the director is modified, the refractive index can in principle be tuned between the ordinary (n_o) and the extraordinary (n_e) index values [99]. However, the low available volume fraction of an opal ($f_{\text{LC}}^{\text{opal}} = 0.26$) strongly limits director reorientation within the tetrahedral and octahedral interstitial sites, resulting in reduced dielectric anisotropy. Moreover, the topology of conventional inverse shell opals hinders complete infiltration with LC molecules, due to small interconnections (necking regions) between the spherical chambers. As a consequence, Bragg peak tuning with temperature or an applied electric field is severely limited. On the other hand, inverse opals provide a larger void volume for LC infiltration and higher dielectric contrast, resulting in a wider Bragg peak and complete PBGs, depending upon the backbone refractive index n_{BB} [101–107]. As discussed earlier, high-quality inverse shell opals can be fabricated by ALD of a high-index material onto the dielectric spheres to the maximum infiltration of 86% of the opal interstitial volume. Removing the original dielectric template results in an inverse structure with ~0.776 air volume fraction. The inverse opal can then be infiltrated with liquid crystal to a filling fraction of $f_{LC} = 0.74$. In this configuration, the trapped tetrahedral and octahedral air pockets within the backbone contain the remaining air volume fraction, $f_{air} = 0.036$. Since f_{LC} is increased from 0.26 in an infiltrated opal to 0.74 in an infiltrated inverse opal, it results in additional flexibility to tune n_{eff}, given by

$$n_{eff} = \left(f_{BB} n_{BB}^2 + f_{LC} n_{LC}^2 + f_{air} n_{air}^2 \right)^{1/2}, \tag{15.5}$$

where f_{BB} is the backbone volume fraction of 0.224. In addition, the volume fractions are related to each other by

$$f_{air} + f_{LC} + f_{BB} = 1. \tag{15.6}$$

It is clear from Eqs. (15.5) and (15.6) that maximum tunability will be attained when the LC (or any EO material) volume fraction is maximal (or conversely backbone volume fraction is minimal). Therefore, and as mentioned earlier, because LP and NCP inverse structures offer lower backbone volume fractions compared to regular inverse opals, they clearly provide the best geometrical host for EO infiltration and dynamic tunability. Experimentally, a mechanically stable LP inverse opal with ultralow backbone volume fraction of 0.05 was fabricated through the SL ALD technique [38]. In addition, it is noteworthy that such structures not only ease the EO penetration into the thin film but also favor complete infiltration within.

In order to assess the dynamic optical properties of these composite materials, we have to investigate all possible configurations and identify which structures provide the greatest PBG tunability. The potential of EO infiltrated LP and NCP inverse architectures to obtain dynamically tunable PBG properties has been explored by Gaillot *et al.* with LC molecules and lanthanum-modified lead zirconate titanate (PLZT) [108, 109]. This work was also in good agreement with an earlier experimental study by Graugnard *et al.* on LC infiltrated LP TiO_2 inverse opals, where they demonstrated a large 20 nm band gap shift under a 25 kHz electric field in the near-infrared region [110] as shown in Figure 15.21a.

In this work, the effects of both hydrophilic and hydrophobic surfaces were investigated to ascertain the effect of surface pinning of the LC molecules on the

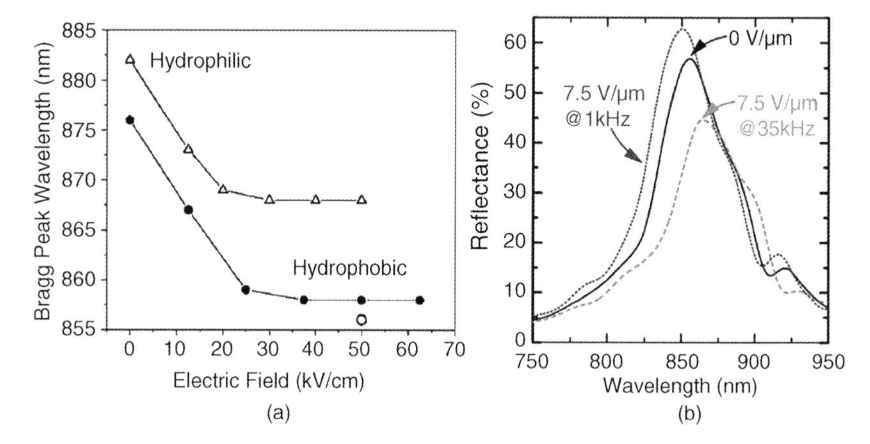

Figure 15.21 (a) Bragg peak position versus applied electric field at 1 kHz for hydrophilic (open triangles) and hydrophobic (filled circles) samples. The single open circle is the peak position for the hydrophobic sample with 50 V applied at 25 kHz. After Ref. [110]. (b) Spectral dependence of Bragg reflectance for no applied field (solid curve) at 1 kHz (dotted curve) and 35 kHz (dashed curve) for square wave electric field amplitude of 7.5 V/m. Blue and red Bragg peak shifts yield an overall shift of 19 nm. After Ref. [111].

inverse opal template. Hysteresis was observed for the hydrophobic treated sample but absent in the hydrophilic sample, which can be understood in terms of the surface pinning of the LC molecules to the TiO_2 surface, which produces a strong restoring force that reorients the LC molecules when the applied bias is removed.

An additional functionality was introduced in a sequel paper by the same group, where they studied the dynamic optical properties of LP TiO_2 inverse opals infiltrated with dual-frequency LC molecules [111]. The dual-frequency liquid crystal MLC-2048 (EMD Chemicals, Inc.) employed in this work displays a birefringence that reverses sign above a critical frequency for an applied electric field. Under low-frequency bias, a blueshift was induced in the inverse opal optical properties. However, under high-frequency bias, a redshift was produced in the optical properties. A total Bragg peak tuning range of 19 nm was achieved comprising a 6 nm blueshift at 1 kHz and a 13 nm redshift at 35 kHz, as shown in Figure 15.21b.

A significant application of infiltrated opals was reported in 2000 by Puzzo *et al.* for low-power full-color reflective displays [112, 113]. To fabricate these devices, an opal was first formed using 270 nm diameter silica spheres and infiltrated with polyferrocenylsilane (PFS) and polymerized. A HF etch was then used to remove the silica temples to form an inverse opal (Figure 15.22). This polymer/electrolyte structure swells or shrinks under the application of an electric field. As shown in Figure 15.23, a large degree of tuning was obtained when the electric field was applied such that the reflectivity characteristics of the opal change from blue to green and then to red and/or deep red. However, the process was found to be rather slow and must be increased significantly for display applications.

Finally, we encourage the reader to look at the more complete review on tunable opals and inverse opals structures recently published by Aguirre *et al.* [114].

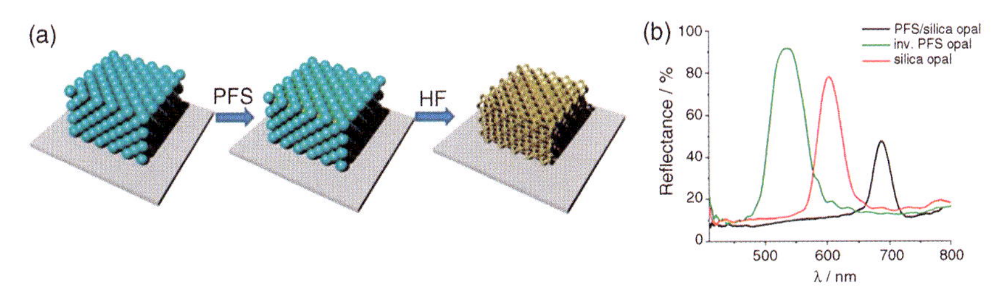

Figure 15.22 (a) Schematic of opal formation, PFS infiltration into opal, and removal of opal to form an inverse opal of PFS. (b) Resulting change in optical reflectance during these processes. After Ref. [112].

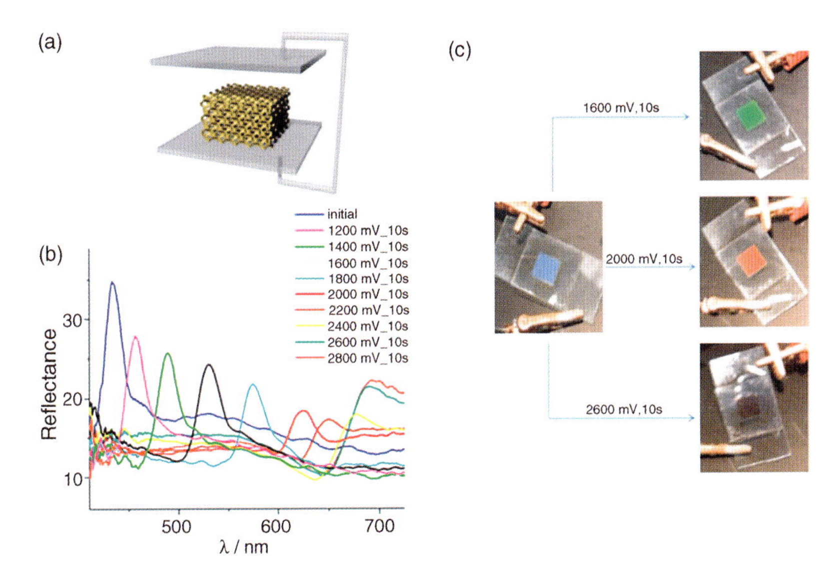

Figure 15.23 (a) Representation of an electrochemical cell fabricated and (b) effects of electrical activation of the active inverse opal on optical properties. (c) Change of color response upon application of an applied voltage. For an opal showing a blue reflectivity, the color can be progressively tuned to green and red and a deeper red by applying increased applied voltages. After Ref. [113].

15.6 Summary

The theoretical and experimental properties of opal-based PCs have been intensively investigated over the years, and have led to a progressive and increasing understanding of their properties and potential applications of 3D optical structures in general. Also, the interaction between theory and experiment has led to new growth and fabrication protocols. Atomic layer deposition has been shown to play a significant role in these developments by providing a perfect complement to template formation leading to the enhancement of optical properties, such as larger photonic band gaps, and the static and dynamic tuning of optical properties with potential in future device applications

In particular, unique structures such as multilayer depositions have been demonstrated and shown to lead to a modification of optical reflectance and photoluminescence properties. In addition, sacrificial layer atomic layer deposition (SL-ALD) enables significant structural basis modification. Using this technique, new structures and templates have been formed by the use of conformal backfilling, a case in point being the formation of large-pore opals that can then be used for infiltration. Conformal backfilling has enabled the static tuning >600 nm of the photonic band gap and for liquid crystal infiltration dynamic tuning of ~ 20 nm has been measured with potentially a tuning range of >80 nm being possible. In addition, conformal thin film models have been shown to accurately explain the observed data and these results have been conformed by higher level simulations.

The technique of double templating demonstrates the advantage of converting a structure from a material with a low melting point to one with a high melting point and/or from a low-index to a high-index material or one with a different functionality. The technique enables faithful structural or topographical replication and predictable dielectric modifications that can be used to modify optical properties [115]. These attributes have been demonstrated in the modification and fabrication of unique two-dimensional photonic crystal structures and the tuning of optical cavities [116] and cloaking structures [117]. These results demonstrate that ALD has very significant applications to a variety of nanophotonics devices that will continue to be exploited and extended to new applications. In fact, ALD has been used to give insight and to modify the optical properties of biotemplates in butterflies [74, 118–120] and diatoms [121].

The ALD fabrication technique is applicable to a wide range of complex two- and three-dimensional nanostructures and materials, which can range from oxides to semiconductors and metals, and because of its low-temperature attributes is even capable of replicating the morphology of tissue [122]. This feature is key and potentially enables in the near future the deposition of both organic materials and polymers. The combination of inorganic and organic materials is already being developed [123] and adaption of these advances to ALD offers many exciting new developments that are beneficial not only to classical photonics but also to emergent fields where optically active and functionalized nanostructures are now considered very appealing and therefore strongly desired.

Acknowledgments

The authors thank Dr. Elton Graugnard, Dr. Jeffrey King, Dr. Curtis Neff, and Dr. T. Yamashita for their many discussions and suggestions. We are particularly indebted to Dr. Elton Graugnard for his development of the ALD systems at Georgia Tech and for his insights into this technology. This work was supported in part by a MURI grant from the Army Research Office.

References

1 Pieranski, P. (1983) Colloidal crystals. *Contemp. Phys.*, **24**, 25–73.

2 Rayleigh, L. (1887) On the maintenance of vibrations by forces of double frequency, and on the propagation of waves through a medium endowed with a periodic structure. *Philos. Mag.*, **24**, 145–159.

3 Iler, R.K. (1979) *The Chemistry of Silica: Solubility, Polymerization, Colloid and Surface Properties and Biochemistry of Silica*, John Wiley & Sons, Inc., New York.

4 Yablonovitch, E. (1987) Inhibited spontaneous emission in solid-state physics and electronics. *Phys. Rev. Lett.*, **58**, 2059–2062.

5 John, S. (1987) Strong localization of photons in certain disordered dielectric superlattices. *Phys. Rev. Lett.*, **58**, 2486–2489.

6 Yariv, E. and Yeh, P. (1984) *Optical Waves in Crystal: Propagation and Control of Laser Radiation*, John Wiley & Sons, Inc., New York.

7 Gaillot, D., Yamashita, T., and Summers, C.J. (2005) Photonic band gaps in highly conformal inverse-opal based photonic crystals. *Phys. Rev. B*, **72**, 205109–205111.

8 Blanco, A., Chomski, E., Grabtchak, S., Ibisate, M., John, S., Leonard, S.W., Lopez, C., Meseguer, F., Míguez, H., Mondla, J.P., Ozin, G.A., Toader, O., and van Driel, H.M. (2000) Large-scale synthesis of a silicon photonic crystal with a complete three-dimensional bandgap near 1.5 micrometres. *Nature*, **405**, 437–440.

9 Joannopoulos, J.D., Meade, R.D., and Winn, J.N. (1995) *Photonic Crystals: Molding the Flow of Light*, Princeton University Press, Princeton, NJ.

10 Joannopoulos, J.D., Johnson, S.G., Winn, J.N., and Meade, R.D. (2008) *Photonic Crystals: Molding the Flow of Light*, 2nd edn, Princeton University Press, Princeton, NJ.

11 Busch, K. and John, S. (1998) Photonic band gap formation in certain self-organizing systems. *Phys. Rev. E*, **58**, 3896–3908.

12 Lodahl, P., Floris van Driel, A., Nikolaev, I.S., Irman, A., Overgaag, K., Vanmaekelbergh, D., and Vos, W.L. (2004) Controlling the dynamics of spontaneous emission from quantum dots by photonic crystals. *Nature*, **430**, 654–657.

13 Ho, K.M., Chan, C.T., and Soukoulis, C.M. (1990) Existence of a photonic gap in periodic dielectric structures. *Phys. Rev. Lett.*, **65**, 3152–3155.

14 Sozuer, H.S., Haus, J.W., and Inguva, R. (1992) Photonic bands: convergence problems with the plane-wave method. *Phys. Rev. B*, **45**, 13962.

15 Yablonovitch, E., Gmitter, T.J., and Leung, K.M. (1991) Photonic band structure: the face-centered-cubic case employing non-spherical atoms. *Phys. Rev. Lett.*, **67**, 2295–2299.

16 Lin, S.Y., Fleming, J.G., Hetherington, D.L., Smith, B.K., Biswas, R., Ho, K.M., Sigalas, M.M., Zubrzycki, W., Kurtz, S.R., and Bur, J. (1998) Three-dimensional photonic crystal operating at infrared wavelengths. *Nature*, **394**, 251–253.

17 Ozbay, E., Michel, E., Tuttle, G., Biswas, R., Sigalas, M., and Ho, K.-M. (1994) Micromachined millimeter-wave photonic band-gap crystals. *Appl. Phys. Lett.*, **64**, 2059–2061.

18 Noda, S., Tomoda, K., Yamamoto, N., and Chutinan, A. (2000) Full three-dimensional photonic bandgap crystals at near-infrared wavelengths. *Science*, **289**, 604–606.

19 Campbell, M., Sharp, D.N., Harrison, M.T., Denning, R.G., and Turberfield, A.J. (2000) Fabrication of photonic crystals for the visible spectrum by holographic lithography. *Nature*, **404**, 53–56.

20 Cumpston, B.H. *et al.* (1999) Two-photon polymerization initiators for three-dimensional optical data storage and microfabrication. *Nature*, **398**, 51.

21 Kennedy, S.R. and Brett, M.J. (2002) Fabrication of tetragonal square spiral photonic crystals. *Nano Lett.*, **2**, 59–62.

22 García-Santamaría, H.T.M.F., Urquía, A., Ibisate, M., Belmonte, M., Shinya, N., Meseguer, F., and López, C. (2002)

Nanorobotic manipulation of microspheres for on-chip diamond architectures. *Adv. Mater.*, **14**, 1144–1147.

23 Xia, Y., Gates, B., and Li, Z.-Y. (2001) Self-assembly approaches to three-dimensional photonic crystals. *Adv. Mater.*, **13**, 409–413.

24 Gratson, G.M., Garcia-Santamaria, F., Lousse, V., Xu, M., Fan, S., Lewis, J.A., and Braun, P.V. (2006) Direct-write assembly of three-dimensional photonic crystals: conversion of polymer scaffolds to silicon hollow-woodpile structures. *Adv. Mater.*, **18**, 461–465.

25 Míguez, H., Meseguer, F., Lopez, C., Mifsud, A., Moya, J.S., and Vazquez, L. (1997) Evidence of FCC crystallization of SiO_2 nanospheres. *Langmuir*, **13**, 6009–6011.

26 Vos, W.L., Megens, M., van Kats, C.M., and Bosecke, P. (1996) Transmission and diffraction by photonic colloidal crystals. *J. Phys.: Condens. Matter*, **8**, 9503–9507.

27 Van Blaaderen, A., Ruel, R., and Wiltzius, P. (1997) Template-directed colloidal crystallization. *Nature*, **385**, 321–324.

28 Woodcock, L.V. (1997) Entropy difference between the face-centred cubic and hexagonal close-packed crystal structures. *Nature*, **385**, 141–143.

29 Bruce, A.D., Wilding, N.B., and Ackland, G.J. (1997) Free energy of crystalline solids: a lattice-switch Monte Carlo method. *Phys. Rev. Lett.*, **79**, 3002–3005.

30 Chabanov, A.A., Jun, Y., and Norris, D.J. (2004) Avoiding cracks in self-assembled photonic band-gap crystals. *Appl. Phys. Lett.*, **84**, 3573–3575.

31 Wong, S., Kitaev, V., and Ozin, G.A. (2003) Colloidal crystal films: advances in universality and perfection. *J. Am. Chem. Soc.*, **125**, 15589–15598.

32 Míguez, H., Chomski, E., Garcia-Santamaria, F., Ibisate, M., John, S., Lopez, C., Meseguer, F., Mondia, J.P., Ozin, G.A., Toader, O., and Van Driel, H.M. (2001) Photonic bandgap engineering in germanium inverse opals by chemical vapor deposition. *Adv. Mater.*, **13**, 1634–1637.

33 Garcia-Santamaria, F., Ibisate, M., Rodriguez, I., Meseguer, F., and Lopez, C. (2003) Photonic band engineering in opals by growth of Si/Ge multilayer shells. *Adv. Mater.*, **15**, 788–792.

34 Graugnard, E., Chawla, V., Lorang, D., and Summers, C.J. (2006) High filling fraction gallium phosphide inverse opals by atomic layer deposition. *Appl. Phys. Lett.*, **89**, 211102–211111.

35 Szekeres, M., Kamalin, O., Schoonheydt, R.A., Wostyn, K., Clays, K., Persoons, A., and Dékány, I. (2002) Ordering and optical properties of monolayers and multilayers of silica spheres deposited by the Langmuir–Blodgett method. *J. Mater. Chem.*, **12**, 3268–3274.

36 Bardosova, M., Hodge, P., Pach, L., Pemble, M.E., Smatko, V., Tredgold, R.H., and Whitehead, D. (2003) Synthetic opals made by the Langmuir–Blodgett method. *Thin Solid Films*, **437** (1–2), 276–279.

37 van Duffel, B., Ras, R.H.A., De Schryver, F.C., and Schoonheydt, R.A. (2001) Langmuir–Blodgett deposition and optical diffraction of two-dimensional opal. *J. Mater. Chem.*, **11**, 3333.

38 Graugnard, E., King, J.S., Gaillot, D.P., and Summers, C.J. (2006) Sacrificial-layer atomic layer deposition for fabrication of nonclose-packed inverse-opal photonic crystals. *Adv. Funct. Mater.*, **16**, 1187–1196.

39 Braun, P.V., Zehner, R.W., White, C.A., Weldon, M.K., Kloc, C., Patel, S.S., and Wiltzius, P. (2001) Epitaxial growth of high dielectric contrast three-dimensional photonic crystals. *Adv. Mater.*, **13**, 721–724.

40 Braun, P.V. and Wiltzius, P. (1999) Electrochemically grown photonic crystals. *Nature*, **402**, 603.

41 Soten, I., Míguez, H., Yang, S.M., Petrov, S., Coombs, N., Tetreault, N., Matsuura, N., Ruda, H.E., and Ozin, G.A. (2002) Barium titanate inverted opals: synthesis, characterization, and optical properties. *Adv. Funct. Mater.*, **12**, 71–77.

42 Holland, B.T., Blanford, C.F., and Stein, A. (1998) Synthesis of macroporous minerals with highly ordered three-dimensional arrays of spheroidal voids. *Science*, **281**, 538–540.

43 Li, B., Zhou, J., Li, Q., Li, L.T., and Gui, Z.L. (2003) Synthesis of $(Pb,La)(Zr,Ti)O_3$

inverse opal photonic crystals. *J. Am. Ceram. Soc.*, **86**, 867–869.

44 Li, B., Zhou, J., Li, L.T., Wang, X.J., Liu, X.H., and Zi, J. (2003) Ferroelectric inverse opals with electrically tunable photonic band gap. *Appl. Phys. Lett.*, **83**, 4704–4706.

45 Yang, P., Deng, T., Zhao, D., Feng, P., Pine, D., Chmelka, B.F., Whitesides, G.M., and Stucky, G.D. (1998) Hierarchically ordered oxides. *Science*, **282**, 2244–2246.

46 Vlasov, Y.A., Nan, Y., and Norris, D.J. (1999) Synthesis of photonic crystals for optical wavelengths from semiconductor quantum dots. *Adv. Mater.*, **11**, 165–169.

47 Wijnhoven, J.E.G.J. and Vos, W.L. (1998) Preparation of photonic crystals made of air spheres in titania. *Science*, **281**, 802–804.

48 Juarez, B.H., Ibisate, M., Palacios, J.M., and Lopez, C. (2003) High-energy photonic bandgap in Sb_2S_3 inverse opals by sulfidation processing. *Adv. Mater.*, **15**, 319–323.

49 Kosaka, H., Kawashima, T., Tomita, A., Notomi, M., Tamamura, T., Sato, T., and Kawakami, S. (1998) Superprism phenomena in photonic crystals. *Phys. Rev. B*, **58**, 10096–10099.

50 Galisteo-López, J.F., Galli, M., Balestreri, A., Patrini, M., Andreani, L.C., and López, C. (2007) Slow to superluminal light waves in thin 3D photonic crystals. *Opt. Express*, **15**, 15342–15350.

51 Fenollosa, R. and Meseguer, F. (2003) Non-close-packed artificial opals. *Adv. Mater.*, **15**, 1282–1285.

52 Míguez, H., Tetreault, N., Yang, S.M., Kitaev, V., and Ozin, G.A. (2003) A new synthetic approach to silicon colloidal photonic crystals with a novel topology and an omni-directional photonic bandgap: micromolding in inverse silica opal (MISO). *Adv. Mater.*, **15**, 597–600.

53 Blanco, A. and López, C. (2006) Silicon onion-layer nanostructures arranged in three dimension. *Adv. Mater.*, **18**, 1593–1597.

54 King, J.S., Graugnard, E., and Summers, C.J. (2005) TiO_2 inverse opals fabricated using low-temperature atomic layer deposition. *Adv. Mater.*, **17**, 1010–1013.

55 Doosje, M., Hoenders, B.J., and Knoester, J. (2000) Photonic bandgap optimization in inverted fcc photonic crystals. *J. Opt. Soc. Am. B*, **17**, 600–606.

56 Suntola, T., Antson, J., Pakkala, A., and Lindfors, S. (1980) Atomic layer epitaxy for producing EL-thin films. Presented at SID International Symposium, San Diego, CA, April 29–May 1, 1980, Digest of Technical Papers.

57 Ritala, M. and Leskelä, M. (2002) *Handbook of Thin Film Materials*, vol. **1**, Academic Press.

58 Juppo, M., Vehkamaki, M., Ritala, M., and Leskelä, M. (1998) Deposition of molybdenum thin films by an alternate supply of $MoCl_5$ and Zn. *J. Vac. Sci. Technol. A*, **16**, 2845–2850.

59 Juppo, M., Ritala, M., and Leskelä, M. (1997) Deposition of copper films by an alternate supply of CuCl and Zn. *J. Vac. Sci. Technol. A*, **15**, 2330–2333.

60 Klaus, J.W., Ferro, S.J., and George, S.M. (2000) Atomically controlled growth of tungsten and tungsten nitride using sequential surface reactions. 5th International Symposium on Atomically Controlled Surfaces, Interfaces and Nanostructures (ACSIN-5), July 6–9, 1999. *Appl. Surf. Sci.*, **162**, 479–491.

61 Klaus, J.W., Ferro, S.J., and George, S.M. (2000) Atomic layer deposition of tungsten nitride films using sequential surface reactions. *J. Electrochem. Soc.*, **147**, 1175–1181.

62 Klaus, J.W., Ott, A.W., Johnson, J.M., and George, S.M. (1997) Atomic layer controlled growth of SiO_2 using binary reaction sequence chemistry. Presented at Chemistry and Physics of Small-Scale Structures, Santa Fe, NM, February 9–11, 1997, 1997 OSA Technical Digest Series, Vol. 2.

63 Ott, A.W., Klaus, J.W., Johnson, J.M., and George, S.M. (1997) Al_2O_3 thin film growth on Si(100) using binary reaction sequence chemistry. *Thin Solid Films*, **292**, 135–144.

64 Romanov, S.G., Maka, T., Torres, C.M.S., Muller, M., and Zentel, R. (2001) Emission in a SnS_2 inverted opaline photonic crystal. *Appl. Phys. Lett.*, **79**, 731–733.

65 Saito, K., Watanabe, Y., Takahashi, K., Matsuzawa, T., Sang, B., and Konagai, M. (1997) Photo atomic layer deposition of transparent conductive ZnO films. *Sol. Energy Mater. Sol. Cells*, **49**, 187–193.
66 Koleske, D.D. and Gates, S.M. (1994) Atomic layer epitaxy of Si on Ge(100) using Si_2Cl_6 and atomic hydrogen. *Appl. Phys. Lett.*, **64**, 884–886.
67 Sugahara, S., Uchida, Y., Kitamura, T., Nagai, T., Matsuyama, M., Hattori, T., and Matsumura, M. (1997) A proposed atomic-layer-deposition of germanium on Si surface. Presented at 1996 International Conference on Solid State Devices and Materials (SSDM'96), Yokohama, Japan, August 26–29, 1996.
68 Becker, J.S., Kim, E., and Gordon, R.G. (2004) Atomic layer deposition of insulating hafnium and zirconium nitrides. *Chem. Mater.*, **16**, 3497–3501.
69 Sumakeris, J., Sitar, Z., Ailey-Trent, K.S., More, K.L., and Davis, R.F. (1993) Layer-by-layer epitaxial growth of GaN at low temperatures. Presented at 2nd International Atomic Layer Epitaxy Symposium, Raleigh, NC, June 2–5, 1992.
70 Morishita, S., Sugahara, S., and Matsumura, M. (1997) Atomic-layer chemical-vapor-deposition of silicon-nitride. Presented at 4th International Symposium on Atomic Layer Epitaxy and Related Surfaces Processes, ALE-4, Linz, Austria, July 29–31, 1996.
71 Graugnard, E., Gaillot, D.P., Dunham, S.M., Neff, C.W., Yamashita, T., and Summers, C.J. (2006) Photonic band tuning in 2D photonic crystal slab waveguides by atomic layer deposition. *Appl. Phys. Lett.*, **89**, 181108–181110.
72 Gaillot, D.P., Graugnard, E.D., Blair, J., and Summers, C.J. (2007) Dispersion control in two-dimensional superlattice photonic crystal slab waveguides by atomic layer deposition. *Appl. Phys. Lett.*, **91**, 181123.
73 Gaillot, D.P., Deparis, O., Wagner, B.K., Welch, V., Vigneron, J.P., and Summers, C.J. (2008) Composite organic/inorganic butterfly scales: producing photonic structures with atomic layer deposition. *Phys. Rev. E*, **78**, 031922.
74 King, J.S., Neff, C.W., Summers, C.J., Park, W., Blomquist, S., Forsythe, E., and Morton, D. (2003) High-filling-fraction inverted ZnS opals fabricated by atomic layer deposition. *Appl. Phys. Lett.*, **83**, 2566–2568.
75 Rugge, A., Park, J.-S., Gordon, R.G., and Tolbert, S.H. (2005) Tantalum(V) nitride inverse opals as photonic structures for visible wavelengths. *J. Phys. Chem. B*, **109**, 3764–4377.
76 King, J.S., Heineman, D., Graugnard, E., and Summers, C.J. (2005) Atomic layer deposition in porous structures: 3D photonic crystals. *Appl. Surf. Sci.*, **244**, 511–516.
77 King, J.S. (2004) Fabrication of opal-based photonic crystals using atomic layer deposition. PhD thesis, Georgia Institute of Technology.
78 Summers, C.J. and Park, W. (2006) Photonic crystal structures. US Patent 6,999,669.
79 Povey, M., Whitehead, D., Thomas, K., Pemble, M.E., Bardosova, M., and Renard, J. (2006) Photonic crystal thin films of GaAs prepared by atomic layer deposition. *Appl. Phys. Lett.*, **89**, 104103.
80 King, J.S., Gaillot, D.P., Graugnard, E., and Summers, C.J. (2006) Conformally back-filled, non-close-packed inverse-opal photonic crystals. *Adv. Mater.*, **18**, 1063–1067.
81 Scharrer, M., Wu, X., Yamilov, A., Cao, H., and Chang, R.P.H. (2005) Fabrication of inverted opal ZnO photonic crystals by atomic layer deposition. *Appl. Phys. Lett.*, **86**, 151113.
82 Scharrer, M., Yamilov A., Wu, X., Cao, H., and Chang, R.P.H. (2006) Ultraviolet lasing in high-order bands of three-dimensional ZnO photonic crystals. *Appl. Phys. Lett.*, **88**, 201103.
83 Withall, R., Martinez-Rubio, M., Fern, G.R., Ireland, T.G., and silver, J. (2003) Photonic phosphors based on cubic Y_2O_3: Tb^{3+} infilled into a synthetic opal lattice. *J. Opt. A*, **5**, 81–85.
84 Velev, O.D., Jede, T.A., Lobo, R.F., and Lenhoff, A.M. (1997) Porous silica via colloidal crystallization. *Nature*, **389**, 447.
85 Sechrist, Z.A., Schwartz, B.T., Lee, J.H., McCormick, J.A., Piestun, R., Park, W., and George, S.M. (2006) Modification of

opal photonic crystals using Al_2O_3 atomic layer deposition. *Chem. Mater.*, **18**, 3562–3570.

86 Braun, P.V. and Wiltzius, P. (1999) Electrochemically grown photonic crystals. *Nature*, **402**, 603.

87 Wijnhoven, J.E.G.J., Zevenhuizen, S.J.M., Hendriks, M.A., Vanmaekelbergh, D., Kelly, J.J., and Vos, W.L. (2000) Electrochemical assembly of ordered macropores in gold. *Adv. Mater.*, **12**, 888–890.

88 Astratov, V.N., Adawi, A.M., Skolnick, M.S., Tikhomirov, V.K., Lyubin, V., Lidzey, D.G., Ariu, M., and Reynolds, A.L. (2001) Opal photonic crystals infiltrated with chalcogenide glasses. *Appl. Phys. Lett.*, **78**, 4094.

89 King, J.S., Graugnard, E., and Summers, C.J. (2006) Photoluminescence modification by high-order photonic bands in TiO_2/ZnS:Mn multilayer inverse opals. *Appl. Phys. Lett.*, **88**, 081109.

90 Blanco, A., Míguez, H., Meseguer, F., Lopez, C., Lopez-Tejeira, F., and Sanchez-Dehesa, J. (2001) Photonic band gap properties of CdS-in-opal systems. *Appl. Phys. Lett.*, **78**, 3181.

91 Gaillot, D.P. and Summers, C.J. (2006) Photonic band gaps in non-close-packed inverse opals. *J. Appl. Phys.*, **100**, 113118–113121.

92 Graugnard, E., Roche, O.M., Dunham, S.N., King, J.S., Sharp, D.N., Denning, R.G., Turberfield, A.J., and Summers, C.J. (2009) Replicated photonic crystals by atomic layer deposition within holographically defined polymer templates. *Appl. Phys. Lett.*, **94**, 263109.

93 John, S. and Busch, K. (1999) Photonic bandgap formation and tunability in certain self-organizing systems. *J. Lightwave Technol.*, **17**, 1931–1943.

94 Busch, K. and John, S. (1999) Liquid-crystal photonic-band-gap materials: the tunable electromagnetic vacuum. *Phys. Rev. Lett.*, **83**, 967–970.

95 Meng, Q.-B., Fu, C.-H., Hayami, S., Gu, Z.-Z., Sato, O., and Fujishima, A. (2001) Effects of external electric field upon the photonic band structure in synthetic opal infiltrated with liquid crystal. *J. Appl. Phys.*, **89**, 5794.

96 Shimoda, Y., Ozaki, M., and Yoshino, K. (2001) Electric field tuning of a stop band in a reflection spectrum of synthetic opal infiltrated with nematic liquid crystal. *Appl. Phys. Lett.*, **79**, 3627.

97 Kang, D., Maclennan, J.E., Clark, N.A., Zakhidov, A.A., and Baughman, R.H. (2001) Electro-optic behavior of liquid-crystal-filled silica opal photonic crystals: effect of liquid-crystal alignment. *Phys. Rev. Lett.*, **86**, 4052–4055.

98 Yoshino, K., Satoh, S., Shimoda, Y., Kajii, H., Tamura, T., Kawagishi, Y., Matsui, T., Hidayat, R., Fujii, A., and Ozaki, M. (2001) Tunable optical properties of conducting polymers infiltrated in synthetic opal as photonic crystal. Presented at International Conference on Science and Technology of Synthetic Metals, Gastein, Austria, July 15–21, 2000.

99 Yoshino, K., Shimoda, Y., Kawagishi, Y., Nakayama, K., and Ozaki, M. (1999) Temperature tuning of the stop band in transmission spectra of liquid-crystal infiltrated synthetic opal as tunable photonic crystal. *Appl. Phys. Lett.*, **75**, 932–934.

100 Khoo, I.-C. (1995) *Liquid Crystals: Physical Properties and Non-Linear Optical Phenomena*, John Wiley & Sons, Inc., New York.

101 Kubo, S., Gu, Z.-Z., Takahashi, K., Ohko, Y., Sato, O., and Fujishima, A. (2002) Control of the optical band structure of liquid crystal infiltrated inverse opal by a photoinduced nematic–isotropic phase transition. *J. Am. Chem. Soc.*, **124**, 10950–10951.

102 Mertens, G., Roder, T., Schweins, R., Huber, K., and Kitzerow, H.-S. (2002) Shift of the photonic band gap in two photonic crystal/liquid crystal composites. *Appl. Phys. Lett.*, **80**, 1885–1887.

103 Mach, P., Wiltzius, P., Megens, M., Weitz, D.A., Lin, K.-H., Lubensky, T.C., and Yodh, A.G. (2002) Electro-optic response and switchable Bragg diffraction for liquid crystals in colloid-templated materials. *Phys. Rev. E*, **65**, 031720.

104 Ozaki, M., Shimoda, Y., Kasano, M., and Yoshino, K. (2002) Electric field tuning of the stop band in a liquid-crystal-infiltrated polymer inverse opal. *Adv. Mater.*, **14**, 514–518.

105 Gottardo, S., Wiersma, D.S., and Vos, W.L. (2003) Liquid crystal infiltration of complex dielectrics. Presented at 6th International Conference on Electrical Transport and Optical Properties of Inhomogeneous Media, ETOPIM 6, Snowbird, UT, July 15–19, 2002.
106 Kubo, S., Gu, Z.-Z., Takahashi, K., Fujishima, A., Segawa, H., and Sato, O. (2005) Control of the optical properties of liquid crystal-infiltrated inverse opal structures using photo irradiation and/or an electric field. *Chem. Mater.*, **17**, 2298–2309.
107 Kubo, S., Gu, Z.-Z., Takahashi, K., Fujishima, A., Segawa, H., and Sato, O. (2004) Tunable photonic band gap crystals based on a liquid crystal-infiltrated inverse opal structure. *J. Am. Chem. Soc.*, **126**, 8314–8319.
108 Gaillot, D.P., Graugnard, E., King, J.S., and Summers, C.J. (2007) Tunable Bragg peak response in liquid crystal infiltrated photonic crystals. *J. Opt. Soc. Am. B*, **24**, 7.
109 Gaillot, D.P. (2007) Optical properties of complex periodic media structurally modified by atomic layer deposition. PhD thesis, Georgia Institute of Technology.
110 Graugnard, E., King, J.S., Jain, S., and Summers, C.J. (2005) Electric-field tuning of the Bragg peak in large-pore TiO_2 inverse shell opals. *Phys. Rev. B*, **72**, R233105.
111 Graugnard, E., Dunham, S.N., King, J.S., Lorang, D., Jain, S., and Summers, C.J. (2007) Enhanced tunable Bragg diffraction in large-pore inverse opals using dual-frequency liquid crystal. *Appl. Phys. Lett.*, **97**, 111101.
112 Puzzo, D.P., Arsenault, A.C., Manners, I., and Ozin, G.A. (2008) Electroactive inverse opal: a single material for all colors. *Angew. Chem., Int. Ed.*, **47**, 943.
113 Arsenault, A.C., Puzzo, D.P., Manners, I., and Ozin, G.A. (2007) Photonic-crystal full-color displays. *Nat. Photon.*, **1**, 468.
114 Aguirre, C.I., Reguera, E., and Stein, A. (2010) Tunable colors in opals and inverse opal photonic crystals. *Adv. Funct. Mater.*, **20**, 2565–2578.
115 Graugnard, E., King, J.S., Gaillot, D.P., and Summers, C.J. (2007) Atomic layer deposition for nano-fabrication of 3D optoelectronic devices. *ECS Trans.*, **3**, 151.
116 Yang, X., Chen, C.J., Husko, C.A., and Wong, C.W. (2007) Digital resonance tuning of high-Q/V_m silicon photonic crystal nanocavities by atomic layer deposition. *Appl. Phys. Lett.*, **91**, 161114.
117 Tamma, V.A., Blair, J., Summers, C.J., and Park, W. (2010) Dispersion characteristics of silicon nanorod based carpet cloaks. *Opt. Express*, **18**, 25746–25756.
118 Summers, C.J., Gaillot, D.P., Crne, M., Blair, J., Park, J.O., Srinivasarao, M., Deparis, O., Welch, V., and Vigneron, J.P. (2010) Investigations and mimicry of the optical properties of butterfly wings. *J. Nonlinear Opt. Phys.*, **19**, 489–501.
119 Crne, M., Sharma, V., Blair, J., Park, J.O., Summers, C.J., and Srinivasarao, M. (2011) Biomimicry of optical microstructures of *Papilio palinurus*. *Europhys. Lett.*, **93**, 14001.
120 Huang, J., Wang, X., and Wang, Z.L. (2006) Controlled replication of butterfly wings for achieving tunable photonic properties. *Nano Lett.*, **6**, 2325.
121 Losic, D., Triani, G., Evans, P.J., Atanacio, A., Mitchell, J.G., and Voelcker, N.H. (2006) Controlled pore structure modification of diatoms by atomic layer deposition of TiO_2. *J. Mater. Chem.*, **16**, 4029–4034.
122 Kemell, M., Pore, V., Ritala, M., Leskelä, M., and Lindén, M. (2005) Atomic layer deposition in nanometer-level replication of cellulosic substances and preparation of photocatalytic TiO_2/cellulose composites. *J. Am. Chem. Soc.*, **127** (41), 14178–14179.
123 Scott, I.D., Seok Jung, Y., Cavanagh, A.S., Yan, Y., Dillon, A.C., George, S.M., and Lee, S.-H. (2011) Ultrathin coatings on nano-$LiCoO_2$ for Li-ion vehicular applications. *Nano Lett.*, **11** (2), 414–418.

16
Nanolaminates

Adriana V. Szeghalmi and Mato Knez

16.1
Introduction

The previous chapters concentrated on the deposition of individual layers by atomic layer deposition (ALD). The ALD process is, however, also a highly efficient technique to produce multicomponent layers of various oxides, nitrides, metals, and/or organic materials through combination of multiple processes involving even molecular layer deposition (MLD). The alternating materials are usually deposited in the same run in order to avoid delays in the production. Commonly, the individual sublayers can be identified with clear boundaries between the components even for layers with thicknesses of about 1–2 nm. However, alloy-type mixed-phase systems can also be obtained. Nanolaminates with repeating structural parameters are of great interest for various branches of materials science and we will focus on these alternating systems in the following chapter.

The interest for such multicomponent systems is broad. On the one hand, the interest is derived from the physical phenomena at the interfaces such as in the case of optical elements. On the other hand, it is interesting to achieve material assemblies with multiple functionalities derived from the individual components such as diffusion barriers combined with electrical properties. There are also further benefits that nanolaminates exhibit over individual layers that can be of particular importance for the production of novel materials [1–3].

16.2
Optical Applications

16.2.1
Interference Optics

Dielectric and metallic coatings are essential for the production of both high-performance optics and optics used in daily life. Their function and quality are determined by the material properties and composition, the micro- and nanostruc-

Atomic Layer Deposition of Nanostructured Materials, First Edition. Edited by Nicola Pinna and Mato Knez.

tural characteristics of the sample, and the optical phenomena occurring within the material and at the materials' interfaces. The ability to understand, analyze, and control these phenomena will allow a production of the desired optical element. Atomic layer deposition offers an excellent opportunity to produce ultrathin, highly uniform, and extremely smooth films, and to tailor the material's refractive index. The conformal growth of ALD films onto complex 3D structures opens routes for defining and mixing the material composition and constructing novel optical elements, which have never before been available.

First optical applications of ALD multilayers or nanolaminates have been envisioned quite early, in parallel to the semiconductor-oriented investigations. Zinc sulfide and zinc selenide thin films have been proposed as antireflective coatings in the infrared (IR) range and as light emitting diodes (LEDs), thin film electroluminescent (TFEL) displays, multilayer dielectric filters, optical phase modulators, and light guides in integrated optics [4]. Shurtleff *et al.* [5] anticipated X-ray optics produced by ALD. With the work published in 1996 by Ritala and coworkers [6], optical thin film applications were demonstrated, and ALD received an international attention in the optical community. The interest for optical coatings produced by ALD has been continuously increasing since then and novel strategies for improved optical elements have been and still are targeted.

Antireflective coatings, high reflective coatings, neutral beam splitters, and Fabry–Perot filters were produced using ZnS and Al_2O_3 as high (H) and low (L) refractive index materials, respectively, having quarter-wave optical thicknesses ($nd = \lambda/4$). The transmission spectrum of a 12-layer (HLHLH-LL-HLHLH/glass) Fabry–Perot bandpass filter is shown in Figure 16.1 [5]. The peak position of the bandpass filter is determined by the thickness of the resonance cavity layer (LL), and it has been found to vary from about 620 to 560 nm in wavelength as a function of the distance from the leading edge of the glass substrate. It is proposed that this is a result of nonuniformities within the coating [5]. The dependence of the transmittance peak

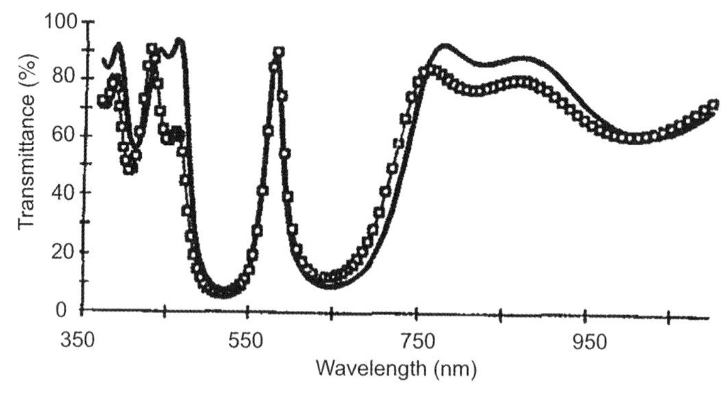

Figure 16.1 Transmittance spectrum (dot: measured; line: target) of a 12-layer Fabry–Perot filter on glass consisting of a high (H)/low (L) refractive index nanolaminate. Reprinted with permission from Ref. [6]. Copyright 1996, Elsevier.

of the filter on the LL thickness also allows to produce filters with various target bandpass positions without significantly affecting the full-width at half-maximum (FWHM) of the peak [7].

Numerous dielectric materials are available by ALD as high and low refractive index multilayer components and the morphology and optical properties of Al_2O_3/TiO_2 nanolaminates have been extensively investigated [7–9] for the visible spectral range. Some fluorides and oxides are advantageous for the production of optical elements for the deep UV region since the required film thicknesses for those application fields are significantly smaller than those for the visible spectrum and compensate for the relatively low film growth rates. Critical issues are the long-term stability and performance under the operating conditions, indicating the necessity to target more robust optical designs.

Historically, interference filters were made of homogeneous layers in a bulk-like fashion, separated by sharp, parallel, and smooth interfaces. Having perfect interfaces was considered to be ideal [5]. With improved production technologies, rugate filters having a gradient in the refraction index became possible. Such rugate filters exhibit superior performance over standard notch filters. A rugate notch filter has a deep, narrow rejection band while also providing a high and flat transmission for the rest of the spectrum. Atomic layer deposition is an excellent technique for tailoring the refractive index [10] with nanoscale laminated films (see Figure 16.2), and thus is

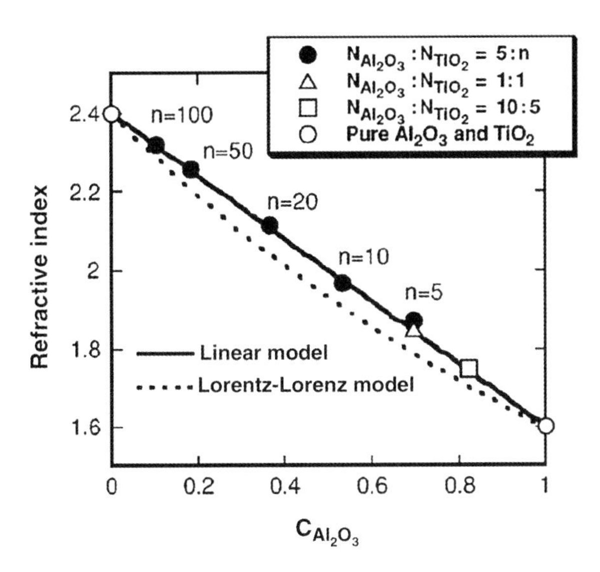

Figure 16.2 Linear dependence of the refractive index in the visible spectral region of a nanolaminate made of high (TiO_2) and low (Al_2O_3) refractive index materials as a function of the volume concentration of the alumina component. A composite material with arbitrary refractive index can be obtained by varying the ratio of the ALD cycles (*N*) of the alternating materials when the individual film thicknesses are much lower than the wavelength of light. Reprinted with permission from Ref. [10]. Copyright 2002, the American Institute of Physics.

ideal for the production of rugate filters. The refractive index of the composite material is fine-tuned with the relative ratio of the individual components. Experimental data showed that the refractive index of an Al_2O_3/TiO_2 composite varies linearly with the volume concentration of the alumina [10]. The composition of the material is precisely controlled by the number of ALD cycles. Since the total thickness of the nanolaminate composite is much less than the wavelength of light, the ALD composite material acts as an optically continuous effective medium. This technique of gradient layers is also of interest for femtosecond laser chirped mirrors [10]. Zaitsu *et al.* [11] have shown that the incorporation of Al_2O_3 into TiO_2 significantly changed the TiO_2 morphology and increased the laser damage resistance of the composite compared to pure TiO_2 films. Especially, amorphous thin films are favorable for a reduction of the laser damage threshold in laser optics, the deposition of which is a strength of the ALD process.

Dielectric ALD nanolaminates also hold an attractive potential for soft and hard X-ray optics. Multilayer mirrors for the X-ray spectral range require short-period nanolaminates, where the random variation between the layers is minimal [12, 13]. Other factors influencing the performance of X-ray optics include the interface roughness and the interdiffusion between the layers [12]. Oxides prevent the formation of alloys or strong interdiffusion at the interfaces, thus enhancing the performance of the optical element. Optics for the "water window" wavelength region (2.33–4.36 nm), between the oxygen and carbon K absorption edges, may particularly benefit from metal oxide-based nanolaminates, since the absorption of oxygen in oxides is minor in this spectral range [14]. Titania/alumina ALD nanolaminates with a bilayer thickness (period) of 4.43 nm achieved a reflectance of over 30% at the wavelength of 2.734 nm and an incidence angle of 71.8°. The number of bilayers was limited to 20, but 60 layers are expected to produce a reflectance of above 50% if the optimal multilayer structure is maintained during the long deposition process. TiO_2/NiO ALD nanolaminates have been proposed as a high reflectance mirror for the "water window" and as a chirped mirror for pulsed attosecond soft X-rays [15]. Chirped mirrors are important for the development of pulsed lasers since they are able to compress a temporal broadening of the pulses. Titanium shows anomalous dispersion around the L absorption edge at 2.733 nm; thus, TiO_2 has a maximum reflection at this wavelength. Soft X-ray mirrors have been demonstrated for TiO_2/ZnO nanolaminates, epitaxially grown on *c*-plane sapphire substrates at 450 °C. The reflectivity was about 30% around 2.7 nm at a grazing angle 2θ of 10° [16]. The same group investigated TiO_2/Al_2O_3 nanolaminates with a reflectivity of approximately 1.54% at a grazing angle 2θ of 29.5° corresponding to an incidence angle of 75.25°. Further research is necessary to attain the theoretical maximum reflectance of nearly 50% at an angle of incidence of 18.2°; however, the experimental data show promise for ALD-based optics in soft X-ray spectroscopy and high-resolution X-ray microscopy, specifically in the "water window," which is essential for an investigation of biomaterials in aqueous media.

Metallic nanolaminates are dominating the production of optics for the hard X-ray spectral region. X-ray reflectivity (XRR) spectra of nanolaminates of alternating W/Al_2O_3 layers with individual thickness as low as 14 and 17 Å, respectively, show

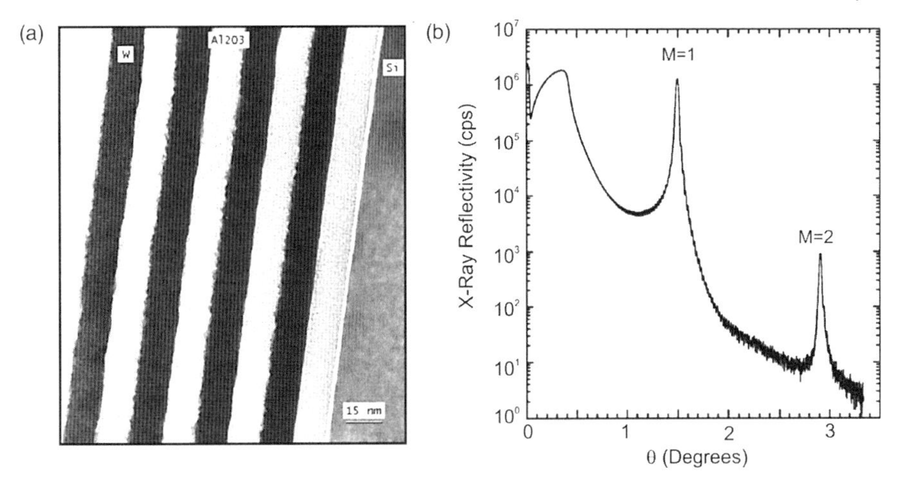

Figure 16.3 Transmission electron micrograph of a 4-bilyer W/Al_2O_3 nanolaminate (a) and the X-ray reflectivity spectrum of a 64-bilayer nanolaminate (b) showing first- and second-order Bragg peaks. Reprinted with permission from Ref. [17]. Copyright 2005, the American Chemical Society.

a first-order Bragg peak at a grazing angle of 1.49° [17] at the 1.54 Å Cu Kα line. The Bragg interference occurs because of the alternating layers being well separated and interdiffusion of the two materials is negligible. XRR and grazing angle X-ray diffraction (GIXRD) are powerful techniques for the characterization of ultrathin materials and provide information on film thickness, roughness, material density, and crystallinity. The density of the alumina ALD nanolayer is significantly lower than that of the crystalline material or thicker Al_2O_3 ALD layers, probably due to incorporation of fluorine [17] at the Al_2O_3/W interface during the WF_6 precursor pulses. Interestingly, transmission electron microscopy images revealed clear differences in the roughness at the interfaces between W and Al_2O_3 (sharp) versus the interfaces between Al_2O_3 and W (rough) (see Figure 16.3). The alumina layers grown onto rougher polycrystalline W layers contribute to a smoothing of the nanolaminate. Thinner W layers benefit more from the smoothing effect as well as layers deposited at lower temperatures [18].

16.2.2 Diffractive Optical Elements

Technological developments in micro- and nanostructuring allow a production of novel and highly interesting optical elements. Among the available coating technologies, ALD will probably play a key role in tailoring the optical properties of nanostructured optics because of its inherent capability to conformally coat high aspect ratio and complex structures. In addition, specific functionalities can be achieved by using composite or layered materials. ALD films can be targeted as functional, buffer, or sacrificial layers. They can fine-tune the optical properties or build novel and temperature-resistive templates for subsequent processing steps [19].

Summers and coworkers thoroughly investigated the use of ALD for optimization of 2D and 3D photonic crystal (PC) structures [19]. Photonic crystals with wide photonic band gaps (PGBs) are highly interesting for low-loss waveguiding, laser optics, metamaterials, light trapping, and controlling the emission of embedded emission sources. Their PBGs strongly depend on the lattice structure and the refractive index contrast both of which can be well tailored by ALD. The infiltration, replication, and conversion of opals with fcc lattice structures and their derivatives demonstrate the large degree of flexibility provided by ALD to tune the optical properties of the opals [19]. The biggest challenge is the growth of conformal coatings within very small pores with diameters of 30–40 nm and a very large surface area. The Bragg peak position and the photoluminescence intensity of an inverted opal consisting of ZnS:Mn and TiO_2 multilayers can be tuned by backfill infiltration of the photonic crystal, meaning that after removal of the template, the occurring hollow area is coated in a subsequent step.

Novel PC structures can be produced when conformal coating is combined with the sacrificial layer selective etching technique. Highly porous materials with a filling fraction as low as 3–5% of the total structure volume were achieved by this technique [19] (see also Chapter 15 for a more detailed discussion of ALD-based photonic crystals). These PCs allow for a shift of the Bragg peak position in a very broad spectral range of about 100% by backfilling with incremental TiO_2 ALD layers.

Subwavelength nanostructured optics show interesting guided-mode resonance (GMR)-type Wood's anomalies with application potential as optical filters and sensors [20]. Such optics is characterized by a narrow band in the transmittance or reflectance spectrum due to the coupling of the incident wavelength to leaky modes guided by the grating. Multilayer ALD coatings were applied to 2D shallow polycarbonate nanostructures to produce tunable GMR filters in the visible spectral range [20]. Figure 16.4 shows a focused ion beam scanning electron micrograph (FIBSEM) and polarized reflectance spectra of the tunable GMR filter as a function of the azimuthal angle of incidence. These systems have great application potential as filters since the peak position of the reflectance/transmittance band can be tuned in a broad spectral range through rotation of the optical element around the axis normal to the substrate without otherwise changing the optical setup. Thus, these elements may be easily integrated in portable, miniaturized optical equipment while their production is facile over large sample areas.

With increasing recognition of the ALD coating technology by the optical community, an intense exchange of know-how is taking place. It is necessary to achieve a common base of understanding to develop novel and improved optics. Experts in optical components and spectroscopy are already contributing and motivating the ALD community. This is also reflected in the development of new ALD materials, such as fluorides for optical applications in the DUV, or improving various aspects of the ALD equipment, such as implementing *in situ* ellipsometry. With new impulses from optics, the future of ALD in optics might turn very bright.

(a)

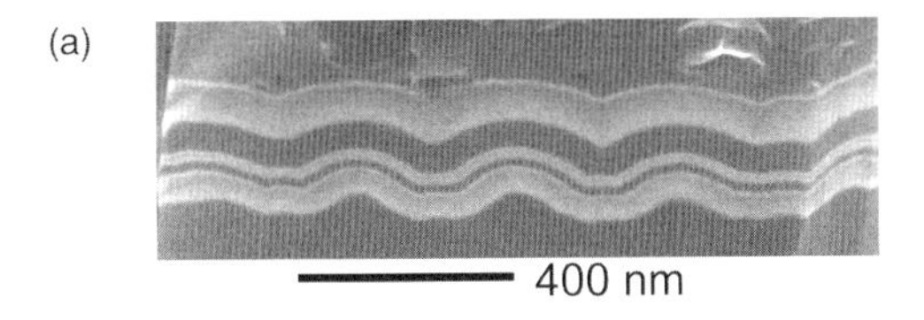

(b)

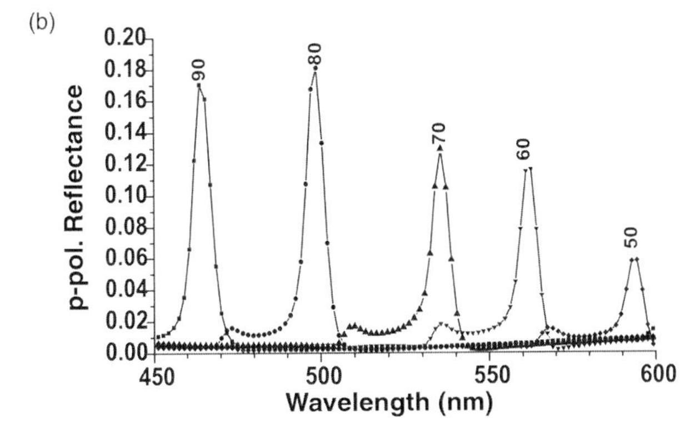

Figure 16.4 Scanning electron micrograph of a nanostructure coated with a five-layer alumina/titania multilayer in cross section prepared with the focused ion beam technique (a). Discrete layers of Al_2O_3 (gray) and TiO_2 (black) can be identified on the polycarbonate substrate. Porous carbon paste was deposited on the ALD multilayer for imaging purposes. (b) Polarized reflectance spectra showing narrowband guided-mode resonance peaks of the optical filter measured at an angle of incidence of 55°. The spectra were recorded at various azimuthal angles by rotating the sample around its axis. Reprinted with permission from Ref. [20]. Copyright 2010, Wiley-VCH Verlag GmbH, Weinheim.

16.3
Thin Film Encapsulation

TFEL devices were among the first industrial applications to incorporate ALD layers. Their performance and reliability strongly depend on the properties and morphology of the ALD films acting as an insulator. The insulator should possess high dielectric permittivity, strong resistance against electrical breakdown, a pinhole-free structure, and thickness uniformity over a large area. Insulator films deposited by ALD can easily fulfill the morphological requirements, and nanolaminates comprising a material with high permittivity and a second material with high electrical stability can improve the luminescent and electrical characteristics of TFELs [21]. The reported charge storage factor for a Ta_2O_5–Al_2O_3 nanolaminate with a relative Ta_2O_5 thickness ratio of 0.67 is about two times larger than that of pure Al_2O_3. The luminescence onset of the nanolaminated device occurs at a lower threshold voltage than that of the Al_2O_3 device resulting in lower power consumption and an improved long-term reliability of luminance [21]. The electrical characterization indicates that the nanolaminated device is more favorable for the space charge generation within the ZnS:Tb phosphor layer than with an Al_2O_3 insulator, resulting in an improved performance.

Nanolaminate structures produced by atomic layer deposition have also been demonstrated as highly effective encapsulation materials for polymer-based photovoltaic cells (PVCs) and organic light emitting diodes (OLEDs) [22, 23]. The lifetime of these devices could be extended by ultrathin ALD coatings of alternating Al_2O_3/HfO_2 or Al_2O_3/ZrO_2 layers because of the capability of ALD to grow defect-free inorganic films at low temperatures.

There are several aspects identified as important for the encapsulation: (i) The changes in the morphology of the coated organic material, induced during the ALD process, will affect the performance of the device. Therefore, the optimal deposition temperature for the ALD deposition must be determined. Optimal thermal annealing during the coating process induces aggregation of the organic domains that facilitates the transport of charge carriers. (ii) The nucleation of the ALD films on the active polymer surface might be hindered resulting in poor encapsulation at the device edges and rapid degradation of the device. Prolonged exposure of the substrate to the ALD precursors within the first cycles can reduce the nucleation delay on the active layers. (iii) Increasing the hydrophobicity of the coating material will reduce the hydrolysis of the capping layer and the degradation of the active layer. Al_2O_3/HfO_2 nanolaminates have greatly increased the resistance to hydrolysis of the encapsulation layer since HfO_2 is more hydrophobic than Al_2O_3 and served as a moisture barrier [22]. It is worth noting that the mentioned nanolaminate is a more effective moisture barrier than a two-layer coating, which is probably related to the crystallinity of the deposited HfO_2 films. The formation of large grain boundaries is effectively suppressed in the nanolaminates with thinner sublayers. (iv) Improved mechanical stability of the ALD layer is essential for long-time performance specifically if flexible devices are targeted. Chang *et al.* applied a UV curable resin to improve the mechanical stability of the nanolaminate encapsulated PVC even reaching the reference values of a reference PVC without encapsulation, but with the advantage of encapsulation [22].

Another interesting approach is the application of nanolaminates as diffusion barriers for gases. One example of such an approach is the encapsulation of OLEDs with a nanolaminate consisting of Al_2O_3 and ZrO_2 [23]. Although this topic was discussed in Chapter 6 of this book and further details can be extracted from there, it may be mentioned at this point that the nanolaminates showed much higher performance for blocking gas diffusion and better chemical stability than neat Al_2O_3 coatings. In addition, the nanolaminates show a much better compatibility with flexible electronics different from glass lids that are commonly used for encapsulation.

Hybrid inorganic–polymer multilayers are expected to be mechanically even more stable and hydrophobic. Such systems allow an increase of the elasticity and flexibility of organic optoelectronic devices. MLD in combination with ALD [24, 25] was applied to produce organic–inorganic hybrid nanolaminates for the production of elastic, flexible, and reliable materials. In this way, the durability and the blocking properties of the coatings were also improved in direct comparison to pure inorganic layers.

Ultrathin hybrid nanolaminates are rapidly produced and can be easily integrated into the manufacturing line of OLEDs and PVCs. Detailed studies of the optical, dielectric, and mechanical properties of such hybrid nanolaminates are still scarce, and intense studies are necessary to optimize the device performance and the long-time stability.

Organic–inorganic superlattices were produced by combining alkylsiloxane self-assembled multilayers (SAMs) grown by MLD with ultrathin TiO_2 layers [26]. The MLD cycle consisted of (i) adsorption of the alkylsiloxane on hydroxyl-terminated surfaces via hydroxylation of the $SiCl_3$ group in the presence of H_2O, (ii) ozone treatment to convert the vinyl group of the SAM to a carboxylic group, and (iii) activation of the carboxylic acid group to form a highly reactive hydroxyl-terminated surface by a $Ti(O^iPr)_4$ and (iv) a H_2O pulse. The estimated growth per cycle (GPC) of the SAM with fully extended alkyl groups was 12 Å, and the measured self-limited GPC was 11 Å. Cross-sectional transmission electron microscopy images of the superlattices are depicted in Figure 16.5 [26]. The nanolaminates are uniform, smooth, and have sharp interfaces. The hydrophobicity of the superlattice with a SAM top layer can be easily shifted from hydrophobic to hydrophilic by stopping the MLD cycle at the vinyl- or the $Ti(OH)_x$-terminated moiety. This aspect may further improve the long-term resistance of encapsulated materials based on such hybrid superlattices.

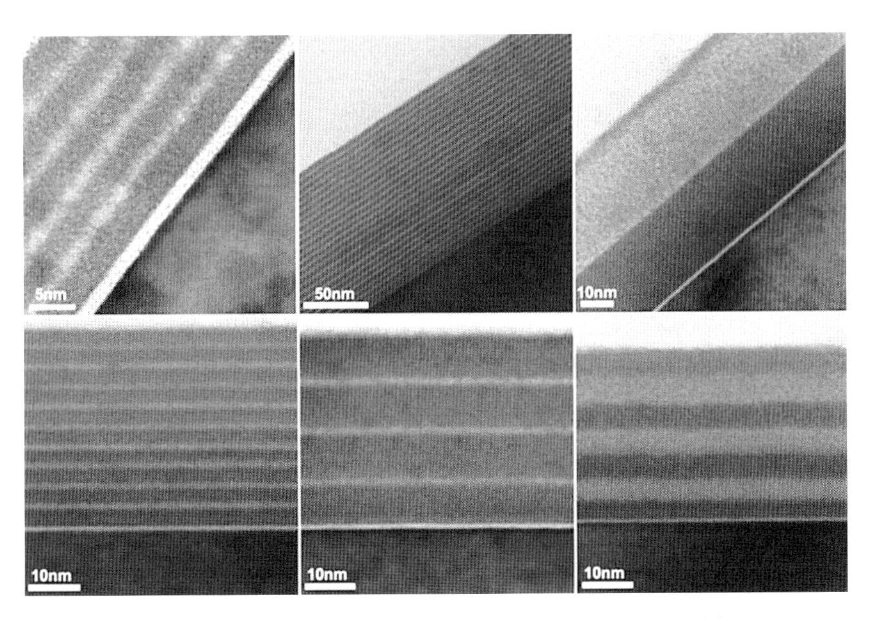

Figure 16.5 Transmission electron micrographs of self-assembled organic multilayer/TiO_2 nanolaminate films made by using MLD with ALD (white is the organic layer, and dark is the titania layer). Reprinted with permission from Ref. [26]. Copyright 2007, the American Chemical Society.

16.4 Applications in Electronics

Nanolaminates deposited by ALD also have another aspect aside from the described optical property improvement of materials. In some previous chapters, the extraordinary importance of ALD for the microelectronics industry has already been discussed. Upon decreasing the feature sizes and structural fashions in electronics, the electronic properties of thin films become increasingly important. Miniaturization in electronics inevitably goes hand in hand with thinner dielectric coatings, which in turn results in a change or loss of the desired electronic properties. One way to solve the problem is to seek for materials that can replace SiO_2 in the form of high-κ dielectrics.

The capacitance of materials is proportional to their dielectric constant and inversely proportional to the material thickness. In order to obtain the appropriate capacitance that is required for modern electronic devices, the thickness of the traditionally used SiO_2 dielectric has to be decreased. This is, however, not as simple. Decreasing the thickness of SiO_2, for example, leads to an occurrence of tunneling effects that below about 2 nm thickness of SiO_2 dominate the leakage current [27, 28]. Aside from the tunneling effects, formation of interface layers of the film with the substrate, the amount of interface state densities, the crystalline or amorphous nature of the dielectric, and so on contribute to the leakage current and play a significant role for the selection of the material to replace SiO_2 in future electronic devices. Among the most promising alternative materials are ZrO_2 and HfO_2 and a significant number of research groups investigate these materials in detail [29]. Commonly, the approaches are based on the formation of thin films of the mentioned and other materials either by means of chemical vapor deposition (CVD) or more recently by ALD. Since even better electrical characteristics are required than those revealed by the pure materials, improvement of the properties is followed by homogeneous or heterogeneous mixtures of materials in order to take advantage of the particular properties of each material. In the following, some approaches towards heterogeneous mixtures, that is, nanolaminate structures of materials designed for applications in electronics, will be discussed.

In order to obtain a suitable material for the replacement of SiO_2, several electronic characteristics of the new material have to be evaluated. Among these important factors are a high dielectric permittivity of the resulting material, a low gate leakage current, a high charge storage factor, and a good long-term stability of the dielectric properties. Aside from the dielectric properties, chemical properties also play an important role as these can strongly affect the dielectric properties. A material that easily crystallizes will contain many grain boundaries that will act as conduction paths and increase the leakage current. Many electronic devices require annealing steps during production; thus, the induced heat might easily lead to crystallization of an initially deposited amorphous material, leading to unwanted increase in the leakage current. Another important factor is the occurrence of mixed interface layers at the boundary of two materials. Such behavior will affect the dielectric properties as the permittivity of such interface mixtures will deviate from the permittivity of the two pure materials.

If we simply take a look at few metal oxides that are commonly deposited by ALD, the above dilemmas will become obvious. Alumina has a large band gap of 8.7 eV and a high crystallization temperature of 900 °C, which keeps the leakage current very low. However, the dielectric constant is low (9.0). Titania has a high dielectric constant (50–80) [30, 31], but a narrow band gap (3.5 eV) and a low crystallization temperature (400 °C), which increases the leakage current induced by tunneling and creates current paths along the grain boundaries [32]. Hafnium oxide is a very popular high-κ material with a dielectric constant of 25 and a band gap of 5.7 eV [31], but suffers from a low crystallization temperature (500 °C) [33]. Further two candidates include zirconium oxide and tantalum oxide with similar considerations to hafnium oxide, albeit with slight variations in the band gap and crystallization temperature.

16.4.1 Dielectric Properties of Inorganic Nanolaminates

One possibility to overcome those shortcomings was followed in the past decade. The idea behind is that creating nanolaminates from two or more alternating materials would have beneficial effects of each applied material to the final stack. In other words, although the dielectric constant of the stack might decrease with respect to the higher-κ component, the leakage current might also decrease because of the contribution of the other component, eventually leading to a compromise material with optimized values towards replacing SiO_2. The experimental optimization of such a nanolaminate is difficult because of a large number of variables that have an influence on the final characteristics of the resulting stack. The variables include the selection of the individual materials, the number of bilayers, the thickness of each layer, thickness variations among the individual layers, and the choice of the initial and terminal layers. Considering that even the simplest case with two metal oxides and no thickness gradients of the materials within a stack allows a huge number of variations, predictions of the dielectric behavior become very difficult.

Among the above-mentioned materials, ALD of titania and alumina are the most widespread and best understood processes. Their complementary electronic characteristics could be beneficially used maintaining a high dielectric constant from the titania and inserting thin alumina layers for decreasing the leakage current. Indeed, a series of works has been published on that system, but we will concentrate on few examples that explain the common trend: Jögi *et al.* [34] reported on Al_2O_3–TiO_2–Al_2O_3 nanolaminates deposited by thermal ALD, while Jeon *et al.* [35] deposited the inverse sequence (TiO_2–Al_2O_3–TiO_2) by plasma-enhanced ALD (PEALD). In the first case, two different compositions of trilayers were evaluated, a 4–6–4 nm trilayer and a 3–2–3 nm trilayer each with respect to alumina, titania, and alumina. The measurements showed that the leakage currents were lower with the thinner layers, even decreasing after annealing. The annealing step apparently improves the quality of the films, presumably removing impurities such as carbon, chlorine, and hydrogen. However, the leakage current even after annealing became about two orders of magnitude worse than that with pure alumina. Similar leakage currents were observed with the inverted stack produced by PEALD. The leakage current was measured to be 5.11×10^{-5} A/cm^2 at 1 MV/cm. The trilayer in this case consisted of 4–1–4 nm titania, alumina, and titania. It, however, required additional optimization of the films. Crystallization of one titania layer from amorphous to anatase and the other titania layer to rutile improved the dielectric constant from 35.1 (found with amorphous titania) to 48. A slight increase in the thickness of the sandwiched alumina layer from 1 to 1.25 nm finally showed a trilayer stack with a dielectric constant of 44 and a leakage current of 6.46×10^{-7} A/cm^2 at 1 MV/cm, which appears very promising from an application point of view. The observation is quite interesting since in an earlier report by Kim *et al.* [36] the minimal thickness of alumina in order to prevent a current flow through the nanolaminates amounted to 2.5 nm. The synthesized nanolaminate, however, consisted of multiple sublayers (see the cross-sectional TEM micrograph in Figure 16.6) instead of only three and it was processed by thermal ALD implying that the alumina contains more impurities than the alumina deposited by PEALD. In the context of alumina/titania nanolaminates, it might also be worth mentioning that in multilayer stacks with total thicknesses of 150 nm and sublayer thicknesses below 0.5 nm an unusual increase in dielectric constants has been observed [37]. This giant dielectric constant reached values of around 1000 for frequencies below 10^2 Hz. The authors propose a Maxwell–Wagner-type dielectric relaxation to be the mechanism behind the giant dielectric constant. The charge carriers responsible for the relaxation are proposed to be oxygen vacancies within the defective sublayers. Oxygen atoms, being highly mobile charge carriers, would diffuse from a TiO_2 layer to an Al_2O_3 layer, which was also observed by elemental mapping. Namely, the oxygen concentration within the Al_2O_3 sublayers was higher in comparison to the concentration in TiO_2 sublayers, although from the theoretically expected stoichiometry it should be opposite. The resulting accumulated

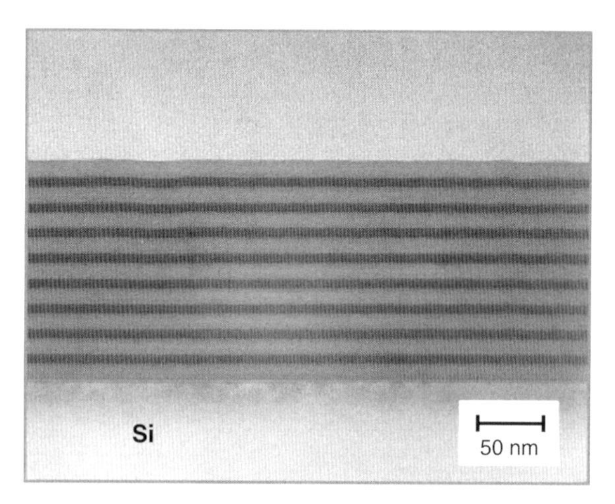

Figure 16.6 Cross-sectional TEM image of an Al_2O_3–TiO_2 nanolaminate with 17 alternating sublayers. Reprinted with permission from Ref. [36]. Copyright 2005, Elsevier.

surface charges at the interfaces of the two dielectrics are proposed to induce the Maxwell–Wagner relaxation.

The presumably most prominent high-κ material is hafnia. Aside from the investigations on the pure material and mixed phases that can be found in the literature, a large number of manuscripts have been published that are related to nanolaminates. Similar to the titania–alumina system, laminate structures with hafnia and alumina have been intensively investigated. In such systems, the reported values for the leakage current are in the range of 4×10^{-7} to 6×10^{-8} A/cm^2 [38, 39] and the dielectric constants reach 12.6–14.6 [40, 41]. The laminate structures in those works consist of rather thin sublayers. The thicknesses of the sublayers range from subnanometers to only few nanometers in thickness, sometimes even pulsing single pulses of one material. Such a thin layer should not act as a separate blocking layer, but the very small amount of added alumina should suppress the crystal formation in the hafnia during the deposition and the subsequent annealing process. Indeed, with very thin alumina layers, a crystallization of the HfO_2 could be effectively suppressed [41]. However, it was also observed that decreasing a sublayer thickness to 1 nm resulted in a dielectric constant of the laminate that is close to that of a mixed phase of both materials [41] leading to the question whether or not one can really obtain discrete sublayers in those thickness regimes. At the interface of two dielectrics, a thin mixed phase of the two materials can form, which is, however, dependent on the mobility of the ions. Even if those mixed layers are extremely thin, upon shrinking the sublayer thicknesses to 1 nm or less, the mixed phases might dominate the structure and the electrical properties, thus the above observation appears reasonable. Another question is whether or not one can still talk about laminates if only one or very few deposition cycles of one material are applied. Considering the fact that the alumina or hafnia processes are not ideal and that a monolayer growth per cycle is not obtained, applying few cycles of each precursor might easily lead to a mixed phase instead of a laminate structure and no discrete sublayers are obtained. This effect is even emphasized with nonideal substrates since many materials show growth inhibition during the first few cycles (see previous chapters of this book). On the other side, PEALD might overcome those problems at least to a certain degree.

Aside from the mentioned two systems (TiO_2/Al_2O_3 and HfO_2/Al_2O_3), nanolaminates were deposited from various combinations of Al_2O_3, TiO_2, HfO_2, Ta_2O_5, ZrO_2, Nb_2O_5, SiO_2, and La_2O_3 [42–53]. In most of those combinations, the leakage currents reach values between 10^{-4} and 10^{-7} A/cm^2 at their best and strongly depend on the used material combination. However, values of 3×10^{-8} A/cm^2 have also been measured at 2 MV/cm for the Ta_2O_5/HfO_2 system, which is the lowest value observed among the nanolaminates [50]. The leakage current densities versus the applied electric field strength of those structures are shown in Figure 16.7. The dielectric constants of the nanolaminates strongly depend on the components of the stacked layers. For example, nanolaminates containing tantalum oxide have shown permittivities of around 30 [43, 44], and with niobium oxide values of 38 [47] and 54 [46] have been obtained.

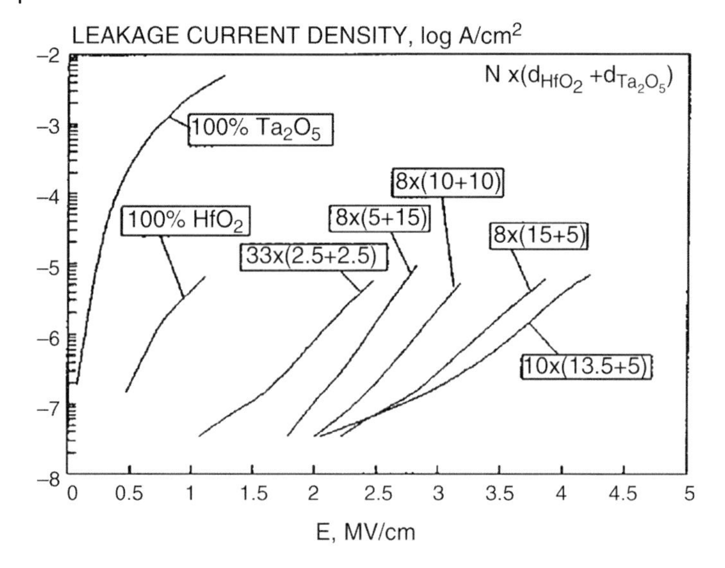

Figure 16.7 Dependence of leakage current density on the applied electric field strength in single Ta_2O_5 and HfO_2 films and HfO_2–Ta_2O_5 nanolaminates. The labels describe the sample configuration $N(d_{HfO_2} + d_{Ta_2O_5})$, where N is the number of bilayers, and d_{HfO_2} and $d_{Ta_2O_5}$ are thicknesses of single HfO_2 and Ta_2O_5 layers, respectively, expressed in nanometers. Reprinted with permission from Ref. [50]. Copyright 1996, the American Institute of Physics.

16.4.2
Dielectric Properties of Organic–Inorganic Nanolaminates

The recent development of molecular layer deposition (see Chapter 5) implies that laminate structures of inorganic and organic materials might also be interesting candidates for dielectrics. Indeed, first reports with such materials appeared recently in the literature, combining alkylsiloxane multilayers with TiO_2 [26] or Al_2O_3 [54] and polyimide with Ta_2O_5 [24]. In particular, the latter combination exhibited quite interesting dielectric properties. The films were composed of 10 alternating layers of polyimide and tantalum oxide and the ratios of the two materials were varied (see Figure 16.8 for a micrograph of the structure). The interesting observation is that with more vol% of polyimide the laminate can reach very low leakage current densities in the range of 10^{-8} A/cm^2 at 1 MV/cm. This is comparable to the best inorganic nanolaminates observed so far (see above). However, the permittivity of this material is comparatively low and amounts to about 5. With an opposite ratio of the constituent materials, the permittivity raises to 10, but on the expense of the leakage current, which becomes about one order of magnitude worse. Nevertheless, the values are still outperforming those of the single materials by one to two orders of magnitude, which appears very promising for a future development. An additional fact to be noted here is the mechanical properties of the organic–inorganic laminates. The resulting stack

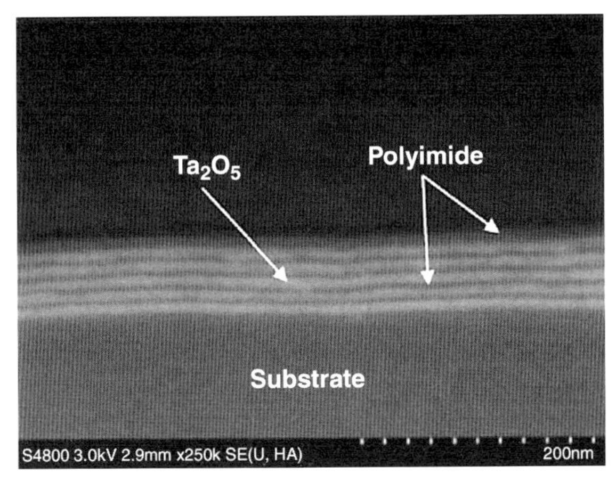

Figure 16.8 FESEM image of a polyimide (PI)/Ta_2O_5 nanolaminate. The target structure is $5 \times (10\,nm\ PI + 10\,nm\ Ta_2O_5)/Si$ with the dark layers being PI. Reprinted with permission from Ref. [24]. Copyright 2009, Wiley-VCH Verlag GmbH, Weinheim.

gains softness and ductility with increasing amount of polyimide, which makes such materials very interesting for applications in flexible electronics.

Given the many possible variation parameters, that is, the composition of the sublayers, the thicknesses, the sequences, and so on, numerous more combinations for the synthesis of nanolaminates are possible. Although already much work has been published, the field still shows good potential for further optimization and offers chances to find the ideal nanolaminate with very low leakage currents and very good dielectric constants. Hybrid materials add more flavor to the dielectric nanolaminates and expand the variation possibilities significantly.

16.4.3
Applications for Memories

Another important aspect of dielectric nanolaminates is their application in memory devices. The next generation of nonvolatile memories is expected to fulfill a number of requirements, including a faster programming/erasing speed, longer retention times, which include 10^6 programming/erasing cycles, 10 years charge retention at 85 °C, and a programming voltage below 5 V [55]. A very attractive design for flash memories is the polysilicon (or metal)/SiO_2/Si_3N_4/SiO_2/semiconductor structure (SONOS), where Si_3N_4 and SiO_2 are used as a charge storage layer and a blocking layer, respectively. The program/erase speed and the charge retention are, however, not sufficiently good. A possibility to improve the characteristics of the flash memories lies in the replacement of either or both of those layers for more efficient charge storage and/or blocking, some of which have been attempted with nanolaminates. The most prominent materials for the laminates are Al_2O_3 in combination with HfO_2 or NiO [55–59]. The results obtained from the published experiments vary, strongly depending on the composition and structure of the nanolaminates. A direct

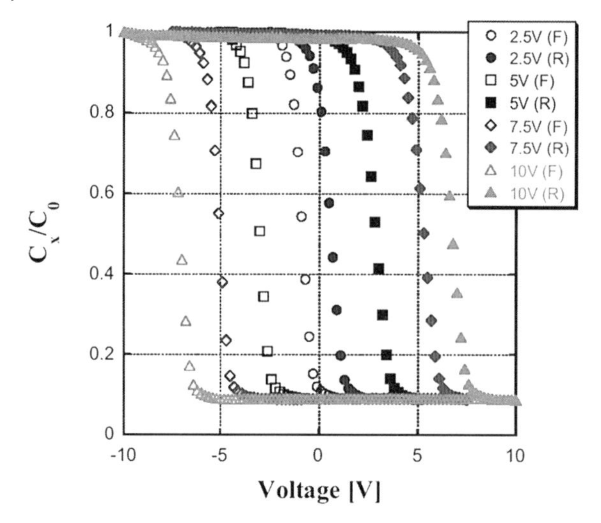

Figure 16.9 *C–V* measurements of a Si/Al_2O_3 (3 nm)/NiO (5 nm)/Al_2O_3 nanolaminate. (F) and (R) indicate the forward and reverse sweeping directions, respectively. Reprinted with permission from Ref. [55]. Copyright 2010, the Electrochemical Society.

comparison of the charge trapping behavior of those materials is not straightforward. While one measure is the size of the so-called memory window, that is, the width of the hysteresis curve of the capacitance upon sweeping the gate voltage, this curve often depends on the extension of the applied sweeping gate voltage. Nevertheless, it may be mentioned at this point that the investigated memory prototypes containing Al_2O_3 and HfO_2 nanolaminates showed memory windows of 1.45 and 2.4 V upon programming at +12 V and erasing at −12 or −14 V, respectively [56, 57]. Nanolaminates consisting of HfO_2 and HfAlO instead of Al_2O_3 showed much larger memory windows, namely about 10 V at a sweeping gate voltage of ±15 or ±16 V [58, 59]. This increase in the memory window is attributed to the formation of HfAlO nanocrystals upon annealing and the ability of those crystals to store charge. The largest memory window, however, was observed with Al_2O_3/NiO/Al_2O_3 nanolaminates. With a sweeping gate voltage of ±15 V the memory window reached 13.8 V [55] (see Figure 16.9), which is explained with a large amount of defects within the NiO in the form of cation vacancies. Those vacancies act as charge traps that upon trapping/detrapping are programmed and erased.

16.5
Copper Electroplating Applications

Semiconductor integrated circuits based on copper interconnections with improved performance and less energy consumption are currently being developed. Decreasing feature sizes of copper-based chips, however, pose severe difficulties for the electrochemical copper deposition process. The electroplating usually occurs on a copper seed layer, which is sputter deposited onto a diffusion barrier layer. This barrier layer is necessary to prevent the migration of copper into the substrate material on the expense of the performance. Improved ultrathin diffusion barrier layers with reduced resistivity are needed to allow direct copper electroplating along with an improved adhesion. Metal/metal nitride nanolaminates are considered appropriate barrier layers, whereby ALD is a promising technique to obtain the thinnest possible nanolaminates.

Several nitrides and metals can be deposited by ALD. Ruthenium in combination with TaN and WNC [60, 61] and WNC/TiN nanolaminates [62] were tested as copper diffusion barriers and adhesive layers. In the case of Ru/TaN nanolaminates, the total thickness of the barrier layer is only few nanometers and the materials intermix to form an alloy. The composition of the alloy is easily varied by the number of ALD cycles in a wide range of Ru:Ta ratios (from 1 : 1 to 1 : 30) to find the most suitable composition for copper interconnections. The Ru/TaN mixed-phase composition with Ru:Ta of 12 : 1 and a film thickness of 5 nm effectively suppressed the Cu diffusion up to temperatures of 250 °C. A higher Ru content, however, failed to prevent Cu migration. Direct copper electroplating was possible on the diffusion barrier layer and showed a texture and void-free filling performance comparable to the conventional plating on sputtered Cu seed layers [60]. WNC mixed phases might substitute the TaN component because of their robust diffusion barrier properties, low resistivity, and capability for enhanced copper metallization [61]. The PEALD deposited alloys will allow copper electroplating for next-generation interconnects with trench sizes below 35 nm, but further studies are necessary to achieve diffusion barriers with thicknesses of 2–3 nm for advanced interconnects.

16.6 Solid Oxide Fuel Cells

Energy-related research is one of the central points of the present and future research. Among the many aspects of energy production, fuel cells became increasingly important. Solid oxide fuel cells (SOFCs) are well established. In SOFCs, a solid metal oxide acts as electrolyte and is embedded into the fuel cell as a solid membrane. The conduction of oxygen ions through this membrane, however, requires a high temperature (more than 600 °C) and ways to reduce the high operation temperature are matter of intense investigation. One proposed approach is the reduction of the resistance of the electrolyte by decreasing its thickness. Yttria-stabilized zirconia (YSZ) is a common material used as solid electrolyte and a way to reduce the thickness of the electrolyte was proposed with a nanolaminate deposition of Y_2O_3 and ZrO_2 with subsequent annealing [63]. Especially when bilayer thicknesses of the two materials below 1 nm were selected (total film thickness was 25–35 nm), a substantial interdiffusion of the metal oxides was observed upon annealing. The electrical conductivity of the resulting films increased, which may be attributed to the small grain sizes of YSZ formed after the annealing procedure. The thin sublayers in the initial nanolaminate may suppress the formation of large crystals, which in turn leads to more grain boundaries and interface defects. This strategy appears very promising for future research in the field of SOFCs.

16.7 Complex Nanostructures

The importance of nanostructures for catalytical, optical, optoelectrical, and renewable energy applications is continuously increasing. Although many products already incorporate nanostructures, extensive research is carried out in various application aspects of nanostructures. Intelligent nanostructuring is required for future innovations, which should rely on simple, economical, and flexible fabrication techniques. Figure 16.10a shows a strategy to design templates for nanoimprint lithography.

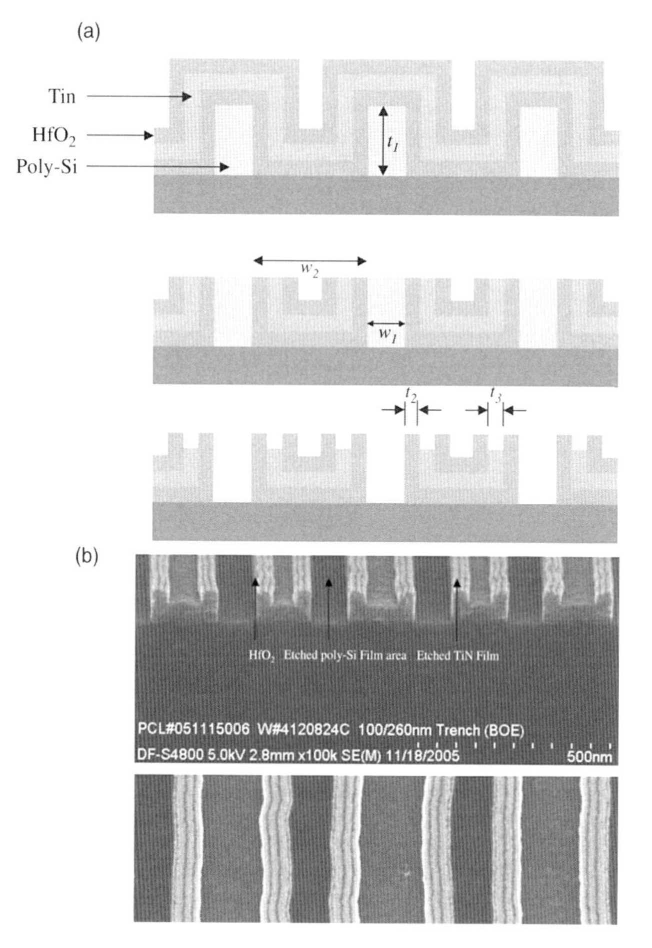

Figure 16.10 Fabrication of high-throughput nanoimprint template by combining ALD with chemical mechanical polishing, and selective wet or dry etch to remove the silicon and the alternate film layer. The line width t_2 and the spacing t_3 are defined with atomic precision by ALD (a). Scanning electron micrograph images in cross section and top view of high-frequency HfO_2 nanostructures (b). Reprinted with permission from Ref. [64]. Copyright 2007, Elsevier.

Based on this processing sequence, combining ALD with chemical mechanical polishing and wet or dry selective etching techniques, Hussain *et al.* demonstrated sub-15 nm line patterns with a 5 nm separation on 200 mm wafers [64]. The ALD coating consisted of alternating TiN and HfO_2 layers. After polishing the horizontal section of the ALD coating, the silicon and TiN components are selectively removed to achieve linear gratings with increased groove density that cannot be fabricated otherwise (see Figure 16.10a). The approach shows a way to cost effectively produce templates for nanoimprint lithography for semiconductor applications.

Nanostructures with increased specific surface area have promising application potential in photocatalysis and as catalytic backbones, as charge collectors in solar cells, and in Li ion batteries. Such structures can be easily prepared by template-assisted coating and selective etching techniques. Inexpensive porous AAO (anodized aluminum oxide) membranes often serve as negative templates for the synthesis of nanowires, nanorods, and nanotubes. ALD is an ideal technique for coating nanomaterials and is exceptional for achieving nanolaminates within high aspect ratio templates such as AAO membranes. Such nanolaminates are important to obtain multiwall tubular structures with wall thicknesses and spacings of a few nm that possess multiple surfaces and an increased effective area of the active element [65]. In further processing steps, solid-state reactions can be easily induced to obtain "microtube-in-microtube" assemblies, wherein the nanotubes are physically separated by a nanoscale gap (Kirkendall effect) [66]. The disadvantage of this technique is that high annealing temperatures must be applied, and the material combinations that undergo nonequilibrium counterdiffusion at the interfaces are limited.

In the case of Ir/TiO_2/cellulose composites, the roughness of the Ir layer additionally contributes to the increase in the specific surface area and the enhancement of their catalytic properties [67]. Successful coating of carbon nanotubes with a multilayer of Al_2O_3/W/Al_2O_3 was presented by Hermann *et al.* [68] The ALD coating is an insulator/metal/insulator sequence building a nanotube coaxial cable with a very high aspect ratio.

16.8 Summary

One key feature of the ALD is the possibility to have multiple processes in a run, allowing for the deposition of multilayers from different materials. The thickness control further enables extremely thin sublayers, even thinner than 1 nm. Having multiple materials deposited in a stack, physical or chemical properties of the individual materials can add value to the final structure and the properties of the stacks can even strongly differ from the corresponding solid solutions or alloys of the same materials with the same thickness. Nanolaminates have a great potential to effectively tune the functionalities of the substrates turning those interesting for optics, electronics, corrosion protection, and as gas diffusion barriers, and could be of immense importance for future technologies as those persistently rush toward smaller feature sizes.

Acknowledgments

The authors greatly acknowledge the financial support by the German ministry of education and research (BMBF) within the project "EFFET" (FKZ 13N9711). M.K. greatly acknowledges the financial support by the BMBF in the frame of the "Nanofutur" program (FKZ 03X5507). A.S. greatly acknowledges the financial support by the German Research Foundation (DFG) within the "Emmy-Noether Program" (SZ 253/1-1). We dedicate our work to the late Professor Ulrich Gösele.

References

1 Manne, S. and Aksay, I.A. (1997) Thin films and nanolaminates incorporating organic/inorganic interfaces. *Curr. Opin. Solid State Mater. Sci.*, **2**, 358–364.

2 Stueber, M., Holleck, H., Leiste, H., Seemann, S., Ulrich, S., and Ziebert, C. (2009) Concepts for the design of advanced nanoscale PVD multilayer protective thin films. *J. Alloys Compd.*, **19**, 3425–3438.

3 Knez, M., Nielsch, K., and Niinistö, L. (2007) Synthesis and surface engineering of complex nanostructures by atomic layer deposition. *Adv. Mater.*, **19**, 3425–3438.

4 Gao, Y.-M., Wu, P., Baglio, J., Dwight, K., and Wold, A. (1989) Growth and characterization of zinc sulfide films by conversion of zinc oxide films with H_2S. *Mater. Res. Bull.*, **24**, 1215–1221.

5 Shurtleff, J.K., Allred, D.D., Perkins, R.T., and Thorne, J.M. (1989) Multilayer X-ray optics produced by atomic layer epitaxy. *Proc. SPIE*, **1159**, 664–673.

6 Riihelä, D., Ritala, M., Matero, R., and Leskelä, M. (1996) Introducing atomic layer epitaxy for the deposition of optical thin films. *Thin Solid Films*, **289**, 250–255.

7 Szeghalmi, A., Helgert, M., Brunner, R., Heyroth, F., Gösele, U., and Knez, M. (2009) Atomic layer deposition of Al_2O_3 and TiO_2 multilayers for applications as bandpass filters and antireflection coatings. *Appl. Opt.*, **48**, 1727–1732.

8 Mitchell, D.R.G., Triani, G., Attard, D.J., Finnie, K.S., Evans, P.J., Barbe, C.J., and Bartlett, J.R. (2006) Atomic layer deposition of Al_2O_3 and TiO_2 thin films and nanolaminates. *Smart Mater. Struct.*, **15**, S57–S64.

9 Abaffy, N.B., Evans, P., Triani, G., and McCulloch, D. (2008) Multilayer alumina and titania optical coatings prepared by atomic layer deposition. *Proc. SPIE*, **7041**, 704109/1–704109/10.

10 Zaitsu, S.-I., Jitusno, T., and Nakatsuka, M. (2002) Optical thin films consisting of nanoscale laminated layers. *Appl. Phys. Lett.*, **80**, 2442–2444.

11 Zaitsu, S.-I., Jitusno, T., Nakatsuka, M., and Yamanaka, T. (2004) Laser damage properties of optical coatings with nanoscale layers grown by atomic layer deposition. *Jpn. J. Appl. Phys. Part 1*, **43**, 1034–1035.

12 Jensen, J.M., Oelkers, A.B., Toivola, R., Elam, D.C., and George, S.M. (2002) X-ray reflectivity characterization of ZnO/Al_2O_3 multilayers prepared by atomic layer deposition. *Chem. Mater.*, **14**, 2276–2282.

13 Szeghalmi, A., Senz, S., Bretscheider, M., Heyroth, F., Gösele, U., and Knez, M. (2009) All dielectric hard X-ray mirror by atomic layer deposition. *Appl. Phys. Lett.*, **94**, 133111/1–133111/3.

14 Kumagai, H., Toyoda, K., Kobayashi, K., Obara, M., and Iimura, Y. (1997) Titanium oxide/aluminum oxide multilayer reflectors for "water-window" wavelengths. *Appl. Phys. Lett.*, **70**, 2338–2340.

15 Masuda, Y., Fujimoto, T., Kumagai, H., and Kobayashi, A. (2008) Titanium oxide/nickel oxide multilayer mirror for attosecond soft X-rays. *Proc. SPIE*, **6890**, 68900T/1–68900T/11.

16 Murata, M., Tanaka, Y., Kumagai, H., Shinagawa, T., and Kobayashi, A. (2010) Atomic layer epitaxy of ZnO and TiO_2 thin

films on *c*-plane sapphire substrate for novel oxide soft X-ray mirrors. *Proc. SPIE*, **7603**, 76030R/1–76030R/10.

17 Sechrist, Z.A., Fabrequette, F.H., Heintz, O., Phung, T.M., Johnson, D.C., and George, S.M. (2005) Optimization and structural characterization of W/Al_2O_3 nanolaminates grown using atomic layer deposition techniques. *Chem. Mater.*, **17**, 3475–3485.

18 Wind, R.W., Fabrequette, F.H., Sechrist, Z.A., and George, S.M. (2009) Nucleation period, surface roughness, and oscillations in mass gain per cycle during W atomic layer deposition on Al_2O_3. *J. Appl. Phys.*, **105**, 074309/1–074309/13.

19 Graugnard, E., King, J.S., Gaillot, D.P., and Summers, C.J. (2007) Atomic layer deposition for nano-fabrication of optoelectronic devices. *ECS Trans.*, **3**, 191–205.

20 Szeghalmi, A., Helgert, M., Brunner, R., Heyroth, F., Gösele, U., and Knez, M. (2010) Tunable guided-mode resonance grating filter. *Adv. Funct. Mater.*, **20**, 2053–2062; Szeghalmi, A., Kley, E.B., and Knez, M. (2010) Theoretical and experimental analysis of the sensitivity of guided mode resonance sensors. *J. Phys. Chem. C*, **114**, 21150–21157.

21 Kim, Y.S. and Yun, S.J. (2005) Nanolaminated Ta_2O_5–Al_2O_3 insulator effect on luminescent and electrical properties of thin-film electroluminescent devices. *Jpn. J. Appl. Phys. Part 1*, **44**, 4848–4854.

22 Chang, C.-Y., Chou, C.-T., Lee, Y.-J., Chen, M.-J., and Tsai, F.-Y. (2009) Thin-film encapsulation of polymer-based bulk-heterojunction photovoltaic cells by atomic layer deposition. *Org. Electron.*, **10**, 1300–1306.

23 Meyer, J., Schneidenbach, D., Winkler, T., Hamwi, S., Weimann, T., Hinze, P., Ammermann, S., Johannes, H.-H., Riedl, T., and Kowalsky, W. (2009) Reliable thin film encapsulation for organic emitting diodes by low-temperature atomic layer deposition. *Appl. Phys. Lett.*, **94**, 233305/1–233305/3.

24 Salmi, L.D., Puukilainen, E., Vehkamaeki, M., Heikkilae, M., and Ritala, M. (2009) Atomic layer deposition of Ta_2O_5/polyimide nanolaminates. *Chem. Vapor Depos.*, **15**, 221–226.

25 George, S.M., Yoon, B., and Dameron, A.A. (2009) Surface chemistry for molecular layer deposition of organic and hybrid organic–inorganic polymers. *Acc. Chem. Res.*, **42**, 498–508.

26 Lee, B.H., Ryu, M.K., Choi, S.-Y., Lee, K.-H., Im, S., and Sung, M.M. (2007) Rapid vapor-phase fabrication of organic–inorganic hybrid superlattices with monolayer precision. *J. Am. Chem. Soc.*, **129**, 16034–16041.

27 Tang, S., Wallace, R.M., Seabaugh, A., and King-Smith, D. (1998) Evaluating the minimum thickness of gate oxide on silicon using first-principles method. *Appl. Surf. Sci.*, **135**(1-4), 137–142.

28 Neaton, J.B., Muller, D.A., and Ashcroft, N.W. (2000) Electronic properties of the Si/SiO_2 interface from first principles. *Phys. Rev. Lett.*, **85**, 1298–1301.

29 Wallace, R.M., McIntyre, P.C., Kim, J., and Nishi, Y. (2009) Atomic layer deposition of dielectrics on Ge and III–V materials for ultrahigh performance transistors. *MRS Bull.*, **43**, 493–503.

30 Shin, H., De Guire, M.R., and Heuer, A.H. (1998) Electrical properties of thin films formed on self-assembled organic monolayers on silicon. *J. Appl. Phys.*, **83** (6), 3311–3317.

31 Wilk, G.D., Wallace, R.M., and Anthony, J.M. (2001) High-κ gate dielectrics: current status and materials properties considerations. *J. Appl. Phys.*, **89** (10), 5243–5275.

32 Ritala, M., and Leskelä, M. (1999) Atomic layer epitaxy: a valuable tool for nanotechnology? *Nanotechnology*, **10**, 19–24.

33 Ikeda, H., Goto, S., Honda, K., Sakashita, M., Sakai, A., Zaima, S., and Yasuda, Y. (2002) Structural and electrical characteristics of HfO_2 films fabricated by pulsed laser deposition. *Jpn. J. Appl. Phys.*, **41**, 2476–2479.

34 Jõgi, I., Kukli, K., Kemell, M., Ritala, M., and Leskelä, M. (2007) Electrical characterization of $Al_xTi_yO_z$ mixtures and Al_2O_3–TiO_2–Al_2O_3 nanolaminates. *J. Appl. Phys.*, **102**, 114114/1–114114/11.

35 Jeon, W., Chung, H.-S., Joo, D., and Kang, S.-W. (2008) $TiO_2/Al_2O_3/TiO_2$ nanolaminated thin films for DRAM capacitor deposited by plasma-enhanced atomic layer deposition. *Electrochem. Solid-State Lett.*, **11**(2), H19–H21.

36 Kim, Y.S. and Yun, S.J. (2005) Nanolaminated Al_2O_3–TiO_2 thin films grown by atomic layer deposition. *J. Cryst. Growth*, **274**, 585–593.

37 Li, W., Auciello, O., Premnath, R.N., and Kabius, B. (2010) Giant dielectric constant dominated by Maxwell–Wagner relaxation in Al_2O_3/TiO_2 nanolaminates synthesized by atomic layer deposition. *Appl. Phys. Lett.*, **96**, 162907/1–162907/3.

38 Kim, S.-H. and Rhee, S.-W. (2006) Cyclic atomic layer deposition of hafnium aluminate thin films using tetrakis (diethylamido)hafnium, trimethyl aluminum, and water. *Chem. Vapor Depos.*, **12**, 125–129.

39 Wu, D., Lu, J., Vainonen-Ahlgren, E., Tois, E., Tuominen, M., Östling, M., and Zhang, S.-L. (2005) Structural and electrical characterization of $Al_2O_3/HfO_2/Al_2O_3$ on strained SiGe. *Solid-State Electron.*, **49**, 193–197.

40 Yang, T., Xuan, Y., Zenlyanov, D., Shen, T., Wu, Y.Q., Woodall, J.M., Ye, P.D., Aguirre-Tostado, F.S., Milojevic, M., McDonnell, S., and Wallace, R.M. (2007) Interface studies of GaAs metal-oxide-semiconductor structures using atomic-layer-deposited HfO_2/Al_2O_3 nanolaminate gate dielectric. *Appl. Phys. Lett.*, **91**, 142122/1–142122/3.

41 Park, P.K., Cha, E.-S., and Kang, S.-W. (2007) Interface effect on dielectric constant of HfO_2/Al_2O_3 nanolaminate films deposited by plasma-enhanced atomic layer deposition. *Appl. Phys. Lett.*, **90**, 232906/1–232906/3.

42 Triyoso, D., Hegde, R.I., Wang, X.-D., Stoker, M.W., Rai, R., Ramon, M.E., White, B.E., and Tobin, P.J. (2006) Characteristics of mixed oxides and nanolaminates of atomic layer deposited HfO_2–TiO_2 gate dielectrics. *J. Electrochem. Soc.*, **153**(9), G834–G839.

43 Kukli, K., Ritala, M., and Leskelä, M. (2000) Low-temperature deposition of zirconium oxide-based nanocrystalline films by alternate supply of $Zr[OC(CH_3)_3]_4$ and H_2O. *Chem. Vapor Depos.*, **6** (6), 297–302.

44 Kukli, K., Ihanus, J., Ritala, M., and Leskelä, M. (1997) Properties of Ta_2O_5-based dielectric nanolaminates deposited by atomic layer epitaxy. *J. Electrochem. Soc.*, **144**(1), 300–306.

45 Zhang, H., Solanki, R., Roberds, B., Bai, G., and Banerjee, I. (2000) High permittivity thin film nanolaminates. *J. Appl. Phys.*, **87**(4), 1921–1924.

46 Kukli, K., Ritala, M., and Leskelä, M. (2001) Development of dielectric properties of niobium oxide, tantalum oxide, and aluminum oxide based nanolayered materials. *J. Electrochem. Soc.*, **148**(2), F35–F41.

47 Kukli, K., Ritala, M., and Leskelä, M. (1999) Properties of atomic layer deposited $(Ta_{1-x}Nb_x)_2O_5$ solid solution films and Ta_2O_5–Nb_2O_5 nanolaminates. *J. Appl. Phys.*, **86**(10), 5656–5662.

48 Duenas, S., Castan, H., Garcia, H., Barbolla, J., Kukli, K., Ritala, M., and Leskelä, M. (2005) Comparative study on electrical properties of atomic layer deposited high-permittivity materials on silicon substrates. *Thin Solid Films*, **474**, 222–229.

49 Kukli, K., Ritala, M., Leskelä, M., Sundqvist, J., Oberbeck, L., Heitmann, J., Schröder, U., Aarik, J., and Aidla, A. (2007) Influence of TiO_2 incorporation in HfO_2 and Al_2O_3 based capacitor dielectrics. *Thin Solid Films*, **515**, 6447–6451.

50 Kukli, K., Ihanus, J., Ritala, M., and Leskelä, M. (1996) Tailoring the dielectric properties of HfO_2–Ta_2O_5 nanolaminates. *Appl. Phys. Lett.*, **68** (26), 3737–3739.

51 Zhang, H. and Solanki, R. (2001) Atomic layer deposition of high dielectric constant nanolaminates. *J. Electrochem. Soc.*, **148** (4), F63–F66.

52 Zhong, L., Daniel, W.L., Zhang, Z., Campbell, S.A., and Gladfelter, W.L. (2006) Atomic layer deposition, characterization, and dielectric properties of HfO_2/SiO_2 nanolaminates and comparisons with their homogeneous mixtures. *Chem. Vapor Depos.*, **12**, 143–150.

53 Maeng, W.J., Kim, W.-H., and Kim, H. (2010) Flat band voltage (V_{FB}) modulation by controlling compositional depth profile in La_2O_3/HfO_2 nanolaminate gate oxide. *J. Appl. Phys.*, **107**, 074109/1–074109/5.

54 Lee, B.H., Lee, K.H., Im, S., and Sung, M.M. (2008) Monolayer-precision fabrication of mixed-organic–inorganic nanohybrid superlattices for flexible electronic devices. *Org. Electron.*, **9**, 1146–1153.

55 Cho, W., Lee, S.S., Chung, T.-M., Kim, C.G., An, K.-S., Ahn, J.-P., Lee, J.-Y., Lee, J.-W., and Hwang, J.-H. (2010) Nonvolatile memory effects of NiO layers embedded in Al_2O_3 high-κ dielectrics using atomic layer deposition. *Electrochem. Solid-State Lett.*, **13**(6), H209–H212.

56 Ding, S.-J., Zhang, M., Chen, W., Zhang, D.W., Wang, L.-K., Wang, X.P., Zhu, C., and Li, M.-F. (2006) High density and program-erasable metal–insulator–silicon capacitor with a dielectric structure of $SiO_2/HfO_2–Al_2O_3$ nanolaminate/Al_2O_3. *Appl. Phys. Lett.*, **88**, 042905/1–042905/3.

57 Ding, S.-J., Zhang, M., Chen, W., Zhang, D.W., and Wang, L.-K. (2007) Memory effect of metal–insulator–silicon capacitor with $HfO_2–Al_2O_3$ multilayer and hafnium nitride gate. *J. Electron. Mater.*, **36** (3), 253–257.

58 Maikap, S., Tzeng, P.-J., Wang, T.-Y., Lee, H.-Y., Lin, C.-H., Wang, C.-C., Lee, L.-S., Yang, J.-R., and Tsai, M.-J. (2007) HfO_2/HfAlO/HfO_2 nanolaminate charge trapping layers for high-performance nonvolatile memory device applications. *Jpn. J. Appl. Phys.*, **46**(4A), 1803–1807.

59 Maikap, S., Tzeng, P.-J., Wang, T.-Y., Lin, C.-H., Lee, L.S., Yang, J.-R., and Tsai, M.-J. (2008) Memory characteristics of atomic-layer-deposited high-κ HfAlO nanocrystal capacitors. *Electrochem. Solid-State Lett.*, **11** (4), K50–K52.

60 Kumar, S., Greenslit, D., Chakraborty, T., and Eisenbraun, E.T. (2009) Atomic layer deposition growth of a novel mixed-phase barrier for seedless copper electroplating applications. *J. Vac. Sci. Technol. A*, **27**, 572–576.

61 Greenslit, D., Kumar, S., Chakraborty, T., and Eisenbraun, E.T. (2008) Investigations of ultrathin Ru-WCN mixed phase films for diffusion barrier and copper direct-plate applications. *ECS Trans.*, **13**, 63–70.

62 Elers, K.-E., Saanila, V., Li, W.-M., Soininen, P.J., Kostamo, J.T., Haukka, S., Juhanoja, J., and Besling., W.F.A. (2003) Atomic layer deposition of W_xN_y/TiN and WN_xC_y/TiN nanolaminates. *Thin Solid Films*, **434**, 94–99.

63 Ginestra, C.N., Sreenivasan, R., Karthikeyan, A., Ramanathan, S., and McIntyre, P.C. (2007) Atomic layer deposition of Y_2O_3/ZrO_2 nanolaminates: a route to ultrathin solid-state electrolyte membranes. *Electrochem. Solid-State Lett.*, **10**(10), B161–B165.

64 Hussain, M.M., Labelle, E., Sassman, B., Gebara, G., Lanee, S., Moumen, N., and Larson, L. (2007) Deposition thickness based high-throughput nano-imprint template. *Microelectron. Eng.*, **84**, 594–598.

65 Bae, C., Yoon, Y., Yoo, H., Han, D., Cho, J., Lee, B.H., Sung, M.M., Lee, M.G., Kim, J., and Shin, H. (2009) Controlled fabrication of multiwall anatase TiO_2 nanotubular architectures. *Chem. Mater.*, **21**, 2574–2576.

66 Peng, Q., Sun, X.-Y., Spagnola, J.C., Saquing, C., Khan, S.A., Spontak, R.J., and Parsons, G.N. (2009) Bi-directional Kirkendall effect in coaxial microtube nanolaminate assemblies fabricated by atomic layer deposition. *ACS Nano*, **3**, 546–554.

67 Leskelä, M., Kemell, M., Kukli, K., Pore, V., Santala, E., Ritala, M., and Lu, J. (2007) Exploitation of atomic layer deposition for nanostructured materials. *Mater. Sci. Eng. C*, **27**, 1504–1508.

68 Hermann, C.F., Fabrequette, F.H., Finch, D.S., Geiss, R., and George, S.M. (2005) Multilayer and functional coatings on carbon nanotubes using atomic layer deposition. *Appl. Phys. Lett.*, **87**, 123110/1–123110/3.

17
Challenges in Atomic Layer Deposition

Markku Leskelä

17.1
Introduction

In atomic layer deposition (ALD), the precursor vapors are introduced to the substrates alternately, one at a time, separated by purging periods with inert gas. The film grows via saturative surface reactions between the incoming precursor and the surface species remaining after treatment with the previous precursor. The reactions should be saturative and no precursor decomposition should take place. Only then is the film growth self-limiting, which is one of the attractive features of ALD [1]. The success of ALD is built on chemistry, and for each film material an appropriate combination of precursor molecules suited to the ALD process must be found. The requirements for the precursors are contradictory: they must be thermally stable and reactive. Decomposition of the precursors, which is often an important reaction mechanism in CVD techniques, results in nonuniform film deposition in ALD.

The main stream in ALD is thermal processing, where activation for the surface reaction is brought about by heat. Thermal processes in some cases require very high temperatures, which makes reactor design expensive and impractical. Therefore, and because many common substrates do not tolerate high temperatures, the tendency is to develop thermal ALD processes that work preferably below 400 °C. Since it is difficult or even impossible to find precursor combinations working at such low temperatures, for example, for non-noble metals or nonmetal elements (Al, Ti, Ta, Si, Ge), nitrides (TaN, Si_3N_4, BN), and refractory materials (carbides, silicides, borides), plasma-enhanced systems have been developed. Plasma ALD is increasingly used but it has its drawbacks, which are limited conformality of the films and more complicated processing.

ALD of thin films has many challenges, such as the following: (i) difficulties in theoretical modeling of the reactions, (ii) simulation of the film growth, (iii) scaling up the processes for large volumes, and (iv) reaching an acceptable process speed. This chapter is limited to materials challenges, and more precisely, challenges in thermal ALD. The various material types, that is, metals, nonmetal elements, binary compounds, and more complex compounds, are each discussed separately. ALD of

Atomic Layer Deposition of Nanostructured Materials, First Edition. Edited by Nicola Pinna and Mato Knez.

organic polymers or hybrid organic–inorganic materials is an emerging area but it is out of the scope of this chapter.

17.2 Metals

Successful ALD reactions are exchange-type reactions. The metal precursor and nonmetal precursor exchange their "ligands." With compound thin film materials, it has been quite straightforward to find this kind of exchange reaction. A particularly universal approach for the growth of oxides, sulfides, and nitrides [2] has been to exploit nonmetal hydrides (H_2O, H_2S, NH_3) with various metal precursor compounds. For metal films, the simplest reaction would be the reduction of metal precursors, for example, with hydrogen. Only a few processes are, however, based on this principle, the most well known being the transition metal (Fe, Co, Ni, Cu) processes utilizing amidinate precursors [3]. The growth rates in these processes remain very low, indicating that the surface chemistry is not favorable. One of the challenges in ALD metal processes is to study the reaction mechanisms *in situ* and understand how a metal precursor can adsorb at the metal surface.

With thermal ALD, noble metals have been deposited using O_2 as the other precursor, together with metal-organic and organometallic compounds [4, 5]. Platinum-group noble metals are exceptions because of their great tendency to form metals instead of compounds. Oxygen as the other reactant confirms this eagerness to form metal and the process is more like burning of the hydrocarbons of the ligands. Noble metal processes have been studied *in situ* using quartz crystal microbalance (QCM) and quadrupole mass spectroscopy (QMS), and the reaction mechanisms are rather well understood [6]. If ozone is used instead of oxygen, noble metal oxides can be deposited at low temperatures ($<200\,^\circ C$), but at higher temperatures this process also produces metal films [7, 8].

W and Mo are examples of metals that have been deposited with an exchange reaction by reducing WF_6 (MoF_6) with silanes and boranes [9]. The reaction mechanism has been verified by QCM and OMS [10] and the versatility of the processes has been reported in several papers and, for example, in depositions of multilayers [11]. Some other metals such as Cu and Ni have been deposited via oxide routes, that is, first an oxide is made with water and then the oxide is reduced to metal using H_2 as the reducing agent [12]. Copper film has been found to be very challenging to deposit by any chemical vapor deposition process. Its adhesion on many different surfaces in microelectronics in a dual-damascene structure was found to be even more challenging. As reviewed by Juppo in 2001 [13] and Kim in 2003 [12], no process ready for application was available at that time. The use of an oxide intermediate step has been found to improve the adhesion and beginning of the growth. Interestingly, the reduction of oxide to metals can be done by alcohols, aldehydes, and carboxylic acids [14]. The discovery of amidinate precursors was also a significant step forward in ALD copper chemistry [3, 15]. The direct reduction of Cu precursors with hydrogen has been a problem but amidinates behave better in this sense. Another possibility with amidinates is to use Cu_3N in an intermediate step; the Cu_3N is reduced to metal above 200 °C [16]. However, the ALD Cu problem is not completely over and new processes are being studied to find methods for large-scale deposition. Interesting results have recently been obtained by reducing Cu dimethylamino-propoxide with diethylzinc at 100–120 °C [17] and catalyzing the hydrogen reduction of $Cu(hfac)_2$ by pyridine at room temperature [18].

Silver and gold are important metals, not only because of their low resistivity but also due to their catalytic and optical properties. The surface plasmon properties of silver have recently received much attention [19]. Silver and gold are very challenging for ALD. They both form compounds with an oxidation state of +1. Coordinative saturation is difficult to achieve since only one anionic ligand can bind to the metal ion. The shielding of the cation must take place with neutral adduct ligands. Their bonding is usually weak and that limits the thermal stability of Ag(I) and Au(I) complexes. However, volatile complexes are known for Ag(I) and Au(I), and they contain β-diketone, cyclopentadienyl, or some other hydrocarbon as the anionic ligand, and trialkylphosphines fill the coordination sphere [20]. In some cases, dimerization allows the desired four coordination to be achieved [21]. The most thermally stable Ag(I) complexes are carboxylates and β-diketonates adducted with alkyl phosphines [22]. Gold also forms complexes with an oxidation state of +3, and for these Au(III) complexes, coordination saturation is easier to achieve [23]. The thermal CVD of Ag and Au films relies on decomposition of the precursors to metals. Different activation methods such as photoactivation by lasers, plasma enhancement, and ion- and electron beam-assisted processes have been used [20]. No reaction-type processes that could be applied to ALD have been reported.

Only two reports have been published on silver ALD and those are based on the hydrogen radical-enhanced reaction of (2,2-dimethylpropionato)silver(I)triphenylphosphine or $Ag(fod)(PEt_3)$ (fod = 2,2,-dimethyl-6,6,7,7,8,8,8-heptafluoro-3,5-octanedionato) [24, 25]. The process itself is rather well suited to ALD and the quality of the silver is reasonable. The process has, however, the normal limitations of radical-enhanced processes.

The electropositive metals (Al, Ti, Ta) are of interest not only because they are conductive, but also because they can be used as adhesion layers and barrier layers in Cu interconnects (Ti, Ta). Thermal ALD of these metals from simple halides and hydrogen requires temperatures that are far too high for the substrates and structures they should be deposited on. Therefore, it was necessary to use hydrogen plasma in their deposition [12]. Both Ti and Ta have been grown from chlorides and hydrogen plasma, while in a report on Al films TMA and hydrogen plasma were used [26–28]. The metal films are not very well characterized. Since they are thin, they oxidize rapidly when exposed to ambient atmosphere. A careful *in situ* capping is needed if these metal films are to be used. Deposition of tantalum films using an exchange reaction between TaF_5 and Si_2H_6 has been attempted [29]. Only fundamentals of the reaction have been studied and it has been concluded that the reaction takes place between these precursors but the film is clearly more silicide than metal.

The report on aluminum ALD with hydrogen plasma and TMA describes deposition of well-conducting films with a high growth rate per cycle. Capping with

TiN was essential to avoid oxidation in air [28]. Using CVD, aluminum films have been grown from different alkyl and hydride (alane) compounds [30]. Triisobutylaluminum decomposes to metallic aluminum at around 200 °C, while with TMA the natural decomposition product is carbide, Al_4C_3 [31, 32]. By increasing the temperature and adding a hydrogen atmosphere or giving extra energy in the form of plasma or photons, it possible to get metallic aluminum from TMA [33]. Alanes, which contain Al−H bond(s), also decompose cleanly to metallic aluminum [34]. Aluminum ALD has not been thoroughly studied, and as seen, however, with hydrogen plasma and alkyl compounds deposition is possible. The use of alanes may also be beneficial in ALD. The background pressure should be low and the reactant and flow gases pure to ensure the formation of metallic aluminum instead of oxide.

There are a number of other electropositive metals such as alkali, alkaline earth, and rare earth metals that are at least as difficult as Al, Ti, and Ta for thermal ALD. The question is if these metals have any use as thin films and thus whether there is any need to deposit these films and conformal thin films. Obviously, the need is limited, but organic LEDs and Li ion batteries are examples where, in the future, electropositive metal films may be needed.

Main group metals on the right-hand side of the periodic table, namely, gallium, indium, thallium, tin, lead, and bismuth, form a group of metals for which ALD processes have not been developed. Their fabrication is probably easier than that of the electropositive metals on the left-hand side of the periodic table. Probably, there are no reports of deposition studies with these metals due to the lack of need for them as thin films. From the electronegativity and reduction potential point of view, these main group metals are very similar to transition metals in the first row. As pointed out above, the first row transition metals are challenging but possible for ALD and the same conclusion can be made for the group 13, 14, and 15 metals.

In conclusion, metal ALD is not yet mature. There is still a need for better processes for first row transition metals such as iron, cobalt, nickel, and copper. Processes for silver and gold do not exist at all. The non-noble metals such as aluminum, titanium, and tantalum are challenging because of their high reduction potential and easy oxidizability. There is a need to make progress in the development of ALD for these metals because of the possible 3D applications that may appear in the future. Many things regarding precursor chemistry can be learned from the CVD literature, but the need for thermally stable precursors is a significant difference between ALD and CVD.

17.3
Nonmetal Elements

ALD of compound films relies on exchange reactions. However, for elements such a universal chemistry does not exist, at least not for thermal ALD. With plasma-enhanced ALD, in principle all elements can be deposited using hydrogen radicals as reducing agents, but at this point it is too early to say how universal this approach will turn out to be. In the area of metalloids and nonmetals, many attempts have been made toward developing ALD of silicon and germanium but no real success has been achieved yet [2]. In addition to Si and Ge, some studies have been done with carbon films. The missing elements in the review of Puurunen [2] are phosphorous, arsenic, antimony, sulfur, selenium, and tellurium. There is a clear motivation to study the group 14 elements (C, Si, Ge), with the other heretofore unstudied nonmetal elements being less interesting. Of those, Sb, Se and Te have applications as thin films.

The ALD processes reported for carbon, silicon, and germanium use hydrogen plasma or temperature modulation (flash heating) to obtain deposition [35–38]. This shows how difficult it is to get the tightly covalently bonded precursors of these elements to adsorb on the surface and react with the other precursor. There have also been attempts to use exchange-type reactions for Si, for example, $Si_2Cl_2H_2 + H_2$ and $Si_2Cl_6 + Si_2H_6$ [39, 40]. The success here has also been limited and the thermodynamic calculations show that with silanes the reactions are uphill but with atomic hydrogen downhill in energy [41]. The chemistry of silicon is very rich and among the thousands of existing compounds it might be possible to find some producing elemental silicon in a thermal process. Germanium, in principle, should be somewhat easier and it has, like silicon, very rich organometallic chemistry. The recent enormous interest in graphene certainly increases the possibilities to study deposition of thin carbon films.

Recently, ALD of tellurides and selenides using alkylsilyls of tellurium and selenium and various metal halides as precursors has been reported [42]. This chemistry was found to work at very low temperatures, that is, below 100 °C, and this was explained in terms of the favorable formation of alkylsilyl halides as volatile by-products. The same principle can be exploited for elemental antimony, and similar dehalosilylation reactions between $(Et_3Si)_3Sb$ and $SbCl_3$ are possible [43]. For the first time, an element other than a metal was deposited by ALD. Pure and conformal thin films of elemental antimony were prepared by thermal ALD. *In situ* reaction mechanism studies showed that the dehalosilylation reactions involved are very efficient in eliminating the ligands from the growing surface, enabling the use of low growth temperatures, that is, down to 95 °C. Various antimony compounds, such as GeSb, Sb_2Te, GaSb, and AlSb, can also be deposited by reacting $(Et_3Si)_3Sb$ with other metal halides or mixing Sb growth cycles with other ALD processes. The new antimony ALD process is a major step forward toward the realization of nonvolatile phase change random access memories (PCRAM) and ALD of III–V compounds.

The success with antimony films may open possibilities for other metalloid films. The requirement is for smart chemistry – as for instance in this case the utilization of Lewis hard acid–hard base, soft acid–soft base principle. Silicon and germanium chemistries are so rich that most probably there exist molecules that can by ligand exchange result in elemental films. The question is, again, about the need for ALD films, and that determines the amount of effort and resources that will be put into this research.

17.4 Binary Compounds

17.4.1 Oxides

Oxides form the most studied material group in ALD. Almost ideal, versatile ALD processes exist for Al_2O_3, TiO_2, and ZnO because of their precursors, which eagerly react with water or other oxygen sources. These favorable reactions make low-temperature processing possible. Due to the need for high dielectric constant oxide films in microelectronics, ALD of ZrO_2 and HfO_2 has been extensively studied and numerous precursors and processes have been developed. ALD of these oxides is used industrially in microelectronics. ALD of niobium, tantalum, and rare earth oxides has also been studied from the high-*k* oxide point of view, but there is still room for further development. Processes to deposit transition metal oxides (Cr, Mn, Fe, Co, Ni, Cu) have been reported [2]. The precursors are β-diketonates, amidinates, or Cp compounds, and the oxygen source is either ozone or water. Although the number of papers on transition metal oxides is not high, the processes are available if needed.

Alkali metal oxide ALD has been very little studied. Li precursors have been synthesized and the ALD reaction between Li(thd) and ozone produces lithium carbonate [44]. The authors speculate that the films they obtained from Li^tOBu and water could be oxide or hydroxide. Although pure lithium oxide was not formed, these lithium precursors can be used in ternary or quaternary compounds, as shown with lithium lanthanate or lithium lanthanum titanate [45]. The alkali metal (Na, K) precursors have been even less studied. In the early 1990s, they were studied in codoping of rare earth-doped ZnS and SrS films. It was shown that their thd precursors are usable in ALD as are double β-diketonate complexes such as $NaPr(tfa)_4$ (tfa = 1,1,1-trifluoro-2,4-pentanedione) [46, 47]. It can be concluded that alkali metal oxides are not very interesting materials and their ALD research will remain minor but as components in ternary compounds alkali metals are important. ALD of Li compounds for batteries is already being done by several groups. It can be expected that materials such as $KNbO_3$ will in the future be deposited by ALD.

Of the alkaline earth oxides, MgO has been the most widely studied in ALD. Cyclopentadienyl compounds and oxygen have commonly been the precursors [48, 49]. MgO is rather stable in air, in contrast to other alkaline oxides, for which the film has to be capped against air exposure, otherwise carbonate forms immediately. Therefore, ALD of CaO, SrO, and BaO has been infrequently studied, but as components in ternary compounds like titanates they are very important. Interestingly, it seems that Ca, Sr, and Ba oxides grow easily on a TiO_2 surface than on their own surfaces and $SrTiO_3$ and $BaTiO_3$ have been successfully grown since the late 1990s [50]. In this process, where cyclopentadienyls and water are used as precursors, the formation of $Sr(OH)_2$ and $Ba(OH)_2$ and the loss of surface reaction control is an issue. The alternation of alkaline earth cyclopentadienyl–water cycles as much as possible with the titanium alkoxide–water cycles removes the problem to a large extent [51]. Bismuth behaves very similarly to alkaline earth metals: the deposition of pure Bi_2O_3 is challenging but using bismuth as component for ternary or quaternary compounds such as $Bi_3Ti_4O_{12}$ or $SrBi_2Ta_2O_9$ makes the introduction of bismuth easier [52, 53]. The best Bi precursor for binary oxide described thus far is Bi $(OCMe_2{}^tPr)_3$ that allows deposition at 150–250 °C using water as the oxygen reagent [54].

Boron oxide ALD films were reported in one study, where B_2O_3 was grown from BBr_3 and water at room temperature [55]. Due to instability in ambient atmosphere, boron oxide may not have many applications but as a component in borates the development of its ALD processes may be useful. In contrast to aluminum oxide, ALD of gallium and indium oxide is difficult, and large-scale application processes do not exist. The alkyl compounds of Ga and In are much less reactive than those of Al. Oxide has been grown from TMIn and water, but cycle times were measured in minutes [56]. $InCl_3$ can be used together with water but the deposition rate remains low, 0.27 Å per cycle [57]. It was noticed subsequently that by increasing the water dose, the growth rate per cycle of In_2O_3 as well as many other oxides could be significantly increased [58]. Since indium oxide is used in ITO as a transparent conducting oxide, in applications relatively thick films are needed. Even the increased growth rate is too low for industrial applications of the $InCl_3 + H_2O$ process. The recently developed InCp + O_3 process shows a clearly higher growth rate but the high price of the In(I)Cp limits its use [59]. The recent report on producing a film of In_2O_3 from $In(acac)_3$ partially solves the problems of $InCl_3$: the deposition temperature is below 300 °C, but the growth rate per cycle remains low [60].

Gallium oxide ALD has been examined to some extent. The growth from $Ga(acac)_3$ and water occurs at 370 °C but the films contain carbon impurities. Better films are obtained if ozone is used as the oxygen source [61]. A change of the Ga precursor to an alkylamido-based compound increases the growth rate in the water process to about 1.0 Å per cycle at 150–275 °C [62]. Ga_2O_3 oxide films have not been grown from trimethylgallium but instead an amine derivative $[(CH_3)_2GaNH_2]_3$ has been used together with oxygen plasma [63]. Pure, amorphous films were grown at 200 °C at a rate of 0.7 Å per cycle. Basic research on gallium oxide has been done but the field is not mature yet giving room for new processes to be developed. The existing research, however, provides a good basis for application-oriented work, if some industrial interest arises.

Silicon dioxide is one of the most important oxides in microelectronics. Its deposition by ALD has been studied since the early 1990s. This was not very successful since finding thermal processes utilizing water as the oxygen source was difficult. $SiCl_4$ reacts with water at high temperature but requires large doses [64]. The reaction temperature can be lowered and reactant dose decreased with a pyridine catalyst [65]. A major advancement in SiO_2 ALD was made with the discovery of a catalytic process where the reaction between tris(*tert*-butoxy)silanol and water was catalyzed by an aluminum oxide layer made by TMA and water [66]. The growth rate in this process is exceptionally high, and even too high for very thin films. At the moment, the most commonly used SiO_2 ALD processes utilize alkylamino silanes (mostly TDMAS = tris(dimethylamino)silane) and ozone [67]. Some carbon remains in the films, but the amount can be decreased by using bis(dimethylamino)silane or post-deposition annealing [68]. Alkylamino silanes react even with hydrogen peroxide and form relatively pure oxide with some carbon and nitrogen residues [69]. In principle, the problems with silicon dioxide have been solved, the major reason being the use of ozone as the oxygen source. ALD of germanium dioxide has been studied to a lesser extent than silicon dioxide. Similar problems to those of SiO_2 have been recognized but use of germanes and ozone could be a solution.

ALD of tin oxide films has attracted quite a lot of interest because of its possible applications as a host material in transparent conducting films. Tin precursors range from halides (chlorides and iodides) to alkyl, alkyl amine, and alkoxide compounds [2]. Antimony is used as a dopant in SnO_2 and therefore pure Sb_2O_5 has not received so much attention. In any case, there is a working process for SnO_2:Sb that is based on metal chlorides and water [70]. The other common dopant for conducting tin oxides is fluorine. Serious ALD studies on SnO_2:F films have not been reported. Several lead precursors have been studied in ALD of lead sulfides, oxides, and ternary oxides. The volatile diethyldithiocarbamate, lead halides, some alkoxides, and thd compounds are all possible precursors. Their reactivity toward H_2S is better than that toward water [71, 72]. The best reaction reported so far for PbO_2 films is that based on tetraphenyl lead and ozone where the oxidation of lead to Pb(IV) occurs [72]. Again, the water process, $Pb(thd)_2 + H_2O$, is not very efficient for binary oxide but seems to work better in the ternary reaction, as exemplified by $PbTiO_3$ [73]. It seems that an ALD process for PbO films is lacking but processes for PbO_2 with ozone and ternary oxides with both water and ozone exist [74, 75].

17.4.2 Nitrides

Transition metal nitride films have been successfully deposited by ALD since the late 1980s using reactions between metal chlorides and ammonia [75]. The most common reaction temperature is 500 °C. This reaction is very suitable for TiN since titanium does not form other nitrides, that is, nitride with an oxidation state of +4. In cases where nitrides are formed at higher oxidation states, ammonia is not a powerful enough reducing agent and, for example, in the case of tantalum, nonconducting Ta_3N_5 is formed instead of TaN [76]. Also in the case of hafnium, formation of nonconducting Hf_3N_4 has been reported [77]. Reasonably conducting NbN and MoN films can be obtained by the reaction between metal chloride and ammonia, although these metals form other nitrides with stoichiometries deviating from 1 : 1 [78, 79]. Additional reducing agent is needed, for example, in the case of Ta, to get conducting nitride films. Zinc has been used in this role and it, causing reduction, also purifies the film by removing traces of chlorine.

Many studies have been carried out to use alkyl amines as precursors for nitride films in order to avoid halide residues and to lower the deposition temperature. The problem with alkyl amines is their thermal instability, and at the low temperatures used reactivity of ammonia is low and impurities remain in the films. It has been shown for TiN that the reaction between alkyl amine and ammonia is not an ideal ALD reaction [80]. Tantalum amido–imido complexes combined with ammonia have been the subject of many studies. The results are similar: in thermal processes, impure nonconducting films are produced, and in plasma-assisted processes better conducting but carbon-contaminated TaN films are obtained [81, 82]. Use of hydrazine yields better films than those obtained with ammonia [83]. With tungsten, however, amido–imido complexes seem to produce well-conducting WN films in a thermal process with ammonia [84]. The use of WF_6 as metal precursor and ammonia as the nitrogen source results, however, in a mixture of W_2N and WN phases and the film exhibits resistivity that is far too high to allow it to be used as a copper barrier layer [85].

Aluminum and especially gallium nitrides are important high band gap semiconducting III–V compounds. As they are used in optoelectronics as emissive layers, they need to be epitaxial layers. The simplest precursors are metal alkyl (methyl) compounds and ammonia. This reaction works for AlN [86] at 325–425 °C but in the case of GaN requires at least 550 °C [87]. Besides trimethyl compounds, other alkyl compounds, alkyl halide compounds, and halides have been studied as gallium precursors in ALE (atomic layer epitaxy, the name used for ALD before mid 1990s) [2]. There is no reason in principle why blue-emitting GaInN could not be deposited by ALE at 600–700 °C other than process time [88], although with a rotating substrate setup, GaAs has been grown at as fast as 2 μm/h [89]. On large scale, blue-emitting GaInN LEDs are fabricated by MOCVD.

Nonmetal elements form important nitrides of which Si_3N_4 and BN are good examples. Both are very challenging for ALD. Si_3N_4 suffers from the same problems as SiO_2: covalent bonds in the silicon precursors are not very reactive toward ammonia. $SiCl_4$ reacts with ammonia at >425 °C, producing Si_3N_4 with a high growth rate per cycle, but long pulse times are needed [90]. Hydrazine is in this case a more efficient nitrogen precursor than ammonia [91]. Use of a plasmaassisted process is beneficial for purity [92]. The challenge in Si_3N_4 ALD is still the need for a fast thermal process, preferably employing ammonia as the nitrogen source. The selection of silicon precursors examined is limited to silane (SiH_4), dichlorosilane, and tetrachlorosilane [2]. A solution probably could be found among more reactive silicon compounds.

Boron nitride depositions have been studied using only the reaction of boron chloride or bromide with ammonia. Films deposited at 750 °C were crystalline and purer than the amorphous films made at 400 °C [93]. Films deposited at low temperatures contain significant amounts of hydrogen, which can be removed by high-temperature annealing [94]. The tight bonding of hydrogen is a problem and even laser-assisted ALD could not reduce the hydrogen content to below 10% [95]. Although BN ALD looks rather straightforward, there is still a need for process development to obtain pure crystalline film at reasonable temperatures. Boron has very rich borane (boron hydride) chemistry but the problem of removal of hydrogen may remain if boranes are used as boron precursors. B–N bond-containing compounds, if volatile enough, could be interesting alternative precursors.

17.4.3
Other III–V Compounds

ALE of III–V semiconductors other than nitrides was extensively studied during 1985–1995. With traditional precursors such as TMA, TMG, TMI, PH_3, and AsH_3, it has been possible to grow AlAs, GaAs, GaAlAs, GaP, InP, and InAs as epitaxial layers [2, 96]. The reaction temperature and pressure are connected so that in ultrahigh vacuum a ML per cycle for GaAs has been achieved at 520 °C, at 20 Torr and 550 °C, and when H_2 carrier gas was used at 600 °C [97–99]. In all cases, the temperature window was narrow. Interestingly, the pulse time of TMG cannot be too long since more than 1 ML per cycle is obtained with longer pulse times indicating decomposition of the precursor [100]. The pulse time is connected to the reaction temperature, being shorter at higher temperatures. The utility of GaAs ALE as well as that of other III–V compounds has been limited due to the relatively narrow operating temperature window, low growth rate, and the unintentional incorporation of high levels of carbon impurity. The monolayer growth mechanisms have been studied by several groups and three different mechanisms have been proposed but no common agreement has been found [96].

The precursors employed in III–V ALE are those known from MOCVD. Besides the trimethyl compounds, other alkyl compounds and chlorides have been tested as precursors for the metals [101, 102]. Arsine has been replaced by $As(NMe_2)_3$ or t-$BuAsH_2$ (t-$BuPH_2$) mainly for safety reasons [103]. With TMG, the reason for the monolayer growth could be blocking by methyl groups preventing multiple adsorption of TMG, but this probably is also the reason for the carbon contamination. With TEG and ^{i}BuGa, carbon contamination can be decreased but 1 ML per cycle is hard to achieve. The results for ALE of other compounds (AlAs, GaP, InP, InAs) are very similar to those found for GaAs [96]. Process temperatures differ but the problems are the same.

No precursor design of group 13 and 15 elements dedicated for ALD (ALE) has been done. Obviously, one reason for the modest results in III–V ALD is the lack of suitable precursors. The question is whether there is a need to deposit III–V compounds only as epitaxial layers, or are polycrystalline III–V films of interest. Substrates have an important role for the structure of the III–V film. If the substrates are three dimensional, ALD (ALE) is a possible fabrication technique. Some examples exist in the literature where V-shaped grooves etched on the [101] surface of GaAs have been used as substrates for GaAs quantum wires embedded in a $GaAs_{1-x}P_x$ layer. The deposition utilizes the selective isotropic or anisotropic deposition on the [101] surface on the bottom of the grooves and on the [104] surfaces on the sidewalls. The selectivity is achieved by controlling the purge times after AsH_3 and PH_3 pulses during GaAs and GaP depositions [105, 106].

17.4.4
Carbides

Research on metal carbide ALD has concentrated on tantalum and tungsten carbides. Often the films contain nitrogen because the precursors are amido–imido complexes [2], and they are fabricated by plasma ALD using hydrogen, nitrogen, or ammonia plasma. The aim of the studies may have been the deposition of pure carbide or nitride films, but unintentional contamination by the other element has occurred. If the aim is to prepare conducting barrier layers, the contamination of nitride film by carbide carbon or vice versa is not critical for its properties. The composition of the films changes depending on the plasma and temperature used. For example, Hossbach *et al.* [107] obtained with H_2/Ar plasma at 250 °C a film with a composition of Ta_2CN, and in XRD the cubic TaC phase was seen. Song and Rhee, on the other hand, used ammonia plasma and saw a mixture of TaN, Ta_3N_5, TaC, and Ta_2O_5 phases, with relative proportions changing depending on temperature and plasma power [108].

Thermal processes for transition metal carbide films have been presented only in patents. Both organic compounds of nonmetal elements (B, Si, P) and metals (Al) can result in conductive carbide-like films [109, 110]. In the case of tungsten, a three-precursor process has been developed for WN_xC_y films, the precursors being WF_6, NH_3, and $B(C_2H_5)_3$. The films deposited at 300–350 °C had a composition of W:N:C = 55 : 15 : 30, as determined by XPS [111]. The films showed excellent barrier properties, especially when nanolaminated with TiN [112].

It can be concluded that pure transition metal carbide films have rarely been deposited by ALD. Most of the films are mixtures of carbides and nitrides. However, the tungsten compound WN_xC_y shows properties acceptable for applications.

All transition metals form carbides. They usually exhibit extreme hardness, outstanding wear resistance, good electric conductivity, and a high melting point. They are used as wear-resistant coatings but in those applications the film thicknesses needed are out of the range feasible for ALD. Perhaps because of this there is little in the literature concerning ALD processes developed for important carbides such as TiC, ZrC, HfC, VC, Cr_7C_3, Cr_2C_3, Co_2C, and so on.

Silicon carbide is the most important nonmetal carbide. ALD of SiC has been studied in 1990s. Silicon precursors were dichlorosilane, diethylsilane, disilane, and the carbon precursor ethylene or acetylene [2]. Film growth has been determined in low-pressure reactors at high temperatures (800–1050 °C) – temperatures where the silicon precursor ($SiCl_2H_2$) is known to decompose [104]. A constant deposition rate was seen up to 900 °C, above which the rate increased rapidly. Hara *et al.* [113] have observed self-limited growth of SiC from Si_2H_6 and C_2H_2 at 1050 °C, the experiments being limited to a few cycles only, while Sumakeris *et al.* were able to grow SiC films at 850 °C from disilane and ethylene [114]. It seems that SiC ALD is possible, although the process and the resulting films have not been studied thoroughly. The high process temperature is a limitation.

Boron carbide, B_4C, is known as one of the hardest materials next to diamond and cubic BN. It is used as a coating for tools. In CVD, films have been fabricated from a mixture of boron halides and methane at 1000–1600 °C [115]. No one has tried to make boron carbide films by ALD.

17.4.5
Silicides and Borides

Metal silicides are interesting conducting or semiconducting materials that have extremely good chemical and thermal resistance. They have been widely used in microelectronics. Silicides are often made by direct reaction between metal and silicon at high temperatures and in an inert atmosphere. In ALD, unintentional formation of silicides has been reported in connection with silane precursors. For example, in a reaction between TaF_5 and Si_2H_6, formation of silicide has been reported [29]. When using silanes as precursors, the formation of silicide has to be taken into account. However, in one report TDMAS was used as a reducing agent in order to process TaN. Significant amounts of silicon remained in the film, but instead of forming conductive tantalum silicide, nonconductive silicon nitride was formed, as determined by XPS [116].

Plasma-enhanced ALD has been used to deposit $CoSi_2$ films. The cobalt precursor was a cyclopentadienyl compound, and plasma ammonia was mixed with silane [117]. If silane is not used, the resulting film is metallic cobalt [118]. The silicide film was nanocrystalline, and if the film was thin, the crystals were separated.

Silicide ALD is a rather virgin research area. The richness of silane chemistry certainly suggests good possibilities for finding thermal processes. Of course, plasma processes are more straightforward and could be developed following the report of Lee and Kim [117].

Metals form borides with two different basic stoichiometries: boron-rich and metal-rich. Of the boron-rich compounds, LaB_6 used as an electron emitter can be mentioned. The metal-rich borides have similar properties to metal silicides – they are inert, hard, and conducting. One of the most interesting borides is the superconducting MgB_2. No boride compounds have been deposited as thin films by ALD. Plasma ALD should be a rather straightforward method, and the versatility of borane chemistry may also enable thermal processing.

17.4.6
Halides

ALD of halide films has focused on fluorides and only one paper on chlorides namely growth of CuCl nanoparticles from [bis(trimethylsilyl)acetylene]-(hexafluoroacetylacetonato)copper(I) and hydrogen chloride has been reported [119]. The first process developed was based on HF that was obtained by thermal decomposition of NH_4F [120]. The aggressive nature of HF, especially toward silicates, makes it a less favorable precursor. The more recent studies of fluorides are based on volatile metal fluoride precursors (TiF_4, TaF_6). They can react with, for example, thd complexes of alkaline earth and rare earth metals and form stable fluorides [121–123]. This exchange reaction produces pure fluoride films with about 1% Ti or Ta impurities. The films are usable in optic applications but better purity would be welcomed. The use of organic or fluorine-containing ligands may be one way to obtain purer fluoride films, as exemplified by Putkonen with CaF_2 [124]. Both $Ca(hfac)_2/O_3$ and $Ca(thd)_2/Hfac/$

O_3 (hfac = 1,1,1,5,5,5-hexafluoro-2,5-pentanedione) pulsing produce CaF_2 films with minimal oxygen impurities.

No bromide, or iodide films have been grown by ALD. However, there are several halide materials (e.g., NaI, TlBr, Tl(Br,I)) used as detector materials. Their preparation as thin films for high surface area detectors, for example, could be beneficial. The chemical approach to these materials could be similar to that for fluorides. The number of halogen-containing organic compounds is enormous giving plenty of possibilities for suitable process chemistry.

17.4.7
Compounds with Oxoanions

Many oxoanions containing a tetrahedral XO_4^{n-} unit (X = Si, P, S, V, Mo, W) form stable solid compounds with metals. Of these, the ALD of silicates is most widely studied because silicates of group 4 and rare earth metals have been considered for use as dielectric layers in silicon-based microelectronics. Silicate films can be grown directly from metal chlorides and silicon alkoxides [125] or by mixing binary processes using a pulse sequence: metal precursor – oxygen precursor – silicon precursor – oxygen precursor. Here a wide variety of precursors have been used, from $HfCl_4$ – $SiCl_4$ – H_2O [126] to different alkyl amines of hafnium and silicon (alkylaminosilanes) and ozone [127]. Combinations of different precursors such as halides and organometallics are also possible. Yet another way to make silicates is the use of a single precursor for the metal and silicon [128]. The advantage of single-source precursors is that the metal:Si ratio is constant. On the other hand, by using different metal and silicon precursors the metal:Si ratio can be varied if needed. The deposition of ternary compounds has challenges of its own and depending on the situation, the growth of layers can be enhanced or quenched by the other component.

In most of the silicate high-*k* dielectric studies, however, the silicate film is formed after interfacial reaction between the metal oxide and SiO_2 on the silicon surface. The higher the annealing temperature, the stronger is the silicate formation. Rare earths, for example, form silicates very easily [129, 130]. In devices, silicate formation is beneficial for leakage current properties but harmful for the dielectric constant.

Solid phosphate materials are interesting for many applications, and for optic, electrical, and biomaterials, ALD of phosphates is possible, and even P_2O_5 can be used as a precursor [131]. Calcium phosphate has been grown from $Ca(thd)_2$ and $(CH_3O)_3PO$, and after rapid thermal annealing the desired hydroxyapatite phase was formed [132]. Yet another phosphate precursor is $(CH_3)_3PO_4$, which has been used for deposition of $LiFePO_4$ and $Ca:LaPO_4$ films [133, 134]. The reaction requires the use of both water and ozone simultaneously as oxygen precursors. These reports show the potential of ALD for producing phosphate films but a lot more research is needed if applications are to be considered.

Vanadium oxide ALD processes have been characterized rather well, the most popular vanadium precursors being $VOCl_3$ and $VO(O^iPr)_3$ [2, 135, 136]. The known vanadium chemistry could enable the processing of vanadate thin films. So far, there are only reports on lithium vanadate films, where the lithium has been added to ALD V_2O_5 films by electrochemistry [137].

The ALD of oxoanion-containing compounds is largely unexplored. For example, no sulfate films have been deposited. One can see a lot of potential in this field.

17.5 Ternary and Quaternary Compounds

The number of ternary compounds studied in ALD is limited but is increasing. The compounds studied are mainly multicomponent oxides and interest in them stems from their electrical, ferroelectric, or magnetic properties. Ternary chalcogenides have recently been studied for phase change memory applications, as mentioned above [42].

There are several ways that ternary compounds can be processed. The most straightforward is that where separate binary deposition cycles are mixed to get the right stoichiometry. Since the growth rates per cycle are different in binary processes, careful adjustment in pulsing is needed [50]. The optimum process temperatures may also be different in binary processes, making the temperature window for ternary processes narrow [52]. The second possibility for deposition of ternary compounds is the use of bimetallic single-source precursors. The approach is common in CVD but has been very rarely studied in ALD, the most well-known case being the deposition of $SrTa_2O_6$ from $SrTa_2(OEt)_{10}dmae_2$ and water [52, 138]. The reason for the limited use of bimetallic precursors is the lack of volatile, thermally stable, and reactive compounds. The third way to obtain ternary compounds is the use of one reaction that introduces the two components into the film, as exemplified above with silicates, where the reaction between metal chlorides and silicon alkoxides produces metal silicates [125].

In transistors and DRAMs, a shift away from SiO_2 to binary high-k oxides such as HfO_2 has already occurred. One possible alternative high-k material type is aluminates, and $LaAlO_3$ has been of particular interest because it makes a stable structure on Si, in contrast to pure La_2O_3 [139]. As a very high-k material, $SrTiO_3$ is attracting constant interest as a possible material for DRAM capacitors. The limited Sr precursor chemistry has slowed industrial use of ALD of $SrTiO_3$. The oldest chemistry is based on Sr cyclopentadienyls with water as the oxygen source [50]. Subsequently, Sr β-diketonates have been used in combination with ozone or oxygen plasma [140, 141]. Mixed rare earth oxides form a third type of ternary oxides receiving attention as possible candidates for high-k materials in microelectronics. The combination of large rare earth ions with small ones can produce separate ternary compounds with a perovskite structure, such as $LaLuO_3$. Ternary rare earth perovskite oxides can be made with both cyclopentadienyl and β-diketonate chemistry, ozone being the oxygen source [142].

Bismuth titanates form a group of ferroelectric materials. In their ALD, the precursor chemistry of bismuth has played a major role [52, 143–145] and so far the most versatile precursor has been $Bi(OCMe_2{}^iPr)_3$ [54]. The quaternary compound

$SrBi_2Ta_2O_9$ is an interesting ferroelectric material. In its ALD process, the bimetallic $SrTa_2(OEt)_{10}dmae_2$ has been utilized [52]. Lead-containing ternary compounds are interesting due to their applications in sensors and transducers and as PZT ferroelectrics. In their ALD, the lead precursor chemistry has been a challenge. $PbZrO_3$ films have been grown using Ph_4Pb, $Zr(thd)_4$, and ozone at around 300 °C, with reasonable results [74, 146].

Magnetic oxides containing transition metals ($LaCoO_3$, $LaNiO_3$) made using thd chelates and ozone as precursors are one type of ternary oxide group that has been of interest during past 15 years [147, 148]. The depositions suffer from low growth rates and annealing is needed to get crystalline films with the desired magnetic and electric properties.

Ternary oxide films are usually deposited at low temperatures because of the low thermal stability of the precursors. Crystalline films are obtained after high-temperature annealing that, for example, in the case of mixed rare earth oxides, requires 800–1000 °C. It would be beneficial from the electrical property point of view to get as-deposited crystalline films, as has been done, for example, by selecting higher temperature tolerant precursors, allowing crystalline $SrTiO_3$ films to be deposited at 370 °C [149].

ALD of ternary compounds is challenging because a larger number of factors affect the process compared to binary compounds: adjustment of the temperature for two or more metal precursors may be difficult and the surface may change significantly during successive stages of growth. The change in the surface can be in some cases beneficial, as seen with silicates and bismuth-containing ternary oxides, where the growth of binary compounds has been difficult.

17.6 Nucleation

Growth of a thin film occurs via several steps, of which nucleation is the first and a very important one. Thereafter growth of the nuclei takes place, the nuclei coalesce, and finally the film thickens, which is usually considered film growth. Nucleation of the film is a complex process and depends on the film material, substrate material, substrate structure, defects on the substrate surface, and of course the deposition conditions, of which the temperature is the most important. In general, nucleation on a substrate material type that is different from the thin film material, that is, ceramic film on metal or polymer or metal film on ceramic or polymer substrate, is always difficult. In ALD research, many examples of nucleation difficulties of metal films on oxide surfaces have been reported in the literature [150]. If the nucleation is slow, tens or hundreds of ALD cycles are needed before any material is deposited on the substrate. This "no growth" period is called the incubation time for film growth. Slow nucleation results in coarsening of the film, and a relatively thick film (often about 5 nm) is needed to fully cover the surface. Another very serious problem has been the nucleation of dielectric oxides on hydrogen-terminated silicon surfaces. For example, in the $ZrCl_4 + H_2O$ process, 50–60 ALD cycles are needed to get a linear growth regime [151].

Nucleation of the films often takes place at defect sites on the substrates. On single crystal substrates, it has been proven many times that terraces and kinks are favored nucleation sites compared to planar crystal surfaces. In ALD films, this has been shown, for example, in deposition of oxide films on carbon materials (carbon nanotubes, graphene) [152, 153]. The film growth starts on defects and a relatively thick film is needed to fully cover the substrate. Adhesion of the film to the substrate may be a problem and depends also on the nucleation. In ALD, it is common to use an adhering interface layer, Al_2O_3, to help in both adhesion and nucleation [154]. Aluminum oxide has been selected for the interfacial layer because of the excellent ALD process between TMA and water. The reaction works on almost all surfaces, including Teflon, although on polymer surfaces the porosity of the material may be the reason for film growth. TMA and water penetrate inside the pores and the removal of water is especially difficult. After a few ALD cycles, the pores are closed and a continuous Al_2O_3 coating is formed [155].

Typical properties of ALD affecting nucleation are low growth temperature, indicating low surface mobility of the species; the exchange reaction between functional surface groups and the incoming precursor, which may further lower the surface mobility; an increase in deposition temperature, which may change the amount and nature of surface groups; an alternate supply of precursors; and purge periods in between cycles, which interrupt growth and give surface species time to move and react. Nucleation during ALD processes has not been studied in very much detail and conclusions from comparisons to other deposition techniques cannot be made. The effects of pulse and purge times on the structure of the film material have been studied in the case of ZnO [156]. That research was not focused on nucleation but gives indications of the effect of purge time on surface mobility and ordering of the structure.

17.7 Conclusions

ALD is still a relatively small research and technology area. The different processes published up to 2010 can still be presented in one review paper with 2000+ references [157]. ALD activities increase constantly and the selection of processes available and materials that are possible to deposit as films continues to expand.

At the moment, the challenges in deposition of materials as thin films by ALD can be divided into different categories:

- materials for which deposition has never been attempted (halides other than fluorides, borides, sulfates);
- materials for which deposition has been attempted but for which serious problems have been encountered (electropositive metals, boron nitride);
- processes needing improvements (many metals and nitrides, ternary compounds);
- process development (fine-tuning) for applications.

Common to the items in the above list as well as ALD in general is the need to develop precursors and find suitable precursor combinations. The challenge is how to increase interest in development of new processes. It seems that ALD process development is done on demand, indicating that more research money is available for applied than basic research.

Another challenge in ALD is that as a chemical technique it is thermodynamically driven. The most stable phase (structure) is formed in the reaction conditions used. Some applied thin film materials (nitrides, carbides) can be metastable, a kind of artificial phase, which may be achievable by PVD techniques but may be impossible by ALD.

The low process temperature necessary due to the low thermal stability of precursors, as mentioned with ternary compounds, often results in amorphous as-deposited films that require annealing after deposition. Another possibility is crystallization of an undesired structure, also requiring further thermal treatments. The formation of a desired phase in as-deposited films can be affected to some extent by precursor selection as shown with ZrO_2 films, where the high-k cubic or tetragonal phases are formed more easily from certain precursors (heteroleptic alkyl amide cyclopentadienyls) than from others (cyclopentadienyls, chloride) [158, 159].

References

1 Ritala, M. and Leskelä, M. (2002) *Handbook of Thin Film Materials*, vol. 1 (ed. H.S. Nalwa), Academic Press, New York, pp. 103–159.

2 Puurunen, R.L. (2005) *J. Appl. Phys.*, **97**, 121301.

3 Lim, B.S., Rahtu, A., and Gordon, R.G. (2003) *Nat. Mater.*, **2**, 749.

4 Aaltonen, T., Alén, P., Ritala, M., and Leskelä, M. (2003) *Chem. Vapor Depos.*, **9**, 45.

5 Aaltonen, T., Ritala, M., Sajavaara, T., Keinonen, J., and Leskelä, M. (2003) *Chem. Mater.*, **15**, 1924.

6 Christensen, S.T. and Elam, J.W. (2010) *Chem. Mater.*, **22**, 2517.

7 Hämäläinen, J., Munnik, F., Ritala, M., and Leskelä, M. (2008) *Chem. Mater.*, **20**, 3564.

8 Hämäläinen, J., Puukilainen, E., Kemell, M., Costelle, L., Ritala, M., and Leskelä, M. (2009) *Chem. Mater.*, **21**, 4858.

9 Klaus, J.W., Ferro, S.J., and George, S.M. (2000) *Thin Solid Films*, **360**, 145.

10 Fabreguette, F.H., Sechrist, Z.A., Elam, J.W., and George, S.M. (2005) *Thin Solid Films*, **488**, 103.

11 Sechrist, Z.A., Fabreguette, F.H., Heintz, O., Phung, T.M., Johnson, D.C., and George, S.M. (2005) *Chem. Mater.*, **17**, 3475.

12 Kim, H. (2003) *J. Vac. Sci. Technol. B*, **21**, 2231.

13 Juppo, M. (2001) Atomic layer deposition of metal and transition metal nitride thin films and in situ mass spectrometry studies PhD thesis, University of Helsinki, http://ethesis.helsinki.fi.

14 Soininen, P.J., Elers, K.-E., Saanila, V., Kaipio, S., Sajavaara, T., and Haukka, S. (2005) *J. Electrochem. Soc.*, **152**, G122.

15 Li, Z.W., Rahtu, A., and Gordon, R.G. (2006) *J. Electrochem. Soc.*, **153**, C146.

16 Li, Z. and Gordon, RG. (2006) *Chem. Vapor Depos.*, **12**, 435.

17 Lee, B.H., Hwang, J.K., Nam, J.M., Lee, S.U., Kim, J.T., Koo, S.-M., Baunemann, A., Fischer, R.A., and Sung, M.M. (2009) *Angew. Chem., Int. Ed.*, **48**, 4536.

18 Kang, S.-W., Yun, J.-Y., and Chang, Y.H. (2010) *Chem. Mater.*, **22**, 1607.

19 Fang, Z.Y., Dai, T., Fu, Q., Zhang, B., and Zu, X. (2009) *J. Microsc.*, **235**, 138.

20 Kodas, T. and Hampden-Smith, M. (1994) *The Chemistry of Metal CVD*, Wiley-VCH Verlag GmbH, Weinheim, pp. 303–327.

21 Adams, S.K., Edwards, D.A., and Richards, R. (1975) *Inorg. Chem. Acta*, **12**, 163.

22 Grodzicki, A., Lakomska, I., Piszczek, P., Szymanska, I., and Szlyk, E. (2005) *Coord. Chem. Rev.*, **249**, 2232.

23 Shibata, S., Iijima, K., and Baum, T.H. (1990) *Dalton Trans.*, 1519.

24 Niskanen, A., Hatanpää, T., Arstila, K., Leskelä, M., and Ritala, M. (2007) *Chem. Vapor Depos.*, **13**, 408.

25 Kariniemi, M.,. Niinistö, J., Hatanpää, T., Kemell, M., Sajavaara, T., Ritala, M., and Leskelä, M. (2011) *Chem. Mater.* **23**, 2901.

26 Rossnagel, S.M., Sherman, A., and Turner, F. (2000) *J. Vac. Sci. Technol. B*, **18**, 2016.

27 Kim, H. and Rossnagel, S.M. (2003) *Thin Solid Films*, **441**, 331.

28 Lee, Y.J. and Kang, S.-W. (2002) *J. Vac. Sci. Technol. A*, **20**, 1983.

29 Lemonds, A.M., White, J.M., and Ekerdt, J.G. (2003) *Surf. Sci.*, **538**, 191.

30 Kodas, T. and Hampden-Smith, M. (1994) *The Chemistry of Metal CVD*, Wiley-VCH Verlag GmbH, Weinheim, pp. 57–86.

31 Bent, B.E., Nuzzo, R.G., and Dubois, L.H. (1989) *J. Am. Chem. Soc.*, **111**, 1634.

32 Suzuki, N., Anayama, C., Masu, K., Tsubouchi, K., and Mikoshiba, N. (1986) *Jpn. J. Appl. Phys.*, **25**, 1236.

33 Gow, T.R., Lin, R., Cadwell, L.A., Lee, F., Backman, A.L., and Masel, R.I. (1989) *Chem. Mater.*, **1**, 406.

34 Gladfelter, W.L., Boyd, D.C., and Jensen, K.F. (1989) *Chem. Mater.*, **1**, 339.

35 Komarov, S.F., Lee, J.J., Hudson, J.B., and D'Evelyn, M.P. (1998) *Diamond Relat. Mater.*, **7**, 1087.

36 Suda, Y., Ishida, M., Yamashita, M., and Ikeda, H. (1994) *Appl. Surf. Sci.*, **82–83**, 332.

37 Akazawa, H. (1997) *J. Cryst. Growth*, **173**, 342.

38 Sugahara, S., Hosaka, K., and Matsumura, M. (1998) *Appl. Surf. Sci.*, **132**, 327.

39 Nishizawa, J., Aoki, K., Suzuki, S., and Kikuchi, K. (1990) *J. Electrochem. Soc.*, **137**, 1898.

40 Kolske, D.D., Gates, S.M., and Beach, D.B. (1992) *J. Appl. Phys.*, **72**, 4073.

41 Gates, S.M. (1992) *J. Phys. Chem.*, **96**, 10439.

42 Pore, V., Hatanpää, T., Ritala, M., and Leskelä, M. (2009) *J. Am. Chem. Soc.*, **131**, 3478.

43 Pore, V., Knapas, K., Hatanpää, T., Sarnet, T., Kemell, M., Ritala, M., Leskelä, M., and Mizohata, K. (2011) *Chem. Mater.*, **13**, 247.

44 Putkonen, M., Aaltonen, T., Alnes, M., Sajavaara, T., Nilsen, O., and Fjellvåg, H. (2009) *J. Mater. Chem.*, **19**, 8767.

45 Aaltonen, T., Alnes, M., Nilsen, O., Costelle, L., and Fjellvåg, H. (2010) *J. Mater. Chem.*, **20**, 2877.

46 Tiitta, M., Leskelä, M., Nykänen, E., Soininen, P., and Leskelä, M. (1995) *Thermochim. Acta*, **256**, 47.

47 Tiitta, M. and Niinistö, L. (1997) *Chem. Vapor Depos.*, **3**, 167.

48 Huang, R. and Kitai, A. (1992) *Appl. Phys. Lett.*, **61**, 1450.

49 Putkonen, M., Sajavaara, T., and Niinistö, L. (2000) *J. Mater. Chem.*, **10**, 1857.

50 Vehkamäki, M., Hatanpää, T., Hänninen, T., Ritala, M., and Leskelä, M. (1999) *Electrochem. Solid-State Lett.*, **2**, 504.

51 Vehkamäki, M., Hatanpää, T., Ritala, M., Leskelä, M., Väyrynen, S., and Rauhala, E. (2007) *Chem. Vapor Depos.*, **13**, 239.

52 Vehkamäki, M., Hatanpää, T., Ritala, M., and Leskelä, M. (2004) *J. Mater. Chem.*, **14**, 3191.

53 Vehkamäki, M., Hatanpää, T., Kemell, M., Ritala, M., and Leskelä, M. (2006) *Chem. Mater.*, **18**, 3883.

54 Hatanpää, T., Vehkamäki, M., Ritala, M., and Leskelä, M. (2010) *Dalton Trans.*, **39**, 3219.

55 Putkonen, M. and Niinistö, L. (2006) *Thin Solid Films*, **514**, 145.

56 Ott, A.W., Johnson, J.M., Klaus, J.W., and George, S.M. (1997) *Appl. Surf. Sci.*, **112**, 2015.

57 Asikainen, T., Ritala, M., and Leskelä, M. (1994) *J. Electrochem. Soc.*, **141**, 3210.

58 Matero, R., Rahtu, A., Ritala, M., Leskelä, M., and Sajavaara, T. (2000) *Thin Solid Films*, **368**, 1.

59 Elam, J.W., Martinson, A.B.F., Pellin, M.J., and Hupp, J.T. (2006) *Chem. Mater.*, **18**, 3571.

60 Nilsen, O., Balasundaraprabhu, R., Monakhov, E.V., Muthukumarasamy, N.,

Fjellvåg, H., and Svensson, B.G. (2009) *Thin Solid Films*, **517**, 6320.

61 Nieminen, M., Niinistö, L., and Rauhala, E. (1996) *J. Mater. Chem.*, **6**, 27.

62 Dezelah, C.L., Niinistö, J., Arstila, K., Niinistö, L., and Winter, C.H. (2006) *Chem. Mater.*, **18**, 471.

63 Shan, F.K., Liu, G.X., Lee, W.J., Lee, G.H., Kim, I.S., and Shin, B.C. (2005) *J. Appl. Phys.*, **98**, 023504.

64 Klaus, J.W., Ott, A.W., Johnson, J.M., and George, S.M. (1997) *Appl. Phys. Lett.*, **70**, 1092.

65 Klaus, J.W., Sneh, O., and George, S.M. (1997) *Science*, **278**, 1934.

66 Hausmann, D., Becker, J., Wang, S., and Gordon, R.G. (2002) *Science*, **298**, 402.

67 Kamiyama, S., Miura, T., and Nara, Y. (2005) *Appl. Phys. Lett.*, **8**, F37.

68 Kamiyama, S., Miura, T., and Nara, Y. (2006) *Thin Solid Films.*, **515**, 1517.

69 Burton, B.B., Kang, S.W., Rhee, S.W., and George, S.M. (2009) *J. Phys. Chem. C*, **113**, 8249.

70 Viirola, H. and Niinistö, L. (1994) *Thin Solid Films*, **251**, 127.

71 Leskelä, M., Niemelä, P., Niinistö, L., Nykänen, E., Soininen, P., Tiitta, M., and Vähäkangas, J. (1990) *Vacuum*, **41**, 1457.

72 Harjunoja, J., Putkonen, M., and Niinistö, L. (2007) *Thin Solid Films*, **497**, 77.

73 Watanabe, T., Hoffmann Eiffert, S., Mi, S.B., Jia, C.L., Waser, R., and Hwang, C.S. (2007) *J. Appl. Phys.*, **101**, 014114.

74 Harjunoja, J., Väyrynen, S., Putkonen, M., Niinistö, L., and Rauhala, E. (2007) *Thin Solid Films*, **253**, 5228.

75 Hiltunen, L., leskelä, M., Mäkelä, M., Niinistö, L., Nykänen, E., and Soininen, P. (1988) *Thin Solid Films*, **166**, 149.

76 Ritala, M., Kalsi, P., Riihelä, D., Kukli, K., Leskelä, M., and Jokinen, J. (1999) *Chem. Mater.*, **11**, 1712.

77 Becker, J.S., Kin, E., and Gordon, R.G. (2004) *Chem. Mater.*, **16**, 3497.

78 Alen, P., Ritala, M., Arstila, K., Keinonen, J., and Leskelä, M. (2005) *J. Electrochem. Soc.*, **152**, G361.

79 Alen, P., Ritala, M., Arstila, K., Keinonen, J., and Leskelä, M. (2005) *Thin Solid Films*, **491**, 235.

80 Elam, J.W., Schuisky, M., Ferguson, J.D., and George, S.M. (2003) *Thin Solid Films*, **436**, 145.

81 Wu, Y.Y., Kohn, A., and Eizenberg, M. (2004) *J. Appl. Phys.*, **95**, 6167.

82 Rayner, G.B. and George, S.M. (2009) *J. Vac. Sci. Technol. A*, **27**, 916.

83 Burton, B.B., Lavoie, A.R., and George, S.M. (2008) *J. Electrochem. Soc.*, **155**, D508.

84 Becker, J. and Gordon, R.G. (2003) *Appl. Phys. Lett.*, **82**, 2239.

85 Elers, K.-E., Saanila, V., Soininen, P.J., Li, W.-M., Kostamo, J.T., Haukka, S., Juhanoja, J., and Besling, W.F.A. (2002) *Chem. Vapor Depos.*, **8**, 1.

86 Riihelä, D., Ritala, M., Matero, R., Leskelä, M., Jokinen, J., and Haussalo, P. (1996) *Chem. Vapor Depos.*, **2**, 277.

87 Karam, N.N., Parodos, T., Colter, P., McNulty, D., Rowland, W., Schetzina, J., El-Masry, N., and Bedair, S.M. (1995) *Appl. Phys. Lett.*, **67**, 94.

88 Boutros, K.S., McIntosh, F.G., Roberts, J.C., Bedair, S.M., Piner, E.L., and El-Masry, N.A. (1995) *Appl. Phys. Lett.*, **67**, 1856.

89 Dip, A., Eldallal, G., Colter, P., Hayafuji, N., and Bedair, S.M. (2002) *Appl. Phys. Lett.*, **62**, 2378.

90 Klaus, J.W., Ott, A.W., Dillon, A.C., and George, S.M. (1998) *Surf. Sci.*, **418**, L14.

91 Morishita, S., Sugahara, S., and Matsumura, M. (1997) *Appl. Surf. Sci.*, **112**, 198.

92 Goto, H., Shibahara, K., and Yokoyama, S. (1996) *Appl. Phys. Lett.*, **68**, 3257.

93 Märtlid, B., Ottoson, M., Heszler, P., Carlsson, J.O., and Larsson, K. (2002) *Thin Solid Films*, **402**, 167.

94 Ferguson, J.D., Weimer, A.W., and George, S.M. (2002) *Thin Solid Films*, **413**, 13.

95 Olander, J., Ottoson, L.M., Heszler, P., Carlsson, J.O., and Larsson, K. (2005) *Chem. Vapor Depos.*, **11**, 330.

96 Jones, A.C. and O'Brien, P. (1997) *CVD of Compound Semiconductors. Precursor Synthesis, Development and Applications*, Wiley-VCH Verlag GmbH, Weinheim, p. 273.

97 Nishizawa, J., Kurabayashi, T., Abe, H., and Nozue, A. (1987) *Surf. Sci.*, **185**, 249.

98 Ozeki, M., Mochizuki, K., Ohtsuka, N., and Kodama, K. (1988) *Appl. Phys. Lett.*, **53**, 1504.

99 Reid, K.G., Urdianyk, M.H., El-Masry, N.A., and Bedair, S.M. (1991) *Mater. Res. Soc. Symp. Proc.*, **222**, 133.

100 Reid, K.G., Myers, A.F., El-Masry, N.A., and Bedair, S.M. (1993) *Thin Solid Films*, **225**, 59.

101 Ohno, H., Ohtsuka, S., Ishii, H., Matsubora, Y., and Hasegawa, H. (1989) *Appl. Phys. Lett.*, **54**, 2000.

102 Usui, A. and Sunakawa, H. (1986) *Jpn. J. Appl. Phys. Part 2*, **25**, L212.

103 Maa, B.Y. and Dapkus, P.D. (1991) *Appl. Phys. Lett.*, **58**, 2261.

104 Nagasawa, H. and Yamaguchi, Y. (1993) *Thin Solid Films*, **225**, 230.

105 Isshiki, H., Iwai, S., Meguro, T., Aoyagi, Y., and Sugano, T. (1994) *J. Cryst. Growth*, **145**, 976.

106 Isshiki, H., Aoyagi, Y., and Sugano, T. (1997) *Appl. Surf. Sci.*, **112**, 122.

107 Hossbach, C., Teichert, S., Thomas, J., Wilde, L., Wojcik, H., Schmidt, D., Adolphi, B., Bertram, M., Muehle, U., Albert, M., Menzel, S., Hintze, B., and Bartha, J.W. (2009) *J. Electrochem. Soc.*, **156**, H852.

108 Song, M.-K. and Rhee, S. (2008) *J. Electrochem. Soc.*, **155**, H823.

109 Elers, K.-E. (2002) Deposition of transition metal carbides US Patent 6,482,262.

110 Härkönen, K., Doczy, M., Lang, T., and Baxter, N.E. (2004) Deposition of carbon and transition metal containing thin films US Patent Application 2004/0208994.

111 Smith, S., Li, W.-M., Elers, K.-E., and Pfeifer, K. (2002) *Microelectron. Eng.*, **64**, 247.

112 Elers, K.-E., Saanila, V., Li, W.-M., Soininen, P.J., Kostamo, J.T., Haukka, S., Juhanoja, J., and Besling, W.F.A. (2003) *Thin Solid Films*, **434**, 94.

113 Hara, S., Meguro, T., Aoyagi, Y., Kawai, M., Misawa, S., Sakuma, E., and Yoshida, S. (1993) *Thin Solid Films*, **225**, 240.

114 Sumakeris, J.J., Rowland, L.B., Kern, R.S., Tanaka, T., and Davis, R.F. (1993) *Thin Solid Films*, **225**, 219.

115 Kräuter, G.E. and Rees, W.S. (1996) *CVD of Nonmetals* (ed. W.S. Rees), Wiley-VCH Verlag GmbH, Weinheim, p. 386.

116 Alen, P., Aaltonen, T., Ritala, M., Leskelä, M., Sajavaara, T., Keinonen, J., Hooker, J.C., and Maes, J.W. (2004) *J. Electrochem. Soc.*, **151**, G523.

117 Lee, H.-R.-B. and Kim, H. (2010) *J. Cryst. Growth*, **312**, 2215.

118 Lee, H.-R.-B., Son, J.Y., and Kim, H. (2007) *Appl. Phys. Lett.*, **90**, 213509.

119 Natrajan, G., Maydannik, P.S., Cameron, D.C., Akopyan, I., and Novikov, B.V. (2010) *Appl. Phys. Lett.* **97**, 241905.

120 Ylilammi, M. and Ranta-aho, T. (1994) *J. Electrochem. Soc.*, **141**, 1278.

121 Pilvi, T., Arstila, K., Leskelä, M., and Ritala, M. (2007) *Chem. Mater.*, **19**, 3387.

122 Pilvi, T., Puukilainen, E., Arstila, K., Leskelä, M., and Ritala, M. (2008) *Chem. Vapor Depos.*, **14**, 85.

123 Pilvi, T., Puukilainen, E., Kressig, U., Leskelä, M., and Ritala, M. (2008) *Chem. Mater.*, **20**, 5023.

124 Putkonen, M. (2009) 9th International Conference on Atomic Layer Deposition, ALD 2009, Monterey, CA, Technical Program & Abstracts, p. 142.

125 Ritala, M., Kukli, K., Rahtu, A., Räisänen, P.I., Leskelä, M., Sajavaara, T., and Keinonen, J. (2000) *Science*, **288**, 319.

126 Delabie, A., Pourtois, G., Caymax, M., De Gendt, S., Ragnarsson, L.-Å., Heyns, M., Fedorenko, Y., Swerts, J., and Maes, J.W. (2007) *J. Vac. Sci Technol. A*, **25**, 1301.

127 Kamiyama, S., Miura, T., and Nara, Y. (2006) *J. Electrochem. Soc.*, **153**, G187.

128 Nam, W.-H. and Rhee, S.-W. (2003) *Electrochem. Solid-State Lett.*, **7**, C55.

129 Kukli, K., Ritala, M., Pilvi, T., Sajavaara, T., Leskelä, M., Jones, A.C., Aspinall, H.C., Gilmer, D.C., and Tobin, P.J. (2004) *Chem. Mater.*, **16**, 5162.

130 Jones, A.C., Aspinall, H.C., Chalker, R.P., Potter, R.J., Kukli, K., Rahtu, A., Ritala, M., and Leskelä, M. (2004) *J. Mater. Chem.*, **14**, 3101.

131 Tiitta, M., Nykänen, E., Soininen, P., Niinistö, L., Leskelä, M., and Lappalainen, R. (1998) *Mater. Res. Bull.*, **33**, 1315.

132 Putkonen, M., Sajavaara, T., Rahkila, T., Xu, L.C., Niinistö, L., and Whitlow, H.J. (2009) *Thin Solid Films*, **517**, 5819.

133 Gandrud, K.B., Pettersen, A., Nilsen, O., and Fjellvåg, H. (2010) BALD and 2nd

GerALD Conference, Hamburg, Conference Program and Abstract Book.
134 Sönsteby, H., Östering, E., Fjellvåg, H., and Nilsen, O. (2010) BALD and 2nd GerALD Conference, Hamburg, Conference Program and Abstract Book.
135 Tsvetskova, M.N., Pak, N.V., Malygin, A.A., and Koltsov, S.I. (1984) *Inorg. Mater.*, **20**, 121.
136 Badot, J.C., Mantoux, A., Baffier, N., Dubrufaut, O., and Lincot, D. (2004) *J. Mater. Chem.*, **14**, 3411.
137 Badot, J.C., Mantoux, A., Baffier, N., Dubrufaut, O., and Lincot, D. (2006) *J. Phys. Chem. Solids*, **67**, 1270.
138 Vehkamäki, M., Ritala, M., Leskelä, M., Jones, A.C., Davies, H.O., Sajavaara, T., and Rauhala, E. (2004) *J. Electrochem. Soc.*, **151**, F69.
139 Nieminen, M., Sajavaara, T., Rauhala, E., Putkonen, M., and Niinistö, L. (2001) *J. Mater. Chem.*, **11**, 2340.
140 Kosola, A., Putkonen, M., Johansson, L.-S., and Niinistö, L. (2003) *Appl. Surf. Sci.*, **211**, 102.
141 Kwon, O.S., Kim, S.K., Cho, M., Hwang, C.S., and Jeong, J. (2007) *J. Electrochem. Soc.*, **152**, C229.
142 Myllymäki, P., Nieminen, M., Niinistö, J., Putkonen, M., Kukli, K., and Niinistö, L. (2006) *J. Mater. Chem.*, **16**, 563.
143 Schuisky, M., Kukli, K., Ritala, M., Hårsta, A., and Leskelä, M. (2000) *Chem. Vapor Depos.*, **6**, 139.
144 Hwang, G.W., Kim, W.D., Min, Y.-S., Cho, Y.J., and Hwang, C.S. (2006) *J. Electrochem. Soc.*, **153**, F20.
145 Vehkamäki, M. (2007) Atomic layer deposition of multicomponent oxide materials PhD thesis, University of Helsinki, http://ethesis.helsinki.fi.
146 Harjunoja, J., Väyrynen, S., Putkonen, M., Niinistö, L., and Rauhala, E. (2007) *Appl. Surf. Sci.*, **253**, 5228.
147 Seim, H., Mölsä, H., Nieminen, M., Fjellvåg, H., and Niinistö, L. (1997) *J. Mater. Chem.*, **7**, 449.
148 Seim, H., Nieminen, M., Niinistö, L., Fjellvåg, H., and Johansson, L.-S. (1997) *Appl. Surf. Sci.*, **112**, 243.
149 Lee, S.W., Kwon, O.S., Han, J.H., and Hwang, C.S. (2008) *Appl. Phys. Lett.*, **92**, 22903.
150 Kukli, K., Ritala, M., Kemell, M., and Leskelä, M. (2010) *J. Electrochem. Soc.*, **157**, D35.
151 Besling, W.F.A., Young, E., Condrad, T., Zhao, C., Carter, R., Vandervorst, W., Caymax, M., De Gendt, S., Heyns, M., Maes, J., Tuominen, M., and Haukka, S. (2002) *J. Non-Cryst. Solids*, **303**, 123.
152 Cavanagh, A.S., Wilson, C.A., Weimer, A.W., and George, S.M. (2009) *Nanotechnology*, **20**, 255802.
153 Xuan, Y., Wu, Y.Q., Shen, T., Qi, M., Capano, M.A., Cooper, J.A., and Ye, P.D. (2008) *Appl. Phys. Lett.*, **92**, 013101.
154 Aaltonen, T., Ritala, M., Tung, Y.-L., Ch, Y., Arstila, K., Meinander, K., and Leskelä, M. (2004) *J. Mater. Res.*, **19**, 3353.
155 Wilson, C.A., Grubbs, R.K., and George, S.M. (2005) *Chem. Mater.*, **17**, 5625.
156 Kowalik, I.A., Guziewicz, E., Kopalko, K., Yatsunenko, S., Wojcik-Glodowska, A., Godlewski, M., Dluzewski, P., Lusakowska, E., and Paszkowicz, W. (2009) *J. Cryst. Growth*, **311**, 1096.
157 Miikkulainen, V., Puurunen, R.I., Ritala, M., and Leskelä, M., *J. Appl. Phys.*, Submitted.
158 Niinistö, J., Kukli, K., Kariniemi, M., Ritala, M., Leskelä, M., Blasco, N., Pinchart, A., Lachaud, C., Laarousi, N., Wang, Z., and Dussarant, C. (2008) *J. Mater. Chem.*, **18**, 5243.
159 Niinistö, J., Kukli, K., Heikkilä, M., Ritala, M., and Leskelä, M. (2009) *Adv. Eng. Mater.*, **11**, 223.

Index

Atomic Layer Deposition of Nanostructured Materials, First Edition. Edited by Nicola Pinna and Mato Knez.

d

t